Ethics and AI Integration Into Modern Classrooms

Mohamed Khaldi
Abdelmalek Essaâdi University, Morocco

IGI Global
Scientific Publishing
Publishing Tomorrow's Research Today

Vice President of Editorial Melissa Wagner
Director of Acquisitions Mikaela Felty
Director of Book Development Jocelynn Hessler
Production Manager Mike Brehm
Cover Design Phillip Shickler

Published in the United States of America by
IGI Global Scientific Publishing
701 East Chocolate Avenue
Hershey, PA, 17033, USA
Tel: 717-533-8845
Fax: 717-533-7115
Website: https://www.igi-global.com E-mail: cust@igi-global.com

Library of Congress Cataloging-in-Publication Data

Names: Khaldi, Mohamed, 1959- editor
Title: Ethics and AI integration into modern classrooms / edited by: Mohamed Khaldi.
Other titles: Ethics and artificial intelligence integration into modern classrooms
Description: Hershey, PA : IGI Global Scientific Publishing, [2025] | Includes bibliographical references and index. | Summary: "This book offers an in-depth exploration of how to integrate artificial intelligence technology into pedagogical practice to revolutionize education"-- Provided by publisher.
Identifiers: LCCN 2025001373 (print) | LCCN 2025001374 (ebook) | ISBN 9798337322629 hardcover | ISBN 9798337322636 paperback | ISBN 9798337322643 ebook
Subjects: LCSH: Artificial intelligence--Educational applications | Artificial intelligence--Moral and ethical aspects
Classification: LCC LB1028.43 .E74 2025 (print) | LCC LB1028.43 (ebook) | DDC 371.33/4--dc23/eng/20250424
LC record available at https://lccn.loc.gov/2025001373
LC ebook record available at https://lccn.loc.gov/2025001374

British Cataloguing in Publication Data
A Cataloguing in Publication record for this book is available from the British Library.

Table of Contents

Detailed Table of Contents

These advancements are not merely incremental but represent a revolutionary shift in how educational content is delivered, accessed, and personalized. AI education tools are designed to enhance learning outcomes by providing innovative solutions that cater to the diverse needs of students, educators, and institutions alike. Moreover, AI-driven tools facilitate the creation of intelligent tutoring systems that provide individualized support to learners. These systems simulate one-on-one tutoring by offering instant feedback, explanations, and additional practice opportunities. For instance, AI can create dynamic simulations and virtual environments that respond to students' actions, offering a more engaging and experiential learning process. . AI-powered language translation tools also facilitate learning in multilingual settings, making it easier for students to access content in their preferred language and ensuring a more inclusive educational environment

Education is a constantly evolving field, where the decisions made by teachers, administrators, and policymakers significantly impact students' academic journey and progress. In this dynamic context, integrating Artificial Intelligence (AI) open new perspectives to support and enhance informed decision-making. Using AI, with its advanced data analysis, predictive modeling, and process automation capabilities, can provide valuable information to education stakeholders, enabling them to make more informed and effective decisions. Assessment plays a crucial role in the educational process, enabling to measure learners' knowledge, skills,

Emotion detection in distance education during online classes has emerged as a critical area of research to enhance student engagement, motivation, and learning outcomes. With the shift to virtual learning environments, understanding students' emotional states has become essential to bridge the gap created by physical absence. This study leverages advanced deep learning techniques, specifically ResNet50, to detect emotional intelligence through facial expressions captured during online classes. By preprocessing facial images and employing robust feature extraction, the model classifies emotional states such as happiness, confusion, boredom, or frustration. The integration of emotional intelligence detection into virtual classrooms enables educators to adapt teaching strategies in real time, fostering a more interactive and supportive learning environment. The proposed system is evaluated on diverse facial expression datasets, showcasing its effectiveness in accurately identifying emotions and contributing to the broader goal of improving student experiences in distance education.

The integration of AI into Learning Management Systems (LMS) communication tools represents a major advancement in online learning. This chapter explores how AI is transforming learner interaction and engagement in distance education, starting with an overview of existing tools like discussion forums and video conferencing. It then highlights AI-driven innovations such as chatbots and virtual assistants for personalized support, sentiment analysis for better moderation, and real-time translation to enhance accessibility. AI-generated personalized recommendations and collaboration tools further enrich the learning experience and reduce educators' workload. Despite these benefits, ethical challenges like privacy and algorithmic biases arise. As AI continues to evolve, it promises to reshape not only communication but also the entire structure of education in the digital.

Preface

The integration of artificial intelligence (AI) into modern classrooms represents a transformative shift in educational landscapes, one that promises to reshape teaching practices and learning experiences across the globe. As educational institutions increasingly turn to AI to enhance pedagogy, it becomes evident that AI has the potential to foster a more personalized, efficient, and inclusive environment. This potential reaches beyond just adapting to the needs of individual learners; it includes offering personalized learning paths tailored to students' specific strengths, weaknesses, and preferences. With AI's ability to adjust content and pace, even students with special needs or those speaking different languages can benefit from educational tools designed specifically for them. Through AI's interactive nature, students receive instant feedback, keeping them engaged while promoting continuous learning.

AI's influence extends to the role of educators as well. By harnessing the power of data analytics, AI provides teachers with real-time insights into student progress, enabling them to make informed decisions about how best to tailor their teaching methods. This capacity to assess and adapt is invaluable in ensuring that both students and teachers remain aligned with the evolving needs of the educational process. However, as with any technological revolution, the integration of AI into classrooms also presents a set of ethical challenges that must be carefully considered. Issues such as privacy, accuracy, accessibility, and fairness must be addressed to ensure that AI's application benefits all students without compromising their rights or the integrity of their learning environment.

This book, *Supporting Personalized Learning and Students' Skill Development with AI*, delves deeply into these topics, exploring how AI can be harnessed to address a wide variety of challenges and opportunities in education. From intelligent tutoring systems that personalize lessons to automating administrative tasks that free up teachers to focus more on instruction, this volume serves as a comprehensive guide to understanding how AI can elevate educational practices. It also examines the ethical considerations and challenges surrounding AI's use in educational settings, highlighting the urgent need for teachers and administrators to be equipped with the necessary skills to effectively and responsibly utilize AI technologies. Covering diverse topics ranging from emotion detection and digital literacy to recommendations systems and autonomous learning, this book is a valuable resource for educators, administrators, policymakers, computer developers, and researchers seeking to understand and implement AI in education.

As the world of education continues to evolve, the integration of AI is not simply an enhancement but an imperative for the future of learning. The pace at which technology advances means that educators must be proactive in embracing these innovations, while simultaneously remaining mindful of the ethical implications and challenges that come with them. By addressing these issues head-on, we can ensure that AI in education remains a tool for good, one that empowers both students and teachers. This book not only provides insights into the current state of AI in education but also offers a roadmap for its future integration, helping all stakeholders navigate the complexities and possibilities of this exciting new frontier.

OVERVIEW OF CHAPTERS

Here is the revised version of the content with added chapter overviews and excluding author names:

Chapter 1: Current Trends in Artificial Intelligence Educational Practices: A Literature Review

This chapter provides a comprehensive exploration of the integration of artificial intelligence (AI) into educational practices. By reviewing recent studies from the last six years, it examines the current innovative applications of AI, effective pedagogical strategies, and the challenges educators face. The aim is to offer a broad understanding of how AI can be used to enhance teaching and learning processes, identifying both the opportunities and obstacles associated with its implementation in educational settings.

Chapter 2: The Impact of Generative AI on Education, a Review of Applications and Challenges

This chapter reviews the use of generative AI in education, highlighting technologies such as GPT, BERT, and other neural networks. It explores how these tools are reshaping traditional teaching methods, course design, and student-teacher interactions. Through an analysis of current academic research, it presents the benefits and challenges of applying generative AI across various academic levels, shedding light on the transformative potential of these technologies in educational environments.

Chapter 3: Tools and Technologies for AI Education

This chapter discusses the revolutionary AI-driven tools and technologies that are transforming educational content delivery, learning personalization, and engagement. It highlights the development of intelligent tutoring systems, dynamic simulations, virtual environments, and multilingual tools. These innovations offer new ways to support diverse student needs and create immersive learning experiences, ultimately improving educational outcomes and fostering more inclusive learning environments.

Chapter 4: AI-Enhanced Education: Revolutionizing Assessments and Informed Decision-Making

This chapter explores how AI is enhancing the educational decision-making process through data analysis, predictive modeling, and automation. It discusses the role of AI in improving assessments, offering teachers and administrators more accurate tools for measuring student progress and identifying areas for intervention. By integrating AI into assessments, this chapter shows how education stakeholders can make more informed, timely decisions to better support students.

Chapter 5: Integrating AI Education in Non-Technical Disciplines, Bridging the Gap Between Theory and Practice: Strategies for AI Integration Across Disciplines

This chapter examines how AI education can be integrated into non-technical disciplines such as social sciences, arts, and humanities. It emphasizes the importance of accessible teaching methods and interdisciplinary collaboration to make AI concepts approachable for non-technical learners. The chapter offers case studies and instructional frameworks that enable students in various fields to understand AI and apply it in ways that enrich their academic and professional practices.

Chapter 6: Artificial Intelligence as a Harbinger of Engagement and Collaboration

This chapter explores AI's role in enhancing learner engagement through personalized learning experiences. It discusses AI applications such as intelligent tutoring systems and adaptive learning paths that cater to individual student needs. Furthermore, the chapter examines AI-driven collaborative tools that promote active participation and critical thinking, highlighting both the potential benefits and ethical considerations of AI in fostering inclusive and effective learning environments.

Chapter 7: Reinventing AI Education: From Collaborative Learning to Real-World Projects

This chapter explores the integration of project-based learning (PBL) and active learning approaches in AI education. By focusing on collaborative and problem-based learning strategies, it emphasizes the importance of hands-on, real-world projects that enhance both practical and theoretical skills. The chapter also addresses the challenges of implementing these methods and provides insights into how they can foster creativity, critical thinking, and student engagement in AI curricula.

Chapter 8: Artificial Intelligence (AI) and Enabling Collaboration in Adaptive Distance Education

This chapter examines how AI can enhance collaboration in distance education, offering personalized support and creating dynamic learning environments. It focuses on AI tools that promote student engagement and streamline collaborative projects, from task management to performance evaluation. By analyzing the evolving role of AI in online learning, this chapter highlights its growing significance in improving student interaction and academic collaboration in distance education settings.

Chapter 9: Reinventing the Online Classroom with AI from Platform Models to the Smart Room

This chapter delves into the impact of AI on online education, particularly through virtual classrooms and smart classrooms. It explores how AI technologies can optimize knowledge delivery, enhance motivation, and personalize learning. The chapter discusses the technological, social, and cognitive dimensions of AI's role in transforming online learning, addressing the infrastructure needed to create interactive, student-centered environments that foster collaboration and self-regulation.

Chapter 10: The Future of Online Learning, Emerging Trends and Innovative Technologies

This chapter explores the emerging trends and innovations shaping the future of online learning. It covers AI-powered adaptive learning systems, immersive virtual and augmented reality experiences, and blockchain technology's potential in academic credentialing. Additionally, the chapter looks at the growing accessibility of mobile learning and microlearning platforms. Together, these advancements promise to make online learning more personalized, engaging, and accessible for learners around the world.

Chapter 11: Enhancing Distance Education Emotion Detection in Online Classrooms Using Deep Learning

This chapter discusses the integration of emotion detection in online classrooms to enhance student engagement and learning outcomes. By leveraging deep learning techniques like ResNet50, the chapter explores how emotion detection through facial expression analysis can enable educators to adapt teaching strategies in real time, improving the learning experience and fostering a more interactive, supportive online environment.

Chapter 12: Artificial Intelligence and Communication in LMS to Enhancing Engagement and Learning

This chapter explores how AI-driven innovations in Learning Management Systems (LMS) are transforming communication and engagement in online education. It highlights AI-powered tools like chatbots, virtual assistants, and real-time translation, which offer personalized support and enhance accessibility. The chapter also addresses ethical concerns such as privacy and algorithmic biases, discussing how AI can reshape the structure of online education to improve both student learning and educator productivity.

Chapter 13: Human-Technology Interaction: Enhancing Education Through AI-Powered Adaptive Learning

This chapter focuses on the role of AI-powered adaptive learning systems in enhancing educational experiences. By addressing ethical issues such as data security and algorithmic bias, it emphasizes the need for transparent, equitable AI systems. The chapter explores the transformative potential of adaptive learning technologies to bridge gaps in accessibility and personalized instruction, advocating for responsible AI practices that align with human-centered educational goals.

Chapter 14: Ethical and Social Issues of AI in Education

This chapter addresses the ethical challenges posed by AI in education, including data privacy concerns, algorithmic bias, and unequal access to technology. It explores how AI can exacerbate educational inequalities and change the traditional teacher-student dynamic. The chapter calls for policies that prioritize ethical considerations, inclusive designs, and fairness to ensure that AI's potential benefits are equitably distributed across diverse educational settings.

Chapter 15: Ethical Integration of Artificial Intelligence in Inclusive Education: Addressing Challenges and Advancing Opportunities for Equitable Learning

This chapter explores how AI can be leveraged to create more inclusive educational environments. It addresses the challenges of algorithmic bias, data privacy, and the digital divide, offering insights into ethical approaches to AI integration. Emphasizing the importance of diversity, equity, and inclusion (DEI), the chapter provides policy recommendations and future research directions to harness AI’s transformative potential while ensuring fairness in educational opportunities.

Chapter 16: Ethical Design Thinking Embedding Moral Decision Making in AI Development Projects

This chapter discusses how AI developers can embed ethical decision-making into their design processes through a design-thinking approach. It explores how empathy, collaboration, and problem-solving can help developers address ethical issues like bias and transparency. By examining case studies and societal impacts, the chapter highlights the importance of designing AI systems that are fair, transparent, and aligned with societal values.

Chapter 17: Doctoral Researchers' Perspectives on Ethical Considerations in Artificial Intelligence in Education through the Lens of the PAPA Framework

This chapter delves into doctoral researchers' perspectives on the ethical considerations surrounding the use of AI in education. It specifically analyzes the PAPA (Privacy, Accuracy, Property, and Accessibility) framework to explore how AI's implementation in educational settings can be ethically managed to ensure fairness and accountability. The chapter offers insights into the evolving ethical landscape and provides recommendations for future research on AI's role in education.

CONCLUSION

The integration of artificial intelligence into the educational system offers immense potential for transforming how we approach teaching and learning. By personalizing educational experiences to meet the unique needs of each learner, AI can help address a broad range of challenges, from catering to students with special needs to providing real-time feedback that keeps learners engaged and motivated. With AI, teachers can also significantly enhance their pedagogical practices by gaining valuable insights into student progress and adapting their teaching strategies accordingly. However, this transformation also requires careful consideration of the ethical challenges that accompany the use of AI, including issues of privacy, fairness, and access. It is essential that these challenges be addressed thoughtfully to ensure that AI benefits all students equitably.

As educators, administrators, and policymakers seek to integrate AI into the classroom, it is critical to provide the necessary training and support for those who will be using these technologies. This includes not only understanding the technical aspects of AI but also being aware of the ethical and practical concerns that accompany its implementation. For AI to truly enhance the educational experience, those who use it must be equipped with the skills and knowledge to navigate its complexities. This book serves as a valuable resource for those seeking to understand AI's role in education and provides a comprehensive overview of the strategies, tools, and frameworks that can be used to integrate AI into pedagogical practice effectively.

Looking ahead, the continued evolution of AI in education will undoubtedly shape the future of teaching and learning in profound ways. As we move forward, it is essential that we remain mindful of the potential risks and challenges while embracing the opportunities that AI offers. By prioritizing ethical considerations, fostering continuous professional development, and ensuring equitable access to AI tools, we can create an educational system that is not only more efficient and personalized but also more inclusive and fair. The promise of AI in education is vast, and with careful thought and strategic implementation, it can be harnessed to create a brighter, more inclusive future for learners worldwide.

Acknowledgment

I would like to begin by expressing my deep gratitude to the management of IGI Global Publishing House, a renowned international academic publishing house specializing in the publication of reference works and scholarly journals, for agreeing to publish this book, which is scheduled for publication in 2025. This confidence shown in my work is a source of great satisfaction and valuable encouragement.

The publication of this book is the result of a collective effort and constant support. This is why my thanks go to all the people who have contributed, directly or indirectly, to the realization of this project. Their involvement, encouragement and sharing have been essential drivers and motivate me to continue my work and to consider other publications. I am particularly touched by the international support I have received, coming from India, Hong Kong, Saudi Arabia, Qatar, Greece, South Africa, and especially from Morocco, my country of origin. This diversity of support is a real wealth and a source of inspiration.

I would like to express a special gratitude to my family, friends and colleagues, whose unwavering support has been a pillar throughout this process. A warm thank you also to my doctoral students, whose constant encouragement and significant contributions have been instrumental in making this project a reality, turning a dream into reality. I also thank all those who believed in this project from the beginning, actively contributing to it and encouraging me to persevere.

Finally, it is with great affection that I dedicate this book to my wife, Malika, whose support and love have been an inexhaustible source of strength. To my children, Inass, Marwane and Maha, who light up my life. To my beloved granddaughters, Malak and Nada, who represent the future. And to my son-in-law, Ismail, whom I consider a member of my family.

Chapter 1

Current Trends in Artificial Intelligence Educational Practices: A Literature Review

Sara Rguig
LLCIS Laboratory, ENSATE, Abdelmalek Essaadi University, Morocco

ABSTRACT

Artificial intelligence continues to advance in many spheres, including education. The integration of Artificial intelligence into educational practices has evidently transformed teaching and learning processes over the decades. In this context, this chapter seeks to explore the dynamic landscape of AI in educational practices, focusing on current innovative applications, effective pedagogical strategies, and the challenges educators face. By reviewing recent studies published within the last six years, this chapter aims to provide a comprehensive understanding of how AI can be effectively integrated into educational settings.

A LITERATURE REVIEW

The term "artificial intelligence" (AI) has evolved significantly since its initiation in the 1950s by John McCarthy at Dartmouth College in the United States. It was during a workshop proposal that McCarthy first coined the term "artificial intelligence" (Russell & Norvig, 2010). AI was defined as the science of making machines smart, focusing on problem-solving and symbolic reasoning. Yet, the advancements in AI have momentously broadened and enhanced the original definition proposed by the computer scientist John McCarthy in 1956. Currently, AI incorporates a wide

DOI: 10.4018/979-8-3373-2262-9.ch001

range of techniques, including machine learning, deep learning, and neural networks, which allow systems to learn from data rather than rely solely on predefined rules. Accordingly, AI has been defined in many different ways; it is that "activity devoted to making machines intelligent, and intelligence is that quality that enables an entity to function appropriately and with foresight in its environment" (Nilsson, 2010, p. 13). On the other hand, Russell and Norvig (2003) describe the term AI as Computational Intelligence in which specific activities, such as playing chess, proving mathematical theorems, composing poetry, and diagnosing illnesses, can be accomplished. Besides, Baker and Smith (2019). provide a broad definition of AI explaining that it does not refer to a single technology. Instead, it is a broad term that encompasses various technologies and methods, including machine learning, natural language processing, data mining, neural networks, and algorithms: "Computers which perform cognitive tasks, usually associated with human minds, particularly learning and problem-solving" (Baker and Smith 2019, p. 10).

Accordingly, the term "artificial intelligence" has evolved from a focus on symbolic reasoning and problem-solving to encompass a broad array of techniques and applications driven by data and learning. As technology continues to advance, the definition and implications of AI will likely continue to change, shaping the future of various industries and every aspect society, including education. Essentially, the integration of AI into educational practices has evidently transformed teaching and learning processes over the decades, including intelligent tutoring systems, teaching robots, learning analytics dashboards, adaptive learning systems, and human-computer interactions (Chen, Xie, & Hwang, 2020). Since its emergence in education, AI has ushered in new opportunities and challenges for educational innovation; in this context, this chapter seeks to explore the dynamic landscape of AI in educational practices, focusing on current innovative applications, effective pedagogical strategies, and the challenges educators face. By reviewing recent studies published within the last six years, this chapter aims to provide a comprehensive understanding of how AI can be effectively integrated into educational settings.

METHODOLOGY

The aim of this study is to explore previous research themes and trends related to AI in education through an extensive review of literature published in reputable journals over the past six years. In order to search and collect relevant journal articles for the current review, we used electronic databases as the most efficient way to begin the literature search; mainly, Wiley Online Library, ScienceDirect, and Google Scholar. We considered also some scientific journals, which scope lies in the field of Artificial Intelligence and education, such as the International Journal of Artificial

Intelligence in Education and Computers and Education. Regarding exclusion and inclusion criteria, only peer-reviewed journal articles published in English were selected for examination in order to ensure the quality of the studies selected. The time period under review is mainly from 2019 to 2024, relevant to the focus of the study. The period was chosen to guarantee the novelty of research results, given the fast advancement in technology and AI in education. The keywords used to search in the literature are: "Artificial intelligence", "AI", "AIED", "Machine intelligence in education", "Online Classroom", "Intelligent Tutoring System", "Personalized learning", "Technology Enhanced Learning", "Virtual Reality", and "Challenges in AIED". A comprehensive full text search was performed within each of the databases to line up with the research themes and inquiries. The search process resulted in 40 articles that were included in the review. The 40 articles were then organized into different parts that correspond to the research objectives, mainly:

- To identify current trends in AI educational practices;
- To highlight innovative AI applications and tools utilized in educational contexts;
- To explore effective pedagogical strategies that incorporate AI to improve learning outcomes;
- To analyze the challenges and considerations for an effective integration of AI in education.

In order to make an adequate presentation of the literature, we suggested to organize the findings into three main themes that guided the study and search:

- Theme 1/ Current Trends in AI Educational Practices: This section delves into pioneering AI technologies currently being employed in educational settings, such as adaptive learning systems, intelligent tutoring systems, data-Driven Insights, virtual reality, and AI-driven assessment tools. It explores also how these tools personalize learning experiences and improve student engagement. A review of user-friendly AI tools that facilitate lesson planning, grading, and classroom management are included, emphasizing their ease of integration into existing curricula.
- Theme 2/ Effective Pedagogical Strategies: This section discusses pedagogical strategies that leverage AI to enhance learning outcomes. Examples include the problem-based learning, blended learning models, flipped classrooms, and collaborative learning environments augmented by AI.
- Theme 3/ Challenges and Considerations: This section addresses the practical challenges educators and institutions face when integrating AI into their teaching practices. Topics may include lack of training, resistance to change,

ethical considerations, data privacy issues, equity and accessibility, and scalability.

CURRENT TRENDS IN AI EDUCATIONAL PRACTICES

In recent years, the amalgamation of artificial intelligence in education has occurred as a transformative trend that has restructured teaching and learning milieus and practices, shifting towards more personalized and data-driven approaches (Akyuz, 2020; Chen, L., Chen, P., & Lin, Z, 2020; Ouyang, F., Zheng, L., & Jiao, P, 2022). One noticeable trend is the rise of personalized learning through adaptive learning technologies and intelligent tutoring systems (Bates, T., Cobo, C., Mariño, O., & Wheeler, S, 2020; Mousavinasab, E., Zarifsanaiey, N., Kalhori, S. R. N., Rakhshan, M., Keikha, L., & Saeedi, M. G.2021; Ahmad, 2021; Judijanto, 2024; and Zawacki-Richter, O., Marín, V.I., Bond, M, 2019. Also, data-driven insights are becoming progressively vital, with learning analytics and predictive analytics that enable instructors to supervise student performance and intervene proactively (Wu, F., Lu, C., Zhu, M. *et al.* 2020). Furthermore, the automation of administrative tasks has modernised old procedures in educational institutions. Today, AI systems can handle grading and administrative support through chatbots, thus allowing educators to focus more on teaching (Chen, L., Chen, P., & Lin, Z, 2020; Judijanto, 2024; Rutner & Scott, 2022). The integration of gamification and simulations has also enhanced the learning experience, making it more interactive and engaging for students (Crompton, H., & Burke, D., 2023; Elaish et al. 2024; Chen, X., Xie, H., Zou, D., & Hwang, G. J, 2020; Huang, J. Salmiza Saleh & Yufei Liu, 2021).

Personalized Learning

Understanding the learning differences among students is the core concept of personalized learning which is crucial for individuals' development. "The basic principle of the formulation of personalized learning is the assertion that each individual is unique and learns differently [...] Personalized learning takes into account multiple variables of learning, ranging from identification of the learning style of people, and clearly, personalized learning has a range of advantages over traditional learning practices" (Akyuz, 2020, p. 954). Due to advancements in technology and changes in lifestyle, these patterns are evolving rapidly, leading to new and innovative learning methods. Personalized learning has recently emerged as the most effective teaching approach that considers all the factors that affect individual learning patterns. It employs various methods and strategies to deliver pertinent information and teach each student in a personalized manner. This approach considers factors such as cultural background, geographic location, individual disabilities, gender, age, and many other elements to create an effective learning style for each person. In fact, personalized learning differs from traditional education, where instructors use the same approach, content, and method for all students. Quite the reverse, personalised learning adapts educational content to individual student needs, preferences, and learning paces. This personalization enhances engagement and improves learning outcomes.

Intelligent Tutoring Systems: Ouyang et al. (2022) and Akyuz (2020) discussed Intelligent Tutoring Systems (ITS) as significant tools in personalized education in today's educational system. They emphasized that ITS control artificial intelligence to provide personalised instruction and feedback, adapting to individual student needs and learning styles. As well, they emphasized the transformative potential of ITS to recognise and address learning gaps, making them valuable for instructors and learners in the educational process. ITS have become a major component of personalized learning; the fusion of intelligent tutoring systems and personalized learning fosters the seamless collaboration and also cultivates student's engagement. Similarly, Mousavinasab et al. (2021) explored the role of (ITS) in enriching educational experiences. According to them, ITS apply adaptive learning technologies to provide personalized feedback and support, which can significantly improve student learning outcomes. The authors highlighted that ITS can analyze student performance data to adjust instructional strategies in real-time, corresponding to students' learning rates and styles. Actually, ITS can tutor students in several ways. They, first, introduce theoretical concepts and provide examples. Then, the system poses questions to the students. It can comprehend the students' responses and assess their understanding, which influences the material presented and the subsequent questions asked. Additionally, students can pose their own questions, and the system is equipped to respond or address issues within the relevant knowledge area. (Ahmad, 2021; Judijanto, 2024). Accordingly, ITS "can be used to simulate one-to-one personal tutoring. Based on learner models, algorithms and neural networks, they can make decisions about the learning path of an individual student and the content to select, provide cognitive scaffolding and help, to engage the student in dialogue" (Zawacki-Richter et al., 2019, p. 4).

The different models commonly integrated into Intelligent Tutoring Systems include the student model, the teacher model, the domain model, and the diagnosis model.The student model encompasses information regarding the student's knowledge level, cognitive abilities, motivation for learning, and preferred learning styles. The teacher model involves assessing the students' current conditions, choosing suitable teaching strategies and methods, and offering support and guidance. The domain model represents the knowledge held by both students and teachers, whereas the diagnosis model assesses errors and shortcomings based on this domain model. (Zawacki-Richter et al. 2019)

These models create a feedback loop where student performance notifies the teacher model, which in turn stimulates instructional strategies. The domain model provides the required content framework, while the diagnosis model ensures that students receive personalised support based on their exclusive learning tracks. In this context, Guan et al. (2020) asserted that "AIEd has demonstrated its potential to help learners identify knowledge gaps and receive specialized support, thereby freeing instructors from daily menial tasks and allowing them to respond to students more effectively" (p. 136). This combined functioning of the models allows ITS to provide a reactive and tailored learning experience with real-time feedback and support, adapting to the individual needs of each learner while supporting educators in their teaching efforts. Actually, Chen, X., Xie, H., & Hwang, G. J, (2020) explored various avenues in this field in their work. For instance, they pointed out Korbit experience; a dialogue-driven intelligent tutoring system that uses machine learning, natural language processing, and reinforcement learning to deliver interactive and personalized learning experiences. To enhance STEM education, Korbit employed a mixed interface that incorporates a range of elements, including multimedia, dialogue-based activities, concept mapping, and gamification. Additionally, the authors emphasized that ITS enhanced with conversational and dialogue capabilities and animated agents, such as chatbots or collaborative robots (cobots), have proven effective in educational settings. Similarly, Pokrivcakova (2019) stressed the role of AI applications in education, including Computer-Assisted Language Learning (CALL), which offers tailored instruction to learners, along with writing and translation supports for language acquisition.

Table 1 presents a comparison of different intelligent tutoring systems (ITS). The evaluation of these Intelligent Tutoring Systems (ITSs) is conducted based on several factors, including the areas they address, the feedback they provide, the policies they implement, the pedagogical agents involved, and the methodology applied to the student model, if applicable (Akyuz, 2020). Actually, there are several specialized systems used across various fields; below are some of the key ones:

Table 1. Specialist areas of ITSs (Akyuz, 2020, p. 967).

	Covered Domains	Feedback	Usage of animated medium	Student Model	Other characteristics
Meta Tutor	Anatomy	Verbal	Yes	Interactive Multimedia	Self-regulatory learning
PAT	Algebra (school level)	Instant (timely)	No	ACT Theory	
ANDES	Physics	Instant	No (use videos for instructions)	Bayesian Network	
Animal Watch	Mathematics	Instant	Yes	Markov Models	
SQL Tutor	SQL Query	Five level feedback	No	CBM (Short-term) Overlay (Long-term)	
ELM-ART	LISP Programming	Instant	No	Collabor ative	Content is displayed as interactive textbooks

Adaptive Learning: Actually, personalized learning is a broadest concept that encompasses various approaches and strategies; adaptive learning is also one of the implementations within that framework. While, ITS emphasizes one-on-one tutoring experiences through AI, adaptive learning focuses on using data to dynamically adjust content and pacing. According to Judijanto (2024, p. 1730), an adaptive learning approach maximizes students' learning potential by offering content tailored to their individual levels of understanding, which in turn enhances their engagement and motivation. "They will not feel too behind or bored with material that is too easy. Instead, they will be challenged according to their abilities, increasing their chances of achieving better learning outcomes".

In the realm of digital education, several initiatives have emerged that leverage adaptive learning technologies to enhance student outcomes. One notable example is Plan Ceibal in Uruguay, recognized as one of the most advanced state agencies dedicated to digital education in the region. A key initiative under Plan Ceibal is the Mathematics Adaptive Platform (PAM), which aligns its content with the national curriculum. This platform provides personalized feedback tailored to each student's skill level, informed by an analysis of their learning experiences. Research has indicated that PAM has positively impacted student learning outcomes (UNESCO, 2019). Another significant example is Daptio, a South African solution based in Cape Town that employs deep analytics and artificial intelligence to deliver personalized learning experiences to teachers, students, and content creators across Africa and other emerging markets. The platform assesses each student's proficiency level and matches them with relevant content, thereby facilitating a more tailored educational experience (UNESCO 2019).

More importantly, adaptative learning systems can create new possibilities for students' assessment. Artificial intelligence technologies "can give quick feedback on student performance, not only in terms of academic knowledge but also in terms of skill proficiency such as problem solving, creativity, and collaboration" (Judijanto, 2024, p. 1724). Accordingly, this approach allows teaching and learning to be more effectively customized to meet each student's unique needs, guaranteeing that every learner receives the necessary support to achieve their full potential. These initiatives illustrate the potential of AI-driven adaptive learning tools to transform educational practices, providing personalized support that can enhance student engagement and academic achievement.

Data-Driven Insights

Data-Driven Insights is another example of AI trends in education. In this approach, AI can analyze data from educational software to identify learning patterns and predict outcomes. Systems that track student engagement and performance provide educators with insights to inform teaching strategies and curriculum interventions. Recent advancements in the field have led to the development of various tools aimed at enhancing educational outcomes and supporting educators. Thus, AI can analyze data from educational software to identify learning patterns and predict outcomes. For instance, PIPE is a deep learning method designed to ease the workload of educators, particularly when assisting novice learners in debugging programs. This model utilizes a representation learning algorithm combined with a multi-label classification algorithm to identify latent features in programming tasks and predict logical errors effectively. In addition to PIPE, there are ongoing efforts to create novel applications that leverage AI technologies to facilitate learning. Ralston, K., Chen, Y., Isah, H., & Zulkernine, F, (2019) developed a voice-interactive and multilingual chatbot that utilizes IBM Watson Assistant, Tone Analyzer, and Language Translator. This chatbot is capable of responding to learners' mood, tone, and language, providing timely support in an engaging manner. The programme shows a promise in addressing examination stress and enhancing student support through data-driven analysis of responses, which demonstrated its effectiveness in addressing emotional and educational needs.

The research by Wu et al. (2020) further illustrates this by demonstrating how machine learning can analyze social media interactions to predict learning success, highlighting the importance of online engagement as a predictor of academic achievement. This study employs various machine learning algorithms to classify Facebook messages, extracting features such as keywords, sentiment, and message length. The findings reveal a significant correlation between the nature of interactions in online discussions and students' learning achievements. These examples illustrate how AI-driven solutions can not only enhance personalized learning experiences, but also support educators in managing their responsibilities more efficiently, ultimately contributing to improved learning outcomes.

Automation of Administrative Tasks

AI has also automated routine tasks in the educational sector, including grading, scheduling, lesson designing, and attendance tracking. This automation allows educators to focus more on teaching and engaging with students; "this enables educators to concentrate on deeper interpersonal relationships with their students, such as providing individualized feedback, facilitating class discussions, and inspiring interest in learning" (Judijanto, 2024, p. 1724). According to Chen et al. (2020), AI has notably enhanced efficiencies in institutional and administrative services, especially in distance and online education. For instance, programs like Knewton alleviate the burden on instructors by offering platforms for providing feedback based on student interactions. Similar conclusions are drawn in other studies, such as those by Rutner and Scott (2022), which highlight that Intelligent Tutoring Systems can perform a range of functions, including grading and feedback provision. Instructors utilizing ITS achieve improved efficiencies in administrative tasks while also fulfilling their core responsibilities of guiding and instructing students.

The arguments presented by Crompton, H., & Burke, D. (2023) further support these findings, indicating that the integration of AI in education fosters effectiveness and efficiency in administrative tasks, particularly in grading student assignments. An examination of today's online learning environments reveals various programs, including TurnItIn, Grammarly, PaperRater and Ecree, that facilitate instructors in performing administrative functions such as plagiarism detection, rating and suggestive grading. Knewton, on the other hand, provides built-in functionalities to assess performance, grade assignments, and offer feedback to promote continuous improvement in learning. More importantly, "ITS analyze and asses the students on the basis of their skills, attributes, and performance which reduces the chances of biased judgments and unfair treatment. ITS provide authentic and reliable results that are based on the individual performance of the student" (Akyuz, 2020, p. 971).

More importantly, AI can play a significant role in assisting instructors with the generation of questions and the creation of multiple-choice tests. For example, researchers developed a natural language processing system capable of automatically generating tests. Their research, which employed a Turing-type evaluation, demonstrated that AI technologies can produce highly realistic short-answer questions.

Overall, AI has significantly reduced the paperwork and workload for instructors, particularly regarding administrative functions. This allows educators to focus on their primary responsibilities, delivering content and materials aligned with institutional or national curricula. While the focus on educational administration was limited in many evaluated articles, those that addressed it provided evidence of improved quality and efficiency in administrative processes and tasks, enhancing the overall effectiveness of educators (Alam, 2022).

Virtual and Augmented Reality

Virtual and augmented reality technologies are increasingly being integrated into educational settings, providing immersive learning experiences that can enhance student engagement and understanding. Within this context, AI assistants play a crucial role in facilitating these experiences, enabling students to explore complex concepts in a more engaging way. Crompton et al. (2023) indicated that AI assistants, referred to by different names such as virtual assistants, intelligent agents, and intelligent tutors, support learners in various capacities. They described how these terms delineate the AI's presentation to users, with some having an anthropomorphic presence, like avatars, while others engage through text prompts. This distinction is particularly relevant in VR and AR environments, where the interaction style can significantly impact user experience.

Moreover, AI assistants can provide real-time support within virtual learning environments. Chatbots and intelligent tutors can engage with students, helping them navigate complex simulations or immersive scenarios. For instance, an AI named Alex functioned as a virtual change agent in a mathematics course, offering diagnostic questions and support tailored to students' needs. This interaction exemplifies how AI can enhance the immersive learning experience by guiding students through the learning process.

Additionally, simulation-based learning has been shown to play a major role in this enhancement. For example, the implementation of simulation-based learning encourages deep learning, which improves the overall learning experience for students. Virtual classrooms, on the other hand, can use virtual technology to: "simulate teaching scenes that are difficult to explain, and for natural phenomena or changes in things that cannot be observed or hard to observe in real life, it can be presented in a smart classroom to create a contextual learning environment for students" (Huang et al, 2021, p. 209). This method makes abstract concepts and theories more intuitive and visual, enhancing students' interest in learning and improving teaching outcomes.

This is supported by Chen, X., Xie, H., & Hwang, G. J, (2020), who observed that virtual reality and simulations foster enriched learning environments, allowing students to explore complex concepts in an interactive manner. These immersive technologies not only engage students more effectively, but also facilitate a deeper comprehension of the material by providing realistic scenarios and hands-on experiences. The integration of AI into machines and robots has led to the creation of powerful instructional tools that improve pedagogical strategies. The writers emphasize that the incorporation of AI principles into robotics has resulted in the creation and implementation of robots as teaching assistants and collaborators, known as cobots, can perform both basic and advanced instructional tasks, including helping students learn to read and pronounce words.

Actually, the application of AI is also integrated in medical education, where virtual reality and simulation allow medical students to experience practical aspects of their training, such as surgical procedures and human anatomy. For instance, Elaish, M.M., Yadegaridehkordi, E. & Ho, YS, (2024) focused on VR's role in medical education, demonstrating how VR simulations improve clinical skills, enhance decision-making, and provide safe environments for practicing procedures. They have also highlighted the VR's effectiveness in teaching complex scientific and mathematical concepts, where simulations can visualize abstract ideas. For example, stimulations allow students to visualize molecular structures or astronomical phenomena, thereby improving comprehension and retention. Additionally, students can conduct virtual experiments in science classes, allowing them to explore concepts interactively. VR training simulations in fields like medicine or engineering provide hands-on experiences that are crucial for skill development.

Oyelere, S.S., Bouali, N., Kaliisa, R, (2020), on the other hand, explored Educational Virtual Reality Games (EVRGs), focusing on their technological, pedagogical, and gaming characteristics. The study identified a predominance of specific VR technologies, particularly the Oculus Rift and HTC Vive, which are commonly used in educational settings for immersive learning experiences, including healthcare, science, and language learning. The interactive nature of these games and the ability to simulate real-world scenarios facilitates deeper understanding and retention of information, leading to improved learning outcomes and enhanced engagement and motivation among students. The combination of AI assistants with VR and AR technologies not only enriches the learning experience, but also ensures that support is personalized and responsive to individual student needs. This innovative approach showcases the potential for AI to transform educational practices in immersive environments. This integration emphasizes the synergy between AI, VR, and AR, highlighting how AI assistants enhance the effectiveness of immersive learning experiences.

The exploration of current trends in AI educational practices reveals a promising landscape characterized by innovative applications that enhance learning experiences. Adaptive learning systems, intelligent tutoring systems, and AI-driven assessment tools are at the forefront of personalizing education and boosting student engagement. Additionally, the availability of user-friendly AI tools simplifies the tasks of lesson planning, grading, and classroom management, making it easier for educators to integrate these technologies into their existing curricula. Together, these advancements not only support educators in their teaching efforts, but also foster a more tailored and effective learning environment for students. Below is a series of AI application used in education cited by Chen, X., Xie, H., & Hwang, G. J, 2020:

Table 2. Examples of AI software tools used in education (Chen, X., Xie, H., & Hwang, G. J, 2020, p.9)

Examples of software tools used to apply AI technologies for educational purposes.

Name of app	Subject	Description	Website
Amira	literacy	Amira is a web-supported, AI-empowered educational application for K-3 literacy learning[1]	[1]https://www.amiralearning.com
Carnegie Learning	math	Carnegie Learning is an AI-empowered app for helping learners cultivate deep conceptual understanding of math[2]	[2]http://www.carnegielearning.com/
Cognii	language	Cognii provides AI-driven educational tools for K-12 education, higher education, and corporate training[3]	[3]http://www.cognii.com/
CueThink	math	CueThink is an AI-driven app for math education that is able to assess learner knowledge based on an engaging peer-to-peer strategy[4]	[4]https://www.cuethink.com/
DreamBox Learning	math	DreamBox Learning is an AI-driven application for math education targeting elementary and middle school students[5]	[5]https://www.dreambox.com/
Edwin	language	Edwin provides AI-empowered services to learn foreign languages[6]	[6]https://edwin.ai
Elemental Path	language	Elemental Path is an AI-driven product for learners to engage with real-time speech-driven communication[7]	[7]https://elementalpath.com/
GlobalEnglish	language	GlobalEnglish is an English learning application based on machine learning technologies[8]	[8]https://globalenglish.com/
Grammarly	language	Grammarly is a machine learning-driven app combined with various NLP technologies for improving writing[9]	[9]https://app.grammarly.com/
iTutorGroup	language	iTutorGroup is an AI-driven application for individualized and personalized language learning[10]	[10]https://www.itutorgroup.com/
KidSense	language	KidSense provides AI-driven applications for children's' language learning[11]	[11]https://kidsense.ai/
Lingos Classroom	language	Lingos Classroom is an AI-driven tool for personalized language learning [12]	[12]https://www.lingos.io/
Linguix	language	Linguix is an AI-driven tool for detecting and resolving spelling and grammar issues in writing[13]	[13]https://linguix.com/
MyGrammarCheck	language	MyGrammarCheck is a grammar checker based on machine learning technologies[14]	[14]https://www.my-grammarcheck.com/
Newsela	language	Newsela is an AI-driven tool for improving reading comprehension with rich article selection on various topics[15]	[15]https://newsela.com/
Noplag App	language	Noplag App is a writing assistance tool for plagiarism checking based on AI technologies [16]	[16]https://noplag.com/
Nuance	language	Nuance provides AI-driven speech recognition tools for students and faculty[17]	[17]http://www.nuance.com/
Pimsleur	language	Pimsleur is an AI-driven language learning app[18]	[18]https://www.pimsleur.com/
Querium	STEM	Querium is an AI-driven application for delivering customizable STEM tutoring courses to high school and college students[19]	[19]http://querium.com/
QuillBot	language	QuillBot is an AI-driven paraphrasing tool[20]	[20]https://quillbot.com/
Quizlet	language, science, math, history, and coding	Quizlet provides services ranging from the learning of foreign language, history, and science to standardized test preparation[21]	[21]https://quizlet.com/
Ready4	language	Ready4 provides services concerning test preparation and college admissions services startup based on AI technologies[22]	[22]http://www.ready4.com/
Rosetta Stone Language Learning	language	Rosetta Stone Language Learning is an augmented reality AI-empowered app for language learning[23]	[23]https://www.rosettastone.com
Symbolab	math	Symbolab is an AI-driven tool for mathematics education[24]	[24]https://www.symbolab.com/
Thinkster Math	math	Thinkster Math provides custom programs for individual learners based on AI technologies[25]	[25]https://hellothinkster.com/
TrueShelf	math	TrueShelf is AI-empowered adaptive math learning software for middle school and high school students[26]	[26]https://trueshelf.com/
VIPKID	language	VIPKID is an AI-empowered personalized language learning tool[27]	[27]http://www.vipkid.com
Yabla	language	Yabla provides personalized course materials based on AI technologies[28]	[28]http://www.yabla.com/
Maths AI App		Maths AI App is an AI-empowered online K-12 mathematics educational tool that is able to provide tailored learning materials[29]	[29]https://www.mathsai.com/
Talk2Learn	Math	Talk2Learn is an open source AI-driven mathematics tutoring tool featuring revolutionary innovations such as an adaptive sequencer and voice recognition[30]	[30]https://www.talk2learn.com/

PEDAGOGICAL STRATEGIES

The integration of artificial intelligence into the educational sphere has, on the other hand, directed the shift of pedagogical practices in online environments, leading to the development of different pedagogical strategies that enhance students' learning outcomes. According to Guan, C., J. Mou, and Z. Jiang, (2020), the instructional design of online environments has increasingly integrated student-led and competency-based pedagogical activities. Models focused on learner-centered instruction have been developed to assist educators and system designers in creating more flexible and self-directed learning spaces, using various theoretical frameworks. These strategies have been shown to positively impact student achievement and engagement, creating more personalized, engaging, and collaborative learning environments. Examples include the problem-based learning, blended learning models, flipped classrooms, and collaborative learning environments augmented by AI.

The Problem-Based Learning

The Problem-Based Learning approach is an instructional strategy designed to help students tackle real-world issues. In problem-based learning, the problems presented are typically ill-structured, mirroring the complexities of real-life situations, and students engage actively in the process of finding solutions. Research indicates that problem-based learning approach is effective in enhancing students' learning experiences. Aslan (2021) examined the effectiveness of problem-based learning in online educational settings. The study focuses on four key aspects: learning achievement, problem-solving skills, communication skills, and student interaction. The author assumes that implementing problem-based learning in live online classes significantly enhances learning outcomes by improving academic achievement, problem-solving skills, communication skills, and student interaction. The article concludes that educators should consider problem-based learning as a valuable approach to foster deeper learning in online education.

Blended Learning Models

Blended learning is a mixed educational approach that combines traditional face-to-face classes and distance learning in e-learning. This hybrid model allows for a mix of in-person and digital interactions, creating a more flexible and engaging learning experience. There isn't a single model of blended learning; instead, there are multiple models available. Instructional designers can select the combination of face-to-face and online elements that they believe is most appropriate for the training program. For instance, some blended courses may focus primarily on e-learning with just a few in-person sessions. Conversely, there are hybrid courses that are predominantly in-person but include some online training modules or activities. AI has notably enhanced efficiencies, especially in distance and online education. For instance, programs like Knewton offers platforms that provides feedback based on student interactions; it provides adaptive learning paths that adapt to individual specific student needs, improving engagement and academic performance. Similarly, Khan Academy has recently emerged as a noteworthy means for educators employing blended learning in their classes (UNESCO, 2019). It is an educational platform that expand face-to-face teaching by proposing instructional videos and exercises that engage students outside the classroom, allowing for more interactive class time. It offers students personalized education that enables them to learn at their own pace, improving their academic performance that can be measured by their professors as well.

Flipped Classrooms

The flipped classroom model changes traditional learning methods by giving students new instructional content to learn at home, while the class session is dedicated to interactive activities with other students. AI tools used in this approach analyze students' comprehension and engagement of pre-class materials, providing effective feedback to teachers that help them adapt in-class activities efficiently. While flipped classrooms and blended learning are related concepts, they differ significantly in their structure and focus. A flipped classroom diverges from the traditional learning model by having students first engage with new content at home through online materials, such as videos or readings, before practicing their understanding in interactive classroom activities. This approach accentuates on maximizing in-class time for active learning and collaboration, with teachers acting as facilitators rather than traditional lecturers. Blended learning, on the other hand, involves a broader range of instructional paradigms that blend face-to-face and online sections, allowing for various arrangements of learning environments. While a flipped classroom is a particular application of blended learning, the latter includes diverse approaches, such as fully online courses with occasional in-person sessions or predominantly in-person classes supplemented by digital resources.

Collaborative Learning Environments

Collaborative learning is another educational approach in online education that points up group work and interaction among students to achieve shared learning goals. In this approach, students work all together in teams, engaging in discussions, problem-solving, and project-based tasks that encourage active participation and cooperation. AI facilitates collaborative learning by connecting students across different locations, developing their skills to work together on projects. According to Zawacki-Richter et al. (2019), AI "can contribute to collaborative learning by supporting adaptive group formation based on learner models, by facilitating online group interaction or by summarising discussions that can be used by a human tutor to guide students towards the aims and objectives of a course" (p.4). Collaborative learning is a powerful educational strategy that not only develops academic outcomes, but also develops essential interpersonal skills through various means, such as critical thinking and communication skills. By working together in groups, students engage in discussions that encourage them to evaluate different perspectives, analyze arguments, and construct well-reasoned responses, thereby fostering critical thinking. Shared problem-solving tasks prompt students to consider multiple solutions, weigh pros and cons, and negotiate the best course of action, further strengthening their analytical abilities. Exposure to diverse viewpoints enriches their understanding and challenges their assumptions, while active listening in collaborative settings enhances their communication skills as they learn to articulate their thoughts and engage meaningfully with peers. Additionally, giving and receiving constructive feedback helps refine their ideas and improve their ability to communicate effectively. Role assignments within collaborative projects promote accountability, requiring students to prepare and present their ideas clearly.

All these pedagogical strategies reveal the transformative potential of AI in education, by personalizing learning, fostering collaboration, and providing educators with actionable insights. Yet, as research continues to evolve, further exploration of these strategies will be crucial in capitalizing on the benefits of AI integration in educational settings that significantly enhance student outcomes. Actually, the articles reviewed reveal a concerning lack of critical reflection on the pedagogical practices and strategies associated with AI applications education. According to Ifenthaler (2020), the future of AI in education presents significant opportunities for transforming learning and teaching, but realizing its full potential requires collaboration among stakeholders, including educators, policymakers, technologists, and researchers to address evolving pedagogical considerations. Hence, while AI offers substantial benefits, there is a pressing need for reliable empirical studies to understand its impact in educational contexts. This was further emphasized by Zawacki-Richter et al. (2019) and Ouyang et al. 2021, who note that system-wide implementations are still lacking and point out significant gaps in the pedagogical implications of AI, especially in higher education. Crompton and Burke 2023, on the other hand, suggested a need for increased educational perspectives from faculty on AI developments and emphasize the importance of sharing findings across disciplines to enhance pedagogical practices. According to Guan et al. (2020), EdTech encompasses not just technology but also the pedagogical aspects of education. In this regard, AIEd research is inherently interdisciplinary, meaning that significant advancements in related fields can lead to a paradigm shift in AIEd research. Hence, Educators must critically reflect on their pedagogical practices to ensure technology aligns with educational goals (Chu, S. T., Hwang, G. J., & Tu, Y. F, 2022).

CHALLENGES

Although AI has the potential to transform education; still, "the application of AI also brings up a number of other challenges, including the potential replacement of the teaching position, disparities in educational quality, and moral dilemmas with data privacy, security, fairness, and openness" (Judijanto et al. 2024, p.1723). Thus, various research teams have carried out research to highlight recurring problems in Artificial Intelligence in education which critically effect the implementations of AI in education, prompting discussions about the responsible use of AI in educational settings (Baker, T., & Smith, L, 2019; Zawacki-Richter et al. 2019; Huang et al. 2021; Akgun, S., & Greenhow, C, 2022; Chiu, T. K., Xia, Q., Zhou, X., Chai, C. S., & Cheng, M, 2023; Li, F., Ruijs, R., & Lu, Y, 2023; Ifenthaler, D., Majumdar, R., Gorissen, P, 2024; Ivanashko, O., Kozak, A., Knysh, T., & Honchar, K, 2024).

According to Akgun et al. 2022, the implementation of AI systems in education raises significant ethical challenges and risks, often contrasting with marketing claims that present these technologies as objective and neutral. One major ethical issue pertains to privacy. Students and teachers often face privacy violations due to excessive personal information shared on online platforms. Despite existing legislation aimed at protecting sensitive data, breaches by AI tech companies exacerbate privacy concerns. While AI systems request user consent for data access, many individuals unknowingly consent to sharing extensive metadata, undermining their agency and privacy. The pressure on students and parents to use these algorithms in educational settings, despite potential privacy infringements, raises ethical questions about choice and autonomy. According to Chiu et al. 2023, educational institutions must navigate regulations surrounding data protection while ensuring that students' personal information is safeguarded against breaches. This challenge necessitates robust data management practices and transparent policies to maintain trust among stakeholders.

Another challenge raised by Akgun et al. 2022; Chiu et al. 2023; and Ifenthaler et al. 2024 is the potential for bias in AI algorithms, which can lead to unfair treatment of students based on race, gender, demographic factors, or socioeconomic status, manifesting in AI-driven platforms. When people generate algorithms, they correspondingly create "a set of data that represent society's historical and systemic biases, which ultimately transform into algorithmic bias. Even though the bias is embedded into the algorithmic model with no explicit intention, we can see various gender and racial biases in different AI-based platforms" (Akgun et al. 2022, p.434). Similarly, Ivanashko et al. (2024, p. 134) stressed the potential for bias, because "if the data on educational process contains biases, AI may produce discriminatory results, affecting certain student groups unfairly". For instance, Akgun et al. (2022) highlighted that existing societal biases are often embedded in AI systems, as seen in language translation tools that reflect gender stereotypes or facial recognition systems that misidentify individuals from marginalized groups. Instances of biased decision-making algorithms can lead to unfair outcomes, as demonstrated by the controversies surrounding standardized assessments in the UK, where algorithmic grading favored students from private schools over those from underrepresented backgrounds. Hence, addressing these ethical dilemmas requires ongoing scrutiny of AI tools and the establishment of guidelines that promote fairness and equity in educational practices (Chiu et al. 2023).

Another significant concern is surveillance. AI tracking systems collect detailed data on student and teacher behaviors, which can be used to predict future actions and learning patterns. While monitoring can serve educational purposes, it often encroaches on privacy and can create an environment of mistrust. Students may feel unsafe to express themselves if they know their actions are being surveyed. Some studies indicate that educators who utilize social networking sites (SNSs) for teaching purposes face various challenges, including issues related to privacy boundaries, the dynamics of friendship and authority, and questions of responsibility and availability. Although monitoring students' activities may be viewed as a necessary part of a teacher's role and a means to address serious online issues like cyberbullying or exposure to inappropriate content, these actions can also be perceived as invasive surveillance, which raises concerns about student privacy (Akgun et al. 2022).

Another issue raised by Chiu et al. (2023); Ifenthaler et al. (2024); and Ivanashko et al. (2024) is the lack of training and professional development. They emphasized the necessity for training educators and administrators on how to effectively use AI technologies and on ongoing professional development to keep up with rapidly evolving AI tools and methodologies. Similarly, Wardat et al. (2024) stated in their study on mathematics teachers' perspectives that many professors feel unprepared to use AI tools effectively due to insufficient training. This is due to a lack of experience and knowledge in working with IT software, devices, and applications. Some teachers might struggle to incorporate AI-based tutoring systems into their lessons because they lack the necessary skills and knowledge. For example, integrating AI tools into existing mathematics curricula can be challenging for some; they find it difficult to align AI applications with specific learning objectives or state standards.

Some educators, on the other hand, may be resistant to adopting AI technologies, preferring traditional teaching methods. This resistance can stem from concerns about AI replacing the teacher's role or doubts about its effectiveness. For example, a teacher might hesitate to use an AI tool for grading or feedback, fearing it could diminish personal interaction with students; in other words, teachers face the challenge of finding the right balance between using AI tools and maintaining personal interaction with students. According to Ifenthaler et al. (2024), this resistance and hesitance to change among educators and institutions to adopt AI technologies are due to fear of job displacement or skepticism about the effectiveness of AI solutions.

In point of fact, the shortage of AI specialists and experts available to train educators on these technologies presents another challenge, as qualified individuals with the necessary skills and experience are limited compared to other sectors. "This is because the development and modernization of AI systems and applications require adaptation to specific data sets and contexts, which demands a unique set of talents and expertise possessed by only a few individuals globally" (Wardat, Y., Tashtoush, M., Alali, R., & Saleh, S, 2024; p. 73). Ultimately, the level of technical proficiency achievable in this area relies heavily on the investment of human capital in the development of such software.

More importantly, according to Huang et al. (2021), the shift in learning styles places greater demands on students' ability to learn autonomously, as the AI-driven educational landscape becomes increasingly student-centered, allowing learners to take the lead in their educational activities. Through intelligent teaching systems, students can create personalized learning plans, independently select materials, manage their learning pace, and engage in collaborative group work. Such individualized approaches require higher levels of self-regulation and self-management from students, prompting educators to focus on nurturing independent learning skills. However, as reliance on AI platforms grows, it is also essential to enhance student communication skills among peers. Increased interactions with machines could hinder social communication abilities, making it vital for AI education initiatives to implement distance education models that prioritize socialization. This would facilitate collaborative projects and social activities, allowing students to engage with classmates from different locations while developing both independent learning and effective communication skills in a technology-driven educational environment

Another emerging challenge is the question of equity and accessibility; essentially, not all students have equal access to AI technologies, leading to disparities in learning opportunities. For instance, students from low-income backgrounds may lack access to devices or reliable internet, making it difficult for them to benefit from AI-enhanced learning tools (Wardat et.al. 2024). Hence, although AI may have a lot of advantages in education, t can also act as a disruptive technology, potentially exacerbating existing inequalities and divides, as marginalized and disadvantaged populations are more likely to be excluded from AI-driven education. This leads to a new form of digital divide: a gap in the ability to utilize data-driven knowledge for informed decision-making. (UNESCO, 2019). Similarly, Huang et al. (2021 p. 212) stressed that developing countries may face the risk of "exacerbating the divisions in education by new technologies. Just as the digital divide has separated those who can access to the Internet from those who cannot, the ever-widening algorithmic divide now threatens to deprive many educational opportunities provided by AI". According to them, most AI algorithms originate from developed countries, which means they may not adequately address the specific conditions and needs of developing nations. So, to leverage AI effectively for enhancing learning, the education sector must overcome substantial challenges, including insufficient basic technology and infrastructure. Equity, accessibility and inclusion must be fundamental principles in the formulation of AI education policies. Consequently, policymakers should consider various questions related to inclusion and equity during the policy development process (Ifenthaler et al. 2024).

The successful integration of AI tools with existing educational frameworks poses a considerable challenge (Ifenthaler et al. 2024). Many institutions may have legacy systems that are not compatible with new AI technologies, leading to disruptions in teaching and learning processes. Effective integration requires careful planning, investment in compatible technologies, and ongoing support for educators during the transition (Chiu et al. 2023). More importantly, challenges exist in scaling AI solutions to accommodate diverse educational contexts and student populations. In fact, what works in one setting may not be easily transferable to another due to differences in resources, student demographics, and institutional goals. Hence, developing scalable AI solutions that can be adapted to various contexts is vital for maximizing their impact on education (Chiu et al. 2023).

The challenges identified underscore the complexities involved in integrating AI into educational settings. Addressing these challenges through targeted research, policy development, and collaboration among stakeholders is essential for unlocking the transformative potential of AI in education. Ongoing efforts to mitigate these challenges will not only enhance the adoption of AI technologies, but also ensure that they contribute positively to educational outcomes for all students.

CONCLUSION

The integration of Artificial Intelligence in education offers significant opportunities to enhance learning experiences, yet it also raises important ethical and practical challenges that must be addressed to ensure its effective implementation. As AI technologies continue to evolve and permeate educational settings, the potential for personalized learning, intelligent tutoring, and adaptive educational tools is immense. Yet, the lack of clear regulatory and policy frameworks governing AI use in educational settings accentuates the need for policymakers to establish guidelines that promote ethical practices while facilitating the integration of technology; Issues such as bias, privacy, and fairness must be carefully considered to mitigate potential negative effects associated with AI deployment in educational environments.

RECOMMENDATIONS

Basically, to navigate these challenges effectively, educational institutions should join forces with policymakers to create vigorous strategies and plans that tackle ethical considerations surrounding AI use. For instance, educational institutions must implement robust data management systems that prioritize the protection of personal information. This may embrace systematic examinations and updates to ensure compliance with data protection regulations. On the other hand, research teams should focus on identifying and mitigating biases in AI algorithms. This can involve developing best practices for data selection and model training to ensure that AI tools provide equitable treatment for all students, regardless of their backgrounds. These procedures should found bases for data privacy, bias mitigation, and transparency in AI-driven decisions. By doing so, stakeholders can ensure that AI applications are implemented responsibly and ethically, fostering trust among educators, students, and parents.

Policymakers must also prioritize equity and accessibility in AI education initiatives. It is essential to ensure that all students, regardless of their socioeconomic status, have access to the technology and infrastructure needed to benefit from AI-enhanced learning, contributing to a more equitable educational landscape. Investments must be made in the necessary technological infrastructure within schools, especially in rural and underserved urban areas. Similarly, policymakers should support and fund initiatives that provide essential technology, such as laptops, tablets, and reliable internet access, particularly for students from low-income families.

More importantly, professional development programs must be launched in an institutional way to train the educational stuff and equip them with the needed skills to successfully integrate AI tools into their teaching methods. However, the trainings should take into consideration the technical skills, as well as the ethical issues of AI in the classroom. In other words, teachers should be empowered to critically evaluate AI tools and understand how to leverage them to enhance learning outcomes while being aware of potential biases and ethical concerns.

More importantly, as AI tools concentrate on student autonomy, educators should emphasize the development of self-regulation and self-management skills. Programs that cultivate independent learning should also include components that foster social interaction and communication among peers. These skills are critical for students to direct their learning time effectively and to take ownership of their educational experiences. For instance, since AI provide personalized feedback and adaptive learning paths, teachers should encourage their students to use these tools to track their own progress, reflect on their learning strategies, and adjust their approaches as required, fostering a sense of agency. Yet, teachers should propose group projects, peer reviews, and collaborative problem-solving activities that not only enhance learning, but also help students develop interpersonal skills.

In view of that, encouraging co-operation and partnership between educators, technologists, and ethicists can lead to more effective and responsible AI applications in education. Interdisciplinary teams can help align AI innovations with pedagogical goals, ensuring that technology is used to enhance learning rather than weaken the process. By combining various viewpoints, these collaborations can promote creative solutions that tackle the challenges of incorporating AI into educational settings.

FUTURE RESEARCH DIRECTIONS

Understanding the long-term impacts of AI on educational outcomes and student development is a critical area for further research. There is an urgent need for reliable empirical studies to assess the impact of AI applications across various educational contexts. Such research should explore both the benefits and drawbacks of AI integration, providing a balanced view that informs educators and policymakers about effective practices and potential risks. For example, longitudinal studies or research design, that involve repeated observations or measurements of the same subjects over an extended period, can help assess how the integration of AI tools impacts student learning, engagement, and achievement over different stages of their education. By collecting data at multiple points, researchers can identify long-term effects, determine the sustainability of educational innovations, and evaluate how students adapt to AI-enhanced learning environments. Applying this method allows researchers to follow and understand changes and developments over time, providing insights into trends, patterns, and causal relationships.

Future research must also address how AI can be leveraged to improve accessibility for all students, particularly those with disabilities or from underserved backgrounds. Scholars, for instance, can investigate how AI tools can be adapted to meet the needs of disabled students or underserved regions and incorporate case studies that highlight successful implementations in diverse cultural settings. These case studies should encompass diverse educational settings, including urban and rural schools, to ensure broad applicability and relevance across different contexts. This approach will help ensure that the application of AI in education is both equitable and inclusive, ultimately broadening its global applicability.

Accordingly, by addressing these areas, educational stakeholders can better align AI innovations with educational goals, ultimately benefiting all students and fostering a more equitable learning environment. The thoughtful integration of AI, guided by ethical considerations and robust research, has the potential to transform education positively, making learning more accessible, personalized, and effective for every learner.

REFERENCES

Ahmad, S. F., Rahmat, M. K., Mubarik, M. S., Alam, M. M., & Hyder, S. I. (2021). Artificial intelligence and its role in education. *Sustainability (Basel)*, 13(22), 2902. DOI: 10.3390/su132212902

Akgun, S., & Greenhow, C. (2022). Artificial intelligence in education: Addressing ethical challenges in K-12 settings. *AI and Ethics*, 2(3), 431–440. DOI: 10.1007/s43681-021-00096-7 PMID: 34790956

Akyuz, Y. (2020). Effects of Intelligent Tutoring Systems (ITS) on Personalized Learning (PL). *Creative Education*, 11(6), 953–978. DOI: 10.4236/ce.2020.116069

Alam, A., Hasan, M., & Raza, M. M. (2022). Impact of artificial intelligence (AI) on education: changing paradigms and approaches. Towards Excellence, 14(1), 281-289. DOI:DOI: 10.13140/RG.2.2.15581.28646/1

Annus, N. (2023). Chatbots in Education: The impact of Artificial Intelligence based ChatGPT on Teachers and Students. *International Journal of Advanced Natural Sciences and Engineering Researches*, 7(4), 366–370. DOI: 10.59287/ijanser.739

Aslan, A. (2021). Problem- based learning in live online classes: Learning achievement, problem-solving skill, communication skill, and interaction. *Computers & Education*, 171, 104237. Advance online publication. DOI: 10.1016/j.compedu.2021.104237

Baker, T., & Smith, L. (2019). Education rebooted? Exploring the future of artificial intelligence in schools and colleges. Retrieved from Nesta Foundation website: https://media.nesta.org.uk/documents/Future_of_AI_and_education_v5_WEB.pdf

Bates, T., Cobo, C., Mariño, O., & Wheeler, S. (2020). Can artificial intelligence transform higher education? *International Journal of Educational Technology in Higher Education*, 17(1), 1–12. DOI: 10.1186/s41239-020-00218-x

Chen, L., Chen, P., & Lin, Z. (2020). Artificial intelligence in education: A review. *IEEE Access : Practical Innovations, Open Solutions*, 8, 75264–75278. DOI: 10.1109/ACCESS.2020.2988510

Chen, X., Xie, H., & Hwang, G. J. (2020). A multi-perspective study on artificial intelligence in education: Grants, conferences, journals, software tools, institutions, and researchers. Computers & Education. *Artificial Intelligence*, 1, 100005. Advance online publication. DOI: 10.1016/j.caeai.2020.100005

Chen, X., Xie, H., Zou, D., & Hwang, G. J. (2020). Application and theory gaps during the rise of artificial intelligence in education. Computers & Education. *Artificial Intelligence*, 1, 100002. Advance online publication. DOI: 10.1016/j.caeai.2020.100002

Chen, X., Xie, H., Zou, D., & Hwang, G. J. (2020). Application and theory gaps during the rise of artificial intelligence in education. Computers & Education. *Artificial Intelligence*, 1, 100002. Advance online publication. DOI: 10.1016/j.caeai.2020.100002

Chiu, T. K., Xia, Q., Zhou, X., Chai, C. S., & Cheng, M. (2023). Systematic literature review on opportunities, challenges, and future research recommendations of artificial intelligence in education. *Computers and Education: Artificial Intelligence*, 4, 100118. DOI: 10.1016/j.caeai.2022.100118

Chu, S. T., Hwang, G. J., & Tu, Y. F. (2022). Artificial intelligence-based robots in education: A systematic review of selected SSCI publications. Computers and education. *Artificial Intelligence*, 3, 100091. DOI: 10.1016/j.caeai.2022.100091

Crompton, H., & Burke, D. (2023). Artificial intelligence in higher education: The state of the field. *International Journal of Educational Technology in Higher Education*, 20(1), 22. Advance online publication. DOI: 10.1186/s41239-023-00392-8

Elaish, M. M., Yadegaridehkordi, E., & Ho, Y. S. (2024). *Publication performance and trends in virtual reality research in education fields: a bibliometric analysis.* Multimed Tools Appl., DOI: 10.1007/s11042-024-19238-0

Gartner. (2019). Hype cycle for emerging technologies. https:// www.gartner.com/en/documents/3956015/hype-cycle-for-emerging-technolo gies-2019.

Guan, C., Mou, J., & Jiang, Z. (2020). Artificial intelligence innovation in education: A twenty-year data-driven historical analysis. *Int. J. Innov. Stud.*, 4(4), 134–147. DOI: 10.1016/j.ijis.2020.09.001

Huang, J., Saleh, S., & Liu, Y. (2021). A Review on Artificial Intelligence in Education, Academic Journal of Interdisciplinary Studies. *Richtmann Publishing Ltd*, 10(3), 206. Advance online publication. DOI: 10.36941/ajis-2021-0077

Ifenthaler, D., Majumdar, R., Gorissen, P., Judge, M., Mishra, S., Raffaghelli, J., & Shimada, A. (2024). Artificial Intelligence in Education: Implications for Policymakers, Researchers, and Practitioners. *Tech Know Learn*, 29(4), 1693–1710. DOI: 10.1007/s10758-024-09747-0

Ivanashko, O., Kozak, A., Knysh, T., & Honchar, K. (2024). The Role of Artificial Intelligence in Shaping the Future of Education: Opportunities and Challenges. *Futurity Education*, 4(1), 126–146. DOI: 10.57125/FED.2024.03.25.08

Judijanto, L., Atsani, R., & Chadijah, S. (2024). Trends in the Development of Artificial Intelligence-Based Technology in Education. International Journal of Teaching and Learning. Vol. 2 No. 6. µ https://injotel.org/index.php/12/article/view/197

Li, F., Ruijs, R., & Lu, Y. (2023). Ethics & AI: A systematic review on ethical concerns and related strategies for designing with ai in healthcare. *AI*, 4(1), 28–53. DOI: 10.3390/ai4010003

Mousavinasab, E., Zarifsanaiey, N., Kalhori, S. R. N., Rakhshan, M., Keikha, L., & Saeedi, M. G. (2021). Intelligent tutoring sys tems: A systematic review of characteristics, applications, and evaluation methods. *Interactive Learning Environments*, 29(1), 142–163. DOI: 10.1080/10494820.2018.1558257

Nilsson, N. J., Hilpisch, Y., Yao, M., Zhou, A., Jia, M., Baesen, B., . . . Verbeke, W. (2010). The quest for ai: A history of ideas and achievements. Erişim adresi: http:// ai. standford. edu/~ nilsson/(Özgün eser 2009 tarihlidir).

Ouyang, F., Zheng, L., & Jiao, P. (2022). Artificial intelligence in online higher education: A systematic review of empirical research from 2011–2020. *Education and Information Technologies*, 27(6), 7893–7925. DOI: 10.1007/s10639-022-10925-9

Oyelere, S. S., Bouali, N., Kaliisa, R., Obaido, G., Yunusa, A. A., & Jimoh, E. R. (2020). Exploring the trends of educational virtual reality games: A systematic review of empirical studies. *Smart Learn. Environ.*, 7(1), 31. DOI: 10.1186/s40561-020-00142-7

Russel, S., & Norvig, P. (2010). *Artificial intelligence- a modern approach*. Pearson Education.

Wardat, Y., Tashtoush, M., Alali, R., & Saleh, S. (2024). Artificial Intelligence in Education: Mathematics Teachers' Perspectives, Practices and Challenges. *Iraqi Journal For Computer Science and Mathematics*, 5(1), 60–77. DOI: 10.52866/ijcsm.2024.05.01.004

Wu, F., Lu, C., Zhu, M., Chen, H., Zhu, J., Yu, K., Li, L., Li, M., Chen, Q., Li, X., Cao, X., Wang, Z., Zha, Z., Zhuang, Y., & Pan, Y. (2020). Towards a new generation of artificial intelligence in China. *Nature Machine Intelligence*, 2(6), 312–316. DOI: 10.1038/s42256-020-0183-4

Zawacki-Richter, O., Marín, V. I., Bond, M., & Gouverneur, F. (2019). Systematic review of research on artificial intelligence applications in higher education – where are the educators? *International Journal of Educational Technology in Higher Education*, 16(1), 39. DOI: 10.1186/s41239-019-0171-0

Chapter 2
The Impact of Generative AI on Education: A Review of Applications and Challenges

Nisserine El Bahri
https://orcid.org/0009-0006-7451-8840
Independent Researcher, Morocco

Zakaria Itahriouan
https://orcid.org/0000-0001-6573-090X
Independent Researcher, Morocco

Anouar Abtoy
https://orcid.org/0000-0002-7955-6719
National School of Applied Sciences, Tetuan, Morocco

ABSTRACT

Education is utilizing generative artificial intelligence more and more, especially with models like GPT, BERT, and other neural networks. Through the automatic generation of information, individualized learning resources, and responses to student inquiries, these technologies are transforming traditional teaching methods, course design, and interactions with learners. Despite the fact that these tools have the potential to drastically change the nature of education, it is crucial to comprehend their existing uses, advantages, and drawbacks. The purpose of this chapter is to present a thorough analysis of current scholarly research on the application of generative AI in education across all academic levels.

DOI: 10.4018/979-8-3373-2262-9.ch002

INTRODUCTION

The integration of generative AI into education is rapidly transforming how learning is shaped, delivered, and experienced. By providing personalized learning, automating repetitive work, and creating dynamic and interactive learning environments, AI technologies are revolutionizing established teaching methods. Emerging technologies such as intelligent tutors (Lin et al., 2023), virtual assistants(Labadze et al., 2023), and content creation platforms (Kaswan et al., 2023) provide educators with new ways to engage with their students and provide them with access to information. These developments are particularly attractive because they have the potential to improve accessibility, adaptability, and inclusion in education for a wide range of demographics and geographies(Andrin et al., 2024).

However, the emergence of generative AI is accompanied by a number of challenges. Important pedagogical and ethical issues arise due to concerns about academic integrity, biases in AI-generated content, and the evolving role of teachers (Meça & Shkëlzeni, 2024). Furthermore, concerns about data privacy, equitable access to AI tools, and the long-term effects of AI on learning outcomes are still evolving and require careful consideration (Stahl, 2021).

This chapter examines the position of generative AI in education today, examining the challenges and revolutionary possibilities it presents. This review aims to give a comprehensive understanding of how AI is changing education by considering important topics such as personalized learning, teacher-student dynamics, ethical issues, and emerging trends. In order to build a more efficient, fair and sustainable learning environment, the paper ultimately seeks to provide insights into how institutions, educators and legislators can embrace the potential of generative AI while addressing its drawbacks.

BACKGROUND

Rapid progresses in generative technologies, particularly those using natural language processing and deep learning (e.g., GPT[1], DALL-E[2]), have unlocked radical possibilities in a number of issues, including education (Bengesi et al., 2024; Vayadande et al., 2024). At the façade of these developments has been personalized learning, a pedagogical approach that transforms educational practices to sustain the needs of each individual. To meet the varying demands of learners, generative AI succeeds in generating dynamic content, adaptive learning pathways, and real-time feedback.

Generative AI

Generative AI is the most rapidly growing field of AI. It can be used to build multiple types of content like text, images, audio and many other types of data. This technology allows to generate data that perfectly resembles that by which the generative AI models were trained, unlike other traditional artificial intelligence algorithms (Feuerriegel et al., 2024; Foster, 2022). The recent methods of machine learning, especially deep neural networks, are used to do this in combination with transformer-based architectures (Harshvardhan et al., 2020).

Well known models include DALL-E for image synthesis (Ko et al., 2023), Jukebox[3] for music (Dhariwal et al., 2020), and OpenAI's GPT series, which give excellent results at natural language creation (Yenduri et al., 2024). These systems can even simulate human conversations, generate graphics of multiple kinds, because they have been trained by large datasets and substantial computational resources. For example, GPT-based models give correct and context-appropriate answers by predicting the next word in a text of a conversation. In a similar way, models applied to image generation use techniques like stable diffusion to transform random noise into complex visual representations controlled by text input.

Due to its revolutionary importance, generative AI has found many prospects and has rapidly developed in several sectors:

- Creative Industries (Amankwah-Amoah et al., 2024; De Cremer et al., 2023): It helps authors and writers to get inspiration or write articles. On the other hand, it can help designers to create works of art as well as musicians to compose songs by offering a new, faster and more innovative approach.
- Healthcare (Meskó & Topol, 2023; Zhang & Kamel Boulos, 2023): Even in this highly sensitive area, generative AI can simulate complex protein structures for drug discovery and can also help create individualized healthcare solutions.
- Marketing (Kshetri et al., 2024; Wahid et al., 2023): Generative AI can create specific content that targets customers in a particular way. It can also help to significantly personalize advertising and even predict consumer activities to improve marketing strategy.
- Technology and Software Development (Ebert & Louridas, 2023; Peng et al., 2023): Currently, generative AI can help with program troubleshooting tasks, suggesting code snippets while coding, which helps speed up software development as in the case of GitHub Copilot.
- Finance & Business (B. Chen et al., 2023; J. Rane et al., 2024): Along with malfunction modeling, generative AI is being explored in finance to improve fraud detection, financial reporting, and economic scenario modeling. Using

big data and non-human assistants, artificial intelligence (AI) applications are helping businesses optimize their services.

- Gaming and Simulation (J. Lee et al., 2023; Lv, 2023): Generative AI is used by video game designers to produce unusual stories, manage dialogues between characters and vast virtual worlds. AI makes games more real by improving game environments and player interactions.
- Scientific Research and Engineering (Alasadi & Baiz, 2023; Qadir, 2023): AI models are used in simulations in the field of exact sciences such as physics and chemistry. The goal is to be able to simulate complex systems, optimize engineering works and create new products.
- Social media and Content Creation (Brüns & Meißner, 2024; Gayam, 2022): Generative AI is currently being used to automatically produce videos, edits and also to automatically generate translation subtitles for online videos based on user preferences. This is also valid for even generating text and sound or at least optimizing it.
- Urban Planning and Architecture (Jiang et al., 2024; Miao et al., 2020): Whether it's construction plans, modeling the environmental effects of construction projects, or even designing sustainable urban developments, generative AI helps architectural specialists highlight the design and visualize complex buildings even before construction.
- Legal and Documentation (Contini, 2024; Macey-Dare, 2023): Generative AI can help legal professionals draft contracts, legal opinions, and documents in general. It can be used to ensure regulatory compliance while automating repetitive activities.

Moreover, several challenges arise for this emerging field. Ethical issues seem important to explore. Generative AI can also be used for things that are undesirable such as deepfake, fake news and other bad practices. Inequalities in the data used in this type of AI can produce biased results and the lack of transparency can clash with principles related to the protection of personal data. The heavy calculations during the training and use of its models result in a very high consumption of energy and this has its consequences on the environment.

In any case, AI can still be considered a revolution that has its advantages and disadvantages, and its development will continue especially thanks to the assets it presents in terms of improving the creativity of human beings.

Educational technology

The advent of technology in the field of education has revolutionized the approaches used to teach whether in the institutional or non-institutional framework. This has allowed us to think of new perspectives in terms of participation, personalization and accessibility. The notion of educational technologies is really very broad including everything from Learning Management Systems to platforms with functionalities based on artificial intelligence (Mangal & Mangal, 2019). The latter are used to enable content sharing, collaboration and many other services for both learners and teachers. Major advancements in this field include:

- Learning Management Systems (LMS) (Cavus, 2015; Turnbull et al., 2020): used to share courses, exercises and any type of content with learners. They can even be used to produce content directly in the platform. Teachers and institutions can use them to have immediate feedback on learners, track learners' progress in content and even perform automatic assessments.
- Massive Open Online Courses (MOOCs) (De Moura et al., 2021; Wei et al., 2021): MOOCs are designed to deliver educational content to a very large audience, allowing learners at even international levels to access courses for which they must travel distances across the globe. Examples of MOOC platforms are Udemy[4], Coursera[5] and edX[6].
- Adaptive learning technologies (Essa et al., 2023; Martin et al., 2020): These types of platforms are designed to provide a unique and personalized learning experience according to the specific needs of the learner. This personalization is impacted by the learner's performance and engagement in these platforms. Examples of personalized learning environments are DreamBox[7] or Knewton[8].
- Collaborative Technologies (Hewson & Chung, 2021; Salih, 2021): These platforms are used to conduct courses in groups while facilitating exchange, discussions and collaborative editing of documents and deliverables. Examples of these platforms are Google Workspace and Microsoft Teams.
- E-assessment and proctoring tools (Gibson & Kuhn, 2024; Iaaly et al., 2024): Another advantage of using technology in education is the ability to conduct tests on educational platforms that offer a multitude of features such as automated correction, authenticity and integrity of questions, and remote access to tests. ProctorU[9] and Respondus[10] are examples of such platforms.
- Virtual Reality (VR) and Augmented Reality (AR) (Al-Ansi et al., 2023; Jang et al., 2021): Virtual reality and augmented reality technologies are increasingly used for learning. They are used in the simulation of real cases, laboratory experiments and virtual educational visits.

- Intelligent tutoring systems (ITS) (Lin et al., 2023; Mousavinasab et al., 2021): Artificial intelligence has made it possible for intelligent tutoring systems to communicate with students in a natural way, automate corrections, and automatically produce grades and comments.
- Gamification (Ofosu-Ampong, 2020; Oliveira et al., 2023): These are serious games where the goal is to learn in a fun way. By implementing rewards or prizes in these games, learners become more motivated to learn and develop positive relationships with the content learned.

For generative AI there are still obstacles despite the extremely revolutionary results achieved. Biases in the decisions made by these technologies, the lack of equal access to new technologies and the adaptation of institutions and teachers are reported challenges. However, educational technologies increasingly influence learning methods confirming that the future will be impacted by generative AI in the short or long term.

TEACHING AND LEARNING MUTATION

With the implementation of personalized learning experiences for students, more flexible and more accessible, generative AI in education promises radical changes in the way teaching and learning are carried out (Chiu, 2024). Learners can use personalized systems at their own pace and to meet their specific needs thanks to AI-powered systems that can create personalized learning paths while relying on student data for (Zawacki-Richter et al., 2019). AI tutors and virtual assistants can help with guidance during training, providing students with additional resources and quick feedback to improve their understanding of difficult topics (Roll & Wylie, 2016).

According to Ivica Pesovski et al. (Pesovski et al., 2024), in a learning system, generative AI can produce diverse learning resources to fit students' preferences. This study demonstrated that matching students' individual learning interests and preferences significantly improved student engagement by implementing AI-generated content that was available in both traditional and popular culture-influenced styles. In addition, the exams automatically generated with generative AI allowed students to properly assess their progress, improving their understanding and the efficiency of the time used for assimilation of the courses.

Zied Bahroun et al. (Bahroun et al., 2023) studied in an in-depth analytical manner the approach according to which generative AI could revolutionize personalized learning environments. The analysis demonstrates how AI-powered learning support systems adjust to live student progress and suggest more resources. This flexibility is crucial to maximize student understanding and engagement, as AI provides tailored training, immediate response, and adaptive testing based on unique learning methods.

In their discussion of AI-generated characters, Pat Pataranutaporn et al. (Pataranutaporn et al., 2021) address ethical issues while highlighting the beneficial uses of these characters in education and wellbeing. The study emphasizes how AI-generated characters can tailor instruction to each student's requirements, interests, and learning preferences. Each learner can receive personalized comments and support from these virtual characters, who can also serve as classmates, mentors, or teachers.

Based on the references cited above, generative AI allows learning to be personalized according to multiple factors (table 1).

Pat Pataranutaporn et al. (Pataranutaporn et al., 2021) focused on AI-generated characters and shed light on ethical issues while highlighting their positive applications in the field of education and well-being. The study shows that content is adapted according to students' interests and preferences thanks to AI-generated characters. These virtual characters can play multiple roles. They can go from being a classmate to a mentor or teacher, often offering tailored support and feedback.

Based on the references mentioned above, generative artificial intelligence can adapt learning to various factors (Table 1).

Table 1. Personalized learning

Factor	Explanation
Analyzing Data to Understand Learner Needs	By evaluating student data to identify individual strengths, limitations, and learning preferences, generative AI personalizes education. It builds a profile that directs content recommendations by gathering information from user interactions, evaluations, and progress.
Customized Content Production	Generative AI adapts lesson difficulty, generates practice questions, and offers resources based on the learner's knowledge level. For instance, AI-powered tutors like Knewton[11] and Carnegie Learning[12] create custom exercises that target specific learning gaps.
Adaptive Learning Paths	AI allows for dynamic curriculum adaptation, enabling individualized learning. By reacting to performance trends, it assists students in concentrating on areas that need work and moving more quickly through material they can grasp more readily.
Instantaneous Feedback and Support	Instantaneous and personalized feedback on tasks is given by generative AI, which aids students in better understanding errors and concepts. In order to promote ongoing learning, virtual AI tutors also provide advice outside of scheduled class times.

ENHANCING ACCESSIBILITY AND INCLUSIVITY

For every student to have the chance to participate completely in the learning process, regardless of their physical, cognitive, or sensory ability, accessibility in education is essential (Pizzol et al., 2021; Ştefan et al., 2021). It entails taking down obstacles and offering tools and assistance so that all students may access, engage in, and benefit from education. On the other hand, fostering a learning environment that recognizes and accommodates each student's unique requirements requires inclusivity in education (Hasson et al., 2024; Moriña et al., 2020; Tamim, 2021). It guarantees that every student has equal access to educational opportunities and can fully engage in the learning process, irrespective of their background, aptitude, gender, or socioeconomic situation.

To enable every student to have the opportunity to fully participate in the learning process, regardless of their physical, cognitive, or sensory abilities, education must be made accessible to everyone (Pizzol et al., 2021; Ştefan et al., 2021). This means removing all barriers and providing tools and support so that all students can access, engage in, and benefit from education. Furthermore, inclusivity is also necessary to enable a new generation of learning that recognizes and adapts to the unique needs of each student (Hasson et al., 2024; Moriña et al., 2020; Tamim, 2021). This principle means that regardless of the student's origin, gender, or economic status, they should have access to education systems and benefit from all related services.

Generative AI promises to improve accessibility and inclusiveness in education. This is conditioned by the development and improvement of flexible and accessible personalized learning environments that can be realized based on the following points:

- Adaptable Materials for Various Learning Styles (Morales-Chan et al., 2024; Tariq, 2024): Students can interact with the content they learn in the way they prefer, benefiting from the ability of generative AI to build adaptive content based on their interactions and the state of the learner. For example, for kinesthetic, visual, or auditory learners, the course can change its format between video, text, and audio. Generative systems can also change the approach used in content based on learner characteristics such as personality (El Bahri et al., 2023, 2024) or emotions (El Bahri et al., 2022).
- Language Translation (S.-M. Lee, 2023): Translation services as generative AI technologies are currently playing a major role in learning. They help learners learn content that is not in their native language. Intelligent tutors can also use translation to give recommendations or offer assistance to the learner.
- Text-to-Speech and Speech-to-Text Capabilities (Matre & Cameron, 2024; Widyana et al., 2022): Generative AI can translate text into speech, which can

help visually impaired students who find it difficult to read. Also, for students who have difficulty hearing, instantaneous transformation of spoken content into text can help them understand it well.

- Virtual Accessibility Tools for Disabilities (Pierrès et al., 2024): There are AI-based solutions that help visually impaired students by describing course content to them. Examples of such applications are Seeing AI[13] and Be My Eyes[14]. In addition, applications like SignAll[15] and Ava[16] translate English into sign language. Deaf students can rely on these solutions to communicate and interact. By improving accessibility aspects, courses become more welcoming for students with such difficulties.
- Reducing Social Anxiety (Archbell & Coplan, 2022): Social anxiety is a phenomenon that can be reduced with the help of technology anxiety (Almaiah et al., 2022; Y. Chen, 2024; Fernández-Batanero et al., 2021). Students who have anxiety in group learning can have specific private environments based on adapted virtual characters generated.
- Remote and Scalable Tutoring (Cortes et al., 2025; Damasceno et al., 2020): Generative AI can provide accessible and on-demand teachers or instructors. Learners who live in remote or underserved areas can currently have access. This reduces inequalities between learners and increases the chances of learners having access to education.

NEW ROLES OF TEACHERS

Generative AI in education is transforming teaching and learning. AI allows teachers to focus on more complex learning objectives by automating repetitive tasks, producing personalized content, and providing real-time feedback. This shift allows teachers to focus on encouraging critical thinking, creativity, and teamwork while teaching students about ethical issues and responsible use of AI. Teachers are increasingly involved as mentors and facilitators as classrooms adopt new technologies, balancing AI-driven efficiency with essential human-centered skills that foster lifelong learning.

Personalized learning

AI-generated real-time learner indicators can be leveraged by teachers to apply specific methods and pedagogies. AI allows teachers to modify their teaching strategies and pace with remarkable precision by tracking each student's learning habits, strengths, and areas for development through real-world data-driven analysis. Based on these technologies, teachers can design courses with varied versions to meet the individual needs of each student, creating a stimulating learning atmosphere where all students can progress well. By encouraging greater engagement and higher learning outcomes, this personalization improves teaching effectiveness.

Content creation

Generative AI greatly assists teachers in creating their educational content, making it easier to create individualized and engaging educational resources. Teachers can save time and effort by using AI technologies to quickly create lesson plans, practice exercises, and quizzes that are customized to each student's needs and skill level (Chan & Tsi, 2024; KHAZANCHI & KHAZANCHI, 2024; LIU & GU, 2024). This allows teachers to create a wide range of educational materials that adapt to different learning preferences and keep students interested and challenged at the appropriate level. AI allows teachers to spend more time on engaging and high-impact educational activities by automating parts of the materials development process, ultimately improving the learning process.

Mentorship

By using generative AI, teachers can focus now on mentoring, building relationships, and providing specialized support to students (Baskara, 2023b; A. S. George, 2023). They can focus on individual student mentoring, encourage participation, and address their specific needs when they have more time to dedicate to routine tasks. Teachers now work as facilitators, promoting human-centered experiences and active learning that AI cannot match (Nyaaba et al., 2024). This mentoring position helps students improve personally and academically, highlighting the crucial role of the teacher in the development of well-rounded individuals.

Increasing Knowledge of AI

AI is already commonplace in education, and teachers play an essential role in educating students about AI, teaching them to use AI technologies judiciously and effectively. It is the educators' responsibility to help students critically evaluate AI-generated content, identify potential biases, and understand the ethical issues related to AI applications (Cope et al., 2021; Holzinger et al., 2018; Minn, 2022). By teaching students these essential skills, educators equip them to function in a digital world where artificial intelligence is ubiquitous. Beyond technical knowledge, this position gives students the discernment necessary to evaluate AI products critically and morally.

Feedback and support

The ability of generative AI to provide immediate feedback on student work enhances learning by allowing students to receive direction while concepts are still fresh in their brains (Bower et al., 2024; Hashem et al., 2024; KHAZANCHI & KHAZANCHI, 2024). AI can respond instantly to tasks or exercises that require simple assessment, freeing up teachers' time to provide more complex feedback and individualized assistance. This allows teachers to focus on helping students develop critical thinking, creativity, and teamwork skills that are essential for solving real-world problems. As a result, teachers can focus their expertise on helping students develop advanced skills and understanding by using AI to handle repetitive feedback tasks.

LIFELONG LEARNING

Life long Learning is revolutionized by increased personalization, accessibility, and flexibility in education offered by generative AI. AI offers relevant knowledge and practical applications that enhance skill development at any stage of life by adapting to the requirements and pace of each learner. This technology fosters a culture of continuous personal and professional development by helping students to pursue self-directed learning and stay updated with constantly evolving information. Artificial intelligence (AI) significantly contributes to sustainability and the commitment to lifelong learning (Asad & Ajaz, 2024; Milana et al., 2024; Tomaszewska, 2023). It is supported by generative AI in several important ways:

- Customized and Flexible learning Experiences (Asad & Ajaz, 2024; Bayly-Castaneda et al., 2024): Generative AI solutions provide a personalized learning experience by adapting to the learner's preferences, pace, and skill level. This flexibility ensures that students receive training materials and content tailored to their needs, allowing them to progress at their own pace and revisit concepts as needed to fully understand them.
- Targeted Practice and Skill Development (Ahn, 2024; Asad & Ajaz, 2024; Kallonas et al., 2024): Generative AI can identify specific areas for growth and suggest exercises based on an analysis of a learner's development. This encourages effective, goal-oriented learning by helping students build on their current knowledge or learn new skills more effectively.
- Availability of Ongoing, Adaptable Education (Rawas & AlSaeed, 2024; Sandhu et al., 2024): Generative AI eliminates the constraints of time and location, providing 24/7 access to excellent educational resources. Access to on-demand educational content allows for flexible education at any stage of

life, helping those who must balance their studies with other commitments, such as work or family.

- Relevant Information and Practical Applications (Bayly-Castaneda et al., 2024; Milana et al., 2024): Depending on a learner's interests or field of study, AI can curate new and updated content to keep them informed of developing topics or industry trends. Some AI systems even replicate real-world situations, giving students hands-on experience and helping them apply their knowledge in real-world scenarios.
- Promoting a Lifelong Learning Culture (Baskara, 2023a, 2024): Generative AI fosters a lifelong learning mindset and continuous development outside the classroom by providing scalable and easily accessible resources. People can develop their knowledge and abilities through this flexible and self-directed learning culture, which fosters professional and personal growth in an interesting and persistent way.

EDUCATIONAL SYSTEMS

With the introduction of powerful tools that improve efficiency, accessibility to high-quality learning materials, and personalization, generative AI is revolutionizing education systems around the world (Łodzikowski et al., 2024; Rajendran, 2023; Wang, 2023). AI helps educational institutions deliver more individualized and effective instruction by facilitating data-driven decision-making, optimizing administrative tasks, and personalizing curricula to meet the needs of individual students. However, as AI is integrated into education systems worldwide, it also raises important questions related to bias, equity, and data privacy, highlighting the need for responsible deployment to ensure that these tools serve all students.

Curriculum Design

Generative AI is transforming curriculum design, enabling education systems to create more individualized and adaptive learning experiences AI (Nithithanatchinnapat et al., 2024; Tariq, 2024). Traditional courses often take a one-size-fits-all approach, whereas AI enables dynamic modifications based on each student's learning preferences, progress, and pace. AI-driven adaptive learning systems can assess student interactions, identify strengths and weaknesses, and quickly modify content to meet each individual's needs. For example, students who are performing well in a subject can be given access to more advanced materials, while those who want more help can receive targeted exercises and explanations to fill learning gaps and build confidence. This flexibility enables students to interact with the material at their own pace and enables teachers to identify and meet learning requirements early on (Anjali & Yogi, 2024; Hashim et al., 2024). Thus, AI-based curriculum design fosters a more inclusive and student-centered teaching approach, which in turn fosters academic motivation and long-term success.

Decision making

By providing comprehensive, real-time information on student performance, engagement patterns, and overall learning trends, generative AI improves data-driven decision-making in education systems (Bukar et al., 2024; Chiu, 2023; Walczak & Cellary, 2023). AI solutions can help educators and administrators identify areas where children are succeeding or struggling through continuous data collection and analysis, enabling early interventions and specialized support. Schools can use this information, for example, to distribute resources more efficiently, by focusing on learning resources or providing more tutoring in areas where students need it most.

In addition, policymakers can use data-driven insights that assess the effectiveness of programs, instructional strategies, and learning environments to guide change and address a range of educational needs (Bukar et al., 2024; Capraro et al., 2024). This approach not only supports personal development, but also allows educational institutions to make deliberate adjustments that improve learning outcomes for entire student populations. Data- and evidence-based decision-making that continually improves to better serve students supports a more flexible, effective, and responsive education system.

Support for educators

By automating tasks that have traditionally taken up a huge portion of teachers' time, generative AI offers teachers significant administrative and instructional support. AI allows teachers to spend more time on interactive, student-centered activities by automating tedious tasks like grading, attendance tracking, and content creation. Teachers can spend less time on manual assessments by using AI to quickly grade multiple-choice exams and provide personalized feedback on student work. AI can also assist with lesson planning, creating personalized learning materials, and making resource recommendations based on curriculum goals and student needs (Chan & Tsi, 2024; Kaplan-Rakowski et al., 2023).

With this support, teachers can spend more time on high-impact learning activities, including facilitating discussions, providing one-on-one mentoring, and fostering students' critical thinking and creativity (Bower et al., 2024; Hashem et al., 2024). When teachers have fewer administrative tasks to complete, they can engage more fully with students and create a more enjoyable and engaging learning environment.

Overall, AI-based administrative assistance makes teachers' jobs easier, improving both teaching standards and overall student outcomes.

Access in Underserved and Remote Locations

Generative AI is crucial for expanding access to education in underserved and remote areas by offering excellent learning resources and support that would not otherwise be available (Capraro et al., 2024; Olusegun, 2024). AI-driven platforms can offer digital courses, interactive exercises, and one-on-one coaching in locations with poor educational infrastructure, providing underprivileged or rural students with the same opportunities as their urban counterparts. For example, by providing virtual classrooms, translation services, and multimedia learning resources, AI-based solutions can eliminate geographical barriers to education.

Additionally, generative AI's ability to modify materials to accommodate a range of language preferences and learning requirements makes it easier for students to interact with content in a way that is both accessible and relevant to their local context (Sandhu et al., 2024; Yu et al., 2024). By eliminating the disparity between underprivileged rural and urban people, providing remote access to a wide range of media, and lowering dependency on physical infrastructure, artificial intelligence (AI) advances educational equity. In the end, generative AI expands educational institutions' student bases, promoting inclusivity and lifelong learning on a world-wide scale.

ETHICAL CONCERNS

While adopting generative AI into education offers many exciting opportunities, there are also significant ethical issues to consider. Concerns around bias, data privacy, and the veracity of student learning surface when AI systems create instructional materials, customize learning routes, and even evaluate student work (JOHRI et al., 2023; N. L. Rane et al., 2023; Williams, 2024). If generative AI models are not properly maintained, they may unintentionally propagate stereotypes, offer biased viewpoints, or handle student data improperly. Ethical concerns are therefore crucial if AI is to provide a fair, inclusive, and transparent learning environment that benefits every student, protects personal privacy, and promotes critical engagement with AI-generated content.

Integrating generative AI into classrooms offers many exciting possibilities, but it also involves important ethical considerations. When AI systems develop instructional materials, personalize learning paths, and even assess student work, questions about bias, data privacy, and the accuracy of student learning arise (JOHRI et al., 2023; N. L. Rane et al., 2023; Williams, 2024). generative AI models can represent biased viewpoints If they are poorly curated, they also can unintentionally promote preconceived notions, or inappropriately manage student data. For AI to provide a transparent, egalitarian, and inclusive learning environment that benefits all students, protects individual privacy, and fosters critical engagement with AI-generated knowledge, ethical questions are therefore crucial.

Bias in AI-Generated Content

Despite the enormous potential of generative AI in education, there is a risk that its content will perpetuate biases (Abdikhakimov, 2023; Fang et al., 2024; Saxena, 2024). These biases, which can be caused by training data reflecting historical preconceptions or cultural imbalances, can lead to biased interpretations of history, culture, and social norms. Biases based on gender, socioeconomic class, culture, and language may emerge, which could alienate diverse groups of students and hinder inclusion. Furthermore, AI may encourage dominant views, obscuring minority perspectives in subjects such as literature or history, and it may favor standard English or certain dialects, which could disadvantage bilingual learners. Additionally, biases in AI-generated feedback and situational examples that inadvertently reinforce specific problem-solving techniques or reinforce societal biases can impact students' views of material and social roles.

Academic Integrity and Plagiarism

By blurring the line between legitimate help and cheating, generative AI can undermine academic integrity by making it easier for students to produce AI-generated assignments that appear innovative (Eke, 2023; Kumar et al., 2024; Plata et al., 2023). The availability of AI-generated content could increase plagiarism as it often evades detection by traditional plagiarism checkers. This reliance on AI can hinder students' ability to think critically, as they may use it to complete assignments rather than engage with the material. Instructors can maintain their integrity and establish a fair and moral learning environment by structuring assignments to encourage critical thinking, establishing clear boundaries for AI use, and using AI detection technology when necessary.

Data privacy

Data privacy concerns arise when generative AI in education collects and stores students' personal data, often from external vendors (Arora & Arora, 2022; Golda et al., 2024; Holmes & Miao, 2023). Unauthorized access, opaque data use, and prolonged retention are risks that could expose student information to abuse. In order to ensure privacy in institutions, educational organizations should select AI systems that give priority to limited data collection, inform stakeholders of their data rights, and establish transparent data retention and audit procedures. These safeguards allow AI to enhance learning securely while protecting student data.

FRAMEWORKS FOR IMPLEMENTING AI INTO TEACHING AND LEARNING

The integration of AI into educational systems, especially in institutions, seems to require a systematic methodology or frameworks to ensure its effectiveness, inclusiveness, and moral relevance. To guide this integration, several frameworks that address ethical, technological, and pedagogical challenges have been developed (Ahshan, 2021; González-Pérez & Ramírez-Montoya, 2022; Strom & Viesca, 2023). These frameworks place a strong emphasis on both the use of AI tools and their compatibility with stakeholder needs and educational objectives. Generative AI as a subfield of AI have to follow directives of these frameworks to make a successful integration into teaching and learning technology.

Holistic AI Integration Frameworks

Holistic AI integration frameworks (Celik, 2023; Chan, 2023; Chiu, 2021; B. George & Wooden, 2023) has the objective to offer a systematic approach to integrate AI into education to ensure that ethical standards and educational goals are met. These frameworks focus on:

- Planning and goal setting: Setting clear goals, such as improving personalized instruction or speeding up administrative processes.
- Infrastructure development: aims to ensure that reliable digital infrastructure, tools, and AI platforms are accessible.
- Teacher training: offering the teachers all the resources they need to use AI successfully, giving special care to ethical considerations and critical analysis.
- Curriculum design: One of the main objectives of this framework is that AI can be used to edit and modify the curriculum to offer to students a variety of personalized content.
- Evaluation and feedback: At this level we have to define accurate metrics that can be used to evaluate the impact of AI on teaching and learning as you refine implementations.

- Implementation that is inclusive and ethical: This principle aims to ensure that AI systems are accessible to all students, fight algorithmic discrimination, and give data security top priority.

Respecting this framework can ensure that we can foster creativity and innovation with low risks while integrating AI in education and this last will be equitable, efficient, and sustainable.

Instructional Design Models for AI Integration

AI technology can be systematically integrated into teaching and learning environments using instructional design models for AI integration (Weng & Chiu, 2023; Xu & Ouyang, 2022). These models ensure that AI tools are successfully applied, tailored to students' needs, and aligned with educational goals. The five processes listed above are part of the well-known instructional design paradigm known as the ADDIE (Analysis, Design, Development, Implementation, and Evaluation) model (Almelhi, 2021; Spatioti et al., 2022). When adapted for educational purposes, the ADDIE paradigm provides a systematic approach to integrating AI tools and systems in a way that promotes learning objectives and ensures positive learning outcomes. The following are essential elements of AI-integrated instructional design models:

- Analysis: The first step in the integration process is to identify specific issues and opportunities where AI can improve education. Therefore, we have to look at issues such as improving feedback systems or taking into account the different abilities of students. The needs of teachers, students, and institutional goals are assessed simultaneously to select the best AI solutions that meet these requirements.
- Design: The next step is to plan AI-enhanced learning experiences. Resources such as flexible platforms, virtual tutors, and specialized material delivery techniques are used in these experiences. Care is taken to ensure that these designs align with curricular standards, learning objectives, and the diverse needs of learners to provide an inclusive and adaptable educational setting.
- • Development: Teachers create lesson plans, assessments, and AI-integrated activities tailored to specific subjects or skill levels at this stage. The primary goal is to use AI capabilities to deliver engaging and dynamic content. Multimedia and interactive elements enhance learning experiences by ensuring that the material is engaging for students and fosters active engagement.
- Implementation: Once developed, AI tools are used in online learning environments (LMS) or in classrooms. To ensure effective use of the technologies, teachers must receive extensive training and strong technical support.

However, throughout this period, student interactions with AI systems are observed in order to address usability or accessibility issues and ensure seamless integration into the educational process.

- The final step is evaluation, which involves assessing the extent to which AI technologies achieve desired outcomes, such as improved student engagement, greater learning effectiveness, and increased teaching productivity. Feedback from teachers and students is collected and analyzed to improve AI applications and teaching strategies, ensuring alignment with learning objectives and encouraging continuous improvement. This iterative approach facilitates the establishment of meaningful and sustainable AI integration in the classroom.

Education becomes more effective, inclusive, and future-oriented through this rigorous approach, which ensures that AI is used in ways that enhance teaching and learning.

DISCUSSION

Generative AI is transforming education with creative methods that improve administrative efficiency, instructional effectiveness, and learning efficiency. With technology to deliver adaptive content, offer on-demand support, and personalize learning experiences, student interactions with learning materials are changing. By tailoring learning paths to each learner's preferences and needs, AI is helping students at all levels of education better manage their educational journey. Additionally, generative AI tools such as virtual tutors, AI-driven content creation platforms, and accessibility technologies can bridge gaps between students with different backgrounds and skill levels, creating a more inclusive and equitable learning environment. This is especially true in underserved or remote locations, where generative AI can give children access to high-quality education they would not otherwise have.

Using generative AI in lesson plans and teaching resources can also help teachers. By automating repetitive tasks like grading and content development, AI allows teachers to focus on other important aspects of education, like fostering creativity and critical thinking and mentoring students. The role of teachers is also changing in the age of AI, they are now facilitators and mentors who help students navigate the digital and ethical landscape of education. Additionally, teachers can tailor their lessons to the unique needs of each student with generative AI, making them more engaging and relevant.

However, the use of generative AI in the classroom raises several ethical questions. The potential for bias in the information generated by AI, which could reinforce stereotypes or exclude different viewpoints, is one of the most pressing concerns. It is essential to put in place protections to prevent historical or cultural biases from affecting the learning process, as AI systems are often trained on large datasets that are likely to contain these biases. Moreover, while AI has multiple benefits, its massive application compromises academic integrity. Generative artificial intelligence, capable of generating information that could be mistaken for original work, poses an unprecedented problem for the academic dignity of teachers. In order to meet this challenge, it is necessary for institutions to establish clear criteria for the moral use of AI, while putting in place mechanisms to identify and prevent its use.

The other major obstacle for generative AI is data protection, since its devices rely on collecting and examining large amounts of information about students in order to adapt learning processes. Educational institutions must ensure that the AI devices they operate comply with privacy standards, considering the crucial value of data protection. Student privacy must be protected through open data policies and informed consent while utilizing the enhanced learning opportunities offered by AI. Education systems must establish policies that protect personal data and specify exactly how student data should be used and stored.

Generative AI has the ability to completely transform education by improving accessibility, efficiency, and personalization of learning; yet, its application requires careful management. Addressing the ethical issues of bias, academic integrity, and data privacy is essential to achieving the full educational benefits of AI. By balancing technological innovation with ethical responsibility, educators, policymakers, and educational institutions can ensure that generative AI contributes to a more sustainable, equal, and inclusive educational future. Successfully integrating AI into education requires both the creation of cutting-edge technologies and a commitment to preserving the values of transparency, fairness, and student rights.

Finally, integrating AI into education requires a comprehensive strategy that considers educational goals, ethical concerns, and technological advances. To maximize the benefits of AI while minimizing its risks, educational institutions must implement frameworks that prioritize openness, diversity, and continuous assessment. These frameworks give educators the tools they need to design engaging and equitable classroom learning spaces.

CONCLUSION

Generative AI is undoubtedly transforming education by creating new opportunities for personalized learning, creativity, and accessibility. Its ability to produce content precisely tailored to each learner's needs, provide immediate feedback, and adapt to those needs has the potential to significantly improve educational outcomes. Furthermore, its application in areas such as multilingual education, accessibility for children with disabilities, and closing educational gaps in disadvantaged communities underscores its importance in advancing inclusion and equity.

However, there are several obstacles to this progress. To ensure that AI tools appropriately benefit students, careful attention is needed to address ethical issues such as algorithm-based discrimination and data protection. In the face of pedagogical changes, such as the transformation of the role of teachers in AI-integrated classrooms, there is a need to intensify student preparation and redouble efforts to balance technology and human connection. Moreover, digital disparity persists as a barrier to the equal implementation of AI, highlighting the need for infrastructure and regulatory measures.

In summary, generative AI has the potential to radically transform the educational field. However, its application requires the support of relevant stakeholders, ethical considerations, and inclusive methods. By addressing these issues proactively, generative AI can harness its potential to establish a more efficient, fair, and sustainable education system for children around the world. To optimize the use of AI in education while minimizing the dangers, it is necessary for institutions to adopt structures that promote transparency, inclusion, and continuous evaluation.

REFERENCES

Abdikhakimov, I. (2023). Legal aspects of AI generated content. International Conference on Legal Sciences, 1(5), 1–17. https://science-zone.org/index.php/conference/article/view/41

Ahn, H. Y. (2024). AI-Powered E-Learning for Lifelong Learners: Impact on Performance and Knowledge Application. *Sustainability (Basel)*, 16(20), 9066. DOI: 10.3390/su16209066

Ahshan, R. (2021). A framework of implementing strategies for active student engagement in remote/online teaching and learning during the COVID-19 pandemic. *Education Sciences*, 11(9), 483. DOI: 10.3390/educsci11090483

Al-Ansi, A. M., Jaboob, M., Garad, A., & Al-Ansi, A. (2023). Analyzing augmented reality (AR) and virtual reality (VR) recent development in education. *Social Sciences & Humanities Open*, 8(1), 100532. DOI: 10.1016/j.ssaho.2023.100532

Alasadi, E. A., & Baiz, C. R. (2023). Generative AI in Education and Research: Opportunities, Concerns, and Solutions. *Journal of Chemical Education*, 100(8), 2965–2971. DOI: 10.1021/acs.jchemed.3c00323

Almaiah, M. A., Alfaisal, R., Salloum, S. A., Hajjej, F., Thabit, S., El-Qirem, F. A., Lutfi, A., Alrawad, M., Al Mulhem, A., Alkhdour, T., Awad, A. B., & Al-Maroof, R. S. (2022). Examining the impact of artificial intelligence and social and computer anxiety in e-learning settings: Students' perceptions at the university level. *Electronics (Basel)*, 11(22), 3662. DOI: 10.3390/electronics11223662

Almelhi, A. M. (2021). Effectiveness of the ADDIE model within an E-learning environment in developing creative writing in EFL students. *English Language Teaching*, 14(2), 20–36. DOI: 10.5539/elt.v14n2p20

Amankwah-Amoah, J., Abdalla, S., Mogaji, E., Elbanna, A., & Dwivedi, Y. K. (2024). The impending disruption of creative industries by generative AI: Opportunities, challenges, and research agenda. In *International Journal of Information Management* (p. 102759). Elsevier., https://www.sciencedirect.com/science/article/pii/S0268401224000070 DOI: 10.1016/j.ijinfomgt.2024.102759

Andrin, G., Kilag, O. K., Groenewald, E., Benitez, J., Dagala, F., & Ubay, R. (2024). Borderless Learning Environments: Impacts on Educational Management Strategies. [IMJRISE]. *International Multidisciplinary Journal of Research for Innovation, Sustainability, and Excellence*, 1(2), 43–49.

Anjali, R. M., & Yogi, M. K. (2024). AIculum-AI-Driven Curriculum Design in Educational Sector. Journal of Big Data Analytics and Business Intelligence, 8–13.

Archbell, K. A., & Coplan, R. J. (2022). Too Anxious to Talk: Social Anxiety, Academic Communication, and Students' Experiences in Higher Education. *Journal of Emotional and Behavioral Disorders*, 30(4), 273–286. DOI: 10.1177/10634266211060079

Arora, A., & Arora, A. (2022). Generative adversarial networks and synthetic patient data: Current challenges and future perspectives. *Future Healthcare Journal*, 9(2), 190–193. DOI: 10.7861/fhj.2022-0013 PMID: 35928184

Asad, M. M., & Ajaz, A. (2024). Impact of ChatGPT and generative AI on lifelong learning and upskilling learners in higher education: Unveiling the challenges and opportunities globally. The International Journal of Information and Learning Technology. https://www.emerald.com/insight/content/doi/10.1108/IJILT-06-2024-0103/full/html

Bahroun, Z., Anane, C., Ahmed, V., & Zacca, A. (2023). Transforming Education: A Comprehensive Review of Generative Artificial Intelligence in Educational Settings through Bibliometric and Content Analysis. *Sustainability (Basel)*, 15(17), 17. Advance online publication. DOI: 10.3390/su151712983

Baskara, F. R. (2023a). Fostering culturally grounded learning: Generative ai, digital storytelling, and early childhood education. International Conference of Early Childhood Education in Multiperspectives, 351–361. https://repository.usd.ac.id/48848/1/10058_351-361__IE__FX.%2BRisang%2BBaskara_3.pdf

Baskara, F. R. (2023b). Navigating pedagogical evolution: The implication of generative AI on the reinvention of teacher education. Prosiding Seminar Nasional Universitas Jabal Ghafur, 2, 409–421. https://repository.usd.ac.id/48711/1/10357_1923-4619-1-SM.pdf

Baskara, F. R. (2024). From AI to We: Harnessing Generative AI Tools to Cultivate Collaborative Learning Ecosystems in Universities. Proceeding International Conference on Learning Community (ICLC), 1(1). https://repository.usd.ac.id/51989/1/11944_27985-76316-1-PB.pdf

Bayly-Castaneda, K., Ramirez-Montoya, M. S., & Morita-Alexander, A. (2024). Crafting personalized learning paths with AI for lifelong learning: A systematic literature review. *Frontiers in Education*, 9, 1424386. https://www.frontiersin.org/journals/education/articles/10.3389/feduc.2024.1424386/full. DOI: 10.3389/feduc.2024.1424386

Bengesi, S., El-Sayed, H., Sarker, M. K., Houkpati, Y., Irungu, J., & Oladunni, T. (2024). Advancements in Generative AI: A Comprehensive Review of GANs, GPT, Autoencoders, Diffusion Model, and Transformers. IEEE Access, 12, 69812–69837. IEEE Access. DOI: 10.1109/ACCESS.2024.3397775

Bower, M., Torrington, J., Lai, J. W. M., Petocz, P., & Alfano, M. (2024). How should we change teaching and assessment in response to increasingly powerful generative Artificial Intelligence? Outcomes of the ChatGPT teacher survey. *Education and Information Technologies*. Advance online publication. DOI: 10.1007/s10639-023-12405-0

Brüns, J. D., & Meißner, M. (2024). Do you create your content yourself? Using generative artificial intelligence for social media content creation diminishes perceived brand authenticity. *Journal of Retailing and Consumer Services*, 79, 103790. DOI: 10.1016/j.jretconser.2024.103790

Bukar, U. A., Sayeed, M. S., Razak, S. F. A., Yogarayan, S., & Sneesl, R. (2024). Decision-making framework for the utilization of generative artificial intelligence in education: A case study of ChatGPT. IEEE Access. https://ieeexplore.ieee.org/abstract/document/10589404/

Capraro, V., Lentsch, A., Acemoglu, D., Akgun, S., Akhmedova, A., Bilancini, E., Bonnefon, J.-F., Brañas-Garza, P., Butera, L., Douglas, K. M., Everett, J. A. C., Gigerenzer, G., Greenhow, C., Hashimoto, D. A., Holt-Lunstad, J., Jetten, J., Johnson, S., Kunz, W. H., Longoni, C., & Viale, R. (2024). The impact of generative artificial intelligence on socioeconomic inequalities and policy making. *PNAS Nexus*, 3(6), pgae191. https://academic.oup.com/pnasnexus/article-abstract/3/6/pgae191/7689236. DOI: 10.1093/pnasnexus/pgae191 PMID: 38864006

Cavus, N. (2015). Distance learning and learning management systems. *Procedia: Social and Behavioral Sciences*, 191, 872–877. DOI: 10.1016/j.sbspro.2015.04.611

Celik, I. (2023). Towards Intelligent-TPACK: An empirical study on teachers' professional knowledge to ethically integrate artificial intelligence (AI)-based tools into education. *Computers in Human Behavior*, 138, 107468. DOI: 10.1016/j.chb.2022.107468

Chan, C. K. Y. (2023). A comprehensive AI policy education framework for university teaching and learning. *International Journal of Educational Technology in Higher Education*, 20(1), 38. DOI: 10.1186/s41239-023-00408-3

Chan, C. K. Y., & Tsi, L. H. (2024). Will generative AI replace teachers in higher education? A study of teacher and student perceptions. *Studies in Educational Evaluation*, 83, 101395. DOI: 10.1016/j.stueduc.2024.101395

Chen, B., Wu, Z., & Zhao, R. (2023). From fiction to fact: The growing role of generative AI in business and finance. *Journal of Chinese Economic and Business Studies*, 21(4), 471–496. DOI: 10.1080/14765284.2023.2245279

Chen, Y. (2024). Effects of technology-enhanced language learning on reducing EFL learners' public speaking anxiety. *Computer Assisted Language Learning*, 37(4), 789–813. DOI: 10.1080/09588221.2022.2055083

Chiu, T. K. F. (2021). A Holistic Approach to the Design of Artificial Intelligence (AI) Education for K-12 Schools. *TechTrends*, 65(5), 796–807. DOI: 10.1007/s11528-021-00637-1

Chiu, T. K. F. (2023). The impact of Generative AI (GenAI) on practices, policies and research direction in education: A case of ChatGPT and Midjourney. *Interactive Learning Environments*, •••, 1–17. DOI: 10.1080/10494820.2023.2172044

Chiu, T. K. F. (2024). Future research recommendations for transforming higher education with generative AI. *Computers and Education: Artificial Intelligence*, 6, 100197. DOI: 10.1016/j.caeai.2023.100197

Contini, F. (2024). Unboxing generative AI for the legal professions: Functions, impacts and governance. IJCA, 15, 1. https://heinonline.org/hol-cgi-bin/get_pdf.cgi?handle=hein.journals/ijca15&section=11

Cope, B., Kalantzis, M., & Searsmith, D. (2021). Artificial intelligence for education: Knowledge and its assessment in AI-enabled learning ecologies. *Educational Philosophy and Theory*, 53(12), 1229–1245. DOI: 10.1080/00131857.2020.1728732

Cortes, K. E., Kortecamp, K., Loeb, S., & Robinson, C. D. (2025). A scalable approach to high-impact tutoring for young readers. *Learning and Instruction*, 95, 102021. DOI: 10.1016/j.learninstruc.2024.102021

Damasceno, A. R. P., Martins, A. R., Chagas, M. L., Barros, E. M., Maia, P. H. M., & Oliveira, F. C. M. B. (2020). STUART: An intelligent tutoring system for increasing scalability of distance education courses. *Proceedings of the 19th Brazilian Symposium on Human Factors in Computing Systems*, 1–10. DOI: 10.1145/3424953.3426640

De Cremer, D., Bianzino, N. M., & Falk, B. (2023). How generative AI could disrupt creative work. *Harvard Business Review*, 13, •••. https://www.daviddecremer.com/wp-content/uploads/HBR2023_How-Generative-AI-Could-Disrupt-Creative-Work-1.pdf

De Moura, V. F., de Souza, C. A., & Viana, A. B. N. (2021). The use of Massive Open Online Courses (MOOCs) in blended learning courses and the functional value perceived by students. *Computers & Education*, 161, 104077. DOI: 10.1016/j.compedu.2020.104077

Dhariwal, P., Jun, H., Payne, C., Kim, J. W., Radford, A., & Sutskever, I. (2020). Jukebox: A generative model for music. arXiv Preprint arXiv:2005.00341. https://assets.pubpub.org/2gnzbcnd/11608661311181.pdf

Ebert, C., & Louridas, P. (2023). Generative AI for software practitioners. *IEEE Software*, 40(4), 30–38. DOI: 10.1109/MS.2023.3265877

Eke, D. O. (2023). ChatGPT and the rise of generative AI: Threat to academic integrity? *Journal of Responsible Technology*, 13, 100060. DOI: 10.1016/j.jrt.2023.100060

El Bahri, N., Itahriouan, Z., Abtoy, A., & Belhaouari, S. B. (2023). Using convolutional neural networks to detect learner's personality based on the Five Factor Model. Computers and Education: Artificial Intelligence, 100163.

El Bahri, N., Itahriouan, Z., Abtoy, A., & Jamil, M. O. (2024). *Personality analysis of students' writing in social media-based learning environments*. IEEE., DOI: 10.1109/ACCESS.2024.3491934

El Bahri, N., Itahriouan, Z., Belhaouari, S. B., & Abtoy, A. (2022). Using Deep Learning to detect Facial Expression from front camera: Towards students' interactions analyze. E3S Web of Conferences, 351, 01032.

Essa, S. G., Celik, T., & Human-Hendricks, N. E. (2023). Personalized adaptive learning technologies based on machine learning techniques to identify learning styles: A systematic literature review. *IEEE Access : Practical Innovations, Open Solutions*, 11, 48392–48409. DOI: 10.1109/ACCESS.2023.3276439

Fang, X., Che, S., Mao, M., Zhang, H., Zhao, M., & Zhao, X. (2024). Bias of AI-generated content: An examination of news produced by large language models. *Scientific Reports*, 14(1), 5224. DOI: 10.1038/s41598-024-55686-2 PMID: 38433238

Fernández-Batanero, J.-M., Román-Graván, P., Reyes-Rebollo, M.-M., & Montenegro-Rueda, M. (2021). Impact of educational technology on teacher stress and anxiety: A literature review. *International Journal of Environmental Research and Public Health*, 18(2), 548. DOI: 10.3390/ijerph18020548 PMID: 33440738

Feuerriegel, S., Hartmann, J., Janiesch, C., & Zschech, P. (2024). Generative AI. *Business & Information Systems Engineering*, 66(1), 111–126. DOI: 10.1007/s12599-023-00834-7

Foster, D. (2022). Generative deep learning. O'Reilly Media, Inc. https://books.google.com/books?hl=fr&lr=&id=Bkq8EAAAQBAJ&oi=fnd&pg=PP1&dq=what+is+generative+ai+text,+photos,+music,+and+other+types+of+data&ots=IN7yE3E6jO&sig=iebuNJWeETo5t2NE9RlKaVbWHy4

Gayam, S. R. (2022). Generative AI for Content Creation: Advanced Techniques for Automated Text Generation, Image Synthesis, and Video Production. *Journal of Science and Technology*, 3(1), 8–38.

George, A. S. (2023). The Potential of Generative AI to Reform Graduate Education. *Partners Universal International Research Journal*, 2(4), 36–50.

George, B., & Wooden, O. (2023). Managing the strategic transformation of higher education through artificial intelligence. *Administrative Sciences*, 13(9), 196. DOI: 10.3390/admsci13090196

Gibson, Y., & Kuhn, R. (2024). E-Proctoring Tools: Is It a Necessary Inconvenience. *Journal of Higher Education Theory and Practice*, 24(7). Advance online publication. https://search.ebscohost.com/login.aspx?direct=true&profile=ehost&scope=site&authtype=crawler&jrnl=21583595&AN=179288786&h=06G9snvdxKAtRstLmOUgJYMRQEtlghOnia6lxYEGGqV2Av%2ByxubrGv7LVoBvmvm%2B1rAMumxU7fcyg3WjzO89kA%3D%3D&crl=c. DOI: 10.33423/jhetp.v24i7.7088

Golda, A., Mekonen, K., Pandey, A., Singh, A., Hassija, V., Chamola, V., & Sikdar, B. (2024). Privacy and Security Concerns in Generative AI: A Comprehensive Survey. IEEE Access. https://ieeexplore.ieee.org/abstract/document/10478883/

González-Pérez, L. I., & Ramírez-Montoya, M. S. (2022). Components of Education 4.0 in 21st century skills frameworks: Systematic review. *Sustainability (Basel)*, 14(3), 1493. DOI: 10.3390/su14031493

Harshvardhan, G. M., Gourisaria, M. K., Pandey, M., & Rautaray, S. S. (2020). A comprehensive survey and analysis of generative models in machine learning. *Computer Science Review*, 38, 100285. DOI: 10.1016/j.cosrev.2020.100285

Hashem, R., Ali, N., El Zein, F., Fidalgo, P., & Khurma, O. A. (2024). AI to the rescue: Exploring the potential of ChatGPT as a teacher ally for workload relief and burnout prevention. *Research and Practice in Technology Enhanced Learning*, 19, •••. https://www.researchgate.net/profile/Othman-Abu-Khurma/publication/374024140_AI_to_the_rescue_Exploring_the_potential_of_ChatGPT_as_a_teacher_ally_for_workload_relief_and_burnout_prevention/links/650a94da82f01628f032ebb7/AI-to-the-rescue-Exploring-the-potential-of-ChatGPT-as-a-teacher-ally-for-workload-relief-and-burnout-prevention.pdf

Hashim, S. N. A., Mariappan, U., Ab Ghani, M. A., Nor, N. M. M., Ensimau, N. K., Yassin, Y. N. H. M., & Abbas, M. S. (2024). From Data To Insights: How AI Is Shaping Evidence-Based Curriculum Design. *Nanotechnology Perceptions*, •••, 1724–1728.

Hasson, L., Keville, S., Gallagher, J., Onagbesan, D., & Ludlow, A. K. (2024). Inclusivity in education for autism spectrum disorders: Experiences of support from the perspective of parent/carers, school teaching staff and young people on the autism spectrum. *International Journal of Developmental Disabilities*, 70(2), 201–212. DOI: 10.1080/20473869.2022.2070418 PMID: 38481460

Hewson, E., & Chung, G. W. (2021). Beyond the VLE: Transforming online discussion and collaboration through Microsoft Teams. Hosted by UNED, Madrid (Spain), 36. https://pureportal.coventry.ac.uk/files/27637722/Proceedings_OOFHEC2019.pdf#page=48

Holmes, W., & Miao, F. (2023). Guidance for generative AI in education and research. UNESCO Publishing. https://books.google.com/books?hl=fr&lr=&id=cKnYEAAAQBAJ&oi=fnd&pg=PA2&dq=data+privacy+generative+ai&ots=wmGadhI2eK&sig=V-VUh9VLCz_tjG4vb7ZXlFr-X88

Holzinger, A., Kieseberg, P., Weippl, E., & Tjoa, A. M. (2018). Current Advances, Trends and Challenges of Machine Learning and Knowledge Extraction: From Machine Learning to Explainable AI. In Holzinger, A., Kieseberg, P., Tjoa, A. M., & Weippl, E. (Eds.), *Machine Learning and Knowledge Extraction* (Vol. 11015, pp. 1–8). Springer International Publishing., DOI: 10.1007/978-3-319-99740-7_1

Iaaly, A., Daou, P., Nakad, M., & Abboud, R. J. (2024). E-assessment and e-proctoring implementation during coronavirus pandemic: A participatory study using socio-technical systems theory. *Education and Information Technologies*. Advance online publication. DOI: 10.1007/s10639-024-13078-z

Jang, J., Ko, Y., Shin, W. S., & Han, I. (2021). Augmented reality and virtual reality for learning: An examination using an extended technology acceptance model. *IEEE Access : Practical Innovations, Open Solutions*, 9, 6798–6809. DOI: 10.1109/ACCESS.2020.3048708

Jiang, F., Ma, J., Webster, C. J., Chiaradia, A. J., Zhou, Y., Zhao, Z., & Zhang, X. (2024). Generative urban design: A systematic review on problem formulation, design generation, and decision-making. *Progress in Planning*, 180, 100795. DOI: 10.1016/j.progress.2023.100795

Johri, A., Lindsay, E., & Qadir, J. (2023). ETHICAL CONCERNS AND RESPONSIBLE USE OF GENERATIVE ARTIFICIAL INTELLIGENCE IN ENGINEERING EDUCATION. SEFI 2023-51st Annual Conference of the European Society for Engineering Education: Engineering Education for Sustainability, Proceedings (2023), 2244–2253.

Kallonas, C., Piki, A., & Stavrou, E. (2024). Empowering professionals: A generative AI approach to personalized cybersecurity learning. 2024 IEEE Global Engineering Education Conference (EDUCON), 1–10. https://ieeexplore.ieee.org/abstract/document/10578894/

Kaplan-Rakowski, R., Grotewold, K., Hartwick, P., & Papin, K. (2023). Generative AI and teachers' perspectives on its implementation in education. *Journal of Interactive Learning Research*, 34(2), 313–338.

Kaswan, K. S., Dhatterwal, J. S., Malik, K., & Baliyan, A. (2023). Generative AI: A Review on Models and Applications. 2023 International Conference on Communication, Security and Artificial Intelligence (ICCSAI), 699–704. https://ieeexplore.ieee.org/abstract/document/10421601/

KHAZANCHI, R., & KHAZANCHI, P. (2024). Generative AI to improve special education teacher preparation for inclusive classrooms. Exploring New Horizons: Generative Artificial Intelligence and Teacher Education, 159.

Ko, H.-K., Park, G., Jeon, H., Jo, J., Kim, J., & Seo, J. (2023). Large-scale Text-to-Image Generation Models for Visual Artists' Creative Works. *Proceedings of the 28th International Conference on Intelligent User Interfaces*, 919–933. DOI: 10.1145/3581641.3584078

Kshetri, N., Dwivedi, Y. K., Davenport, T. H., & Panteli, N. (2024). Generative artificial intelligence in marketing: Applications, opportunities, challenges, and research agenda. In *International Journal of Information Management* (Vol. 75, p. 102716). Elsevier., https://www.sciencedirect.com/science/article/pii/S026840122300097X DOI: 10.1016/j.ijinfomgt.2023.102716

Kumar, R., Eaton, S. E., Mindzak, M., & Morrison, R. (2024). Academic Integrity and Artificial Intelligence: An Overview. In Eaton, S. E. (Ed.), *Second Handbook of Academic Integrity* (pp. 1583–1596). Springer Nature Switzerland., DOI: 10.1007/978-3-031-54144-5_153

Labadze, L., Grigolia, M., & Machaidze, L. (2023). Role of AI chatbots in education: Systematic literature review. *International Journal of Educational Technology in Higher Education*, 20(1), 56. DOI: 10.1186/s41239-023-00426-1

Lee, J., Eom, S.-Y., & Lee, J. (2023). Empowering game designers with generative AI. *IADIS International Journal on Computer Science & Information Systems*, 18(2), 213–230.

Lee, S.-M. (2023). The effectiveness of machine translation in foreign language education: A systematic review and meta-analysis. *Computer Assisted Language Learning*, 36(1–2), 103–125. DOI: 10.1080/09588221.2021.1901745

Lin, C.-C., Huang, A. Y. Q., & Lu, O. H. T. (2023). Artificial intelligence in intelligent tutoring systems toward sustainable education: A systematic review. *Smart Learning Environments*, 10(1), 41. DOI: 10.1186/s40561-023-00260-y

LIU, C. C., & GU, X. Q. (2024). Developing Frames for Change: The Impact Of Generative AI on the Broad Practices of Teacher Educators. Exploring New Horizons: Generative Artificial Intelligence and Teacher Education, 225.

Łodzikowski, K., Foltz, P. W., & Behrens, J. T. (2024). Generative AI and Its Educational Implications. In Kourkoulou, D., Tzirides, A.-O., Cope, B., & Kalantzis, M. (Eds.), *Trust and Inclusion in AI-Mediated Education* (pp. 35–57). Springer Nature Switzerland., DOI: 10.1007/978-3-031-64487-0_2

Lv, Z. (2023). Generative artificial intelligence in the metaverse era. *Cognitive Robotics*, 3, 208–217. DOI: 10.1016/j.cogr.2023.06.001

Macey-Dare, R. (2023). How ChatGPT and generative AI systems will revolutionize legal services and the legal profession. *SSRN*, 4366749, •••. https://papers.ssrn.com/sol3/papers.cfm?abstract_id=4366749. DOI: 10.2139/ssrn.4366749

Mangal, S. K., & Mangal, U. (2019). Essentials of educational technology. PHI Learning Pvt. Ltd. https://books.google.com/books?hl=fr&lr=&id=Sc9upWs956cC&oi=fnd&pg=PR2&dq=education+technology+&ots=cPvedoKDz7&sig=qB-SRjIATUWxq36Ys-lHd3Ga24s

Martin, F., Chen, Y., Moore, R. L., & Westine, C. D. (2020). Systematic review of adaptive learning research designs, context, strategies, and technologies from 2009 to 2018. *Educational Technology Research and Development*, 68(4), 1903–1929. DOI: 10.1007/s11423-020-09793-2 PMID: 32837122

Matre, M. E., & Cameron, D. L. (2024). A scoping review on the use of speech-to-text technology for adolescents with learning difficulties in secondary education. *Disability and Rehabilitation. Assistive Technology*, 19(3), 1103–1116. DOI: 10.1080/17483107.2022.2149865 PMID: 36427182

Meça, A., & Shkëlzeni, N. (2024). Academic Integrity in the Face of Generative Language Models. In Miraz, M. H., Southall, G., Ali, M., & Ware, A. (Eds.), *Emerging Technologies in Computing* (Vol. 538, pp. 58–70). Springer Nature Switzerland., DOI: 10.1007/978-3-031-50215-6_5

Meskó, B., & Topol, E. J. (2023). The imperative for regulatory oversight of large language models (or generative AI) in healthcare. *NPJ Digital Medicine*, 6(1), 120. DOI: 10.1038/s41746-023-00873-0 PMID: 37414860

Miao, Y., Koenig, R., & Knecht, K. (2020). The development of optimization methods in generative urban design: A review. Proceedings of the 11th Annual Symposium on Simulation for Architecture and Urban Design, 1–8. https://www.researchgate.net/profile/Reinhard-Koenig-2/publication/344460745_The_Development_of_Optimization_Methods_in_Generative_Urban_Design_A_Review/links/5f787c8d92851c14bcaa17f2/The-Development-of-Optimization-Methods-in-Generative-Urban-Design-A-Review.pdf

Milana, M., Brandi, U., Hodge, S., & Hoggan-Kloubert, T. (2024). Artificial intelligence (AI), conversational agents, and generative AI: Implications for adult education practice and research. *International Journal of Lifelong Education*, 43(1), 1–7. DOI: 10.1080/02601370.2024.2310448

Minn, S. (2022). AI-assisted knowledge assessment techniques for adaptive learning environments. *Computers and Education: Artificial Intelligence*, 3, 100050.

Morales-Chan, M., Amado-Salvatierra, H. R., & Hernandez-Rizzardini, R. (2024). AI-Driven Content Creation: Revolutionizing Educational Materials. Proceedings of the Eleventh ACM Conference on Learning @ Scale, 556–558. DOI: 10.1145/3657604.3664640

Moriña, A., Sandoval, M., & Carnerero, F. (2020). Higher education inclusivity: When the disability enriches the university. *Higher Education Research & Development*, 39(6), 1202–1216. DOI: 10.1080/07294360.2020.1712676

Mousavinasab, E., Zarifsanaiey, N., & Niakan Kalhori, R., S., Rakhshan, M., Keikha, L., & Ghazi Saeedi, M. (. (2021). Intelligent tutoring systems: A systematic review of characteristics, applications, and evaluation methods. *Interactive Learning Environments*, 29(1), 142–163. DOI: 10.1080/10494820.2018.1558257

Nithithanatchinnapat, B., Maurer, J., Deng, X., & Joshi, K. D. (2024). Future Business Workforce: Crafting a Generative AI-Centric Curriculum Today for Tomorrow's Business Education. *The Data Base for Advances in Information Systems*, 55(1), 6–11. DOI: 10.1145/3645057.3645059

Nyaaba, M., Shi, L., Nabang, M., Zhai, X., Kyeremeh, P., Ayoberd, S. A., & Akanzire, B. N. (2024). Generative AI as a Learning Buddy and Teaching Assistant: Pre-service Teachers' Uses and Attitudes (arXiv:2407.11983). arXiv. http://arxiv.org/abs/2407.11983

Ofosu-Ampong, K. (2020). The Shift to Gamification in Education: A Review on Dominant Issues. *Journal of Educational Technology Systems*, 49(1), 113–137. DOI: 10.1177/0047239520917629

Oliveira, W., Hamari, J., Shi, L., Toda, A. M., Rodrigues, L., Palomino, P. T., & Isotani, S. (2023). Tailored gamification in education: A literature review and future agenda. *Education and Information Technologies*, 28(1), 373–406. DOI: 10.1007/s10639-022-11122-4

Olusegun, J. (2024). Generative AI Solutions for Expanding Access to Higher Education in Developing Regions. https://www.researchgate.net/profile/John-Olusegun-2/publication/384081407_Generative_AI_Solutions_for_Expanding_Access_to_Higher_Education_in_Developing_Regions/links/66ea0191dde50b32587a643e/Generative-AI-Solutions-for-Expanding-Access-to-Higher-Education-in-Developing-Regions.pdf

Pataranutaporn, P., Danry, V., Leong, J., Punpongsanon, P., Novy, D., Maes, P., & Sra, M. (2021). AI-generated characters for supporting personalized learning and well-being. *Nature Machine Intelligence*, 3(12), 1013–1022. DOI: 10.1038/s42256-021-00417-9

Peng, S., Kalliamvakou, E., Cihon, P., & Demirer, M. (2023). The Impact of AI on Developer Productivity: Evidence from GitHub Copilot (arXiv:2302.06590). arXiv. https://doi.org//arXiv.2302.06590DOI: 10.48550

Pesovski, I., Santos, R., Henriques, R., & Trajkovik, V. (2024). Generative AI for Customizable Learning Experiences. *Sustainability (Basel)*, 16(7), 7. Advance online publication. DOI: 10.3390/su16073034

Pierrès, O., Christen, M., Schmitt-Koopmann, F., & Darvishy, A. (2024). Could the Use of AI in Higher Education Hinder Students With Disabilities? A Scoping Review. IEEE Access. https://ieeexplore.ieee.org/abstract/document/10433192/

Pizzol, B., Giannotti, M., & Tomasiello, D. B. (2021). Qualifying accessibility to education to investigate spatial equity. *Journal of Transport Geography*, 96, 103199. DOI: 10.1016/j.jtrangeo.2021.103199

Plata, S., De Guzman, M. A., & Quesada, A. (2023). Emerging research and policy themes on academic integrity in the age of chat GPT and generative AI. *Asian Journal of University Education*, 19(4), 743–758. DOI: 10.24191/ajue.v19i4.24697

Qadir, J. (2023). Engineering education in the era of ChatGPT: Promise and pitfalls of generative AI for education. 2023 IEEE Global Engineering Education Conference (EDUCON), 1–9. https://ieeexplore.ieee.org/abstract/document/10125121/

Rajendran, R. M. (2023). Importance Of Using Generative AI In Education: Dawn of a New Era. *Journal of Science and Technology*, 4(6), 35–44. DOI: 10.55662/JST.2023.4603

Rane, J., Kaya, O., Mallick, S. K., & Rane, N. L. (2024). Impact of ChatGPT and similar generative artificial intelligence on several business sectors: Applications, opportunities, challenges, and future prospects. Generative Artificial Intelligence in Agriculture, Education, and Business, 27–83.

Rane, N. L., Choudhary, S. P., Tawde, A., & Rane, J. (2023). ChatGPT is not capable of serving as an author: Ethical concerns and challenges of large language models in education. *International Research Journal of Modernization in Engineering Technology and Science*, 5(10), 851–874.

Rawas, S., & AlSaeed, D. (2024). ChatGPT: Innovating Lifelong Learning in Higher Education Through Artificial Intelligence and Digital Transformation. In Lytras, M. D., Serban, A. C., Alkhaldi, A., Malik, S., & Aldosemani, T. (Eds.), *Digital Transformation in Higher Education, Part B* (pp. 13–28). Emerald Publishing Limited., DOI: 10.1108/978-1-83608-424-220241002

Roll, I., & Wylie, R. (2016). Evolution and Revolution in Artificial Intelligence in Education. *International Journal of Artificial Intelligence in Education*, 26(2), 582–599. DOI: 10.1007/s40593-016-0110-3

Salih, M. M. (2021). A Comparative Study Between Google Workspace and Microsoft Office 365 Productivity Services in Iraqi Educational Institutions. *International Journal of Humanities and Educational Research*, 3(5), 123–135. DOI: 10.47832/2757-5403.5-3.11

Sandhu, R., Channi, H. K., Ghai, D., Cheema, G. S., & Kaur, M. (2024). An Introduction to Generative AI Tools for Education 2030. Integrating Generative AI in Education to Achieve Sustainable Development Goals, 1–28.

Saxena, A. K. (2024). Quantitative Measurement of Bias in AI-Generated Content: A Comprehensive Narrative Literature Review. 2024 IEEE International Symposium on Technology and Society (ISTAS), 1–5. https://ieeexplore.ieee.org/abstract/document/10732696/

Spatioti, A. G., Kazanidis, I., & Pange, J. (2022). A comparative study of the ADDIE instructional design model in distance education. *Information (Basel)*, 13(9), 402. DOI: 10.3390/info13090402

Stahl, B. C. (2021). Ethical Issues of AI. In Stahl, B. C. (Ed.), *Artificial Intelligence for a Better Future* (pp. 35–53). Springer International Publishing., DOI: 10.1007/978-3-030-69978-9_4

Ştefan, I. A., Hauge, J. B., Sallinen, N., Ştefan, A., & Gheorghe, A. F. (2021). Accessibility and education: Are we fulfilling state of the art requirements? The International Scientific Conference eLearning and Software for Education, 1, 579–587. https://search.proquest.com/openview/cce87013703da339bf74f3361cb90f53/1?pq-origsite=gscholar&cbl=1876338

Strom, K. J., & Viesca, K. M. (2023). Towards a complex framework of teacher learning-practice. In *Non-linear perspectives on teacher development* (pp. 13–28). Routledge., https://www.taylorfrancis.com/chapters/edit/10.4324/9781003372097-2/towards-complex-framework-teacher-learning-practice-kathryn-strom-kara-mitchell-viesca DOI: 10.4324/9781003372097-2

Tamim, T. (2021). Language, Class, and Education: Deconstructing the Centre to Rethink Inclusivity in Education in Pakistan. *Cogent Education*, 8(1), 1897933. DOI: 10.1080/2331186X.2021.1897933

Tariq, M. U. (2024). Generative AI in curriculum development in higher education. In *Impacts of Generative AI on Creativity in Higher Education* (pp. 227–258). IGI Global., https://www.igi-global.com/chapter/generative-ai-in-curriculum-development-in-higher-education/355433 DOI: 10.4018/979-8-3693-2418-9.ch009

Tomaszewska, R. (2023). Andragogy meets ChatGPT in lifelong learning: Exploring opportunities and challenges. 2023 IEEE International Conference on Data Mining Workshops (ICDMW), 868–874. https://ieeexplore.ieee.org/abstract/document/10411582/

Turnbull, D., Chugh, R., & Luck, J. (2020). Learning Management Systems, An Overview. In Tatnall, A. (Ed.), *Encyclopedia of Education and Information Technologies* (pp. 1052–1058). Springer International Publishing., DOI: 10.1007/978-3-030-10576-1_248

Vayadande, K., Pednekar, C. B., Khune, P. A., Prabhavalkar, V. S., & Dange, V. R. (2024). GPT-3- and DALL-E-Powered Applications: A Complete Survey. In Dey, A., Nayak, S., Kumar, R., & Mohanty, S. N. (Eds.), *How Machine Learning is Innovating Today's World* (1st ed., pp. 329–341). Wiley., DOI: 10.1002/9781394214167.ch19

Wahid, R., Mero, J., & Ritala, P. (2023). Written by ChatGPT, illustrated by Midjourney: Generative AI for content marketing. *Asia Pacific Journal of Marketing and Logistics*, 35(8), 1813–1822. DOI: 10.1108/APJML-10-2023-994

Walczak, K., & Cellary, W. (2023). Challenges for higher education in the era of widespread access to generative AI. *Economics and Business Review*, 9(2). Advance online publication. DOI: 10.18559/ebr.2023.2.743

Wang, T. (2023). Navigating Generative AI (ChatGPT) in Higher Education: Opportunities and Challenges. In C. Anutariya, D. Liu, Kinshuk, A. Tlili, J. Yang, & M. Chang (Eds.), Smart Learning for A Sustainable Society (pp. 215–225). Springer Nature Singapore. DOI: 10.1007/978-981-99-5961-7_28

Wei, X., Saab, N., & Admiraal, W. (2021). Assessment of cognitive, behavioral, and affective learning outcomes in massive open online courses: A systematic literature review. *Computers & Education*, 163, 104097. DOI: 10.1016/j.compedu.2020.104097

Weng, X., & Chiu, T. K. (2023). Instructional design and learning outcomes of intelligent computer assisted language learning: Systematic review in the field. *Computers and Education: Artificial Intelligence*, 4, 100117. DOI: 10.1016/j.caeai.2022.100117

Widyana, A., Jerusalem, M. I., & Yumechas, B. (2022). The application of text-to-speech technology in language learning. Sixth International Conference on Language, Literature, Culture, and Education (ICOLLITE 2022), 85–92. https://books.google.com/books?hl=fr&lr=&id=EaykEAAAQBAJ&oi=fnd&pg=PA85&dq=Text-to-Speech++in+education&ots=P6KtZz0QCb&sig=znet6sj4HFVbtmGzxJ4IbJ1qGzY

Williams, R. T. (2024). The ethical implications of using generative chatbots in higher education. *Frontiers in Education*, 8, 1331607. https://www.frontiersin.org/articles/10.3389/feduc.2023.1331607/full. DOI: 10.3389/feduc.2023.1331607

Xu, W., & Ouyang, F. (2022). A systematic review of AI role in the educational system based on a proposed conceptual framework. *Education and Information Technologies*, 27(3), 4195–4223. DOI: 10.1007/s10639-021-10774-y

Yenduri, G., Ramalingam, M., Selvi, G. C., Supriya, Y., Srivastava, G., Maddikunta, P. K. R., Raj, G. D., Jhaveri, R. H., Prabadevi, B., & Wang, W. (2024). Gpt (generative pre-trained transformer)–a comprehensive review on enabling technologies, potential applications, emerging challenges, and future directions. IEEE Access. https://ieeexplore.ieee.org/abstract/document/10500411/

Yu, H., Guo, Y., Yang, H., Zhang, W., & Dong, Y. (2024). Can ChatGPT Revolutionize Language Learning? Unveiling the Power of AI in Multilingual Education Through User Insights and Pedagogical Impact. *European Journal of Education*, 12749. Advance online publication. DOI: 10.1111/ejed.12749

Zawacki-Richter, O., Marín, V. I., Bond, M., & Gouverneur, F. (2019). Systematic review of research on artificial intelligence applications in higher education – where are the educators? *International Journal of Educational Technology in Higher Education*, 16(1), 39. DOI: 10.1186/s41239-019-0171-0

Zhang, P., & Kamel Boulos, M. N. (2023). Generative AI in medicine and healthcare: Promises, opportunities and challenges. *Future Internet*, 15(9), 286. DOI: 10.3390/fi15090286

ENDNOTES

1 https://chatgpt.com/

2 https://openart.ai/

3 https://openai.com/index/jukebox/

4 https://www.udemy.com/

5 https://www.coursera.org/

6 https://www.edx.org/

7 https://www.dreambox.com/

8 https://www.wiley.com/en-us/education/alta

9 https://www.proctoru.com/

10 https://web.respondus.com/

11 https://www.wiley.com/en-us/education/alta

12 https://www.carnegielearning.com/

13 https://www.seeingai.com/

14 https://www.bemyeyes.com/

15 https://www.signall.us/

16 https://fr.ava.me/

Chapter 3
Tools and Technologies for AI Education

Ajay N. Upadhyaya
https://orcid.org/0000-0002-7583-6430
SAL Engineering and Technical Institute, India

Anjum Parvez
Uttaranchal University, Uttarakhand, India

Sandhya Verma
Allahabad University, India

Rokesh Kumar Yarava
Chalapathi Institute of Engineering and Technology, India

Dharmendra Singh
https://orcid.org/0009-0007-4274-6379
School of Management Sciences, Lucknow, India

V. Lakshmi Prasanna
https://orcid.org/0000-0003-0774-3694
Gokaraju Rangaraju Institute of Engineering and Technology, India

P. Selvakumar
https://orcid.org/0000-0002-3650-4548
Nehru Institute of Technology, India

ABSTRACT

These advancements are not merely incremental but represent a revolutionary shift in how educational content is delivered, accessed, and personalized. AI education tools are designed to enhance learning outcomes by providing innovative solutions that cater to the diverse needs of students, educators, and institutions alike. Moreover, AI-driven tools facilitate the creation of intelligent tutoring systems that provide individualized support to learners. These systems simulate one-on-one tutoring by offering instant feedback, explanations, and additional practice opportunities. For instance, AI can create dynamic simulations and virtual environments that respond to students' actions, offering a more engaging and experiential learning process. . AI-powered language translation tools also facilitate learning in multilingual settings, making it easier for students to access content in their preferred language

DOI: 10.4018/979-8-3373-2262-9.ch003

and ensuring a more inclusive educational environment

INTRODUCTION TO AI EDUCATION TOOLS AND TECHNOLOGIES: ENHANCING LEARNING OUTCOMES

These advancements are not merely incremental but represent a revolutionary shift in how educational content is delivered, accessed, and personalized. AI education tools are designed to enhance learning outcomes by providing innovative solutions that cater to the diverse needs of students, educators, and institutions alike. Moreover, AI-driven tools facilitate the creation of intelligent tutoring systems that provide individualized support to learners. These systems simulate one-on-one tutoring by offering instant feedback, explanations, and additional practice opportunities. For instance, AI can create dynamic simulations and virtual environments that respond to students' actions, offering a more engaging and experiential learning process. These technologies enable a more equitable distribution of educational opportunities and help in addressing disparities in access to quality education. AI-powered language translation tools also facilitate learning in multilingual settings, making it easier for students to access content in their preferred language and ensuring a more inclusive educational environment. R, with its statistical prowess, is well-suited for courses focused on data analysis, statistics, and data-driven decision-making, offering students robust tools for understanding and interpreting data. Julia, with its emphasis on performance, is ideal for advanced courses that require high computational power and efficiency, preparing students for roles in research and development where large-scale data processing is essential.The integration of these programming languages into AI education also reflects broader trends in the field, emphasizing the need for versatility and adaptability in the face of rapidly evolving technologies. Educators must consider the strengths and limitations of each language to tailor their teaching strategies and ensure that students are equipped with the skills needed to tackle real-world challenges. In conclusion, Python, R, and Julia each play a significant role in AI education, offering distinct advantages that cater to different aspects of learning and application. Python's simplicity and extensive libraries make it ideal for beginners and general AI development. R's statistical capabilities and data analysis tools provide depth for understanding complex data interactions. Julia's performance and computational efficiency support advanced scientific and AI research. By incorporating these languages into educational curricula, institutions can provide students with a comprehensive and versatile skill set, preparing them to excel in the dynamic and evolving field of AI.

Programming Languages for AI: Python, R, and Julia in Education

Among the most prominent languages for AI, Python, R, and Julia each offer unique advantages that make them valuable tools in the educational sector. Understanding their distinct characteristics and applications can provide educators and learners with the insights needed to leverage these languages effectively for AI education.Python's widespread adoption is supported by a rich set of libraries and frameworks tailored for AI and machine learning, such as TensorFlow, Keras, and scikit-learn (Acikkar, M., et al .,2009). These libraries provide pre-built functions and tools that streamline the development process, enabling learners to experiment with and implement advanced algorithms without needing to write extensive code from scratch. Furthermore, Python's strong community support and abundant online resources, including tutorials, forums, and documentation, facilitate a collaborative learning environment, helping students overcome challenges and expand their knowledge.In educational settings, R's strengths in handling complex data sets and performing sophisticated statistical analyses can be leveraged to teach students about data science, exploratory data analysis, and statistical modeling. R's syntax and functionality are geared towards data scientists and statisticians, and it provides a rich environment for developing and evaluating data-driven models. Its integration with tools like R Markdown allows for the seamless creation of dynamic reports that combine code, results, and narrative, facilitating an in-depth understanding of the analytical process and its outcomes.Julia is a relatively newer language that has been gaining traction for its high-performance capabilities and ease of use in scientific computing. Julia's design emphasizes both speed and productivity, combining the performance of low-level languages with the usability of high-level languages. This makes it particularly valuable for educational contexts where computational efficiency is critical, such as in large-scale simulations or complex AI models. Julia's syntax is similar to other high-level languages, which eases the learning curve for students familiar with Python or R.

AI Development Platforms: TensorFlow, PyTorch, and Keras in the Classroom

In the evolving landscape of artificial intelligence (AI), the choice of development platforms can significantly impact the learning experience and the effectiveness of educational programs. TensorFlow, PyTorch, and Keras are three leading frameworks that have become integral to AI education, each offering unique features and advantages that cater to different learning needs and objectives. Understanding these platforms' strengths and applications can help educators and students select the most suitable tools for their AI coursework and projects.TensorFlow TensorFlow's architecture is designed for both research and production, making it suitable for students at various stages of learning. In the classroom, TensorFlow's scalability and support for distributed computing allow students to experiment with large-scale models and datasets. TensorFlow's TensorBoard provides powerful visualization tools that help students understand model performance and training dynamics, offering insights into metrics such as loss and accuracy through interactive dashboards (Adamson, D., et al, 2014). This capability enhances the learning process by providing real-time feedback and fostering a deeper understanding of how different parameters influence model behavior. This breadth of tools helps students explore different aspects of AI, from building and training models to deploying them in real-world scenarios. TensorFlow's integration with cloud services such as Google Cloud Platform also facilitates hands-on experience with scalable AI solutions, preparing students for practical challenges in industry settings.In the classroom, PyTorch's clear and concise API facilitates a hands-on approach to learning, enabling students to grasp core concepts of deep learning and neural networks with minimal overhead. PyTorch's integration with the popular library torchvision provides ready-to-use datasets and pre-trained models, streamlining the development process and allowing students to focus on applying and understanding advanced AI techniques. PyTorch's growing ecosystem and supportive community contribute to a rich educational experience, with numerous tutorials, examples, and research papers available to enhance learning. The platform's emphasis on research-friendly features also prepares students for advanced studies and innovative projects in AI.Keras, originally developed as a high-level API for building and training deep learning models, is now part of the TensorFlow ecosystem but continues to offer a standalone experience.

In the classroom, Keras' concise and readable syntax helps students quickly grasp the principles of neural networks and deep learning. Keras' modular design allows for easy experimentation with different model components, such as layers, optimizers, and activation functions, fostering an interactive learning environment. Additionally, Keras' integration with TensorFlow enables students to transition seamlessly between high-level model building and lower-level TensorFlow functionalities as they progress in their studies. This gradual introduction to more complex aspects of AI development supports a smooth learning curve and prepares students for diverse applications of AI technology. In conclusion, TensorFlow, PyTorch, and Keras each offer distinct advantages for AI education, catering to different aspects of learning and development. TensorFlow's scalability and comprehensive ecosystem provide a broad toolkit for exploring various AI applications, while PyTorch's dynamic computation graph and user-friendly interface support an iterative and research-oriented approach to learning. Keras, with its high-level abstractions and simplicity, offers an accessible entry point into deep learning. By incorporating these platforms into educational curricula, instructors can equip students with a diverse skill set and practical experience, preparing them for the challenges and opportunities in the rapidly advancing field of artificial intelligence (Agaoglu, M. 2016).

Data Science Tools for AI Education: Jupyter Notebooks, Pandas, and NumPy

In the field of artificial intelligence (AI) education, effective data science tools are crucial for managing, analyzing, and visualizing data. Jupyter Notebooks, Pandas, and NumPy are three foundational tools that play an integral role in AI and data science curricula. Each tool offers unique functionalities that enhance the learning experience by facilitating hands-on experimentation, data manipulation, and mathematical computation. Understanding these tools' roles and capabilities is essential for educators and students aiming to leverage data science effectively in AI education.jupyter Notebooks have become a cornerstone in both data science and AI education due to their interactive and user-friendly environment. Jupyter Notebooks allow students to create and share documents that contain live code, equations, visualizations, and narrative text. This integrated approach supports an iterative learning process where students can write and execute code in real time, observe the results immediately, and document their findings alongside their code. This format encourages exploratory learning and provides a platform for students to develop, test, and refine AI models in a highly interactive manner (Babić, I. D. 2017).The ability to combine code execution with narrative explanations and visualizations makes Jupyter Notebooks particularly effective for teaching complex AI concepts. Students can document their thought processes, annotate code snippets with explanations, and create visualizations to better understand data patterns and model performance. This approach promotes a deeper understanding of both the theoretical and practical aspects of AI. Their compatibility with numerous data science libraries and frameworks further enhances their utility, allowing students to integrate and utilize various tools and techniques within a single notebook environment.Pandas Pandas simplifies data cleaning, transformation, and aggregation, enabling students to perform a wide range of data operations with minimal code. Its functionality is crucial for tasks such as merging datasets, filtering data, handling missing values, and summarizing data statistics.In educational settings, Pandas helps students develop a strong foundation in data preprocessing and analysis, which are critical skills in AI development. The library's integration with other scientific computing tools, such as SciPy and Matplotlib, further extends its functionality, providing students with a comprehensive environment for numerical and graphical analysis.Integrating Jupyter Notebooks, Pandas, and NumPy into AI education creates a cohesive learning experience where students can effectively manage, analyze, and visualize data. Jupyter Notebooks offer an interactive platform for coding and documentation, Pandas facilitates efficient data manipulation and analysis, and NumPy supports numerical computations and array operations. Together, these tools provide a robust foundation for students to develop and apply AI techniques, enhancing their ability to handle real-world data challenges and build effective AI models.In conclusion, Jupyter Notebooks, Pandas, and NumPy are essential tools in data science and AI education, each offering unique functionalities that contribute to a comprehensive learning experience. Jupyter Notebooks provide an interactive and integrated environment for coding and documentation, Pandas enables efficient data manipulation and analysis, and NumPy supports advanced numerical computations. By incorporating these tools into educational curricula, educators can equip students with the practical skills and knowledge needed to excel in the field of AI and data science, preparing them for the demands of the modern data-driven world.

Machine Learning Libraries and Frameworks: Scikit-learn, OpenCV, and NLTK

Scikit-learn, OpenCV, and NLTK are three prominent tools that offer specialized functionalities for different aspects of machine learning. Each of these libraries provides unique capabilities that enhance the educational experience by facilitating hands-on learning and practical applications of machine learning techniques. This consistency allows students to focus on understanding and experimenting with different algorithms without being overwhelmed by intricate implementation details.In educational settings, Scikit-learn is particularly valuable for teaching fundamental machine learning concepts. Its straightforward API facilitates the development and evaluation of models, enabling students to quickly apply algorithms to real-world datasets (Bahadır, E. 2016). Scikit-learn's built-in functions for model evaluation and selection, such as cross-validation and performance metrics, help students understand how to assess model performance and improve their predictive accuracy. The library's documentation and community support further enhance the learning experience, providing students with ample resources to troubleshoot issues and deepen their understanding of machine learning principles.OpenCV, or Open Source Computer Vision Library, is another essential tool, particularly for applications involving image processing and computer vision. OpenCV offers a vast array of functions and algorithms for processing and analyzing visual data, making it a valuable resource for students interested in image-related AI projects (Ramakrishnan T., et al.,2022).In summary, Scikit-learn, OpenCV, and NLTK each play a significant role in machine learning education, offering specialized tools for different domains within the field. Scikit-learn's user-friendly interface and comprehensive suite of machine learning algorithms make it ideal for teaching core concepts and building predictive models.

OpenCV's capabilities in image processing and computer vision provide valuable experience in working with visual data. NLTK's tools for natural language processing enable students to explore and analyze textual data. By incorporating these libraries and frameworks into educational curricula, instructors can provide students with a diverse set of skills and practical experience, preparing them for various applications and challenges in the rapidly advancing field of machine learning.

AI-Powered Learning Management Systems: Personalized Education Platforms

AI-powered LMS platforms leverage sophisticated algorithms and data analytics to tailor educational content and support to individual learners, thereby addressing diverse learning needs and optimizing the educational journey. These systems go beyond traditional LMS functionalities by incorporating adaptive learning technologies, predictive analytics, and intelligent tutoring systems to create a more personalized and effective learning environment.AI-powered platforms, however, use data-driven insights to customize learning paths based on each student's unique profile. By analyzing data such as past performance, engagement metrics, and learning preferences, these systems can recommend tailored content, adjust the difficulty of assignments, and provide targeted feedback (Calvo, R. A., et al.,2011). This personalization ensures that each student receives the support they need to succeed, fostering a more engaging and effective learning experience.Adaptive learning technologies are a key feature of AI-powered LMS platforms. These systems dynamically adjust the content and learning activities based on real-time analysis of student interactions and performance. Predictive analytics is another significant aspect of AI-powered LMS. Additionally, predictive analytics can provide insights into overall class performance, helping educators make data-informed decisions and improve instructional strategies.Intelligent tutoring systems (ITS) are an innovative component of AI-powered LMS platforms, offering personalized assistance akin to one-on-one tutoring. These systems simulate human tutoring by providing instant feedback, explanations, and guidance based on the student's interactions with the content. ITS can engage students in interactive problem-solving, offer hints and solutions, and adapt their teaching strategies based on the learner's responses. This personalized support enhances the learning experience by providing immediate assistance and helping students overcome challenges more effectively.

The integration of AI in LMS also extends to administrative and operational efficiencies. Automated grading and assessment tools reduce the burden on educators by providing quick and accurate evaluations of student submissions (Camacho, D. et al.,2007). AI-driven analytics can streamline administrative tasks, such as scheduling and resource allocation, allowing educators to focus more on instructional activities. Additionally, AI can assist in content creation and curation by identifying high-quality educational resources and suggesting relevant materials for inclusion in the curriculum.Despite these advancements, the deployment of AI-powered LMS platforms also raises important considerations regarding data privacy, ethical use, and accessibility. Ensuring the security and confidentiality of student data is paramount, and institutions must establish robust safeguards to protect sensitive information. Ethical use of AI involves transparency in how algorithms make decisions and ensuring that these systems do not reinforce biases or inequalities. Additionally, efforts must be made to ensure that AI-powered tools are accessible to all students, including those with disabilities or limited technological resources.

Virtual and Augmented Reality in AI Education: Immersive Learning Experiences

Virtual Reality provides a fully immersive digital environment where students can engage with educational content in a simulated 3D space. In AI education, VR can be used to create dynamic simulations that allow students to explore and interact with AI models and algorithms in a virtual setting. For example, VR can visualize neural network architectures, enabling students to see how data flows through layers and how various components interact within a deep learning model. This hands-on approach helps demystify complex concepts and fosters a deeper understanding of AI systems (DeCarlo, P., et al.,2010). Students can manipulate virtual objects, conduct experiments, and observe outcomes in real-time, providing a more intuitive grasp of abstract concepts.In addition to visualizing AI models, VR can facilitate immersive training environments for practical applications. For instance, VR simulations can replicate real-world scenarios where AI is applied, such as autonomous vehicle navigation or robotics operations. By interacting with these simulations, students can gain practical experience in troubleshooting and optimizing AI systems, preparing them for real-world challenges. VR also offers opportunities for collaborative learning, where students can work together in a shared virtual space to solve problems or develop projects, enhancing teamwork and communication skills.Augmented Reality, For example, students can use AR to visualize data sets and AI algorithms in real-time as they study, overlaying interactive graphs and models onto physical worksheets. This integration of virtual elements with physical objects helps contextualize theoretical knowledge and provides a more engaging way to interact with content.AR can also support experiential learning by providing real-time feedback and guidance during hands-on activities. For example, in a classroom where students are building AI models or programming robots, AR can offer step-by-step instructions and overlay helpful hints on their workspace. This real-time assistance helps students learn by doing and reinforces concepts as they apply them in practical scenarios (Vijayakumar G, et al., 2024). Furthermore, AR can enhance field trips or external learning activities by providing contextual information and interactive elements related to AI applications in various industries. AR can enhance remote learning by providing interactive and contextualized content that students can access from their own devices, ensuring a more connected and engaging learning experience (Delen, D. 2010).Despite the promising potential of VR and AR in AI education, there are several considerations to address. The cost of implementing VR and AR technologies can be significant, requiring investment in hardware and software as well as ongoing maintenance. Additionally, educators need to be trained in effectively integrating these technologies into their teaching practices and developing relevant content. Ensuring that VR and AR experiences are accessible to all students, including those with disabilities, is also crucial to promoting equitable learning opportunities.By providing dynamic simulations, interactive content, and experiential learning opportunities, these technologies enhance student engagement and understanding of complex AI concepts. As VR and AR continue to evolve, their integration into educational curricula will likely drive further innovations and improvements in how AI is taught and experienced.

Natural Language Processing Tools for AI Education: Text Analysis and Generation

In the context of AI education, NLP tools play a critical role in teaching students the intricacies of language models, text analysis, and text generation. These tools offer students practical experience with technologies that are increasingly integral to various applications, from chatbots and virtual assistants to sentiment analysis and automated content creation.Text Analysis is a fundamental aspect of NLP that involves extracting meaningful information from textual data. NLTK offers a comprehensive suite of functions for processing and analyzing text, including text classification, sentiment analysis, and topic modeling (Easterday, M. W. et al,2018). Its extensive documentation and educational resources make it an excellent choice for introducing students to NLP concepts and techniques. SpaCy, on the other hand, provides a more streamlined and efficient approach to text analysis, with robust support for modern NLP tasks such as dependency parsing and word embeddings.In the classroom, these tools enable students to work with real-world text data, develop custom text-processing pipelines, and gain hands-on experience with fundamental NLP techniques. For example, students might use these tools to analyze sentiment in social media posts, extract key information from news articles, or classify text into different categories. The ability to perform and visualize text analysis provides valuable insights into the structure and meaning of language, laying the groundwork for more advanced NLP applications.The interactive nature of text generation projects allows students to see the impact of their work in real time, enhancing their understanding of model behavior and performance.Hugging Face's Transformers library is an example of a powerful tool that facilitates text generation and other NLP tasks. This library provides pre-trained models and easy-to-use APIs for working with transformer-based architectures, making it accessible for educational purposes. Students can leverage these models to generate text, perform language translation, and implement question-answering systems. The library's extensive documentation and community support further assist in the learning process, providing resources and examples to guide students through the development of NLP applications.The integration of text analysis and generation tools into AI education also supports the development of critical skills such as data preprocessing, model evaluation, and algorithmic thinking. Students learn to preprocess and clean text data, select appropriate models and algorithms for specific tasks, and evaluate the performance of their models using various metrics. These skills are essential for working with real-world NLP applications and understanding the limitations and challenges associated with language technologies.Despite the advantages of NLP tools, several considerations must be addressed in educational contexts. Ensuring that students have access to computational resources capable of running sophisticated NLP models is essential, as some models require significant processing power (Feghali, T., Zbib, I., et al, 2011).

Computer Vision Tools for AI Education: Image and Video Analysis

By leveraging these tools, educators can provide students with hands-on experience in developing and applying computer vision technologies.Image Analysis forms a core component of computer vision, involving the extraction and interpretation of meaningful information from still images. Key tasks in image analysis include image classification, object detection, and image segmentation. Libraries such as OpenCV, PIL (Python Imaging Library), and scikit-image are fundamental tools for image processing and analysis (Garcia-Gorrostieta, J. M.,et al.2018). OpenCV, an open-source library developed by Intel, provides a comprehensive set of tools for a variety of image processing tasks, including feature detection, image transformation, and object tracking. Its extensive functionality and optimization for real-time performance make it a valuable resource for both educational and practical applications.In the classroom, OpenCV allows students to engage with fundamental image processing techniques such as filtering, edge detection, and morphological operations. Students can work on projects that involve processing images to identify objects, analyze patterns, or enhance image quality. For instance, students might use OpenCV to develop algorithms for detecting faces in images or segmenting regions of interest. The practical experience gained from these projects helps students understand the underlying principles of computer vision and apply them to real-world scenarios.Both TensorFlow and PyTorch offer pre-trained models and transfer learning capabilities, allowing students to build and fine-tune models for specific image analysis tasks with relatively little computational effort.

TensorFlow, developed by Google, provides the TensorFlow Object Detection API and densorFlow Hub, which offer pre-trained models and tools for various image analysis tasks.. PyTorch, developed by Facebook, also supports CNNs through its Torchvision library, which includes pre-trained models and datasets for image analysis. PyTorch's dynamic computation graph and user-friendly interface facilitate experimentation and model development, making it an excellent choice for educational purposes.Video Analysis extends the principles of image analysis to temporal sequences, introducing additional complexities such as motion tracking, action recognition, and video classification. Tools for video analysis often build on image processing techniques, incorporating temporal information to understand movement and changes over time. Libraries such as OpenCV and frameworks like TensorFlow and PyTorch also support video analysis, providing tools for processing video frames, tracking objects, and detecting activities. In educational settings, video analysis projects can involve developing systems for real-time motion detection, activity recognition, or video summarization. For example, students might create applications that analyze sports footage to track player movements or detect unusual activities in surveillance videos (Gough, D., et al.,2017). These projects help students understand the challenges of working with temporal data and develop solutions that integrate image and video processing techniques. For instance, autonomous vehicles use computer vision for object detection and lane tracking, while medical imaging applications use it for diagnostic purposes and treatment planning. Despite the promising potential of computer vision tools, several challenges must be addressed in educational contexts. Ensuring that students have access to adequate computational resources is essential, as deep learning models and video processing tasks can require significant processing power. Additionally, data privacy and ethical considerations related to computer vision, such as the potential for bias in image recognition systems, must be discussed to promote responsible use of technology. In summary, computer vision tools for image and video analysis are integral to AI education, providing students with practical experience and insights into advanced visual processing techniques.

Robotics and AI: Hands-on Learning with Robotics Platforms

The convergence of robotics and artificial intelligence (AI) represents a significant advancement in both fields, offering students a rich, hands-on learning experience that integrates physical and digital technologies. Robotics platforms equipped with AI capabilities provide an ideal environment for students to explore complex concepts, develop practical skills, and apply theoretical knowledge in a tangible, interactive way. These platforms enable students to design, program, and control robots that can perform a variety of tasks, from simple movements to advanced autonomous operations, bridging the gap between classroom learning and real-world applications.Robotics platforms offer a wide range of tools and features that facilitate experiential learning in AI and robotics. Educational robots such as LEGO Mindstorms, VEX Robotics, and Thymio provide accessible entry points for students to engage with robotics and programming. These platforms often come with user-friendly interfaces, modular components, and comprehensive curricula that support learning across different age groups and skill levels (Hall Jr., O. P., et al.,2008). For example, LEGO Mindstorms integrates a programmable brick with motors and sensors, allowing students to build and program robots for various tasks. The platform's visual programming environment and extensive educational resources make it suitable for introducing foundational robotics concepts and fostering creativity in problem-solving.VEX Robotics offers a more advanced platform with a focus on STEM education and competition. The VEX IQ and VEX V5 systems provide a versatile set of tools for constructing and programming robots, enabling students to engage in both individual and collaborative projects. VEX Robotics emphasizes hands-on learning and real-world applications, with a curriculum that covers topics such as mechanical design, sensor integration, and autonomous behavior (Zhu, W., et al.,2015). ROS enables students to work with complex algorithms, integrate sensors and actuators, and develop advanced robotic behaviors. Its modular design and active community support offer valuable resources for learning and experimentation. NVIDIA Jetson is a powerful platform designed for AI and robotics applications, featuring high-performance computing capabilities and support for deep learning frameworks (Iglesias, A., et al.,2009). Jetson modules, such as the Jetson Nano and Jetson Xavier, provide students with the computational power needed to implement and run AI algorithms on robots. With Jetson, students can develop applications such as object detection, speech recognition, and autonomous navigation, leveraging the platform's capabilities to explore cutting-edge AI technologies in a robotics context.

Programming and control are critical aspects of robotics education, and platforms like Arduino and Raspberry Pi provide versatile tools for developing and controlling robotic systems. Arduino offers a range of microcontrollers and sensors that can be used to build and program robots, while Raspberry Pi provides a compact, powerful computing platform for running more complex software and interfacing with various sensors and actuators. Hands-on learning with robotics platforms offers several educational benefits. By engaging with physical robots, students gain practical experience in designing, building, and programming robotic systems. This experiential learning helps solidify theoretical concepts and fosters a deeper understanding of how AI and robotics technologies work together (Yang,.F. 2007). Additionally, educators need to provide guidance and support to help students navigate the complexities of robotics programming and AI integration. Addressing these challenges requires careful planning and resource allocation to ensure that all students can benefit from the opportunities offered by robotics education.In conclusion, robotics platforms (Jackson, M. et al.,2015) equipped with AI capabilities offer a powerful and immersive learning experience in AI education. The integration of advanced tools and techniques further enhances the learning experience, preparing students for future careers and innovations in these rapidly evolving fields.

AI Education Software: Simulation, Modeling, and Animation Tools

AI education is greatly enhanced by the integration of sophisticated software tools for simulation, modeling, and animation, which provide students with interactive and immersive learning experiences. These tools allow students to visualize complex AI concepts, experiment with different algorithms and scenarios, and gain practical insights into how AI systems operate. By leveraging simulation, modeling, and animation software, educators can create dynamic learning environments that bridge theoretical knowledge and practical application, making abstract concepts more tangible and comprehensible.Simulation software is crucial in AI education as it allows students to model and interact with virtual environments that replicate real-world scenarios (Weston-Sementelli, J. et al.,2018). These simulations can range from simple models to complex systems, providing students with a platform to test and evaluate AI algorithms in a controlled setting. MATLAB/Simulink, for example, offers powerful tools for simulation and model-based design, enabling students to develop and analyze dynamic systems and control algorithms. Its visual interface allows users to build complex models by connecting blocks representing various components and systems, making it easier to understand and manipulate complex interactions.In the context of AI, MATLAB/Simulink can be used to simulate various scenarios such as autonomous vehicle navigation, robotics control systems, and sensor fusion. Students can experiment with different algorithms, adjust parameters, and observe the effects in real-time, gaining valuable insights into the performance and behavior of AI systems (Jain, G. P., et al, 2014). This hands-on approach helps in understanding the practical implications of theoretical concepts and allows for iterative learning and refinement of algorithms. educational purposes and allowing students to quickly develop and test models of complex systems.

Animation tools further enhance AI education by providing visual representations of algorithms and data. These tools help students visualize the inner workings of AI models and understand the dynamics of data processing and decision-making. Blender, a powerful open-source 3D modeling and animation software, can be used to create animations that illustrate AI concepts such as neural network training, reinforcement learning (Kalz, M., et al.,2008), and data visualization. Blender's extensive capabilities allow students to develop custom animations that demonstrate how AI systems learn and adapt, providing a deeper understanding of the underlying processes.Unity, a popular game development platform, is also widely used in AI education for its simulation and visualization capabilities. Unity supports the development of interactive 3D environments and offers tools for integrating AI algorithms into simulations. Students can create virtual worlds, design and program AI agents, and test their behavior in dynamic settings. Unity's versatility makes it an excellent tool for developing projects such as autonomous robots, intelligent game characters, and interactive educational simulations.Google Colab and Jupyter Notebooks are additional tools that facilitate interactive learning through coding and visualization. Google Colab provides a cloud-based environment for running Python code and working with AI libraries such as TensorFlow and PyTorch. It enables students to collaborate on projects, share code, and visualize results without the need for local computing resources. Jupyter Notebooks, similarly, offer an interactive interface for writing and executing code, visualizing data, and documenting results. These tools are valuable for exploring machine learning algorithms, analyzing datasets, and presenting findings in an accessible format.The integration of these simulation, modeling, and animation tools into AI educatin offers several benefits (Welham, D.2008).

However, there are challenges associated with using these tools in education. Ensuring that students have access to the necessary software and hardware resources can be a significant consideration, particularly for advanced simulation and modeling tools that may require substantial computational power (Kao, Chen, et al.,2010). Additionally, educators need to provide appropriate guidance and support to help students effectively use these tools and integrate them into their learning experiences. Developing effective instructional materials and curricula that leverage these tools is also essential for maximizing their educational impact.In conclusion, AI education software for simulation, modeling, and animation plays a crucial role in enhancing learning experiences and understanding of complex AI concepts. By providing interactive and immersive environments for exploring algorithms, testing scenarios, and visualizing data, these tools bridge the gap between theory and practice. The use of simulation platforms like MATLAB/Simulink, modeling tools like AnyLogic and NetLogo, and animation software like Blender and Unity enables students to develop practical skills, gain insights into AI systems, and apply their knowledge to real-world problems. Addressing the associated challenges and ensuring effective implementation will further enhance the educational benefits of these powerful tools.

Cloud-Based AI Education Platforms: Scalable and On-Demand Resources

Cloud-based AI education platforms have transformed the landscape of learning by offering scalable and on-demand resources that cater to diverse educational needs. These platforms leverage cloud computing technology to provide an array of tools, materials, and instructional methods accessible anytime, anywhere, facilitating a flexible learning environment. One of the most significant advantages of cloud-based AI education platforms is their scalability. They can accommodate a large number of users simultaneously without compromising performance, which is particularly beneficial for institutions and organizations with fluctuating or growing user bases. This scalability ensures that educational resources are consistently available, regardless of the number of learners accessing them.The on-demand nature of these platforms means that learners can engage with educational content at their own pace, fitting their studies around other commitments (Li, X. 2007). Such platforms also facilitate collaborative learning environments, where users can interact with peers, participate in discussion forums, and work on group projects in real-time, regardless of geographical locations. This connectivity fosters a global learning community and encourages diverse perspectives and ideas.Data analytics is another powerful feature of cloud-based AI education platforms. By tracking user interactions and performance, these platforms can generate insights into learning trends, identify areas where students may be struggling (Vlugter, P.,et al.,2009). Furthermore, many of these platforms operate on a subscription-based or pay-as-you-go model, making them accessible to a wider audience, including individuals and organizations with limited budgets.In summary, cloud-based AI education platforms represent a significant advancement in the field of education, offering scalable, flexible, and personalized learning experiences (Lodhi, P., et al.,2018). Their ability to deliver on-demand resources, integrate multimedia content, support collaboration, and leverage data analytics enhances the effectiveness of educational programs and addresses the diverse needs of learners.

AI Education Hardware: Specialized Computer Systems and Devices

AI education hardware encompasses specialized computer systems and devices designed to enhance the learning and application of artificial intelligence technologies. These systems and devices are tailored to meet the demanding computational requirements of AI, providing educators and students with the tools needed to explore, experiment, and innovate in the field of AI. As AI continues to evolve and integrate into various sectors, the role of dedicated hardware in education becomes increasingly crucial (Ma, H., et al.,2015). One prominent category of AI education hardware is Graphics Processing Units (GPUs). GPUs are critical for handling the parallel processing needs of AI algorithms, particularly those involved in deep learning and neural networks.AI education hardware also includes high-performance computing clusters and cloud-based resources (McNamara, D. S., 2015). These clusters consist of multiple interconnected computers working together to solve complex problems and run large-scale AI simulations. In educational contexts, they provide a platform for collaborative projects and research, enabling students and faculty to perform extensive computations and analyze substantial amounts of data. Cloud-based solutions further extend these capabilities by offering scalable computing power and storage on-demand, allowing educational institutions to access advanced AI resources without the need for significant upfront investment in physical hardware.Interactive learning devices, such as AI-powered robots and smart sensors, also play a role in AI education hardware. These devices offer hands-on experiences for students to apply AI concepts in real-world scenarios (Samarakou, M., et al.,2015).

Best Practices for Selecting AI Education Tools and Technologies

Selecting the right AI education tools and technologies is crucial for fostering effective learning environments and ensuring that students gain relevant and practical skills in artificial intelligence. Adhering to best practices when choosing these tools can significantly enhance educational outcomes, optimize resource use, and prepare learners for future challenges in the AI field. Here are several key considerations and best practices for selecting AI education tools and technologies Define Educational Objective Before selecting any tools or technologies, clearly outline the educational goals and objectives. Determine what specific AI concepts, skills, or applications the tools should support. This could range from basic machine learning principles to advanced deep learning techniques (Neumann, W.L.2007). Aligning the tools with educational objectives ensures that they effectively meet the needs of both instructors and students, facilitating targeted learning experiences.Evaluate

Technological Capabilities Assess the technological capabilities of potential AI tools. Ensure they support the required computational power, scalability, and compatibility with existing systems. Tools should be able to handle the complexity of AI tasks, such as data processing, model training, and simulation. For instance, if the focus is on deep learning, tools with robust GPU support and high-performance computing capabilities are essential.Consider User-Friendliness Choose tools that are user-friendly and accessible to the target audience. Educational tools should have intuitive interfaces, clear documentation, and supportive learning resources. This is especially important for beginners or those with limited experience in AI. User-friendly tools reduce the learning curve and help students focus on mastering AI concepts rather than struggling with complex software.Ensure Compatibility and Integration Verify that the selected tools are compatible with existing educational technologies and systems. Compatibility with various operating systems, software platforms, and programming languages is crucial for seamless integration. Additionally, consider whether the tools can integrate with other educational resources, such as learning management systems (LMS) and collaboration platforms, to provide a cohesive learning experience.Assess Support and Community Evaluate the level of support and community engagement available for the tools. Strong support from developers and an active user community can be invaluable for troubleshooting (Nguyen, J., et al.,2018), sharing best practices, and accessing additional resources. Tools with comprehensive support, forums, and user groups can enhance the learning experience and provide assistance when challenges arise.Review Cost and Licensing Consider the cost of the tools and their licensing arrangements. Many AI tools have varying pricing models, including free, open-source, subscription-based, or one-time purchase options. Evaluate the cost relative to the budget and the value provided. For educational institutions, exploring discounts, educational licenses, or grants can also be beneficial in managing expenses.Prioritize Security and Privacy Ensure that the tools adhere to robust security and privacy standards (Oztekin, A. 2016). Data protection is crucial, especially when dealing with sensitive information or student data. summary, selecting AI education tools and technologies requires careful consideration of educational objectives, technological capabilities, user-friendliness, compatibility, support, cost, security, and long-term viability (Paquette, L., et al.,2015). By following these best practices, educators and institutions can choose tools that effectively support learning, foster engagement, and prepare students for the dynamic field of artificial intelligence.

The Future of AI Education Tools and Technologies - Trends and Innovations

The future of AI education tools and technologies is poised for transformative advancements, driven by emerging trends and innovations that promise to reshape how artificial intelligence is taught and learned. As AI continues to evolve, educational tools and technologies are adapting to offer more sophisticated, interactive, and personalized learning experiences. Here are some key trends and innovations shaping the future of AI education Enhanced Personalization through Adaptive Learning By dynamically adjusting content and difficulty levels, adaptive learning platforms provide targeted support, helping students grasp complex AI concepts more effectively and at their own pace. This personalization extends beyond traditional classroom settings, enabling remote and self-directed learning to be more engaging and effective.For instance, AR applications could visualize neural network architectures or simulate data analysis processes in real-time (Pedró, F., et al.,2019). VR, on the other hand, can create fully immersive environments where students can experiment with AI algorithms, participate in virtual labs, or explore complex datasets interactively. These technologies provide hands-on experiences that enhance comprehension and retention of abstract AI principles.Development of AI-Powered Educational Assistants The rise of AI-powered educational assistants, such as chatbots and virtual tutors, is transforming the way students interact with learning materials. They also facilitate continuous learning by providing assistance outside traditional classroom hours, making educational support more accessible and responsive.Expansion of Open Educational Resources (OER) The proliferation of open educational resources is democratizing access to high-quality AI education materials. OER platforms offer free or low-cost educational content, including textbooks, tutorials, and interactive tools, which can be easily shared and adapted. Innovations in OER are enabling collaborative development of AI educational resources, where educators and researchers contribute to and refine content collectively. This trend promotes equity in education by making advanced AI learning materials accessible to a broader audience, including those in under-resourced or remote areas.

Advances in AI-Driven Content Creation and Curriculum Development

AI technologies are also transforming content creation and curriculum development. AI-driven tools can assist educators in designing and updating course materials, generating interactive simulations, and creating customized assessments (Quixal, M., 2016). These tools can analyze trends in student performance and learning outcomes to suggest improvements or new content areas. By automating routine tasks, educators can focus more on engaging students and addressing complex educational needs, leading to more dynamic and responsive AI curricula.Emphasis on Ethical AI and Responsible Use As AI education grows, there is a heightened focus on teaching ethical AI practices and responsible use of technology. Future AI education tools will likely incorporate modules on AI ethics, fairness, and transparency, helping students understand the societal implications of AI systems. Educational technologies will emphasize the importance of developing AI applications that are inclusive, equitable, and aligned with ethical standards. This approach ensures that future AI professionals are equipped with the knowledge to create technology that benefits society and minimizes potential harms.Integration of AI in Data-Driven Decision Making AI tools are increasingly being used to analyze educational data and inform decision-making processes. By leveraging AI to analyze and interpret educational data, institutions can enhance the effectiveness of their AI education programs and support student success.Growth of Collaborative Learning Platforms Collaborative learning platforms, supported by AI technologies, are fostering interactive and team-based learning experiences (Raju, D., et al.,2015).

These platforms enable students to work together on AI projects, share insights, and solve problems collectively. AI facilitates collaboration by providing real-time feedback, coordinating group activities, and managing project workflows. Such platforms promote peer learning and simulate real-world AI development environments, where teamwork and communication are essential.Enhanced Accessibility through Assistive Technologies The future of AI education tools includes a strong emphasis on accessibility and inclusivity. Assistive technologies powered by AI are making learning more accessible for students with disabilities. Features such as speech-to-text, text-to-speech, and intelligent navigation aids ensure that educational resources are usable by a diverse range of learners. These technologies help remove barriers to learning and provide equitable access to AI education for all students. Continuous Evolution and Lifelong Learning AI education tools will increasingly support lifelong learning and professional development. As AI technology evolves, continuous education becomes essential for staying current with new advancements (Ramírez, J., et al.,2018). Tools and platforms will offer ongoing learning opportunities, certifications, and professional development programs tailored to various stages of a career in AI. This approach helps professionals adapt to technological changes and maintain their expertise in a rapidly advancing field.In summary, the future of AI education tools and technologies is characterized by innovations that enhance personalization, immersion, accessibility, and collaboration. As AI continues to advance, these trends and innovations will shape how AI is taught, learned, and applied, preparing students and professionals for a dynamic and evolving landscape. By embracing these developments, educational institutions can provide effective and engaging AI education that meets the needs of a diverse and global learner base.

CONCLUSION

In conclusion, the tools and technologies for AI education represent a dynamic and rapidly evolving landscape that is fundamentally reshaping how artificial intelligence is taught and learned. As AI continues to advance and integrate into various sectors, the educational tools and technologies supporting its study are becoming increasingly sophisticated and diverse. The integration of adaptive learning systems, augmented reality (AR), and virtual reality (VR) provides immersive and personalized learning experiences that enhance students' understanding of complex AI concepts. These technologies offer interactive environments where learners can visualize and experiment with AI principles, thereby gaining practical insights that traditional methods may not offer.AI-powered educational assistants and chatbots play a crucial role in offering personalized support and instant feedback, making learning more accessible and responsive to individual needs. These tools facilitate continuous learning and ensure that students receive the guidance they need, even outside of regular classroom settings. Similarly, the expansion of open educational resources (OER) democratizes access to high-quality AI materials, enabling a broader audience to benefit from cutting-edge content and tools. The collaborative nature of OER promotes collective development and refinement of educational resources, contributing to a more inclusive and equitable learning environment. Moreover, advancements in AI-driven content creation and curriculum development are transforming how educational materials are designed and updated. These innovations allow educators to tailor content to current trends and student needs, enhancing the relevance and effectiveness of AI education. The growing emphasis on ethical AI and responsible technology use ensures that future AI professionals are well-versed in the societal implications of their work, promoting the development of technology that aligns with ethical standards and benefits society as a whole. Additionally, collaborative learning platforms supported by AI technologies foster teamwork and peer learning, simulating real-world AI development environments and preparing students for collaborative work in the industry.Assistive technologies enhance accessibility and inclusivity, ensuring that all learners, regardless of their abilities, can engage with AI education. Features such as speech-to-text and intelligent navigation aids help remove barriers to learning, providing equitable access to educational resources. As AI education tools continue to evolve, there is also a strong emphasis on supporting lifelong learning and professional development. Continuous education and certification programs tailored to various stages of a career in AI help professionals stay current with technological advancements and maintain their expertise. In essence, the future of AI education tools and technologies is characterized by a commitment to innovation, personalization, and inclusivity.

REFERENCES

Acikkar, M., & Akay, M. F. (2009). Support vector machines for predicting the admission decision of a candidate to the School of Physical Education and Sports at Cukurova University. *Expert Systems with Applications*, 36(3 PART 2), 7228–7233. DOI: 10.1016/j.eswa.2008.09.007

Adamson, D., Dyke, G., Jang, H., & Rosé, C. P. (2014). Towards an agile approach to adapting dynamic collaboration support to student needs. *International Journal of Artificial Intelligence in Education*, 24(1), 92–124. DOI: 10.1007/s40593-013-0012-6

Agaoglu, M. (2016). Predicting instructor performance using data mining techniques in higher education. *IEEE Access : Practical Innovations, Open Solutions*, 4, 2379–2387. DOI: 10.1109/ACCESS.2016.2568756

Babić, I. D. (2017). Machine learning methods in predicting the student academic motivation. *Croatian Operational Research Review*, 8(2), 443–461. DOI: 10.17535/crorr.2017.0028

Bahadır, E. (2016). Using neural network and logistic regression analysis to predict prospective mathematics teachers' academic success upon entering graduate education. *Educational Sciences: Theory & Practice*, 16(3), 943–964. DOI: 10.12738/estp.2016.3.0214

Calvo, R. A., O'Rourke, S. T., Jones, J., Yacef, K., & Reimann, P. (2011). Collaborative writing support tools on the cloud. *IEEE Transactions on Learning Technologies*, 4(1), 88–97. https://www.learntechlib.org/p/73461/. DOI: 10.1109/TLT.2010.43

Camacho, D., & Moreno, M. D. R. (2007). Towards an automatic monitoring for higher education learning design. *International Journal of Metadata, Semantics and Ontologies*, 2(1), 1. DOI: 10.1504/IJMSO.2007.015071

DeCarlo, P., & Rizk, N. (2010). The design and development of an expert system prototype for enhancing exam quality. *International Journal of Advanced Corporate Learning*, 3(3), 10–13. DOI: 10.3991/ijac.v3i3.1356

Delen, D. (2010). A comparative analysis of machine learning techniques for student retention management. *Decision Support Systems*, 49(4), 498–506. DOI: 10.1016/j.dss.2010.06.003

Easterday, M. W., Rees Lewis, D. G., & Gerber, E. M. (2018). The logic of design research. Learning. *Research and Practice*, 4(2), 131–160. DOI: 10.1080/23735082.2017.1286367

Feng, S., Zhou, S., & Liu, Y. (2011). Research on data mining in university admissions decision-making. *International Journal of Advancements in Computing Technology*, 3(6), 176–186. DOI: 10.4156/ijact.vol3.issue6.21

Garcia-Gorrostieta, J. M., Lopez-Lopez, A., & Gonzalez-Lopez, S. (2018). Automatic argument assessment of final project reports of computer engineering students. *Computer Applications in Engineering Education*, 26(5), 1217–1226. DOI: 10.1002/cae.21996

Gough, D., Oliver, S., & Thomas, J. (2017). *An introduction to systematic reviews* (2nd ed.). SAGE.

Hall, O. P.Jr, & Ko, K. (2008). Customized content delivery for graduate management education: Application to business statistics. *Journal of Statistics Education : An International Journal on the Teaching and Learning of Statistics*, 16(3). Advance online publication. DOI: 10.1080/10691898.2008.11889571

Hew, K. F., Lan, M., Tang, Y., Jia, C., & Lo, C. K. (2019). Where is the "theory" within the field of educational technology research? *British Journal of Educational Technology*, 50(3), 956–971. DOI: 10.1111/bjet.12770

Iglesias, A., Martinez, P., Aler, R., & Fernandez, F. (2009). Reinforcement learning of pedagogical policies in adaptive and intelligent educational systems. *Knowledge-Based Systems*, 22(4), 266–270. https://e-archivo.uc3m.es/bitstream/handle/10016/6502/reinforcement_aler_KBS_2009_ps.pdf?sequence=1&isAllowed=y. DOI: 10.1016/j.knosys.2009.01.007

Jackson, M., & Cossitt, B. (2015). Is intelligent online tutoring software useful in refreshing financial accounting knowledge? *Advances in Accounting Education*, 16, 1–19. DOI: 10.1108/S1085-462220150000016001

Jain, G. P., Gurupur, V. P., Schroeder, J. L., & Faulkenberry, E. D. (2014). Artificial intelligence-based student learning evaluation: A concept map-based approach for analyzing a student's understanding of a topic. *IEEE Transactions on Learning Technologies*, 7(3), 267–279. DOI: 10.1109/TLT.2014.2330297

Jonassen, D., Davidson, M., Collins, M., Campbell, J., & Haag, B. B. (1995). Constructivism and computer-mediated communication in distance education. *American Journal of Distance Education*, 9(2), 7–25. DOI: 10.1080/08923649509526885

Kalz, M., van Bruggen, J., Giesbers, B., Waterink, W., Eshuis, J., & Koper, R. (2008). A model for new linkages for prior learning assessment. *Campus-Wide Information Systems*, 25(4), 233–243. DOI: 10.1108/10650740810900676

Kanipriya, M., Sugumaran, D., Pandiyarajan, V., Yogananda, B. S., Selvakumar, P., & Mohan, S. (2024, January). Analysis of Electromyography Signals for Control of Mechanical Prosthesis using Machine Learning Techniques. In 2024 2nd International Conference on Intelligent Data Communication Technologies and Internet of Things (IDCIoT) (pp. 1059-1067). IEEE.

Kao, G. Y. M., Chen, K. C., & Sun, C. T. (2010). Using an e-learning system with integrated concept maps to improve conceptual understanding. *International Journal of Instructional Media*, 37(2), 151–162.

Li, X. (2007). Intelligent agent-supported online education. *Decision Sciences Journal of Innovative Education*, 5(2), 311–331. DOI: 10.1111/j.1540-4609.2007.00143.x

Lodhi, P., Mishra, O., Jain, S., & Bajaj, V. (2018). StuA: An intelligent student assistant. *International Journal of Interactive Multimedia and Artificial Intelligence*, 5(2), 17–25. DOI: 10.9781/ijimai.2018.02.008

Ma, H., & Slater, T. (2015). Using the developmental path of cause to bridge the gap between AWE scores and writing teachers' evaluations. *Writing & Pedagogy*, 7(2), 395–422. DOI: 10.1558/wap.v7i2-3.26376

McNamara, D. S., Crossley, S. A., Roscoe, R. D., Allen, L. K., & Dai, J. (2015). A hierarchical classification approach to automated essay scoring. *Assessing Writing*, 23, 35–59. DOI: 10.1016/j.asw.2014.09.002

Neumann, W. L. (2007). *Social research methods: Qualitative and quantitative approaches*. Pearson.

Nguyen, J., Sánchez-Hernández, G., Armisen, A., Agell, N., Rovira, X., & Angulo, C. (2018). A linguistic multi-criteria decision-aiding system to support university career services. *Applied Soft Computing*, 67, 933–940. DOI: 10.1016/j.asoc.2017.06.052

Nicholas, D., Watkinson, A., Jamali, H. R., Herman, E., Tenopir, C., Volentine, R., Allard, S., & Levine, K. (2015). Peer review: Still king in the digital age. *Learned Publishing*, 28(1), 15–21. DOI: 10.1087/20150104

Oztekin, A. (2016). A hybrid data analytic approach to predict college graduation status and its determinative factors. *Industrial Management & Data Systems*, 116(8), 1678–1699. DOI: 10.1108/IMDS-09-2015-0363

Paquette, L., Lebeau, J. F., Beaulieu, G., & Mayers, A. (2015). Designing a knowledge representation approach for the generation of pedagogical interventions by MTTs. *International Journal of Artificial Intelligence in Education*, 25(1), 118–156. https://www.learntechlib.org/p/168275/. DOI: 10.1007/s40593-014-0030-z

Pedró, F., Subosa, M., Rivas, A., & Valverde, P. (2019). *Artificial intelligence in education: Challenges and opportunities for sustainable development*. UNESCO.

Quixal, M., & Meurers, D. (2016). How can writing tasks be characterized in a way serving pedagogical goals and automatic analysis needs? *CALICO Journal*, 33(1), 19–48. DOI: 10.1558/cj.v33i1.26543

Raju, D., & Schumacker, R. (2015). Exploring student characteristics of retention that lead to graduation in higher education using data mining models. *Journal of College Student Retention*, 16(4), 563–591. DOI: 10.2190/CS.16.4.e

Ramakrishnan, T., Mohan Gift, M. D., Chitradevi, S., Jegan, R., Hency Jose, P. S., Nagaraja, H. N., & Hailegiorgis, S. M. (2022). Study of numerous resins used in polymer matrix composite materials. *Advances in Materials Science and Engineering*, 2022(1), 1088926.

Samarakou, M., Fylladitakis, E. D., Früh, W. G., Hatziapostolou, A., & Gelegenis, J. J. (2015). An advanced eLearning environment developed for engineering learners. *International Journal of Emerging Technologies in Learning*, 10(3), 22–33. DOI: 10.3991/ijet.v10i3.4484

Vijayakumar, G., Rajkumar, M., Rajiv Chandar, N., Selvakumar, P., & Duraisamy, R. (2024). Muniyandi Rajkumar, Rajiv Chandar N, Selvakumar P and Ramesh Duraisamy, (2024). Environmentally friendly TDS removal from waste water by electrochemical ion exchange batch-type recirculation (EIR) technique. *Environmental Science. Water Research & Technology*, 10(4), 826–835. DOI: 10.1039/D3EW00793F

Vlugter, P., Knott, A., McDonald, J., & Hall, C. (2009). Dialogue-based CALL: A case study on teaching pronouns. *Computer Assisted Language Learning*, 22(2), 115–131. DOI: 10.1080/09588220902778260

Yang, F., Wang, M., Shen, R., & Han, P. (2007). Community-organizing agent: An artificial intelligent system for building learning communities among large numbers of learners. *Computers & Education*, 49(2), 131–147. DOI: 10.1016/j.compedu.2005.04.019

Zhu, W., Marquez, A., & Yoo, J. (2015). "Engineering economics jeopardy!" Mobile app for university students. *The Engineering Economist*, 60(4), 291–306. DOI: 10.1080/0013791X.2015.1067343

Chapter 4
AI-Enhanced Education: Revolutionizing Assessments and Informed Decision-Making

Najoua Hrich
https://orcid.org/0000-0001-7013-8185
Regional Center of Education and Training Professions, Institutions for Higher Executive Training, Morocco

Mohamed Azerki
https://orcid.org/0009-0009-7397-3034
Regional Academy of Education and Training, Ministry of National Education Preschool and Sports, Morocco

Mohamed Khaldi
https://orcid.org/0000-0002-1593-1073
Ecole Normale Sperieure de Tetouan, Abdelmalek Essaadi University, Morocco

ABSTRACT

Education is a constantly evolving field, where the decisions made by teachers, administrators, and policymakers significantly impact students' academic journey and progress. In this dynamic context, integrating Artificial Intelligence (AI) open new perspectives to support and enhance informed decision-making. Using AI, with its advanced data analysis, predictive modeling, and process automation capabilities, can provide valuable information to education stakeholders, enabling them to make more informed and effective decisions. Assessment plays a crucial role in the educational process, enabling to measure learners' knowledge, skills, and progress. Through these assessments, teachers and decision-makers can make informed decisions on the pedagogical direction, identification of individual student needs, and implementation of appropriate educational policies. In this context, the integration of AI into assessment processes provides new opportunities to enhance

DOI: 10.4018/979-8-3373-2262-9.ch004

their efficiency, objectivity, and relevance.

INTRODUCTION

Educational assessment plays a fundamental role in education systems worldwide, serving as the main mechanism for measuring learners' knowledge, skills, and progress (Beaupied, 2009). It is not only a tool for verifying learning achievements, but also an essential lever for guiding pedagogical action and adapting teaching practices to students' needs. Teachers use assessment results to identify gaps in students' understanding, adjust their teaching methods, and propose personalized learning paths (Sayac, 2017), (Sayac, 2018). This process is not limited to classroom assessment, but also supports the development of educational interventions that can better meet students' diverse needs.

For policymakers and administrators, assessment plays a key role in assessing the effectiveness of school programs, teaching methods and educational reforms. It provides an evidence based foundation for shaping and refining educational policies (OECD, 2020). Data collected through assessments can highlight areas in need of improvement, guide resource allocation and inform decisions about the future direction of education systems. In this way, educational assessment is an essential pillar in ensuring that education systems remain responsive to the evolving needs of society and the labor market.

However, despite its importance, traditional assessment methods, such as standardized tests and written examinations, have several inherent limitations. These approaches do not always capture the full range of learners' skills and knowledge, particularly when it comes to measuring higher-order reasoning or problem-solving abilities. In many cases, traditional assessments are limited in their ability to measure practical, creative and social skills, which are increasingly seen as essential for success in the 21st century. Moreover, traditional assessment methods can be biased and inequitable, disadvantaging certain groups of students, such as those from marginalized communities or those with learning difficulties. This can lead to distorted results that do not truly reflect a learner's abilities, compromising the fairness and reliability of educational assessments (Swiecki et al., 2022a).

In addition, traditional assessment methods impose a considerable administrative burden on teachers and administrators. The process of designing and administering assessments, scoring responses and analyzing results can be time-consuming and resource intensive. Teachers are often required to manually grade large quantities of tests, leaving little time for other crucial tasks, such as individualized instruction, professional development or engagement with students on a deeper level. This time-consuming nature of assessments also presents a challenge for administrators, who must ensure that assessments are valid, reliable and well aligned with program objectives. Given these challenges, traditional assessment systems are increasingly seen as inadequate to meet the demands of modern education.

It is in this context that innovation in educational assessment becomes necessary. To overcome these limitations, there is an urgent need to explore new approaches that are more effective, inclusive and representative of diverse learning styles. One of the most promising innovations in this field is the integration of AI into educational assessment processes. AI technologies, which include machine learning, natural language processing and predictive modeling, have the potential to revolutionize the way assessments are designed, administered and analyzed (Hrich et al., 2023).

AI offers several key advantages that can help resolve the limitations of traditional assessment methods. Firstly, AI can automate the analysis of assessment items, identifying patterns and correlations in the data that might be overlooked by human assessors. This can help ensure that assessments are more accurate and freer from bias. For example, AI can detect redundant or poorly worded questions, ensuring that each assessment item serves its specific purpose of measuring a precise skill or knowledge. AI can also detect bias in assessment questions, ensuring that assessments are fair and just for all students, regardless of background or ability (Moses, 2017), (Zanon et al., 2016).

Next, AI can enhance the adaptive nature of assessments, making them more personalized and responsive to learners' individual needs (Liu & Baucham, 2023). Traditional assessments often take a uniform approach, which doesn't always reflect the diversity of students' abilities. In contrast, AI can analyze learners' performance in real time and adjust the difficulty level of the assessment, accordingly, offering a more tailored experience that challenges each learner at their own level. This can improve the accuracy of assessments and provide more detailed information on students' strengths and areas for improvement (Brown, 2022).

In addition, AI can increase the efficiency of the assessment process by automating routine tasks such as marking and generating feedback. This eases the administrative burden on teachers, enabling them to focus more on teaching and interacting with students. AI can also provide instant feedback to learners, enabling them to identify errors and improve their understanding of content. This is particularly useful in formative assessments, where the aim is to support ongoing learning rather than simply awarding a grade.

AI also offers the potential to improve the reliability and validity of assessments. Using data-driven techniques, AI can analyze large quantities of assessment data, enabling the identification of trends and patterns that may not be immediately apparent. This enables more accurate and consistent scoring, as AI systems are less prone to errors and inconsistencies that can arise during manual correction. What's more, AI can be used to perform item analysis, helping to determine which questions are most effective at measuring the skills or knowledge being targeted. This ensures that assessments are not only accurate, but also well aligned with curriculum objectives and learning outcomes (Gardner et al., 2021).

Integrating AI into educational assessment offers considerable scope for transforming the way we assess learners. By leveraging the capabilities of AI, educational systems can move beyond traditional approaches, creating more personalized, effective and equitable assessments. This shift to AI-powered assessments could help meet growing demands for more comprehensive, accurate and inclusive assessment practices in education. However, it's important to stress that implementing AI in educational assessment also raises important ethical and practical considerations, such as ensuring data confidentiality, avoiding bias in AI algorithms and maintaining transparency in the decision-making process (Brown, 2022).

This chapter will explore these dynamics in more detail, providing a comprehensive review of AI applications in the educational assessment process. We'll analyze how AI can help optimize item analysis, improve the relevance of multiple-choice tests, and explore other innovative uses of AI in assessment.

THE EDUCATIONAL ASSESSMENT PROCESS

The Educational Assessment Cycle encompasses all the necessary steps for implementing and analyzing an evaluation system, from test design to decision-making (Hrich et al., 2021). This process gathers data on learners' skills and knowledge to refine and improve teaching practices (Newton, 2007). It begins with defining assessment objectives, which guide test creation. These tests must be designed rigorously to reliably and validly measure the targeted skills. Next, assessments are administered to learners, often in a controlled setting to ensure fairness and result reliability. Once the assessments are completed, results are collected, analyzed, and interpreted. Utilizing these results not only assesses student performance but also provides insights for improving teaching methods and adjusting curricula to meet identified needs (IEA,2023).

Assessment extends beyond merely grading learners; it also includes evaluating the tools and processes used to ensure their relevance and effectiveness. In essence, the educational assessment process is a continuous cycle, leveraging feedback to continually optimize teaching and learning (Europe & ALTE,2012). This process comprises the following steps illustrated in figure 1.

Figure 1. Educational Assessment Process

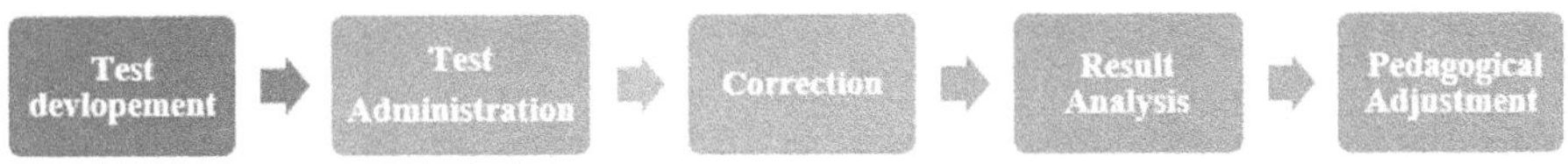

Test Development

The assessment Process begins with the design of learning objectives and assessment questions. This initial phase involves defining the skills or knowledge to be measured and designing the items (questions, tasks, exercises) that reliably gather this information. This phase lays the foundations for the test development process, ensuring coherence between educational objectives and assessment content.

The second phase, Examination Specifications, clarifies the purpose of the assessment, the target audience, and the content areas to be covered. The types of items and cognitive levels required are also defined to ensure that the assessment meets educational objectives and inclusiveness criteria.

Then, in the Item Creation phase, the editors develop questions according to defined specifications, ensuring relevance, accuracy and fairness. They ensure that the items accurately reflect the target skills and respect the defined assessment formats, to guarantee reliable measurement of learning.

The next phase, Pilot Testing/Experimentation, involves administering a preliminary version of the test to a representative sample of the target population. This stage identifies potential problems (ambiguities, inappropriate difficulty levels) and collects essential data for adjusting the items. Statistical analysis is carried out to assess the performance of each question, in terms of reliability and validity.

Based on the results obtained, an item revision is carried out. The items are adjusted to improve their clarity and relevance, based on the feedback collected during the testing phase. Only items that meet the quality criteria are retained for further processing.

Finally, the Final Test Construction and Administration phase involves assembling the selected items to create a coherent and valid assessment. The final test is then administered to the target population. The results obtained help to inform educational decisions, by providing information on the quality of the learning process and the effectiveness of the test. To illustrate the process of developing an assessment in clear steps, here is a suggested diagram of the stages. The diagram in figure 2 organizes the various phases of the test development Process (Lane et al., 2015).

Figure 2. Diagram of the development process for an evaluation test

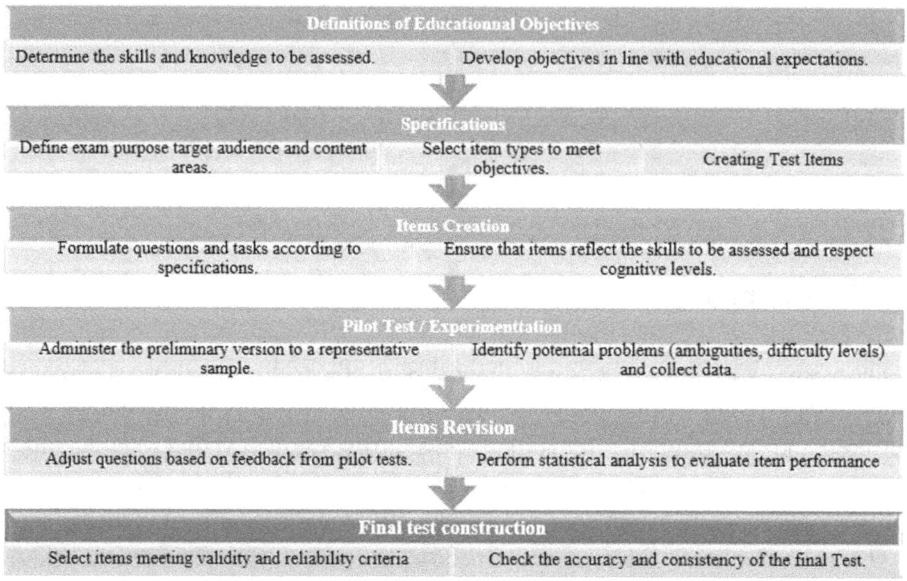

Each stage is interdependent and contributes to the validity, reliability and relevance of the final test. This cyclical process also gathers information that can be used to improve future assessments (Haladyna & Rodriguez, 2013).

Test Administration

This stage involves administering the assessment to the students, whether in the form of MCTs, essays, projects or oral exams. To ensure a smooth process, test administration requires meticulous logistical preparation. The organization of the space is a key element: it is crucial to check the layout of the rooms, the arrangement of the benches and the circulation of candidates in order to guarantee an environment conducive to concentration and compliance with instructions. The premises must be well lit, properly ventilated and accessible to the disabled, where appropriate.

Secondly, materials management plays a key role in this preparation. It's crucial to ensure that all the necessary aids are available: exam papers, pens, calculators (if authorized), time clocks to keep track of time, and monitoring devices (cameras or other control systems). For digital exams, a thorough check of computer and audiovisual equipment is essential.

At the same time, an administrative team must be mobilized to manage the registration of candidates, the delivery of invitations, identification at the entrance, and the follow-up of absences or unforeseen events. The presence of a logistics coordinator throughout the event is essential to quickly resolve any problems that may arise, whether technical or related to the specific needs of the participants.

In short, the key to logistical preparation is anticipation and effective coordination between all teams to ensure a calm, orderly environment conducive to a successful exam (*Elaborer et faire passer des tests - Cadre européen commun de référence pour les langues (CECR)* - www.coe.int, n.d.).

Correction

Once the assessment has been completed, the next step is to correct the students' answers a crucial phase in the assessment process. However, this stage is not without its complex challenges. In particular, the subjectivity inherent in certain types of tests, such as essays or projects, can make evaluation more delicate. Graders must be scrupulous in applying precise, uniform criteria to ensure impartial and fair marking. Furthermore, the workload can be particularly overwhelming, especially when dealing with a large number of papers. Rigorous time management is therefore essential to meet deadlines while ensuring thorough and conscientious marking. Human error, such as omissions or misunderstandings in the interpretation of answers, is also a major risk, hence the importance of rigor in the marking process, and sometimes even the use of double marking to avoid excessive subjectivity. Last but not least, the management of contestations or complaints from candidates represents another major challenge, which can put the credibility of the correction process to the test. Transparency, as well as clear and precise communication of evaluation criteria, are therefore essential to prevent any ambiguity and ensure that complaints are dealt with fairly and equitably.

Results Analysis

Analyzing exam results is a crucial step in the evaluation process, enabling us to assess both student performance and the overall effectiveness of the system in place. This involves an in-depth examination of results, at both individual and group level, in order to identify significant trends. Using statistical tools such as averages, standard deviations and distribution curves, we can measure the dispersion of scores and identify any anomalies, such as an excessive concentration of results around certain values. The use of analysis of variance, for example, enables us to verify the existence of notable differences between groups of students, depending on their specialization or level of preparation. In addition, qualitative analysis of responses, based on rigorously defined evaluation grids, provides a detailed assessment of candidates' strengths and weaknesses, enabling us to pinpoint skills acquired or lacking.

With this in mind, dedicated IT tools such as Excel, SPSS or R greatly facilitate large-scale data processing, offering the possibility of generating visual dashboards - graphs, histograms, box plots - for clear and efficient interpretation of results ("Classical Item & Test Analysis Spreadsheet CITAS," n.d.), (Excel Spreadsheets for Classical Test Analysis, n.d.), (IRT Illustrator – Psychmeasurement Systems, n.d.). This method also makes it possible to identify potentially problematic questions, whose difficulty or ambiguous wording may have impaired student performance. However, results analysis is not limited to an examination of overall performance: it also includes a detailed statistical analysis of the test items, in order to measure the relevance and validity of each question. The difficulty index, which evaluates the percentage of students who answered a question correctly, and the discrimination index, which measures the ability of each item to distinguish good performers from poor performers, are invaluable tools. A question with a low discrimination index, answered correctly by a majority of students, may not effectively differentiate candidates' skill levels, and would need to be revised to better assess students' specific abilities.

The use of advanced statistical theories, such as Item-Response Theory (IRT), makes it possible to model candidates' probability of success with greater precision, depending on their level of competence and the difficulty of the questions (Embretson & Reise, 2013) . These tools make it possible to identify items that are too easy, too difficult, or poorly formulated, all of which could skew the overall assessment (Maclean et al., 2005). Ultimately, this detailed analysis of results, enhanced by statistical and qualitative tools, is not only a means of improving the validity of assessments, but also a lever for adapting teaching practices, by identifying areas where students encounter recurring difficulties (Moses, 2017). The analysis of exam results, as a rigorous process based on scientific tools, thus contributes to improving the quality of teaching and guiding pedagogical reforms, to better meet the specific needs of students (Janssen et al., 2014).

Feedback and Pedagogical Adjustment

Once the exam results have been analyzed, it is crucial to adjust the assessment based on the trends identified. If certain questions were particularly challenging for a large number of students, this may indicate that the wording was ambiguous or that the content was not sufficiently covered in class. In such cases, it is important to review the clarity and relevance of the questions to prevent future confusion. Similarly, the balance between different types of questions (multiple-choice, open-ended, practical exercises) should be evaluated: an assessment that focuses too much on a single format may not accurately reflect the full range of skills intended. The analysis of results also helps to detect biases, such as questions that are too easy or too difficult, which could distort the evaluation of students' competencies. These adjustments help make the exam fairer and more aligned with the course's learning objectives. Thus, the analysis of results serves not only to improve teaching but also to make the assessment more equitable and more representative of students' knowledge and skills (Hrich et al., 2019).

At the same time, the results should be used to adjust teaching. If common trends emerge, such as recurring errors on certain questions, this may indicate a need to revisit certain aspects of the curriculum or teaching methods. The instructor could organize targeted revision sessions or adapt their pedagogy to better meet the students' needs. Finally, the analysis of results can also help assess the relevance of the evaluation criteria and adjust the examination modalities so that they are more closely aligned with the learning objectives. This process of feedback and continuous adjustment ensures a constant improvement in the learning experience (Lee, 2023).

The process of an educational assessment including result analysis and pedagogical adjustment, can be significantly enhanced by artificial intelligence. By automating grading, providing detailed analysis of performance and items, and offering personalized feedback, AI speeds up the evaluation process while improving its accuracy and relevance (Yeung, 2019). However, to maximize its effectiveness, it is essential to maintain a balance between automation and human expertise, particularly in the phases of interpreting results and making pedagogical adjustments.

ARTIFICIAL INTELLIGENCE FOR ASSESSMENT

The integration of AI into the educational assessment process represents a significant advance, particularly in the results analysis and pedagogical adjustment phases. Thanks to its automation capabilities, AI can not only speed up the marking of exams, but also offer in-depth analyses of student performance and the questions themselves, enabling more accurate and relevant assessment.

Automated marking, particularly for MCTs and structured exercises, reduces the time spent on assessment, freeing up teachers for more strategic tasks. In addition, AI is able to identify recurring trends in student responses, spot anomalies or potential biases in questions, and generate detailed reports on student performance, enabling finer-grained analysis of results. For example, it can calculate difficulty and discrimination indices for each question, enabling tests to be adjusted so that they are more balanced, and representative of the skills targeted.

Artificial intelligence (AI) is increasingly integrated into various aspects of educational assessment. It helps improve the effectiveness and fairness of assessments by reducing human bias, enhancing objectivity and providing personalized learning experiences for learners. It can analyze large quantities of data in real time, identify trends or specific difficulties, and provide recommendations tailored to individual needs.

Another key function of AI lies in providing personalized feedback. Artificial intelligence can generate feedback tailored to each student, based on their specific errors, strengths and areas requiring improvement. This feedback, often automated and instantaneous, enables students to become aware of their mistakes and readjust their learning autonomously, while benefiting from more targeted attention.

However, while AI offers considerable advantages, its integration into the assessment process must not be at the expense of human expertise. Teachers play an essential role in interpreting results, in particular to adjust assessment criteria, review the relevance of questions, and adapt pedagogical strategies. Indeed, AI is no substitute for human understanding of learning contexts and students' individual needs. For example, the interpretation of results must not be based on numbers alone: teachers need to be able to understand why a question was problematic for a large number of students, or why a student failed a particular task.

To maximize the effectiveness of AI in educational assessment, it is therefore crucial to maintain a balance between automation and human intervention. AI can provide powerful tools for analyzing and monitoring performance, but the teacher remains indispensable for ensuring the match between assessments, pedagogical objectives and student needs. The successful integration of AI into educational assessment requires a harmonious collaboration between technology and pedagogical expertise, guaranteeing a more effective and equitable learning experience.

AI in Test Development

The development of assessment tests is a critical aspect of educational assessment, and with the advent of AI, new opportunities are emerging to enhance and streamline this process (Peng et al., 2022). AI is increasingly playing a central role in assessment design, notably through deep learning models like long short-term memory networks (LSTMs). These technologies enable the automation and optimization of several key stages in test creation, from refining test items to detecting potential redundancies or inconsistencies between questions, thereby ensuring the validity and reliability of assessments.

For instance, LSTMs can be employed to identify repetitive or ineffective distractors in Multiple-Choice Tests (MCTs) by capturing contextual dependencies across test items over extended sequences. AI also facilitates the personalization of assessments by enabling the creation of adaptive tests that adjust difficulty levels in real-time based on the candidate's performance. Table 1 examines the integration of AI at each stage of assessment test development, highlighting the most promising approaches and applications that are transforming test design to be more accurate, relevant, and equitable.

Table 1. AI for development test process

Phase	AI Approaches	AI Algorithms	Description
Item Edition	Natural Language Processing (NLP)	Word Embeddings (Word2Vec, GloVe)	Word2Vec is a neural network model used to produce word embeddings by learning distributed representations of words based on their context in a continuous vector space (Yilmaz & Toklu, 2020).
			GloVe constructs word embeddings based on global word-word co-occurrence statistics and factorizes a co-occurrence matrix to generate embeddings capturing both local and global semantic relationships (Pennington et al., 2014).
		Bidirectional Encoder Representations from Transformers (BERT)	BERT is used to automatically generate items from source texts by capturing the semantic and contextual relationships between words.(Devlin et al., 2019)
		Generative Pre-trained Transformer (GPT)	GPT is used to generate items using a pre-trained language model that generates coherent and relevant text (H. Zhang et al., 2023).
Pilot test	Recommendation Systems	Neural Networks	Neural network-based recommendation systems analyze participants' performance and recommend the following items based on their skills and preferences (S. Zhang et al., 2020).
	Adaptive Algorithms	Item Response Theory	IRT is used to model the probability of a participant answering an item correctly, based on his or her skills and the characteristics of the item. Adaptive algorithms dynamically adjust the test according to participants' responses (Yeung, 2019).
Item Revision	Neural Networks	Convolutional Neural Networks	Convolutional neural networks detect potentially biased items by analyzing item characteristics and comparing them to predefined norms (Cao et al., 2021).
		Autoencoders	Autoencoders are used to detect similarities between items and identify groups of items with similar characteristics, facilitating item review and categorization (Charte et al., 2018).

Automated Correction & Automated Grading

Artificial intelligence is playing an increasingly important role in the correction of assessment items, whether MCTs or open-ended questions. In the case of MCTs, AI enables fast, accurate and objective automated correction, which is particularly useful in large-scale examinations. AI algorithms compare students' answers with predefined correct answers and instantly generate statistics on overall performance, such as the rate of correct answers or areas of difficulty. This speed and ability to process large amounts of data means immediate feedback to students, optimizing their learning. In addition, the use of AI guarantees objectivity, eliminating the subjective biases of human graders, such as the halo effect, which can influence assessment based on non-academic criteria. In addition, AI offers the possibility of analyzing recurring errors, enabling teachers to better target their interventions and adapt teaching to students' specific needs.

However, correcting open-ended questions remains a more complex challenge, due to the intrinsic subjectivity and diversity of possible answers. In this field, AI, thanks to natural language processing, is able to evaluate more qualitative criteria, such as the relevance of arguments, the structure of reasoning and the quality of expression. This enables faster, more consistent correction than that carried out manually, minimizing the risk of human bias and errors due to fatigue. However, technology still has its limits, particularly when it comes to assessing creativity and critical thinking, elements that are often difficult for algorithms to capture. What's more, the quality of correction depends largely on the data on which systems are trained, and biases can emerge if these datasets are incomplete or undiversified. In short, while AI represents a significant advance in the automation of scoring, it is crucial to maintain a balance between automation and human expertise, especially for assessments requiring a nuanced or subjective understanding of responses.

Result Analysis

AI is transforming assessment analysis, not only by making it possible to process large quantities of data on student performance, but also to analyze test items in depth(Rodriguez, 2013). Using deep learning algorithms, AI can detect recurring trends and patterns in student responses, identify common difficulties with certain concepts, or spot items that may be poorly designed, too difficult or misinterpreted. This analysis of items enables us to better understand their validity and reliability, and to adjust assessments according to the results obtained (Wright, 2008). AI can also predict future student performance by analyzing past results and identifying patterns of success or failure. These predictions enable teachers to provide targeted support, identifying in advance students likely to struggle. In addition, AI tracks the evolution of students' skills over time, measuring progress and evaluating the impact of teaching methods. It also helps identify specific areas where students need extra support, based on how they respond to different types of items, enabling learning paths to be adjusted. In this way, AI, by combining assessment and item analysis, makes the correction process more precise, rapid and objective, while also offering the best possible support to students.

In summary, the integration of AI into the educational assessment process offers great promise for improving the efficiency and accuracy of assessments. By automating and improving every step of the test development process, AI can help educators make more informed decisions and improve educational outcomes. Future research should explore the implications of AI integration, including ethical considerations and the need for robust validation methods to ensure AI-based assessments are fair and effective.

Feedback & Decision Making

Artificial intelligence (AI) plays a crucial role in assessment decision-making, providing automated, data-driven solutions to improve the quality, efficiency and fairness of assessment processes. In educational and professional contexts, AI can be used to analyze candidates' performance in real time, considering various factors such as answers to questions, time spent on each task or behavioral trends. Thanks to machine learning algorithms, AI can provide personalized recommendations, such as adapting the difficulty level of questions according to previous answers (adaptive testing) or detecting potential biases in assessments.

One of the major advantages of AI in assessment decision-making is its ability to process large amounts of data and perform complex analyses in a short space of time. For example, in large-scale exams, AI can identify patterns in candidates' answers, detect anomalies or errors in test construction, or adjust scoring criteria to better reflect the skills being assessed. It can also play a key role in predicting a candidate's future performance, using predictive models that consider their learning history or past assessments.

In addition, AI enables more objective decision-making by eliminating the human biases that could affect the evaluation of responses. It can thus offer a fairer, more accurate assessment, based on rigorous, quantifiable criteria. Integrating AI into the design and administration of assessments therefore offers opportunities to make decisions more informed, transparent and tailored to the individual needs of candidates, while optimizing the resources of educators and institutions.

To summarize, the table below presents various AI applications in the field of educational assessment. Each feature illustrates how advanced technologies can transform assessment and learning practices, bringing efficiency, personalization and security. From automating grading to detecting cheating, personalizing learning paths and analyzing student responses, AI enables a better understanding of learners' strengths and weaknesses, while improving the reliability and validity of tests. This technological evolution thus promises to optimize education, by making assessments fairer, faster and more tailored to students' individual needs.

Table 2. Application of AI for Educational Assessment

Feature	Description
Automated scoring	AI automates the grading of specific assessments, like multiple-choice tests or essays. Through analyzing the language and content of a submission, AI algorithms assess the structure, grammar, and other elements of a student's writing, ultimately saving time and enhancing the fairness and consistency of evaluation (Mizumoto & Eguchi, 2023).
Feedback generation	refers to providing feedback to students based on their assessment performance. AI can generate personalized feedback for each student, which can help them identify their strengths and challenges and subsequently improve their learning outcomes (Lee, 2023).
Personalized learning	AI is also used to personalize learning and assessment experiences. AI algorithms analyze students' data to identify strengths and weaknesses and recommend learning resources or assessment tasks that target improvement areas (Huang et al., 2023).
Cognitive diagnosis	using AI algorithms to analyze student responses to assessment items and diagnose their cognitive strengths and weaknesses. This approach goes beyond traditional Item Response Theory (IRT) models, which focus on measuring overall ability or proficiency in a particular subject (Swiecki et al., 2022b).
Cheating detection	AI is employed to identify cheating in e-assessments, including analyzing examinee behavior and recognizing cheating patterns, such as identical responses or abnormal response times (Wongvorachan, 2023).
Test security	AI enhances the security of the assessment by detecting and preventing fraud, such as impersonation, hacking, or distribution of exam materials (Kurdi et al., 2020).
Item analysis and selection	AI algorithms assist in this process by analyzing large amounts of data and identifying patterns and relationships that may not be immediately apparent to human evaluators. One example of using AI in analyzing items is using machine learning algorithms to identify which test items are most effective in differentiating between high-performing and low-performing students by helping to identify too easy or complex items, remove them from future assessments, and determine which items are most effective at measuring specific learning outcomes. AI can also be used in item selection, which involves selecting a set of items from a more significant item pool to create an assessment test. AI can assist in this process by selecting items that are well-matched to the intended learning outcomes and that maximize the information obtained from the assessment; this can help improve the validity and reliability of the evaluation and provide a more accurate image of student learning (Martínez-Plumed et al., 2019).

AI IN THE EDUCATIONAL ASSESSMENT PROCESS: CASE STUDIES AND BEST PRACTICES

Large Scale Competitive Exams

In general, large-scale competitive examinations are designed to assess the knowledge and skills of candidates with a view to their admission to training establishments or recruitment for professional positions. These examinations, generally organized at national or regional level, are extremely competitive, with a large number of candidates vying for a limited number of places or positions. In order to select the most qualified individuals, it is crucial that these tests distinguish between candidates according to their level on the criteria assessed. It is therefore essential to use items with high discriminating power, i.e. capable of differentiating candidates precisely and finely according to the criterion measured.

However, when the number of items to be processed is considerable, test review and analysis quickly become complex and difficult to manage. In this context, manual management of tests and responses becomes a major challenge, due to the volume and complexity of the data to be analyzed. Identifying redundancies, inconsistencies and errors becomes particularly arduous, which can adversely affect the quality and accuracy of the evaluation.

In addition, MCTs are the most widely used in large-scale competitive examinations, whether administered face-to-face or remotely. A particularity of MCTs is that candidates can sometimes answer correctly by sheer luck. For example, in a test with two options, only one of which is correct, the probability of answering correctly by chance is 50%. It is therefore essential to consider the impact of these random responses. This highlights the need for in-depth analysis of candidates' answers.

Finally, when the number of items in a competition is particularly high, item review becomes an essential step. It enables designers to assess the complexity and scope of the evaluation by examining each item and its distractors. The identification of redundancies then becomes all the more crucial, as item diversity and quality are essential to ensure a fair and equitable assessment of candidates' skills.

In what follows, we present two applications of artificial intelligence that address the peculiarities and challenges outlined above and aim to improve the quality and relevance of multiple-choice tests administered in large-scale competitions.

Application of LSTM for redundancy detection in MCTs

The use of deep learning models to detect redundancies in texts represents a significant advance in the field of Automatic Natural Language Processing (Minaee et al., 2021). These models exploit deep neural architectures to extract relevant features from textual data, enabling them to identify similarities and repetitions within textual content. Among the architectures commonly used for this task are recurrent neural networks (RNNs), convolutional neural networks (CNNs) and transformers.

Recurrent Neural Networks

RNNs are suitable for modeling sequential data, which makes them relevant for detecting repetitive patterns in texts. For example, an RNN model can be used to identify repeated phrases or passages throughout a document (Fu et al., 2017).

Convolutional Neural Networks

CNNs extract local features from text data and can detect similarities between text passages by analyzing local patterns. For example, a CNN can be used to identify similar sentences in a set of documents (Wang et al., 2021).

Transformers:

Transformers have revolutionized the field of NLP by introducing attention mechanisms. They can consider relationships between words across the whole text, making it easier to detect redundancy over long distances. For example, a transformer model can be used to identify repeated passages of text in different parts of a document (Dai et al., 2020).

These deep learning models offer different but complementary approaches to detecting redundancies in texts, by analyzing word sequences, local features and long-distance relationships between words (Hrich & al, 2024), (Hrich & al, 2024a) .

The growing adoption of multiple-choice tests in educational examinations and competitions has considerably transformed the assessment of student learning. This development responds to the need to assess a large number of candidates while measuring a wide range of skills objectively and efficiently. However, this transition is not without its challenges, particularly when it comes to designing questions that are relevant and free from redundancy.

In competitive examinations, the number of items specified is often very high, sometimes reaching 100 to 120 items, each with four to five distractors. This abundance of items increases the risk of redundancy, which can compromise the validity and reliability of assessments, and cause problems during examination taking.

Faced with this challenge, contest organizers and educators have developed an emerging practice called "test piloting", where test designers appoint a group of experts to assess the quality and validity of questions. Despite this method of validation, there remains a risk that certain redundancies will escape the vigilance of these experts.

To overcome this problem and help improve item quality by identifying and removing duplications, (Hrich, Azekri, & Khaldi, 2024a)have developed a deep learning model specifically designed to detect redundancies in items and distractors. This model uses a Long Short-Term Memory neural network, a variant of Recurrent Neural Networks (RNN), which is particularly well suited to capturing long-term dependencies in data sequences, making it an ideal choice for the redundancy detection task.

As part of the development of this model, the use of LSTMs was favored due to their particular effectiveness in detecting redundancies associated with long-term dependencies present in the data. As an advanced variant of RNN, LSTMs are designed so that relevant information is better captured and retained over long sequences, making them particularly suited to complex tasks such as redundancy detection in MCTs involving a large number of items and distractors.

In comparison, CNNs are considered less effective for this specific task due to their inability to perceive the global information needed to detect textual redundancy. Although CNNs can traditionally handle sequences of data, they often prove ineffective when faced with data characterized by long-term dependencies, due to gradient vanishing problems. This makes CNNs less suitable for detecting complex redundancies requiring in-depth contextual understanding.

As a result, LSTMs are particularly well suited to detecting item redundancy and distractors in MCTs, thanks to their ability to capture long-term dependencies within textual data. In such a context, where each test item may contain several distractors, it becomes crucial to identify redundant or similar items that may undermine the validity of the assessment. LSTMs make it possible to efficiently process long sequences of text, both the questions themselves and the proposed answers, by analyzing the subtle relationships between the various elements of the test. Unlike other models, such as CNNs, which are limited in their ability to handle long-range dependencies, LSTMs are able to identify repetitive patterns or ill-conceived distractors, even when these are far apart in the text. An LSTM model has therefore been developed and tested specifically for this task, enabling automatic detection of redundancies within items and distractors, and thus improving overall test quality. By optimizing the item selection and revision process, LSTMs contribute to a more accurate and fair assessment of candidates' skills.

Artificial Intelligence Item Analysis Tool For Educational Assessment

The ever-increasing number of candidates for examinations and competitive exams, combined with technological developments in assessment methods, has led many educational and professional institutions to widely adopt MCTs. Faced with a massive influx of participants, this approach is emerging as a practical and efficient solution to complex logistical challenges. MCTs offer a standardized method of assessment, with multiple-choice questions or closed exercises, enabling automated correction, which considerably reduces the administrative burden and frees up time for teachers and administrators. This approach also facilitates rapid, impartial and objective assessment of candidates' skills, while meeting the strict deadlines imposed in time-limited testing contexts.

However, although the adoption of MCTs is seen as a pragmatic response to the need to manage a large number of candidates, this method raises a number of concerns, particularly with regard to the quality of the assessment and the fairness of the results. Indeed, one of the major challenges of MCTs lies in their potential bias, notably due to the effect of chance. Candidates may, simply by chance, select the correct answer to a question about which they have no prior knowledge. This phenomenon calls into question the validity of the results, as it can give an erroneous state of candidates' actual mastery of knowledge. What's more, the standardized nature of these tests, while effective, doesn't always capture the full wealth of candidates' skills, particularly in areas where critical thinking and creativity are essential.

This is where the test development process comes into its own. To ensure that the test is truly representative of the target skills and can offer a reliable assessment, several steps need to be followed. The pre-test phase, also known as the pilot test, is crucial to this process. This preliminary test enables the test items (questions, exercises, etc.) to be tested on a representative sample of the target population before final implementation. The aim is to assess the quality of the items, in particular their difficulty, discriminating power and internal consistency (Janssen et al., 2014). Based on feedback from participants and analysis of the results, certain items may be revised, modified or deleted to improve the validity and reliability of the test (Wright, 2008).

During the pre-test, the results are analyzed using various statistics to determine the difficulty and discrimination indices of the items (Downing, 2010) . The difficulty index measures the proportion of candidates who answered a question correctly, while the discrimination index evaluates an item's ability to distinguish between high and low performers (Shrock & Coscarelli, 2010). A question that is too easy, and answered correctly by the majority of candidates, may not be relevant for assessing the targeted skill. Conversely, a question that is too difficult may not be effective in differentiating between different skill levels. Statistical analysis of items also enables us to detect potential problems, such as ambiguous or poorly worded questions, which could mislead candidates or distort results.

The integration of a pre-test into the design of MCTs enables the quality of each item to be validated from different angles, ensuring that each question is relevant, valid and discriminating. This in-depth analysis process guarantees a fair and accurate assessment of candidates' knowledge. Once this validation phase is complete, any necessary adjustments are made before the final test is implemented.

However, while pre-testing is an effective method of ensuring test quality, its implementation can pose logistical challenges, particularly in contexts where test confidentiality is paramount, such as professional certification exams. In such cases, setting up experimental groups to test items may be difficult, if not unfeasible. The need to preserve the integrity of the test while obtaining accurate data for its improvement makes this approach complex.

This is where AI comes into play, offering an innovative solution for optimizing the results analysis and item revision process. AI, in particular, enables statistical post-test analysis of results without the need for pre-testing with experimental groups. Using machine learning algorithms (Hrich, Azekri, Elhaddouchi, et al., 2024), AI can analyze candidates' responses in depth, calculate difficulty and discrimination indices for each item, and adjust scores in real time. This approach bypasses logistical constraints and saves precious time, as results can be analyzed in a matter of seconds, unlike traditional methods which take several days or weeks. What's more, AI makes it possible to apply more complex and detailed analysis methods, processing large quantities of data simultaneously, which would be difficult to achieve manually.

(Hrich, Azekri, & Khaldi, 2024b) And (Hrich, Azekri, Elhaddouchi, et al., 2024) uses artificial intelligence to perfect the analysis of results and the item revision process, providing an innovative solution that overcomes the logistical limitations of traditional methods. The proposed model combines proven statistical techniques with artificial neural networks (ANNs), to optimize post-test analysis without the need for pre-testing with experimental groups. AI makes it possible to calculate difficulty (P-index) and discrimination (D-index) indices for each item, and to adjust candidates' scores.

Machine learning algorithms offer the ability to process huge volumes of data in a matter of seconds, significantly reducing lead times compared with conventional methods, which can take days or even weeks. The integration of ANNs gives the model great flexibility, as they are able to learn complex, non-linear relationships between item characteristics, enabling high-quality items to be distinguished and non-discriminatory ones to be eliminated.

The complementarity between traditional statistical analysis and the advanced modeling capabilities of ANNs illustrates the transformative potential that AI brings to evaluation processes, enabling more sophisticated, rapid and accurate analyses. By enabling the simultaneous processing of vast quantities of data, this model refines item selection and enhances score reliability, providing a robust and scalable solution for the assessment and analysis of results in the field of education.

CHALLENGES AND PERSPECTIVE

Challenges

The integration of artificial intelligence (AI) into educational assessment presents promising prospects, but also raises significant challenges. Firstly, ensuring the reliability and validity of algorithms is essential to guarantee the accuracy and relevance of the analyses provided by AI. If models are not properly tested, they risk providing biased or inaccurate assessments, which could compromise the fairness of results for learners. Algorithmic bias is another major challenge: the training data used to train these systems may contain cultural, social or linguistic biases, leading to unintended discrimination. For example, learners from different cultures or environments could be assessed unfairly due to biases inherent in the algorithms, undermining the objectivity and fairness of assessments.

In addition, the use of AI in assessment requires rigorous management of privacy and personal data. AI systems collect sensitive data on students' performance, learning behaviors and even errors, posing a risk to data confidentiality. Complying with standards and regulations, such as the General Data Protection Regulation (GDPR) in Europe, is becoming an imperative for institutions. Furthermore, resistance to change is an obstacle often encountered when introducing new technologies. Some teachers and learners may perceive AI as intrusive or complex, which can hinder its adoption. To ensure smooth integration, awareness-raising and training initiatives need to be put in place to help users familiarize themselves with these tools and understand their benefits.

From a technical point of view, the integration of AI requires specialized skills and can generate high costs, not only for the development of systems, but also for their maintenance and regular updates. Institutions must therefore be prepared to make long-term investments so that these technologies can evolve to meet the needs of educational assessment. Another critical issue concerns the transparency and explicability of AI models, especially those based on deep learning, which are often complex “black boxes”. In order to justify the results obtained and guarantee their acceptability, it is crucial that teachers and administrators can understand and explain AI decisions.

Finally, adaptability to diverse educational contexts represents a final challenge, as AI systems need to be flexible to adjust to different environments, learning cultures and pedagogical objectives. This need for customization requires models to be able to adjust to the specificities of each institution, an aspect that is still difficult to achieve universally. In conclusion, even if AI brings significant advances in educational assessment, these challenges show the importance of vigilance, ethics and caution for a successful integration that truly benefits education.

Perspectives

The integration of AI into educational assessment is opening up significant prospects, profoundly transforming current assessment methods and paving the way for more personalized, accurate and inclusive practices. One of the main opportunities lies in the personalization of assessments. AI makes it possible to create adaptive assessments, adjusting the level of difficulty in real time according to learners' performance. This enables a more individualized approach, where each student can progress at their own pace, maximizing their potential and engagement in learning.

AI also offers powerful tools for improving test quality by automating the creation and revision of questions. Using advanced models such as neural networks or LSTMs, it is possible to detect redundancies, inconsistencies and inefficient distractions in tests, ensuring fairer, more balanced assessments. This automation also helps reduce the workload for teachers, who can concentrate more on essential pedagogical aspects. Furthermore, AI can enrich assessments by offering in-depth analyses of results, identifying not only academic skills, but also cross-curricular skills such as critical thinking, problem-solving and creativity, which are essential for learners' success in the modern world.

In addition, AI can help reduce bias in assessment by ensuring objective marking of work, thus minimizing unconscious influences from markers. This is particularly relevant in multicultural contexts, where cultural and linguistic biases can influence assessments. By enhancing objectivity, AI makes assessments fairer for all learners, whatever their background or context.

AI also promotes continuous assessments through machine learning, enabling students' progress to be tracked and analyzed dynamically. Rather than being limited to final exams, AI enables data to be captured in real time, offering immediate feedback to teachers and students. This helps to quickly identify areas for improvement and make pedagogical adjustments during learning, supporting more responsive and relevant monitoring.

One of the most promising future prospects is predictive analytics, which uses data to anticipate student performance trajectories and potential needs. This would enable teachers to intervene earlier in the learning process, further personalizing educational pathways. However, successful adoption of these perspectives requires investment in technological infrastructure, training for teachers and rigorous ethical frameworks to guarantee confidentiality and respect for personal data.

In short, AI opens up fascinating possibilities for rethinking educational assessment. It transforms assessments to make them more flexible, tailored and focused on the needs of individual learners, while improving the equity and efficiency of the education system as a whole. AI in educational assessment is thus a powerful lever for modernizing education, making it more relevant to the challenges of the XXI century.

CONCLUSION

Artificial intelligence applied to educational assessment represents a breakthrough with great potential, although it also poses significant challenges. One of its main assets is its ability to automate marking and analyze results in detail, saving teachers a considerable amount of time. With automatic correction systems, particularly in quantitative subjects, AI frees teachers from repetitive tasks, enabling them to devote more time to pedagogical support. What's more, the analyses produced by AI provide precise information on students' strengths and weaknesses, making it easier to adapt teaching methods to individual needs.

The speed of this feedback also benefits students, who receive almost immediate feedback. This reactivity boosts motivation and encourages more engaging learning. However, these advantages come with significant limitations. For example, algorithms can introduce biases if they are not carefully calibrated, risking compromising the fairness of assessments and perpetuating inequalities. On the other hand, overreliance on AI could reduce the essential human interaction between teachers and students, risking dehumanizing learning.

The use of AI in assessment also raises questions of security and confidentiality. The collection and storage of personal data makes it crucial to protect this information from any risk of cyber-attack. Another major challenge is managing cheating: AI technologies, while facilitating assessment, can also be abused. Some students could use sophisticated tools to bypass control systems, threatening the integrity of assessments.

In the future, AI could extend its role to assess more complex skills, such as creativity or problem-solving, thus contributing to a more global vision of education. For this evolution to be beneficial, it is essential to put in place strict ethical and safety standards to frame the use of AI, to ensure constant monitoring of assessment systems, and to develop effective means to prevent and detect cheating in order to preserve the credibility of assessments.

In conclusion, AI represents an extraordinary opportunity to personalize assessment, reduce teacher workload, and provide rapid analysis. However, its use must remain balanced and cautious. Technology can never replace the human dimension of education. The future of AI in educational assessment will depend on our ability to make it a complementary tool, respecting ethics, data confidentiality and the integrity of pedagogical practices to ensure inclusive learning that respects the unique needs and strengths of every student.

REFERENCES

Beaupied, A. (2009). L'évaluation par les compétences. *Idées économiques et sociales, 155*(1), 71–77. DOI: 10.3917/idee.155.0071

Brown, G. T. L. (2022). The past, present and future of educational assessment: A transdisciplinary perspective. *Frontiers in Education*, 7, 1060633. Advance online publication. DOI: 10.3389/feduc.2022.1060633

Cao, S., Sun, X., Bo, L., Wei, Y., & Li, B. (2021). *BGNN4VD*: Constructing Bidirectional Graph Neural-Network for Vulnerability Detection. *Information and Software Technology*, 136, 106576. DOI: 10.1016/j.infsof.2021.106576

Charte, D., Charte, F., García, S., del Jesus, M. J., & Herrera, F. (2018). A practical tutorial on autoencoders for nonlinear feature fusion: Taxonomy, models, software and guidelines. *Information Fusion*, 44, 78–96. DOI: 10.1016/j.inffus.2017.12.007

Classical Item & Test Analysis Spreadsheet CITAS. (n.d.). *Assessment Systems.* https://assess.com/citas/

Dai, Z., Lai, G., Yang, Y., & Le, Q. (2020). Funnel-Transformer: Filtering out Sequential Redundancy for Efficient Language Processing. *Advances in Neural Information Processing Systems*, 33, 4271–4282.

Devlin, J., Chang, M.-W., Lee, K., & Toutanova, K. (2019). *BERT: Pre-training of Deep Bidirectional Transformers for Language Understanding* (arXiv:1810.04805). arXiv. https://doi.org//arXiv.1810.04805DOI: 10.48550

Downing, S. M. (2010). Test Development. In Peterson, P., Baker, E., & McGaw, B. (Eds.), *International Encyclopedia of Education* (3rd ed., pp. 159–165). Elsevier., DOI: 10.1016/B978-0-08-044894-7.00270-0

Elaborer et faire passer des tests—Cadre européen commun de référence pour les langues (CECR)—Www.coe.int. (n.d.). Cadre européen commun de référence pour les langues (CECR). Retrieved November 6, 2024, from https://www.coe.int/fr/web/common-european-framework-reference-languages/developing-tests-examining

Embretson, S. E., & Reise, S. P. (2013). *Item Response Theory*. Psychology Press. DOI: 10.4324/9781410605269

Excel Spreadsheets for Classical Test Analysis. (n.d.). Retrieved January 4, 2024, from https://languagetesting.info/statistics/excel.html

Fu, X., Ch'ng, E., Aickelin, U., & See, S. (2017). *CRNN: A Joint Neural Network for Redundancy Detection* (arXiv:1706.01069). arXiv. /arXiv.1706.01069DOI: 10.1109/SMARTCOMP.2017.7946996

Gardner, J., O'Leary, M., & Yuan, L. (2021). Artificial intelligence in educational assessment: 'Breakthrough? Or buncombe and ballyhoo?'. *Journal of Computer Assisted Learning*, 37(5), 1207–1216. DOI: 10.1111/jcal.12577

Haladyna, T. M., & Rodriguez, M. C. (2013). *Developing and Validating Test Items.* Routledge. DOI: 10.4324/9780203850381

Hrich, N., Azekri, M., Elhaddouchi, C., & Khaldi, M. (2024). Enhancing Educational Assessments: Score Regeneration through Post-Item Analysis with Artificial Intelligence. *International Journal of Information and Education Technology, 14*(10), 1414–1420. *Scopus.* Advance online publication. DOI: 10.18178/ijiet.2024.14.10.2172

Hrich, N., Azekri, M., & Khaldi, M. (2023). ARTIFICIAL INTELLIGENCE FOR EDUCATIONAL ASSESSMENT. *ICERI2023 Proceedings*, 2120–2124. 16th annual International Conference of Education, Research and Innovation. DOI: 10.21125/iceri.2023.0598

Hrich, N., Azekri, M., & Khaldi, M. (2024a). *Application of LSTM for Redundancy Detection in MCTS: Enhancing Test Precision | IIETA*. DOI: 10.18280/isi.290430

Hrich, N., Azekri, M., & Khaldi, M. (2024b). Artificial Intelligence Item Analysis Tool for Educational Assessment: Case of Large-Scale Competitive Exams. *International Journal of Information and Education Technology (IJIET)*, 14(6), 822–827. DOI: 10.18178/ijiet.2024.14.6.2107

Hrich, N., Lazaar, M., & Khaldi, M. (2019). Improving Cognitive Decision-Making into Adaptive Educational Systems through a Diagnosis Tool based on the Competency Approach. [iJET]. *International Journal of Emerging Technologies in Learning*, 14(07), 07. Advance online publication. DOI: 10.3991/ijet.v14i07.9870

Hrich, N., Lazaar, M., & Khaldi, M. (2021). Assessment Process-Reflection on Pedagogical Practices and Integration of Technologies in Education. In *Promoting Positive Learning Experiences in Middle School Education* (pp. 42–65). IGI Global., DOI: 10.4018/978-1-7998-7057-9.ch003

Huang, A. Y. Q., Lu, O. H. T., & Yang, S. J. H. (2023). Effects of artificial Intelligence–Enabled personalized recommendations on learners' learning engagement, motivation, and outcomes in a flipped classroom. *Computers & Education*, 194, 104684. DOI: 10.1016/j.compedu.2022.104684

IRT Illustrator – Psychomeasurement Systems. (n.d.). Retrieved January 4, 2024, from https://itemanalysis.com/irt-illustrator/

Janssen, G., Meier, V., & Trace, J. (2014). Classical test theory and item response theory: Two understandings of one high-stakes performance exam. *Colombian Applied Linguistics Journal*, 16(2), 167. DOI: 10.14483/udistrital.jour.calj.2014.2.a03

Kurdi, G., Leo, J., Parsia, B., Sattler, U., & Al-Emari, S. (2020). A Systematic Review of Automatic Question Generation for Educational Purposes. *International Journal of Artificial Intelligence in Education*, 30(1), 121–204. DOI: 10.1007/s40593-019-00186-y

Lane, S., Raymond, M. R., & Haladyna, T. M. (2015). *Handbook of Test Development*. Routledge. DOI: 10.4324/9780203102961

Lee, A. V. Y. (2023). Supporting students' generation of feedback in large-scale online course with artificial intelligence-enabled evaluation. *Studies in Educational Evaluation*, 77, 101250. DOI: 10.1016/j.stueduc.2023.101250

Liu, Y., & Baucham, M. (2023). AI Technology: Key to Successful Assessment. In *Handbook of Research on Redesigning Teaching, Learning, and Assessment in the Digital Era* (pp. 304–325). IGI Global. DOI: 10.4018/978-1-6684-8292-6.ch016

Maclean, R., Watanabe, R., & Baker, R. Boediono, Cheng, Y. C., Duncan, W., Keeves, J., Mansheng, Z., Power, C., Rajput, J. S., Thaman, K. H., Alagumalai, S., Curtis, D. D., & Hungi, N. (Eds.). (2005). *Applied Rasch Measurement: A Book of Exemplars: Papers in Honour of John P. Keeves*. Springer Netherlands. DOI: 10.1007/1-4020-3076-2

Martínez-Plumed, F., Prudêncio, R. B. C., Martínez-Usó, A., & Hernández-Orallo, J. (2019). Item response theory in AI: Analysing machine learning classifiers at the instance level. *Artificial Intelligence*, 271, 18–42. DOI: 10.1016/j.artint.2018.09.004

Minaee, S., Kalchbrenner, N., Cambria, E., Nikzad, N., Chenaghlu, M., & Gao, J. (2021). *Deep Learning Based Text Classification: A Comprehensive Review* (arXiv:2004.03705). arXiv. https://doi.org//arXiv.2004.03705DOI: 10.48550

Mizumoto, A., & Eguchi, M. (2023). Exploring the potential of using an AI language model for automated essay scoring. *Research Methods in Applied Linguistics*, 2(2), 100050. DOI: 10.1016/j.rmal.2023.100050

Moses, T. (2017). A Review of Developments and Applications in Item Analysis. In Bennett, R. E., & von Davier, M. (Eds.), *Advancing Human Assessment: The Methodological, Psychological and Policy Contributions of ETS* (pp. 19–46). Springer International Publishing., DOI: 10.1007/978-3-319-58689-2_2

Newton, P. E. (2007). Clarifying the purposes of educational assessment. *Assessment in Education: Principles, Policy & Practice*, 14(2), 149–170. DOI: 10.1080/09695940701478321

OECD. (2020). *Improving Governance with Policy Evaluation: Lessons From Country Experiences*. Organisation for Economic Co-operation and Development. https://www.oecd-ilibrary.org/governance/improving-governance-with-policy-evaluation_89b1577d-en

Peng, C., Zhou, X., & Liu, S. (2022). An Introduction to Artificial Intelligence and Machine Learning for Online Education. *Mobile Networks and Applications*, 27(3), 1147–1150. DOI: 10.1007/s11036-022-01953-3

Pennington, J., Socher, R., & Manning, C. (2014). Glove: Global Vectors for Word Representation. *Proceedings of the 2014 Conference on Empirical Methods in Natural Language Processing (EMNLP)*, 1532–1543. DOI: 10.3115/v1/D14-1162

Rodriguez, T. M. H., & Michael, C. (2013). *Developing and Validating Test Items*. Routledge., DOI: 10.4324/9780203850381

Sayac, N. (2017). Lucie Mottier-Lopez, Évaluations formative et certificative des apprentissages. Enjeux pour l'enseignement. *Recherches & Educations*, 28, 28. Advance online publication. DOI: 10.4000/ree.5958

Sayac, N. (2018). MERLE Pierre. Les pratiques d'évaluation scolaire. Historique, difficultés, perspectives. *Revue française de pédagogie.Revue Française de Pédagogie*, 204(204), 204. Advance online publication. DOI: 10.4000/rfp.8553

Shrock, S. A., & Coscarelli, W. C. (2010). Criterion-Referenced Measurement. In Peterson, P., Baker, E., & McGaw, B. (Eds.), *International Encyclopedia of Education* (3rd ed., pp. 31–35). Elsevier., DOI: 10.1016/B978-0-08-044894-7.00240-2

Swiecki, Z., Khosravi, H., Chen, G., Martinez-Maldonado, R., Lodge, J. M., Milligan, S., Selwyn, N., & Gašević, D. (2022a). Assessment in the age of artificial intelligence. *Computers and Education: Artificial Intelligence*, 3, 100075. DOI: 10.1016/j.caeai.2022.100075

Swiecki, Z., Khosravi, H., Chen, G., Martinez-Maldonado, R., Lodge, J. M., Milligan, S., Selwyn, N., & Gašević, D. (2022b). Assessment in the age of artificial intelligence. *Computers and Education: Artificial Intelligence*, 3, 100075. DOI: 10.1016/j.caeai.2022.100075

Wang, Z., Li, C., & Wang, X. (2021). *Convolutional Neural Network Pruning with Structural Redundancy Reduction* (arXiv:2104.03438). arXiv. /arXiv.2104.03438DOI: 10.1109/CVPR46437.2021.01467

Wongvorachan, T. (2023). *Cheating Detection in Tests: A Systematic Review*. OSF. DOI: 10.31234/osf.io/uc5hj

Wright, R. (2008). *Educational Assessment: Tests and Measurements in the Age of Accountability*. SAGE Publications., DOI: 10.4135/9781483329673

Yeung, C.-K. (2019). *Deep-IRT: Make Deep Learning Based Knowledge Tracing Explainable Using Item Response Theory* (arXiv:1904.11738). arXiv. https://doi .org//arXiv.1904.11738DOI: 10.48550

Yilmaz, S., & Toklu, S. (2020). A deep learning analysis on question classification task using Word2vec representations. *Neural Computing & Applications*, 32(7), 2909–2928. DOI: 10.1007/s00521-020-04725-w

Zanon, C., Hutz, C. S., Yoo, H., & Hambleton, R. K. (2016). An application of item response theory to psychological test development. *Psicologia: Reflexão e Crítica : Revista Semestral do Departamento de Psicologia da UFRGS*, 29(1), 18. Advance online publication. DOI: 10.1186/s41155-016-0040-x

Zhang, H., Song, H., Li, S., Zhou, M., & Song, D. (2023). A Survey of Controllable Text Generation Using Transformer-based Pre-trained Language Models. *ACM Computing Surveys, 56*(3), 64:1-64:37. DOI: 10.1145/3617680

Zhang, S., Yao, L., Sun, A., & Tay, Y. (2020). Deep Learning based Recommender System: A Survey and New Perspectives. *ACM Computing Surveys*, 52(1), 1–38. DOI: 10.1145/3285029

Chapter 5
Integrating AI Education in Non-Technical Disciplines:
Bridging the Gap Between Theory and Practice

Omar Abdennour
https://orcid.org/0000-0002-2216-0991
Higher Normal School of Tétouan, Morocco

Hassane Kemouss
https://orcid.org/0000-0002-9664-3275
Higher Normal School of Tétouan, Morocco

Mohamed Khaldi
https://orcid.org/0000-0002-1593-1073
Higher Normal School of Tétouan, Morocco

ABSTRACT

As artificial intelligence (AI) continues to transform various sectors, the need for a comprehensive understanding of AI principles and applications has become increasingly vital, even in non-technical disciplines. This paper explores the integration of AI education into fields such as humanities, social sciences, and business, emphasizing the importance of interdisciplinary learning. We discuss the unique challenges and opportunities presented by this integration, including the development of critical thinking skills, ethical considerations, and the enhancement of creativity through AI tools. By examining case studies and educational frameworks, we highlight effective strategies for curriculum design that foster AI literacy among students in

DOI: 10.4018/979-8-3373-2262-9.ch005

non-technical areas. Our findings suggest that equipping future professionals with AI knowledge not only prepares them for a rapidly evolving job market but also empowers them to address complex societal issues through innovative solutions. Ultimately, this research advocates for a paradigm shift in education that embraces AI as a foundational component across all disciplines, ensuring that all students can thrive in an increasingly digital world.

I. INTRODUCTION

The swift progression of artificial intelligence (AI) technologies has infiltrated a multitude of sectors, emphasizing the critical need for a comprehensive understanding of AI principles and their applications that extends beyond conventional technical domains. As AI persistently transforms industries and alters societal dynamics, it becomes progressively essential for students in non-technical fields—such as the humanities, social sciences, and business—to attain proficiency in AI literacy (Uppal, 2022). This incorporation transcends the mere acclimatization of students to technological instruments; it signifies a revolutionary paradigm in education that cultivates critical analysis, ethical discernment, and innovative problem-solving capabilities (Hamal et al., 2022). By investigating the distinctive challenges and prospects that emerge from the integration of AI education within these disciplines, this manuscript endeavors to elucidate effective strategies for curriculum design and pedagogical frameworks that can bolster interdisciplinary learning. Through the investigation of illustrative case studies, we shall exhibit how AI instruments can be employed to tackle intricate issues and enhance educational outcomes ("Integration of AI in Learning," 2022). Ultimately, this inquiry advocates for a fundamental shift in educational methodologies, endorsing AI as a pivotal element that equips all students to adeptly navigate and flourish in an increasingly digital and interconnected milieu.This transformative approach not only emphasizes the importance of technological proficiency but also fosters critical thinking and creativity, essential skills for future leaders in a rapidly evolving world (Kumar et al., 2023).The swift and remarkable progress of artificial intelligence (AI) technologies has infiltrated a multitude of sectors and industries, thereby emphasizing the critical need for a comprehensive and extensive understanding of AI principles, concepts, and practical applications that extend well beyond the conventional boundaries of traditional technical fields and disciplines. As artificial intelligence continues to fundamentally transform various industries and exert a significant influence on the dynamics of society, it becomes increasingly essential and crucial for students who are pursuing studies in non-technical disciplines—such as the humanities, social sciences, and business—to develop and cultivate a solid foundation of AI literacy and knowledge

(Mulianingsih et al., 2020). This integration of AI education into diverse fields is not simply focused on familiarizing students with various technological tools and platforms; rather, it symbolizes a groundbreaking and transformative approach to education that promotes the development of critical thinking abilities, ethical reasoning skills, and creative problem-solving capabilities among learners. By delving into the distinctive challenges and unique opportunities that emerge from the incorporation of AI education within these varied disciplines, this paper seeks to illuminate and emphasize effective strategies for curriculum design as well as pedagogical frameworks that can significantly enhance and enrich interdisciplinary learning experiences (Alam et al., 2022). Through a thorough exploration of relevant case studies, we will illustrate and demonstrate how AI tools and technologies can be effectively utilized to tackle complex issues and enhance educational outcomes for students. Ultimately, this research advocates for a significant and impactful paradigm shift in educational practices, encouraging the promotion of AI as a foundational and essential component that empowers all students to successfully navigate and thrive in an increasingly digital, interconnected, and technologically advanced world (Role of AI in Education, 2023). By integrating AI into the curriculum, educators can foster critical thinking skills and creativity, preparing students to tackle real-world challenges with innovative solutions (Kennedy, 2023).

A. Growing Relevance of AI Beyond STEM

As we progress and move deeper into an era that is increasingly dominated and shaped by the advancements and innovations of artificial intelligence, it becomes increasingly clear that the significance and relevance of AI stretch far beyond the traditional boundaries and confines of the disciplines that are commonly categorized under the umbrella of science, technology, engineering, and mathematics, collectively referred to as STEM (Schumann et al., 2020). In a diverse array of fields, including but not limited to social sciences, the arts, and various sectors of business, artificial intelligence emerges not merely as a conventional tool that individuals might use, but rather as a remarkable and transformative force that fundamentally reshapes existing methodologies, enhances the creative processes that individuals engage in, and profoundly influences the decision-making processes that are critical to operations and strategies in these areas.This transformative impact of AI fosters innovative collaborations across disciplines, enabling professionals to harness its capabilities for more effective problem-solving and enhanced creativity in their respective fields (AI in Education, 2022).As a result, organizations are increasingly seeking to integrate AI-driven solutions into their workflows, recognizing the potential for increased efficiency and groundbreaking advancements that can arise from this synergy between technology and human expertise.The integration of AI not only streamlines operations but also empowers individuals to push the boundaries of what is possible, leading to unprecedented breakthroughs and a reimagining of traditional roles within organizations.This shift towards AI-driven methodologies is prompting a cultural transformation within workplaces, where adaptability and continuous learning become essential traits for success in an ever-evolving landscape.

In social sciences, for instance, AI can analyze vast amounts of data to uncover patterns in human behavior, offering insights that were previously unattainable. This capability allows researchers to tackle complex societal issues with greater depth and precision. Similarly, in the arts, AI tools are being utilized to create innovative works, challenging traditional notions of creativity and authorship. In business, AI-driven analytics empower organizations to make informed decisions, optimize operations, and enhance customer experiences (Zhou & Lawless, 2015).

Figure 1. Layered Pyramid of AI Impact

The integration of AI education in these non-technical fields is essential for preparing students to engage with these evolving landscapes. By understanding AI principles and applications, students can become adept at leveraging these technologies to solve real-world problems, thereby enhancing their employability and impact in their respective fields (Schumann et al., 2020).

This foundational knowledge not only equips them with valuable skills but also fosters a mindset of adaptability and innovation that is crucial in an increasingly automated world.

B. Aim of the Chapter

The primary aim of this chapter is to provide a comprehensive framework for educators and institutions seeking to effectively incorporate AI education into non-technical curricula. This framework will outline strategies for curriculum design that foster AI literacy among students in disciplines such as humanities, social sciences, and business (How & Hung, 2019). By integrating AI concepts into these diverse fields, educators can empower students to critically engage with technology and understand its implications on society, ultimately preparing them for the challenges of a rapidly evolving job market.This chapter will also explore practical examples and case studies that illustrate successful integration of AI education, highlighting best practices and potential pitfalls to avoid in the process.By showcasing these real-world applications, educators can gain valuable insights into how to tailor their approaches and enhance student learning outcomes in an increasingly digital landscape (Sakalle et al., 2021). Furthermore, the chapter will discuss strategies for developing interdisciplinary curricula that not only incorporate AI tools but also promote ethical considerations and responsible use of technology in various professional contexts.

We will explore pedagogical approaches that encourage critical thinking, ethical reasoning, and creative problem-solving, all of which are crucial in an AI-centric world. Additionally, we will highlight case studies that demonstrate successful integration of AI tools in non-technical education, showcasing best practices and innovative methodologies (Eguchi, 2021). These examples will serve as a foundation for educators seeking to implement similar strategies in their own classrooms, ultimately fostering an environment where students are equipped to navigate and thrive in a technology-driven future.By emphasizing collaboration between educators and industry professionals, we can further enhance the relevance of these curricula, ensuring that students are not only knowledgeable about AI but also prepared to address the ethical dilemmas and societal impacts associated with its use.

Ultimately, this chapter advocates for a paradigm shift in educational practices, emphasizing the need for a holistic approach that recognizes AI as a foundational component across all disciplines. By equipping students with the necessary knowledge and skills, we empower them to navigate the complexities of an increasingly digital and interconnected world, ensuring they can thrive and contribute meaningfully to society (Sadiku et al., 2021).This transformative educational framework not only fosters critical thinking and problem-solving abilities but also encourages students to engage in interdisciplinary projects that reflect real-world challenges, thereby bridging the gap between theory and practice.Such an approach prepares students to become innovative leaders who can adapt to rapid technological advancements while addressing ethical considerations and societal needs in their future careers.

Figure 2. Roadmap of AI Integration in Curricula

II. THEORETICAL FRAMEWORK

As we delve deeper into the integration of AI education in non-technical disciplines, it is crucial to consider the ethical implications that accompany this technological shift. The introduction of AI tools in fields such as humanities and social sciences not only enhances analytical capabilities but also raises concerns regarding data privacy, algorithmic bias, and the potential for misuse of AI-generated content (Kujundziski & Bojadjiev, 2024). Educators must therefore prioritize ethical reasoning within their curricula, equipping students with the ability to critically assess the impact of AI on society and individual lives. Furthermore, fostering interdisciplinary collaboration among departments can lead to richer educational experiences, allowing students to engage with diverse perspectives on how AI influences cultural narratives and societal norms (Berendt et al., 2020). By integrating these ethical considerations and collaborative approaches into AI literacy programs, institutions can better prepare students to navigate the complexities of an AI-infused world while promoting responsible innovation.

A. Definition of AI Literacy:

AI literacy refers to the understanding and ability to engage with artificial intelligence technologies, including their principles, applications, and implications. It encompasses knowledge about how AI systems work, the ability to critically evaluate AI-generated information, and the skills to use AI tools in various contexts.This foundational knowledge is essential for fostering an informed citizenry that can participate in discussions about the ethical, social, and economic impacts of AI technologies. By equipping students with AI literacy, educational institutions empower them to make informed decisions and contribute positively to a society increasingly shaped by technology. As AI continues to evolve and integrate into everyday life, fostering a curriculum that prioritizes AI literacy will ensure that future generations are not only consumers of technology but also informed contributors who can navigate the complexities of an AI-driven world.This proactive approach will enable students to critically evaluate AI applications, understand their implications, and advocate for responsible use in various sectors.This emphasis on AI literacy will also prepare students to address challenges such as bias in algorithms, data privacy concerns, and the ethical considerations surrounding automation and job displacement.

B. Importance in Non-Technical Education:

In the realm of non-technical academic disciplines, which encompass areas such as the humanities, social sciences, and the field of business, the acquisition of AI literacy is undeniably critical for a multitude of reasons that cannot be overlooked:

1. **Critical Thinking:** The development of AI literacy significantly contributes to the enhancement of critical thinking skills in students, as it equips them with the necessary tools to thoroughly analyze and interpret the data and insights generated by artificial intelligence in a manner that is both effective and insightful (Long et al., 2023).
2. **Ethical Reasoning:** Gaining a comprehensive understanding of artificial intelligence prompts students to deeply consider the ethical implications that arise within this context, including vital issues related to privacy and inherent biases, which are of paramount importance in our contemporary society that is increasingly driven by data.
3. **Creativity Enhancement:** The integration of AI tools has the remarkable potential to significantly augment and enhance creativity among students, thereby facilitating their ability to explore and devise innovative and effective solutions to complex problems that they may encounter within their respective fields of study (Holmes et al., 2021).
4. **Career Readiness:** In light of the fact that artificial intelligence is poised to have a profound and lasting impact across a wide array of industries, possessing a solid foundation of AI literacy not only improves students' employability but also adequately prepares them to adeptly navigate the challenges and opportunities presented by a job market that is evolving at an unprecedented pace (Zhang et al., 2022).
5. **Societal Engagement:** The empowerment that comes with AI literacy enables students to actively engage with and address the pressing societal challenges that confront us today, thereby fostering and promoting informed citizenship within

a world that is increasingly influenced and shaped by artificial intelligence. By understanding the ethical implications and potential consequences of AI technologies, students can contribute meaningfully to discussions surrounding privacy, security, and equity in an increasingly digital society (Kasinidou, 2023). This active participation not only enhances their critical thinking skills but also encourages collaborative problem-solving as they work alongside diverse communities to create innovative solutions that benefit society as a whole.

C. Educational Theories: TPACK Framework

The TPACK (Technological Pedagogical Content Knowledge) framework emphasizes the integration of technology, pedagogy, and content knowledge in education. It provides a comprehensive approach to understanding how these three components interact to enhance learning experiences.

1. **Technological Knowledge (TK):** Within the realm of artificial intelligence education, this particular concept pertains to the comprehensive understanding and familiarity with the various AI tools, systems, and technologies that are readily accessible for application in disciplines that do not primarily focus on technical skills or engineering-related knowledge (Eguchi, 2021).
2. **Pedagogical Knowledge (PK):** This area encompasses a variety of effective and innovative teaching strategies and methodologies that are specifically designed to foster student engagement and stimulate critical thinking skills, implementing approaches such as project-based learning and collaborative activities that seamlessly integrate the utilization of artificial intelligence into the educational experience (Sánchez & Lama, 2009).
3. **Content Knowledge (CK):** This term refers to the extensive subject matter expertise and deep understanding that exists within various non-technical academic disciplines, wherein the principles and applications of artificial intelligence can be leveraged to significantly enhance educational outcomes, enabling students to engage in activities such as analyzing social trends or creating unique digital art forms (Kang et al., 2022).

By integrating these three areas, educators can design curricula that not only teach AI concepts but also develop critical skills and ethical reasoning necessary for students to thrive in an AI-centric world.

D. Rogers' Diffusion of Innovation Model: Adoption Trends and Barriers

Rogers' Diffusion of Innovation Model describes how, why, and at what rate new ideas and technology spread. In the context of integrating AI education into non-technical fields, several trends and barriers can be identified:

Adoption Trends:

1. Within the realm of education, there exists a distinct group of trailblazers known as innovators and early adopters, which encompasses a variety of educators and institutions that are currently leading the charge by actively exploring and experimenting with various artificial intelligence tools and methodologies that can be integrated into their curricula in order to improve educational outcomes and student engagement (Flores-Vivar & García-Peñalvo, 2023).
2. As compelling success stories continue to surface and gain recognition within the educational community, an increasing number of educators are beginning to recognize the significant value that AI education practices can bring to their teaching methods, leading them to adopt these innovative strategies in order to enhance the overall learning outcomes and experiences of their students (Celik, 2022).
3. Conversely, within the broader educational landscape, there are segments of the population, specifically the late majority and laggards, who may exhibit a degree of resistance or reluctance when it comes to embracing artificial intelligence technologies, a hesitance that can often be attributed to either a fundamental lack of understanding regarding the capabilities and benefits of AI or a perception that the integration of such advanced tools into their educational practices is overly complex and daunting (Rajeswari & Purushothaman, 2023)

E. Barriers to Adoption

1. The absence of adequate training opportunities and resources available for educators may lead them to feel ill-equipped and unqualified to effectively teach complex artificial intelligence concepts that are increasingly becoming essential in today's technological landscape (Druga et al., 2022).
2. Certain educational institutions may exhibit a hesitance or sluggishness in their efforts to update and modernize their curricula, or they might also lack the necessary support and infrastructure required to successfully incorporate and integrate innovative new technologies into their teaching methodologies (Ouyang & Jiao, 2021).

3. The presence of ethical dilemmas, particularly those relating to algorithmic bias and the protection of data privacy, can significantly impede the widespread adoption of artificial intelligence tools within educational settings, as educators may harbor apprehensions regarding the potentially adverse consequences that could arise from their use (Conley, 2023). Addressing these challenges requires a comprehensive approach that includes not only curriculum reform but also training for educators and robust policies to ensure ethical standards are upheld in the implementation of AI technologies.

Figure 3. Combined TPACK and Rogers' Model

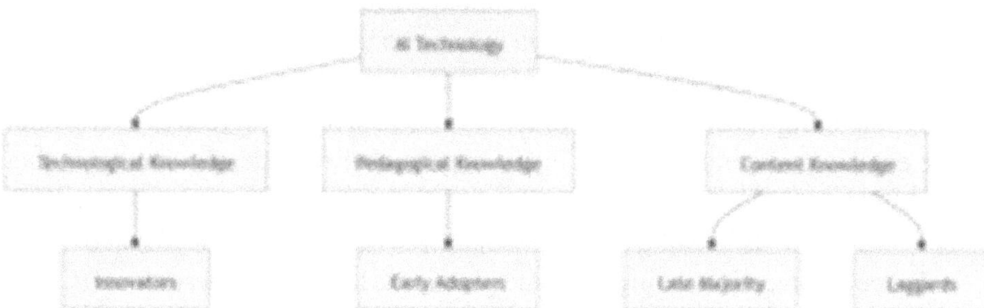

F. Ethical AI Considerations

When integrating AI education into non-technical disciplines, several ethical considerations must be addressed:

1. **Algorithmic Bias:** Artificial Intelligence systems, which are designed to analyze vast amounts of data and make decisions, have the potential to reinforce and perpetuate the existing biases that are often embedded in the training data from which they learn, thereby resulting in outcomes that can be deemed unfair and discriminatory towards certain groups of people. Therefore, it is imperative for educators to impart to their students the essential skills needed to recognize, identify, and critically evaluate these biases, thereby fostering a culture that promotes responsible and ethical use of AI technologies in various facets of life (Xu, 2020).
2. **Data Privacy:** As society increasingly becomes dependent on data for various applications and services, it is of utmost importance for individuals to comprehend the profound implications associated with the collection, usage, and

sharing of personal data, particularly in the context of artificial intelligence. Consequently, students should be educated about the existing data privacy laws, regulations, and ethical practices regarding data management, equipping them with the knowledge necessary to navigate the intricate and often challenging landscape of AI technology in a responsible manner (Kousa & Niemi, 2022).

3.Societal Impact: The advent of artificial intelligence has far-reaching implications for society at large, affecting a multitude of aspects ranging from employment opportunities and job markets to the nature of social interactions and relationships among individuals. In light of these significant changes, educators have the critical responsibility to encourage and facilitate discussions that delve into how AI technology shapes and influences cultural narratives, societal norms, and the everyday lives of individuals, thereby preparing students to engage thoughtfully and meaningfully with these complex and pressing issues (Bu, 2022).

By fostering a curriculum that emphasizes critical thinking and ethical considerations, educators can empower students to navigate the evolving landscape of AI technology while encouraging them to become proactive contributors to society's dialogue on its implications.

Addressing these ethical considerations, educators can ensure that students not only become proficient in AI technologies but also develop a strong ethical framework for their application in society.

Figure 4. Funnel of Ethical AI Awareness

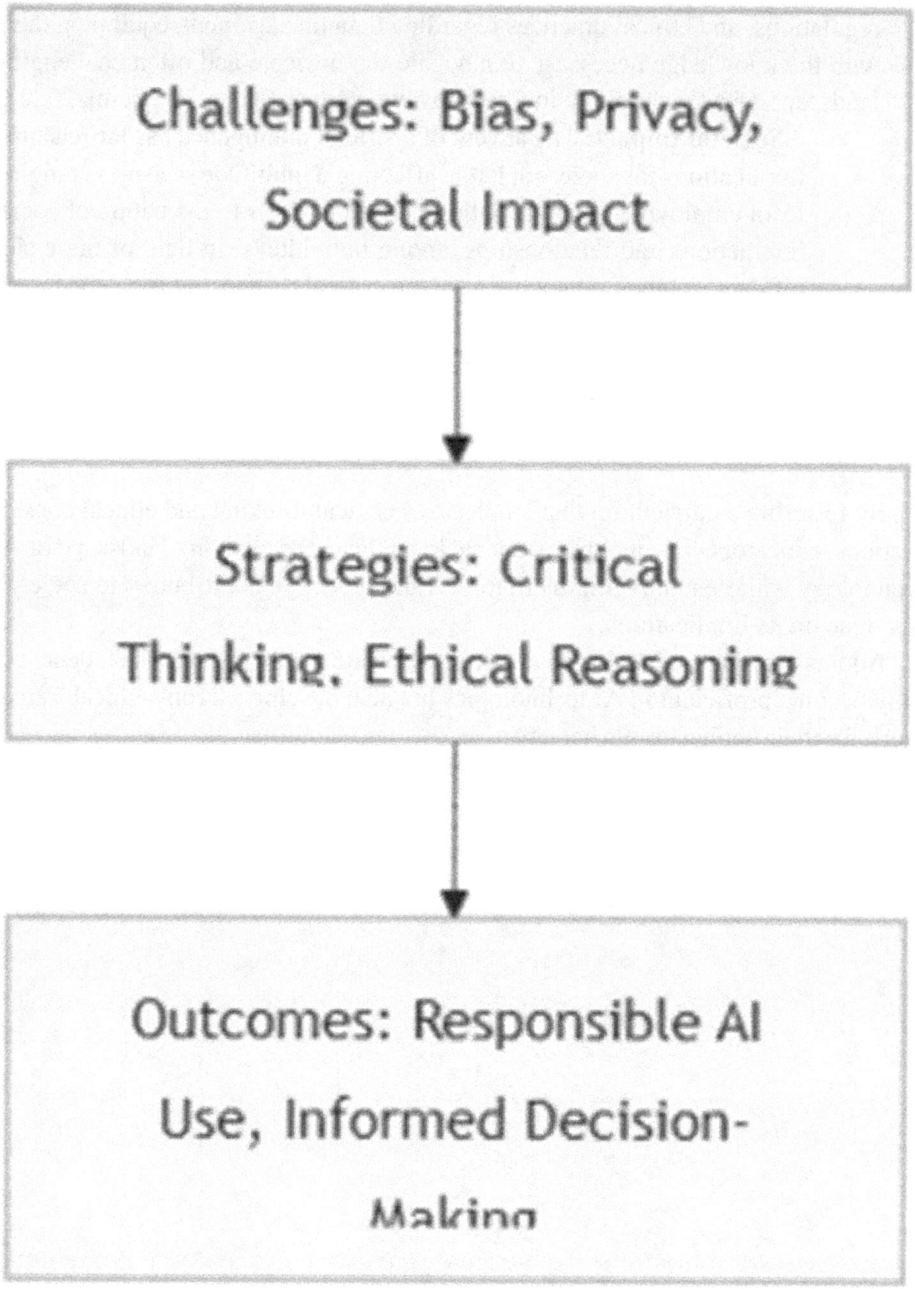

III. CASE STUDIES

Real-World Examples Showcasing Success Stories in AI Integration

1. **Business Ethics Courses:** Analyzing AI Ethical Dilemmas

One notable example of integrating AI education into business ethics is the course offered at Stanford University, which focuses on the ethical implications of AI technologies. In this course, students analyze real-world dilemmas such as bias in facial recognition systems. For instance, the course discusses how AI algorithms used in facial recognition have been found to exhibit significant racial and gender biases, leading to wrongful identifications and ethical concerns about surveillance. By engaging with these case studies, students learn to critically evaluate the ethical ramifications of AI technologies and develop frameworks for responsible AI deployment in business practices (Matias & Zipitria, 2023).This hands-on approach not only equips students with the necessary skills to navigate complex ethical landscapes but also encourages them to advocate for transparency and fairness in AI systems as they enter the workforce.This commitment to ethical considerations in AI is essential, as the technology continues to evolve and permeate various sectors, potentially impacting millions of lives.

2. **Creative Arts:** Using GPT Tools for Storytelling and DALL-E for Art Creation

In the realm of creative arts, institutions like the University of Southern California (USC) have embraced AI tools like GPT-3 for storytelling and DALL-E for visual art creation. In a recent project, students utilized GPT-3 to generate narrative ideas and dialogues for a short film, allowing them to explore new storytelling techniques and enhance their creative processes. Additionally, students in visual arts courses used DALL-E to create unique artwork based on textual prompts, challenging traditional notions of authorship and creativity. These experiences not only foster creativity but also encourage students to critically assess the role of AI in artistic expression (Tiwari, 2023)and its potential to reshape the future of creative industries, prompting discussions about the ethical implications and boundaries of machine-generated content in art.These discussions are crucial as they navigate the balance between human creativity and technological advancement, ultimately shaping how future generations will engage with both art and artificial intelligence.As technology continues to evolve, educational institutions are increasingly integrating AI tools into their curricula, preparing students for a future where collaboration between human creativity and artificial intelligence becomes the norm.

3. Environmental Studies: Leveraging Machine Learning for Climate Predictions

A successful integration of AI in environmental studies can be seen in the work conducted by the University of California, Berkeley, where researchers leverage machine learning algorithms to predict climate patterns and assess renewable energy adoption. For example, their AI models analyze vast datasets to forecast climate changes, helping policymakers make informed decisions regarding environmental conservation and energy management. Additionally, students in the Environmental Science program engage in projects that utilize AI to analyze the effectiveness of various renewable energy sources, contributing to sustainable practices and innovative solutions for climate-related challenges (Srinivasan & Uchino, 2021). This collaborative effort not only enhances academic understanding but also fosters a culture of innovation that empowers future leaders to tackle pressing environmental issues through data-driven strategies.This multidisciplinary approach ensures that students are equipped with the necessary skills to address complex environmental problems while promoting a deeper awareness of the intersection between technology and sustainability.This emphasis on practical experience and interdisciplinary learning prepares graduates to become proactive agents of change in their communities, advocating for policies that prioritize ecological balance and social responsibility (Piskopani et al., 2023) .while also encouraging the development of green technologies that can further mitigate the impacts of climate change and enhance energy efficiency.

These real-world examples illustrate the transformative potential of integrating AI education across various non-technical disciplines. By examining ethical dilemmas in business, enhancing creativity in the arts, and addressing critical environmental issues, students are equipped with the knowledge and skills necessary to navigate an increasingly AI-driven world. Through these success stories, we can see the profound impact of AI literacy on preparing future professionals to engage thoughtfully and innovatively with the complexities of their respective fields.

IV. PRACTICAL STRATEGIES

As we consider the broader implications of integrating AI education into non-technical disciplines, it is essential to address the need for continuous professional development among educators themselves. This training not only equips teachers with the necessary skills to effectively incorporate AI tools in their curricula but also fosters a culture of innovation within educational institutions. By participating in workshops and collaborative projects that emphasize the ethical use of AI and its societal impact, educators can better facilitate discussions around algorithmic bias and data privacy, ensuring that students are prepared to navigate these challenges responsibly (Roopaei & Dehbozorgi, 2024). Moreover, as educators become more proficient in employing AI technologies, they can model effective practices that inspire students to engage critically with AI applications, ultimately enhancing both teaching and learning outcomes across various fields.

Simplifying AI Concepts

A. Relatable Examples:

1. **IBM Watson:** Think of IBM Watson as a super-smart assistant that can help businesses make better decisions. For instance, it can analyze customer feedback from social media to determine what people love or dislike about a product, much like how a friend might tell you their opinions on a movie you're considering watching (Fisher et al., 2012). This ability to process vast amounts of information quickly allows companies to tailor their products and services to meet customer needs more effectively, fostering a deeper connection between consumers and brands.
2. **Teachable Machine:** Imagine you have a pet that you want to train to recognize different objects. Teachable Machine is like a digital version of that training process. You can show it pictures of cats and dogs, and it learns to identify them, just like how you would teach your pet to recognize a ball or a stick (Tanveer et al., 2020)

. This tool uses machine learning to adapt and improve its recognition capabilities over time, making it an invaluable resource for educators and developers alike who seek to create interactive experiences that engage users effectively.

B. Breaking Down Technical Jargon:

- **Natural Language Processing (NLP):** This is just a fancy term for how computers understand and interact with human language. For example, when you use a voice assistant like Siri or Alexa to ask about the weather, NLP helps the device understand your question and respond appropriately (Florea & Radu, 2019). By leveraging NLP, developers can create applications that not only comprehend user input but also generate contextually relevant responses, enhancing the overall user experience and making technology more accessible to everyone.Understanding the nuances of NLP allows for the development of more sophisticated algorithms that can analyze sentiment, recognize intent, and even translate languages in real-time, further bridging the gap between humans and machines.This progression in NLP technology is paving the way for advancements in artificial intelligence, enabling machines to engage in more natural and intuitive conversations with users.

- **Machine Learning:** Think of it as teaching a computer to learn from experience. For example, if you show a computer many pictures of apples and oranges, it learns to tell them apart, just like how you learned the difference between the two fruits when you were young (Roy, 2020). This process of learning from data is fundamental to creating intelligent systems that can adapt and improve over time, ultimately leading to advancements in various applications such as image recognition, natural language processing, and autonomous vehicles.These advancements not only enhance the capabilities of machines but also open up new possibilities for innovation across industries, driving efficiency and transforming how we interact with technology on a daily basis.

Project-Based Learning

C. Practical Projects:

1. **Analyzing Social Media Trends:** Students can use NLP tools to analyze tweets or Instagram posts about a trending topic. They can identify common sentiments (positive, negative, neutral) and present their findings, similar to how a journalist might report on public opinion (Zhou & Lawless, 2015).
2. **Building a Chatbot:** Students can create a simple chatbot using natural language processing techniques, allowing them to understand user queries and provide appropriate responses, thereby gaining hands-on experience with AI technologies in a practical context.3. Developing a Recommendation System: Students can design and implement a recommendation system that suggests products, movies, or music based on user preferences and behavior, enhancing their understanding of machine learning algorithms and data analysis techniques.4. Analyzing Social Media Data: Students can explore social media platforms to gather data on user interactions and trends, applying statistical methods to analyze engagement metrics and draw insights that could inform marketing strategies or content creation.

3. **Predicting Economic Shifts:** Students can work on a project where they analyze historical economic data using AI tools to forecast future trends. For instance, they could look at past unemployment rates and economic indicators to predict how the economy might change, much like how weather forecasters use data to predict the weather (Maitri, 2019). This project would not only enhance their analytical skills but also provide valuable insights into the complexities of economic dynamics and the role AI plays in decision-making processes.These projects encourage students to develop a deeper understanding of data interpretation and its practical applications, preparing them for careers in fields such as marketing, economics, and data science.By engaging in such projects, students can cultivate critical thinking and problem-solving abilities that are essential for navigating the increasingly data-driven landscape of today's job market.As a result, students will be better equipped to analyze trends and make informed decisions that can positively impact their future careers and the organizations they work for.

Interdisciplinary Collaboration

D. Facilitating Teamwork:

- **Collaborative AI Projects:** Pair technical students (like computer science majors) with non-technical students (like business or sociology majors). For example, a technical student can handle the coding aspect of an AI project, while the non-technical student can focus on understanding the societal implications of the AI's findings. Together, they can create a project that not only demonstrates technical skills but also addresses real-world issues, such as how AI can impact job markets or enhance customer experiences (Pugalenthi et al., 2021)

. This approach not only enriches the learning experience for both students but also fosters a deeper understanding of how interdisciplinary perspectives can lead to innovative solutions in AI applications and their societal ramifications.By collaborating, these students can bridge the gap between technology and its ethical considerations, ensuring that AI development is not only efficient but also responsible and inclusive.Such projects can empower students to critically analyze the potential biases in AI algorithms and advocate for fairness, transparency, and accountability in technology deployment (Sandu & Gide, 2019).

Figure 5. Cycle of AI Literacy Implementation

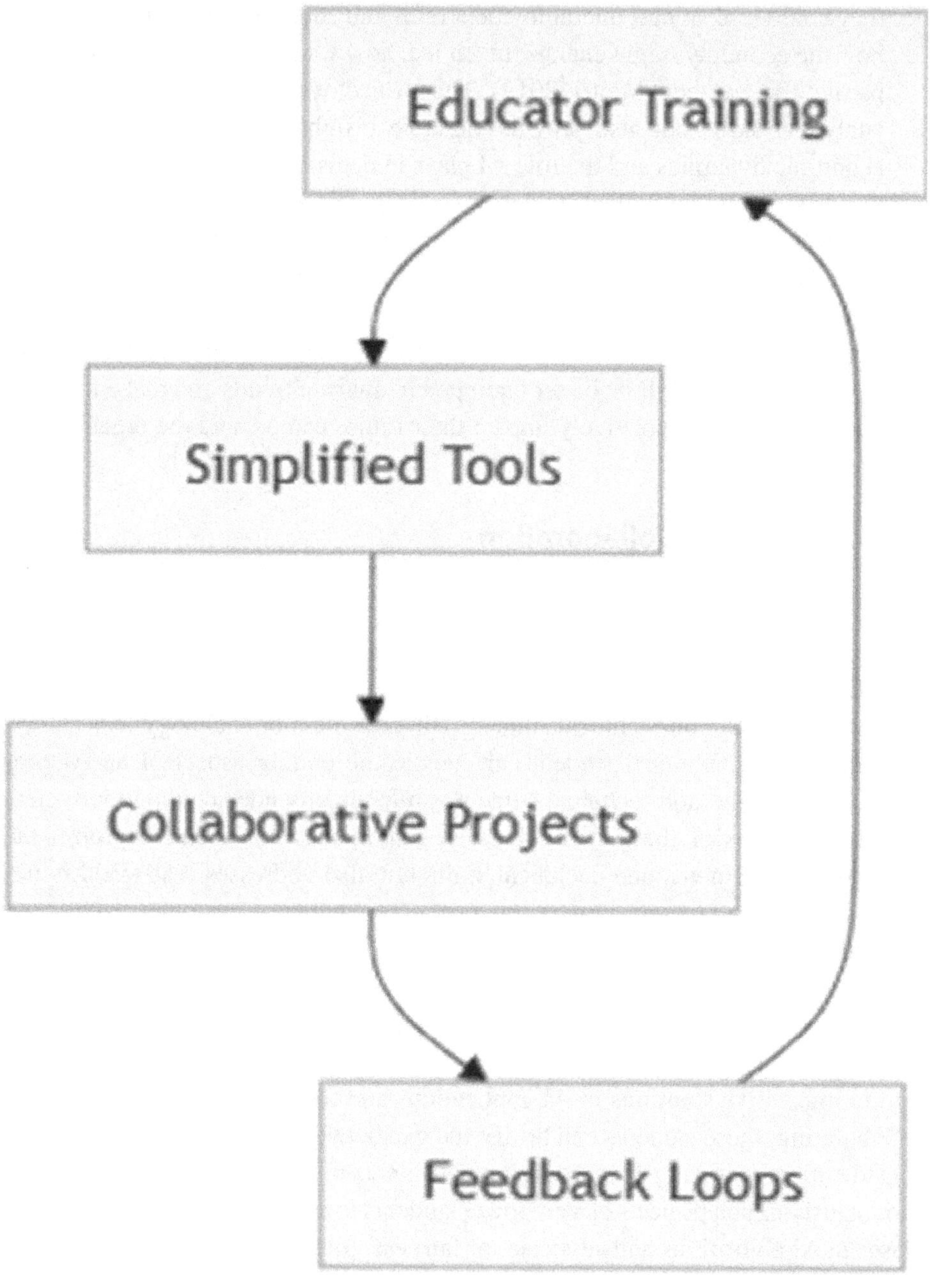

By incorporating these strategies, students can gain a deeper understanding of AI concepts in a relatable and practical manner, preparing them for a future where AI plays a critical role in various fields.

V. CHALLENGES AND SOLUTIONS

As the integration of AI tools in psychology curricula advances, it is also essential to consider the implications for student mental health and well-being. The use of AI-driven assessments and interventions can provide valuable support; however, educators must remain vigilant about the potential risks associated with over-reliance on technology in sensitive areas such as mental health. For instance, while chatbots like Woebot offer immediate assistance, they cannot fully replicate the empathetic understanding of a human therapist, which may lead to feelings of isolation if students rely solely on digital interactions (Muhammad & Matilda, 2024). Thus, fostering a balanced approach that combines traditional therapeutic practices with innovative AI applications will be crucial in preparing students not only to utilize these technologies effectively but also to appreciate the irreplaceable value of human connection in psychological care.This hybrid model can empower students to seek help when needed while ensuring they also develop strong interpersonal skills and emotional intelligence, essential components of overall mental well-being.By integrating both digital and human resources, educational institutions can create a comprehensive support system that addresses the diverse needs of students, promoting resilience and emotional growth in an increasingly complex world.

Challenges and Solutions in Integrating AI Education into Non-Technical Disciplines

A. Challenges:

1. Limited Educator Expertise in AI:

Numerous educators who are primarily engaged in fields that are not centered around technical subjects may find themselves lacking the essential knowledge and skills required to competently and effectively impart crucial concepts related to artificial intelligence, which consequently results in a significant gap in the level of AI literacy among students who are eager to learn and understand these important advancements in technology2. Insufficient Curriculum Resources: The existing educational materials often do not include comprehensive AI content tailored for non-technical disciplines, making it difficult for educators to find suitable resources that bridge the knowledge gap and engage students in meaningful ways.This lack of accessible resources not only hinders educators' ability to teach AI concepts effectively but also limits students' exposure to the interdisciplinary applications of artificial intelligence, which are increasingly relevant in today's job market (Dekker et al., 2020). As a result, educational institutions must prioritize the development of inclusive curricula that incorporate AI literacy across various disciplines, ensuring all students have the opportunity to engage with and benefit from these transformative technologies.By fostering an environment where AI is integrated into diverse fields of study, educators can equip students with essential skills that enhance their employability and prepare them for a rapidly evolving workforce.

2. Resistance to Integrating Technology in Traditional Disciplines:

It is entirely possible that there exists a sense of hesitance or reluctance among faculty members and educational institutions alike when it comes to the adoption of artificial intelligence tools and methodologies, a feeling that likely arises from a strong preference for and attachment to traditional teaching approaches, as well as a prevailing skepticism regarding the actual relevance and applicability of artificial intelligence in fields that are considered to be non-technical or less reliant on technology (Dignum, 2021).This resistance can hinder the potential for innovation in educational practices, ultimately affecting students' readiness to thrive in a technology-driven job market.Embracing artificial intelligence in education could provide transformative opportunities for personalized learning, enabling educators to tailor their teaching strategies to meet the diverse needs of students and better prepare them for future challenges.By integrating AI tools into the curriculum, educators can enhance engagement and foster critical thinking skills, equipping students with the competencies necessary to navigate an increasingly complex world.Integrating AI not only streamlines administrative tasks but also offers valuable insights into student performance, allowing for data-driven decision-making that can further enhance educational outcomes (Huang, 2023).This approach encourages a more adaptive learning environment where feedback is immediate and personalized, ultimately fostering a culture of continuous improvement among both students and educators.

3. Ethical and Societal Concerns:

The emergence and widespread integration of artificial intelligence, or AI, into various sectors of society raises a plethora of significant and complex ethical questions that demand our attention and scrutiny, including but not limited to pressing issues related to the privacy and security of personal data, the inherent biases that may exist within algorithms used to make decisions, and the broader societal consequences that arise from the deployment of AI technologies in our everyday lives, all of which necessitate meticulous and thoughtful consideration, especially within educational environments where the implications can be profound and far-reaching (Sethi et al., 2020). As educators and policymakers grapple with these challenges, it becomes essential to establish frameworks that promote ethical AI usage while fostering an environment of transparency and accountability in the development and implementation of these technologies.Such frameworks should not only prioritize the protection of student data but also encourage critical thinking and digital literacy among learners, equipping them with the skills necessary to navigate an increasingly complex technological landscapeIn addition, ongoing professional development for educators will be vital to ensure they are well-equipped to integrate AI tools effectively and responsibly into their teaching practices (Akgun & Greenhow, 2021). This commitment to professional growth will empower educators to harness the full potential of AI, enabling personalized learning experiences that cater to diverse student needs while maintaining ethical standards.

B. Solutions:

1. Teacher Training Workshops on AI Basics:

Implementing comprehensive and thoughtfully designed professional development programs that are specifically tailored to focus on the essential and foundational knowledge of artificial intelligence can significantly empower educators, allowing them to cultivate a greater sense of confidence and competence in their ability to effectively teach complex AI concepts, which in turn leads to a substantial enhancement in the overall delivery of the curriculum (Gupta & Bhaskar, 2020).In addition to these workshops, collaborative learning communities can be established where educators share best practices and resources, fostering a culture of continuous improvement and innovation in teaching methodologies.By leveraging technology and online platforms, these communities can extend their reach, enabling educators from diverse backgrounds to connect, collaborate, and learn from each other's experiences, ultimately enriching the educational landscape.This collaborative approach not only enhances individual teaching practices but also contributes to a collective knowledge base that can drive systemic change within educational institutions, ensuring that all students receive high-quality instruction in artificial intelligence.This emphasis on collaboration and resource sharing empowers educators to stay updated with the latest advancements in AI, equipping them to better prepare students for a rapidly changing technological landscape (Nguyen et al., 2022) and fostering a culture of lifelong learning among teachers, which is essential for adapting to new challenges and opportunities in the field of education.

2. **Simplified Platforms for AI Experimentation**:

Providing both educators, who play a crucial role in shaping the minds of future generations, and students, who are eager to learn and explore new technologies, with user-friendly artificial intelligence tools and platforms that are designed to be accessible and intuitive can significantly facilitate opportunities for hands-on experimentation and practical learning experiences, thereby making it much simpler to seamlessly integrate artificial intelligence into a diverse array of academic disciplines without necessitating extensive technical expertise or advanced knowledge in the field of technology (Bhimdiwala et al., 2021). By fostering an environment where both educators and students can engage with AI technologies, institutions can cultivate a culture of innovation and creativity that empowers learners to harness the potential of these tools in their academic pursuits.This collaborative approach not only enhances the learning process but also prepares students for a future where AI will play an increasingly vital role in various industries, ensuring they are well-equipped to thrive in an evolving job market.As educational institutions adapt to these changes, they must prioritize training and support for faculty, enabling them to effectively incorporate AI into their teaching methodologies.This investment in professional development is crucial for maintaining the relevance of educational practices and ensuring that instructors are confident in guiding students through the complexities of AI technologies (Mujiono, 2023). By fostering a culture of continuous learning and adaptation, schools can create an environment where both educators and students embrace innovation, ultimately leading to improved educational outcomes and better preparedness for the challenges of tomorrow's workforce. Emphasizing collaboration between educators, industry leaders, and technology experts will further enhance the integration of AI in curricula, allowing for a more comprehensive understanding of its applications and implications in various fields.

3. Regular Student Feedback to Refine Teaching Methodologies:

Creating a systematic and continuous feedback loop with students, wherein their thoughts and opinions are regularly gathered and analyzed, can significantly assist educators in gaining a deeper understanding of how effective their various teaching methods truly are while simultaneously allowing them to pinpoint specific areas that may require improvement or adjustment. This ongoing and iterative process, characterized by its cyclical nature, has the potential to result in a more captivating and pertinent educational experience focused on artificial intelligence, one that is specifically tailored to meet the diverse needs and interests of each individual student (Zhou & Lawless, 2015)

. By fostering an environment where students feel comfortable sharing their insights, educators can adapt their strategies to enhance engagement and comprehension, ultimately promoting a more inclusive learning atmosphere that embraces the rapid advancements in technology and AI.

By strategically tackling and confronting these multifaceted challenges with carefully designed and focused solutions, educational institutions and organizations can significantly enhance their ability to equip and prepare students who are pursuing non-technical disciplines with the necessary skills and knowledge, enabling them to not only survive but also to truly thrive in a world that is becoming increasingly dominated and influenced by advancements in artificial intelligence technology.This proactive approach not only fosters resilience among students but also encourages a mindset of continuous learning, ensuring that they remain adaptable and competitive in an ever-evolving job market.Such initiatives can lead to the development of interdisciplinary programs that bridge the gap between technical and non-technical fields, fostering collaboration and innovation across various sectors.

VI. CONCLUSION

The integration of artificial intelligence (AI) education into non-technical disciplines represents a critical evolution in contemporary educational practices, responding to the pervasive influence of AI across diverse fields. As outlined in this research, the necessity for AI literacy extends beyond traditional STEM domains, encompassing areas such as humanities, social sciences, and business. This paper has demonstrated that incorporating AI education not only enhances students' critical thinking, ethical reasoning, and creativity but also prepares them for the complexities of a rapidly changing job market. Through the examination of real-world case studies and effective curriculum strategies, it is evident that fostering interdisciplinary learning can empower students to address societal challenges and innovate solutions that leverage AI technologies. Furthermore, the ethical considerations surrounding AI, including algorithmic bias and data privacy, underscore the importance of equipping students with the tools to engage thoughtfully with these issues. Ultimately, this research advocates for a paradigm shift in educational methodologies, emphasizing the need for a holistic approach that recognizes AI as a foundational component across all disciplines, thus ensuring that future professionals are well-prepared to thrive in an increasingly digital and interconnected world.This shift not only enhances students' technical skills but also cultivates critical thinking and ethical reasoning, enabling them to navigate the complexities of AI's impact on society effectively.This comprehensive educational framework will empower students to become informed citizens who can contribute positively to discussions about technology and its implications, fostering a generation that prioritizes ethical considerations in innovation.By integrating AI into the curriculum, educators can create a dynamic learning environment that encourages collaboration and creativity, ultimately preparing students to tackle real-world challenges with confidence and foresight.

REFERENCES

Akgun, S., & Greenhow, C. (2021).*Artificial intelligence in education: Addressing ethical challenges in K-12 settings.*DOI: 10.1007/s43681-021-00096-7

Alam, A., Hasan, M., & Raza, M. M. (2022). *Impact of artificial intelligence (ai) on education: changing paradigms and approaches.* Towards Excellence., DOI: 10.37867/TE140127

Berendt, B., Littlejohn, A., & Blakemore, M. (2020). AI in Education: Learner Choice and Fundamental Rights. *Learning, Media and Technology*, 45(3), 312–324. Advance online publication. DOI: 10.1080/17439884.2020.1786399

Bhimdiwala, A., Neri, R. C., & Gomez, L. M. (2021). Advancing the Design and Implementation of Artificial Intelligence in Education through Continuous Improvement. *International Journal of Artificial Intelligence in Education.* Advance online publication. DOI: 10.1007/s40593-021-00278-8

Bu, Q. (2022). *Ethical Risks in Integrating Artificial Intelligence into Education and Potential Countermeasures.* Science Insights., DOI: 10.15354/si.22.re067

Celik, I. (2022). Towards Intelligent-TPACK: An empirical study on teachers' professional knowledge to ethically integrate artificial intelligence (AI)-based tools into education. *Computers in Human Behavior.* Advance online publication. DOI: 10.1016/j.chb.2022.107468

Conley, J. (2023).*Harnessing Artificial Intelligence for Innovation in Education.* DOI: 10.1007/978-981-19-9201-8_8

Dekker, I., Dekker, I., de Jong, E. M., Schippers, M. C., De Bruijn-Smolders, M., De Bruijn-Smolders, M., Alexiou, A., Alexiou, A., & Giesbers, B. (2020). Optimizing Students' Mental Health and Academic Performance: AI-Enhanced Life Crafting. *Frontiers in Psychology*, 11, 1063. Advance online publication. DOI: 10.3389/fpsyg.2020.01063 PMID: 32581935

Dignum, V. (2021). The role and challenges of education for responsible AI. *London Review of Education*, 19(1). Advance online publication. DOI: 10.14324/LRE.19.1.01

Druga, S., Otero, N., & Ko, A. J. (2022, July 7). The Landscape of Teaching Resources for AI Education.*Annual Conference on Innovation and Technology in Computer Science Education.* DOI: 10.1145/3502718.3524782

Eguchi, A. (2021a). *AI-Powered Educational Robotics as a Learning Tool to Promote Artificial Intelligence and Computer Science Education.* DOI: 10.1007/978-3-030-82544-7_26

Eguchi, A. (2021b). *AI-Robotics and AI Literacy.* DOI: 10.1007/978-3-030-77022-8_7

Fisher, D. H., Dilkina, B., Eaton, E., & Gomes, C. P. (2012, July 22). Incorporating computational sustainability into AI education through a freely-available, collectively-composed supplementary lab text. *National Conference on Artificial Intelligence.*

Florea, A. M., & Radu, S. (2019, May 28). Artificial Intelligence and Education. *International Conference on Control Systems and Computer Science.* DOI: 10.1109/CSCS.2019.00069

Flores-Vivar, J. M., & García-Peñalvo, F. J. (2023). *Reflections on the ethics, potential, and challenges of artificial intelligence in the framework of quality education (SDG4).* Comunicar., DOI: 10.3916/C74-2023-03

Gupta, K. P., & Bhaskar, P. (2020). *Inhibiting and Motivating Factors Influencing Teachers' Adoption of AI-Based Teaching and Learning Solutions: Prioritization Using Analytic Hierarchy Process.* DOI: 10.28945/4640

Hamal, O., El Faddouli, N., Alaoui Harouni, M. H., & Lu, J. (2022). Artificial Intelligent in Education. *Sustainability (Basel)*, 14(5), 2862. Advance online publication. DOI: 10.3390/su14052862

Harry, A. (2023). Role of AI in Education. [INJURITY]. *Interdiciplinary Journal and Hummanity*, 2(3), 260–268.

Holmes, W., Porayska-Pomsta, K., Holstein, K., Sutherland, E., Baker, T., Shum, S. B., Santos, O. C., Rodrigo, M. T., Cukurova, M., Bittencourt, I. I., & Koedinger, K. R. (2021). *Ethics of AI in Education: Towards a Community-Wide Framework.* DOI: 10.1007/s40593-021-00239-1

How, M. L., & Hung, W. L. D. (2019). Educing AI-Thinking in Science, Technology, Engineering, Arts, and Mathematics (STEAM) Education. *Education Sciences*, 9(3), 184. Advance online publication. DOI: 10.3390/educsci9030184

Huang, J. (2023). *Artificial Intelligence in Higher Education.* Advances in Computer and Electrical Engineering Book Series., DOI: 10.4018/978-1-6684-6937-8.ch014

Integration of AI in Learning. (2022). *Advances in Higher Education and Professional Development Book Series.* DOI: 10.4018/978-1-6684-4083-4.ch014

Kasinidou, M. (2023, June 29). AI Literacy for All: A Participatory Approach. *Annual Conference on Innovation and Technology in Computer Science Education*. DOI: 10.1145/3587103.3594135

Kennedy, D. R. (2023).*Inteligencia Artificial, como base de un modelo de curriculum learning para el desarrollo de competencias digitales en e-learning*. DOI: 10.24275/uama.5813.9684

Kousa, P., & Niemi, H. (2022). AI ethics and learning: EdTech companies' challenges and solutions. *Interactive Learning Environments*. Advance online publication. DOI: 10.1080/10494820.2022.2043908

Kujundziski, A. P., & Bojadjiev, J. (2024). *Artificial Intelligence in Education*. Advances in Educational Technologies and Instructional Design Book Series., DOI: 10.4018/979-8-3693-4310-4.ch001

Kumar, D., Haque, A., Mishra, K., & Islam, F. (2023).*Exploring the Transformative Role of Artificial Intelligence and Metaverse in Education: A Comprehensive Review*. DOI: 10.56294/mr202355

Long, D., Roberts, J., Magerko, B., Holstein, K., DiPaola, D., & Martin, F. (2023, April). Ai literacy: Finding common threads between education, design, policy, and explainability. In Extended Abstracts of the 2023 CHI Conference on Human Factors in Computing Systems (pp. 1-6).

Maitri, A. (2019).*Rancang Bangun Chatbot sebagai Pusat Informasi Lembaga Kursus dan Pelatihan Menggunakan Pendekatan Natural Language Processing*.

Matias, A., & Zipitria, I. (2023).*Promoting Ethical Uses in Artificial Intelligence Applied to Education*. DOI: 10.1007/978-3-031-32883-1_53

Muhammad, F., & Orji, C. M. (2024). REVOLUTIONIZING EDUCATION IN THE DIGITAL ERA: THE ROLE OF AI IN PROMOTING INCLUSIVITY, EQUALITY, AND ETHICAL INNOVATION. *The American Journal of Social Science and Education Innovations*, 6(09), 151–157.

Mujiono, M. (2023). *Educational Collaboration : Teachers and Artificial Intelligence*. Jurnal Kependidikan., DOI: 10.33394/jk.v9i2.7801

Mulianingsih, F., Anwar, K., Shintasiwi, F. A., & Rahma, A. J. (2020). Artificial Intellegence dengan pembentukan nilai dan karakter di bidang pendidikan. *IJTIMAIYA: Journal of Social Science Teaching*, 4(2), 148–154.

Nguyen, A., Ngo, H., Hong, Y., Dang, B., & Nguyen, B.-P. (2022). Ethical principles for artificial intelligence in education. *Education and Information Technologies*. Advance online publication. DOI: 10.1007/s10639-022-11316-w PMID: 36254344

Ouyang, F., & Jiao, P. (2021). Artificial intelligence in education: The three paradigms. *Computers and Education: Artificial Intelligence*, 2, 100020.

Piskopani, A.-M., Chamberlain, A., & Holter, C. T. (2023, July 11).*Responsible AI and the Arts: The Ethical and Legal Implications of AI in the Arts and Creative Industries*. DOI: 10.1145/3597512.3597528

Pugalenthi, R., Chakkaravarthy, A. P., Ramya, J., Babu, S., & Krishnan, R. R. (2021). Artificial learning companionusing machine learning and natural language processing. *International Journal of Speech Technology*, 24(3), 553–560. Advance online publication. DOI: 10.1007/s10772-020-09773-0

Rajeswari, P., & Purushothaman, S. (2023). Artificial Intelligence in Education. *Edu Spectra*, 5, 47.

Roopaei, M., & Dehbozorgi, N. (2024).*AI in Education: Crafting Policies for Tomorrow's Learning Landscape*. DOI: 10.1109/ISEC61299.2024.10665282

Roy, M. (2020). AI Intervention in Education Systems of India: An Analysis. *Solid State Technology*.

Sadiku, M. N., Musa, S. M., & Chukwu, U. C. (2022). Artificial intelligence in education. iUniverse.

Sakalle, A., Tomar, P., Bhardwaj, H., & Sharma, U. (2021). Impact and Latest Trends of Intelligent Learning With Artificial Intelligence. In Impact of AI Technologies on Teaching, Learning, and Research in Higher Education (pp. 172-189). IGI Global.

Sandu, N., & Gide, E. (2019, September). Adoption of AI-Chatbots to enhance student learning experience in higher education in India. In 2019 18th international conference on information technology based higher education and training (ITHET) (pp. 1-5). IEEE.

Schumann, C. A., Reuther, K., Tittmann, C., Clauß, A. M., & Kauper, J. (2020, June). Impact of AI application on digital education focused on STE (A) M. In EDEN Conference Proceedings (No. 1, pp. 153-161).

Sethi, K., Sharma, A., Chauhan, S., & Jaiswal, V. (2020). Impact of social and cultural challenges in education using AI. In *Revolutionizing Education in the Age of AI and Machine Learning* (pp. 130–151). IGI Global.

Srinivasan, R., & Uchino, K. (2021, March). Biases in generative art: A causal look from the lens of art history. In *Proceedings of the 2021 ACM Conference on Fairness, Accountability, and Transparency* (pp. 41-51).

Tanveer, M., Hassan, S., & Bhaumik, A. (2020). Academic Policy Regarding Sustainability and Artificial Intelligence (AI). *Sustainability (Basel)*, 12(22), 9435. Advance online publication. DOI: 10.3390/su12229435

Tiwari, R. (2023). *Ethical And Societal Implications of AI and Machine Learning*. Indian Scientific Journal Of Research In Engineering And Management., DOI: 10.55041/IJSREM17519

Uppal, M. (2022).*AI in education*. DOI: 10.4324/9780429329067-5

Xu, L. (2020, December 11).*The Dilemma and Countermeasures of AI in Educational Application*. DOI: 10.1145/3445815.3445863

Zhang, H. Z., Lee, I., Ali, S., DiPaola, D., Cheng, Y., & Breazeal, C. (2022). Integrating Ethics and Career Futures with Technical Learning to Promote AI Literacy for Middle School Students: An Exploratory Study. *International Journal of Artificial Intelligence in Education*. Advance online publication. DOI: 10.1007/s40593-022-00293-3 PMID: 35573722

Zhou, M. Y., & Lawless, W. F. (2015). An overview of artificial intelligence in education. Encyclopedia of Information Science and Technology, Third Edition, 2445-2452.

Chapter 6
Artificial Intelligence as a Harbinger of Engagement and Collaboration

Soufiane Ouariach
https://orcid.org/0000-0001-5869-9027
Abdelmalek Essaadi University, Morocco

Fatima Zahra Ouariach
Abdelmalek Essaadi University, Morocco

Mohamed Khaldi
https://orcid.org/0000-0002-1593-1073
Abdelmalek Essaadi University, Morocco

ABSTRACT

Artificial intelligence (AI) in education is at the top of the agenda for education today in developing learners' AI knowledge, skills, attitudes, and values in the 21st century. This chapter examines the transformative impact of artificial intelligence (AI) on education, with a focus on enhancing learner engagement. It explores the key elements of engagement—behavioral, emotional, and cognitive—and their correlation with improved educational outcomes. The chapter delves into AI applications, such as intelligent tutoring systems and adaptive learning paths, highlighting their role in providing personalized and interactive learning experiences. Additionally, it discusses AI-driven feedback systems and collaborative tools that foster active participation and critical thinking. While AI offers significant potential to revolutionize education, the chapter also addresses the ethical considerations involved in preserving the human aspect of teaching. Through a balanced analysis, this chapter underscores the importance of integrating AI to create inclusive, effective learning

DOI: 10.4018/979-8-3373-2262-9.ch006

environments.

INTRODUCTION

Involving learners has become a key factor in academic success. In today's classrooms, engaging learners is essential to achieving better results. Learner engagement can be defined in several important ways. Firstly, there is the behavioral aspect, i.e. active participation. Emotional engagement takes into account key elements such as interest and a sense of belonging. Finally, there is cognitive engagement, i.e. the attention they pay to the lessons. The technological revolution underway promises to redefine the way learners learn and engage with their educational journey. It is propelled by the advent of artificial intelligence (AI), which is transforming education in the 21st century. It has always been a major challenge for teachers to engage learners, given that learner engagement contributes to successful knowledge acquisition. It covers the cognitive, affective, and behavioral involvement of learners in their educational process. AI emerges here as a useful tool for encouraging this last decisive point. Comprehensive research by Zawacki-Richter et al (2019) has shown how AI is changing various aspects of higher education. Their analysis reveals that thanks to AI, it is possible to accelerate massive personalization to meet the particular needs of each learner. Among these techniques is personalization with intelligent tutoring systems. Holmes et al (2019) argue that these systems can track learners' progress, identify their particular difficulties in real time and adapt content accordingly. This bespoke approach not only improves academic results but also boosts their self-esteem. AI is also revolutionizing assessment methods. Chassignol et al (2018) describe AI-based assessment systems that can handle open-ended responses offering more freedom for learners to express themselves and assess their understanding in more subtle ways. In addition to personalization and assessment, artificial intelligence makes immersive learning possible. From adaptive educational games to virtual reality environments, these technologies uniquely hold learners' attention. These approaches work effectively with learners who have a special need in their studies, underlining the potential of this technology to encourage its adoption for all learners. But the integration of AI into education also raises a crucial question. How do we preserve the human element that lies at the heart of education?

This last point is highlighted by Luckin et al (2016). In their view, AI should not replace teachers but rather reinforce them. The aim is to free educators from repetitive tasks so that they can focus on the more human aspects of teaching, such as inspiration, guidance, and individualized support.

This chapter delves into the transformative impact of artificial intelligence on education, beginning with an exploration of learner engagement. It examines the key elements that define engagement in online environments, the correlation between engagement and improved educational outcomes, and the psychological and social factors influencing student commitment. Building on this foundation, the chapter then introduces AI applications in education, focusing on personalized learning through intelligent tutoring systems and adaptive learning paths. The discussion progresses to AI's role as a catalyst for interactive, evaluative, and collaborative learning, highlighting conversational bots, AI-driven feedback systems, and AI-enhanced collaboration tools. Throughout, the chapter emphasizes how these AI technologies can enhance student engagement and learning experiences. Finally, it addresses the crucial ethical considerations surrounding AI in education, providing a balanced view of the potential benefits and challenges of integrating AI into learning environments. The aim of this chapter is to explore how artificial intelligence (AI) can transform learner engagement and improve educational outcomes. By analyzing the behavioral, emotional and cognitive aspects of engagement, the authors will see how AI personalizes learning to meet individual needs, improves assessment methods and delivers immersive experiences. The challenge is to understand how AI can support, rather than replace, the crucial role of teachers, focusing on autonomy and inspiration.

THE IMPORTANCE OF LEARNER ENGAGEMENT IN LEARNING

Redefining Engagement in Modern Education

The 21st century has ushered in unprecedented technological advancements that are fundamentally reshaping every aspect of society, with education standing at the forefront of this transformation. The emergence of artificial intelligence, machine learning, and adaptive technologies has created new possibilities for teaching and learning that were unimaginable just a few decades ago. However, to fully understand the impact and potential of these technological innovations in education, first, it's important to understand the essential foundations of effective learning and learner engagement.

The educational landscape is experiencing unprecedented change. Technology, particularly artificial intelligence (AI), is revolutionizing traditional teaching methods and learning environments. However, before delving into the transformative role of AI in education, it is essential to understand the fundamental concepts of learner engagement and its significance in the educational process.

The way in which education is conceived is changing. Schools are increasingly realizing that it is important to keep learners interested and involved if they are to learn. When learners are engaged, they do better. Engaging learners is essentially about making them want to learn, not by letting them sit back, but by making them feel part of the process. At its core, engagement means that teachers and learners must be actively involved in learning during the course (Ouariach et al., 2023). This involves behaviors such as participation, but also how they feel emotionally and how they think about and interact with the material. All of these elements help them to absorb and retain more of what they learn. Research shows that learner engagement is an essential prerequisite for optimal and deep learning (Barkoukis et al. 2014; Skinner, 2016; Skinner, Zimmer-Gembeck & Connell, 1998). The more they take an interest in the lessons and reflect on them, the more they get out of them. Learner engagement can be defined by several key elements, including behavioral, emotional, and cognitive engagement in the learning process. Understanding the psychological and social factors that influence learner engagement, such as motivation, self-efficacy, peer relationships, and the classroom environment, is essential for educators wishing to create inclusive and stimulating learning spaces. In this section, the authors answer three fundamental questions that influence learner engagement. In doing so, as human beings, It's not always easy to understand things on paper, so it's a matter of getting to know each factor. Therefore, the author will present these factors by answering three fundamental questions before moving on to the main picture.

What are the key elements that define learner engagement in an online environment?

Learner engagement in a classroom is the result of several different elements working together to make learning truly interactive and useful. A lot of it is about engaging learners' curiosity and making them want to explore topics on their own. Fried and Fried (2015) argue that we identify our passions in our field of study to distinguish what we are passionate about from what we are less passionate about. When learners have a natural interest in a topic, they begin to explore it more deeply and ask good questions that help them broaden what they learn from it. Their curiosity drives them to learn more and more. Reschly and Christenson (2012) argue that engagement mediates the relationship between context and learner development. A large part of engaging learners is to pique their interest, to help them become invested in the material so that they proactively seek ways to understand it even more. This kind of engagement and interaction is very important to get the most out of their education. This is called demonstrating natural curiosity. Technology, used effectively, can help to stimulate curiosity and interest, as well as facilitate and maintain voluntary engagement. In addition, technology can play a key role in triggering and addressing the personal, situational, and contextual factors that promote autonomy and competence and engender deep, active learning (Arnone et al., 2011).

Online communication tools encompass instant messaging, forums, chat rooms, email, blogs, wikis, webinars, and course management systems. These tools provide numerous opportunities to promote dialogue and collaboration among teachers, learners, and other stakeholders (Ouariach et al., 2024). Online learning has established itself as an inclusive educational model that enhances access to higher education and supports the development of digital skills (Zahra et al., 2024).

Practical activities are also very important for maintaining learners' interest, as they enable learners to participate actively in their learning. Learners become more involved in the subject when they interact directly with it. Participation in class discussions is another important sign of learner engagement (Frymier & Houser, 2016). Not only does it encourage interaction between the group and the class, but it also fosters a collaborative learning environment where different ideas can be shared and explored. When learners actively contribute to lessons, ask questions, and solve problems together, they become more engaged and reinforce what they are studying. As a general rule, the most motivated learners have a higher level of motivation, which makes them more open to assimilating new information (Filgona et al., 2020). This strengthens their ability to retain what they learn in the long term.

Engagement is also pivotal when learners see how relevant the material is to their lives and future careers. The ability to link courses to real-world applications bridges the gap between theory and practice. Connectivism takes advantage of these digital tools to facilitate learning and create opportunities for collaboration and engagement (Ouariach & Khaldi, 2024). Making sure learners understand how they will use their knowledge outside of school makes courses more understandable. Showing learners why what they are studying is meaningful and useful impresses them more (Tutos'Me, 2024). This kind of contextual learning is essential if education is to have an impact. Ultimately, creating a learning environment that truly engages learners and encourages them to participate is very important to keeping children engaged. When the classroom encourages active participation, learners are more focused and better able to grasp the material. If the atmosphere encourages learners to pay attention and get involved, they will understand what they are being taught much better. Teachers who succeed in designing lessons and structuring their classrooms in such a way as to captivate learners and encourage them to participate are the ones who achieve concrete results. An interactive and stimulating environment in which learners want to participate is the key to engagement, and engagement is the key to learning outcomes. Now that having answered the first question, let's move on to the second, which is this:

How does learner engagement correlate with improved educational outcomes?

There are significant correlations between student engagement and better educational outcomes - on multiple levels. Firstly, when teachers adapt and improve their teaching styles that leads directly to boosting engagement from the students. It creates a more dynamic classroom environment where everyone is more inclined to participate and feel interested. That increased engagement is so crucial because it's associated with students having higher motivation levels too. More engaged students are willing to put in extra effort with their studies (Petiot, 2023). Additionally, when students are receptive to different teaching methods, they tend to feel more satisfied that their educational needs are being met. That higher sense of fulfillment only serves to further amplify their level of engagement in the lessons. There are significant correlations between student engagement and better educational outcomes across multiple levels. When teachers adapt and improve their teaching styles, it leads directly to boosting engagement from students, creating a more dynamic classroom environment where everyone is more inclined to participate and feel interested (Fredricks et al., 2019). This increased engagement is crucial because it's associated with students having higher motivation levels. More engaged students are willing to put in extra effort with their studies (Lei et al., 2018). Additionally, when students are receptive to different teaching methods, they tend to feel more satisfied that their educational needs are being met. This higher sense of fulfillment further amplifies their level of engagement in lessons (Bond et al., 2020). Research has shown that increased learner engagement is linked to improved academic performance across various subjects. When students are actively engaged in their learning process, they tend to retain information more effectively and demonstrate a deeper understanding of complex concepts (Zhoc et al., 2019). Engagement also fosters the development of critical thinking skills. As students become more involved in classroom discussions and activities, they are more likely to question, analyze, and synthesize information. This active participation encourages higher-order thinking, which is essential for academic success and future career readiness. In summary, taking the time to grasp how learner engagement interacts with academic performance means educators can develop more successful teaching tactics. Approaches that consider individual needs and what inspires each student can only serve to benefit all kinds of learners. If schools focus on sparking kids' interest and helping every child feel involved, motivated, and supported in their studies, it'll lead to learning gains across the board no matter a student's background or situation. Ensuring classrooms provide engaging learning experiences tailored for different types of kids is key to promoting academic achievement for students from all walks of life and turning straight to the last question, which plays an important role in the closing of the circle.

What psychological and social factors influence a student's level of commitment?

There's more to student engagement than just teaching styles. A lot of psychological and social factors play a significant role in how involved someone feels in their studies too. An important part is how deep that engagement is it shows how well the student feels connected to their school and academic environment. When a student's own goals and beliefs align well with what their university promotes, it helps them integrate more and ultimately be more engaged. Also, short and medium-term projects seem to positively impact engagement levels. It seems hands-on tasks can strengthen that psychological bond to the learning process. Time management skills and organizational strategies are also key for adapting to university demands (Pirot & De Ketele, 2000). To understand what motivates students, there are a whole range of psychological and social factors that determine the extent to which students engage with their education. At the core is intrinsic drive - learners are more likely to stick with it when they genuinely love their classes and find learning personally fulfilling. That lines up with this theory from Ryan and Deci in 2020 about how independence, skill building, and connections nurture natural inspiration. Closely tied to motivation is feeling welcomed in the academic scene. Having positive relationships with classmates, teachers, and the school overall can majorly boost commitment. Strayhorn's (2012) work pointed out how crucial creating an inclusive environment is, especially for underrepresented groups. Belief in one's abilities, or self-efficacy, also massively affects dedication. Research building on Bandura's theories shows how views of potential shape involvement and grit. Students with higher self-efficacy aim higher and push through challenges better than their peers (Schunk & DiBenedetto, 2020).

Goals are indispensable, are they focused on learning for its own sake or demonstrating being better than others? Learners steering by self-improvement tend to dive deeper into their work and stick with it longer term. Senko et al. (2011) distinguish between mastery goals (focused on learning and improvement) and performance goals (focused on demonstrating ability relative to others). Students who adopt mastery goals often show deeper engagement and commitment to their studies.

Social circumstances surrounding each student impact commitment levels a lot as well. Encouragement from home, group-class, and teachers supplies motivation and means for sustained participation, especially important for first-gen college learners. Roksa and Kinsley's (2019) work emphasizes the importance of social support, particularly for first-generation college students. The way students see themselves as learners, or their academic identity, also influences dedication. Was et al. (2009) explored this dynamic and how it influences choices and performances. Those with more confidence hold steady even when the going gets tough. Students who strongly identify as capable learners are more likely to remain committed to their studies, even when faced with challenges. Perceptions of education's usefulness also significantly shape motivation and commitment. Eccles and Wigfield (2020) addressed how seeing relevancy to goals encourages learners to stay focused. They found when students see clear connections between their studies and future goals, they are more likely to remain committed.

Cultural factors add another layer of complexity to student commitment. Guiffrida et al. (2013) explores how cultural background and values can influence academic motivation and engagement. Understanding and accommodating diverse cultural perspectives is crucial for fostering commitment among all students.

Finally, the overall institutional climate significantly impacts student commitment. Museus et al. (2017) emphasize the importance of creating a supportive and inclusive campus environment that promotes student success and engagement.

Now it's clear how different factors influence learner engagement. On the other hand, there is a type of technology born to influence the learner, this double-edged technology has the potential to change the way learners engage with e-learning. This double-edged technology has the potential to change the way learners engage with e-learning. It brings enormous potential to the online session, but it also poses many challenges. In the following section, the authors address the issue of artificial intelligence in education.

ARTIFICIAL INTELLIGENCE: THE NEW FRONTIER OF EDUCATIONAL INNOVATION

Beyond Traditional Learning Boundaries

Artificial Intelligence (AI) in education refers to the integration of AI technologies and applications into teaching and learning processes. The utilization of AI in education has gained much attention due to its potential to transform traditional educational practices and enhance learning outcomes. In particular, the technological field of AI has made significant strides with a range of educational applications including personalized learning, adaptive assessment, and intelligent coaching systems (Zawacki-Richter et al., 2019). Within educational technology, Artificial Intelligence in Education (AIEd), is emerging as a promising field with new opportunities for enriching the educational experience. Deep integration between artificial intelligence and education has led to innovative tools and approaches that are remaking the education landscape. Educational models have changed because of AI-based technologies, leading to such things as smart teaching systems or smart classrooms (Zhu, 2024). By harnessing AI, educational institutions can make their teaching methods more effective, improve academic performance and offer learners personalized, autonomous learning experiences (Wang et al., 2023). The application of artificial intelligence in education aims to promote learners' cognitive abilities, encourage innovation, improve resource utilization, and support the overall development of education (Evenddy, 2024). One of the major benefits of integrating AI applications into educational systems is an improvement in the level of learning. AI technologies could provide individualized educational experiences that meet students' personal needs, resulting in increased participation and better knowledge retention (Alhumaid et al., 2023). In addition, AI support tools could assist teachers in their academic as well as administrative work; this strengthens the management of education to increase its optimal efficiency (Ahmad et al., 2022). By including AI technology in the education sector, it would be possible for learners to master knowledge much more precisely and receive training that would enable them to improve their specific skills (Hua, 2022). In addition, the use of AI in education goes beyond conventional classrooms and extends to specialized areas such as teaching music. Technology-based intelligent educational design with AI can facilitate the integration of IT and music education thus enhancing the learning experience in this field (Dai, 2021). Similarly, in civic education teaching, developing precise teaching methods using AI can address the relevance of civic education and improve the quality and efficiency of teaching practices. The technologies of AI provide an opportunity for exploring new ways of reforming psychology and civic education teaching in universities by deploying technological innovations to enhance teaching methods and learner engagement (Han & Gong, 2022). Ultimately, responsibly using AI's abilities for assessment and content delivery seems poised to benefit both educators and learners. Its insights could free up teachers to focus more on interactive, project-based learning too. As long as it's implemented thoughtfully with student development in mind, AI may end up elevating education overall.

ARTIFICIAL INTELLIGENCE APPLICATIONS FOR PERSONALIZED LEARNING

Intelligent tutoring systems

Intelligent Tutoring Systems (ITS) represent a significant advance in the use of artificial intelligence in education. These systems have been developed to provide tailored and adaptive teaching, mimicking the approach of a human tutor. They use artificial intelligence methods to analyze learners' responses, identify their strengths and weaknesses, and adjust the content and pace of learning accordingly. According to research by VanLehn et al (2019), they have developed an intelligent orchestration system called FACT. As learners solve difficulties in small groups, it advises the teacher which groups to visit and what to say. Based on the results of more than 50 iterative development trials, it appears that FACT did not lead to improvements in either collaboration or productivity. However, with a few tweaks to FACT, it has managed to increase wrestling productivity, but increasing collaboration remains an unresolved issue. In the context of educational recommender systems, one study suggested a new measure of fairness that ensures equal learning opportunities among learners, respecting a new set of educational principles as well as they suggested a post-processing method that reorganizes the recommended courses initially returned by an algorithm to optimize equal learning opportunities while preserving personalization (Marras et al., 2022). This study aims to optimize recommendation systems to improve the quality and equity of online education. This could have a significant impact on the learning experience of millions of users of online educational platforms, which could have consequences for learner engagement.

An important change in ITS is the increasing adoption of machine learning to dynamically adjust content. According to Kulik and Fletcher (2016), modern information and communication technologies can assess learners' performance in real-time and immediately adjust the difficulty and type of content presented, thus maintaining optimal levels of engagement. In this meta-analysis, in 46 (or 92%) of the 50 controlled evaluations, learners who received intelligent tutoring performed better than learners in traditional classes, and the performance improvement was significant enough to be considered meaningful in 39 (or 78%) of the 50 studies. Across the 50 studies, the median SE was 0.66, which represents a moderate to significant effect for social science studies. This corresponds approximately to an improvement in test performance from the 50th to the 75th percentile.

Thanks to advances in artificial emotional intelligence, ITS was able to respond more effectively to learners' emotional states. The Philippines used Physics Playground to study the positive or negative links between various affective states and learning, such as frustration, the affective state that expresses annoyance and dissatisfaction. After using a standardized system for coding learners' emotions during play and a task (puzzle) different from the one described in the main text, Banawan, Rodrigo, and Andres (2015) observed that, after engaged concentration (79% of the affective states observed), frustration (8%) was the second most experienced state among the eight monitored and the only one to have a statistically significant relationship with achievement - a negative relationship. According to the studies, frustration does not manifest itself in the same way as confusion, which can be positively associated with success once it has been resolved. It also highlights the context of engagement, with Filipino learners not experiencing the same level of enjoyment as American learners using Physics Playground (D'Mello, 2021). Modern ITS has incorporated personalization based on large-scale learning analytics. According to Holmes et al (2022), the use of predictive analytics allows ITS to predict learners' needs and suggest preventive measures.

ITS is one of the advances in Information and Communication Technology (ICT) used in the learning process, and gamification can overcome monotony, lack of interest or motivation, as well as monotony when using ITS. In this study, the use of STI with Gamification is referred to as STI + G. Many previous researches have associated STI with Gamification. For example, Pereira et al (2019) applied STI + G in medical education, demonstrating that game elements such as levels, progress bars, avatars, and leaderboards can be employed in this approach. Based on their evaluation, Pereira and colleagues (2019) demonstrated that ITS + G facilitates interaction between the user and the system. According to Tan and Cheah (2021), ITS + G has been shown to increase learner motivation and engagement. ITS + G also offers learners a positive learning experience.

In summary, ITS continues to evolve rapidly, offering more and more possibilities for improving learner engagement in online environments. Their ability to offer personalized, adaptable, and emotionally intelligent education positions them as valuable tools for education in the future.

Adaptive learning paths

The role of adaptive learning paths is crucial in improving learner engagement by tailoring educational experiences to individual needs and preferences. According to Gligorea (2023), these tailored approaches improve learning performance, increase learner engagement, and promote long-term learning in a flexible and accessible way. Adaptive learning systems, by offering tailor-made learning elements based on learners' specific needs, encourage engagement by providing content that matches learners' interests and skills (Liu et al., 2019). In addition, adaptive learning contexts are naturally conducive to learner motivation, which has a beneficial effect on their level of engagement (Papamitsiou et al., 2020). According to Cheung et al. (2021), these systems offer a wide range of media, real mentoring, and easy access to information, which encourages higher engagement, access to knowledge, experience sharing, and supervision. According to Tan (2024), adaptive learning focuses on content according to learners' skills and weaknesses, emphasizing areas that need improvement while reducing the time spent on mastered concepts, and encouraging engagement and motivation. In revolution, adaptive learning systems are constantly searching for the best learning approaches that match learners' preferences, knowledge levels, and historical data, leading to improved performance and learner satisfaction (Shou et al., 2020). By gathering information about learners' past knowledge and academic performance, adaptive systems anticipate and adapt learning paths to ensure equitable education and encourage engagement (Palanisamy et al., 2021). According to Wang et al (2021), the tailored services offered by adaptive learning pathways, which take into account learners' knowledge, context, and preferences, encourage greater participation through self-control and improved academic performance.

In a study carried out by El-Sabagh (2021) to examine the influence of an online learning environment, they were adapted to learners' learning styles on their engagement. He used a quasi-experimental design with 118 learners divided into an experimental group (60 learners) and a control group (58 learners). Key findings include: Engagement was significantly improved in the experimental group compared to the control group. Engagement was assessed in terms of skills, participation, results, and emotional involvement. The results showed a statistically significant improvement in engagement levels in the experimental group.

Adaptive learner performance recommendation systems (SPR) are being developed through reinforcement learning (RL) in the academic field. According to another research by Amine et al (2023), a reinforcement learning agent acquires the ability to improve recommendations based on the user's actions and feedback. The suggested framework develops tailored advice by taking into account learners' goals, preferences, and academic context. It uses Q-learning to propose learning programs that take into account various elements such as sequential behavior, learning style, activities, materials, levels of difficulty, feedback, preferences, skills, and level of knowledge. In terms of average success rate and lowest average turnaround time, Q-learning with 500 iterations was more effective than Q-learning with fewer iterations. With this method, learning experiences can be tailored to the needs and characteristics of each learner, which significantly improves learner engagement and performance compared with traditional methods.

In summary, adaptive learning methods play a crucial role in improving learner engagement and outcomes. There is strong evidence from the studies presented that these systems, by adjusting content and learning methods according to individual needs, provide a tailored and effective educational experience. The analysis conducted by El-Sabagh (2021) provides strong empirical evidence of the effectiveness of these methods, demonstrating a significant improvement in engagement within the experimental group. In addition, research by Amine et al (2023) into reinforcement learning-based learner performance recommender systems offers promising new opportunities to further develop these adaptive approaches. The growing importance of adaptive learning in the contemporary educational landscape is highlighted by these results.

ARTIFICIAL INTELLIGENCE AS A CATALYST FOR INTERACTIVE, EVALUATIVE, AND COLLABORATIVE LEARNING

Artificial intelligence (AI) is revolutionizing the educational landscape in profound and multidimensional ways. This transformation revolves around several fundamental aspects that are redefining the learning experience. Firstly, the personalization of learning has improved dramatically, thanks to adaptive algorithms that analyze each student's performance, learning style and rate of progress in real time. These intelligent systems automatically adjust the difficulty of exercises, suggest relevant complementary resources and create tailor-made learning paths. Traditional assessments are also transformed by AI, which now enables detailed analyses of student responses, identifying precisely the areas requiring attention and providing instant, constructive feedback. This speed of feedback enables learners to adjust their study methods in real time and progress more effectively. AI also fosters collaborative learning by creating sophisticated virtual environments where students can interact, share knowledge and work on joint projects, overcoming traditional geographical barriers. Teachers also benefit from these technological advances, having powerful analysis tools at their disposal to monitor their students' progress and quickly identify those who require special attention.

What's more, AI makes it possible to automate certain repetitive administrative tasks, freeing up time for the human interactions essential to learning. AI technologies are also introducing new forms of engagement through virtual and augmented reality, educational chatbots and interactive simulations, making learning more immersive and stimulating. This digital revolution in education is not limited to the mere digitization of content, but represents a fundamental transformation in the way an education is conceived and delivered, paving the way for a more accessible, equitable and effective education for all. In the next section, the authors will examine how AI plays a crucial role in these three fundamental aspects, offering innovative solutions that redefine the traditional educational experience.

Conversational Bot: Catalyst for Interaction

The rapid progress of chatbot technology is a striking example of this (Adamopoulou & Moussiades, 2020). Chatbots, also known as conversational agents, allow people to interact with computers through natural language, using natural language processing (NLP) technology (Bradeško & Mladenić, 2012). The use of educational chatbots is increasing due to their ability to offer an affordable method to engage learners and provide a personalized learning experience (Benotti et al., 2018). Chatbots interact using elements such as text, speech, images, gestures, and other means of communication to help learners' complete educational tasks (Kuhail et al., 2022). When interacting with learners, chatbots have assumed different functions such as teaching agents, peer agents, educational agents, and motivational agents (Chhibber & Law, 2019). Learning agents play the role of human teachers and can give instructions, illustrate examples, ask questions (Wambsganss et al., 2020), and give immediate feedback (Kulik & Fletcher, 2016). They can increase learner engagement and motivation, improve learning outcomes, and increase teacher support (Srinivasan, 2024). Generative artificial intelligence (AI) technology can help transform educational activities and interactively guide both learners and teachers (Ilieva, 2023). Focusing on the psychological aspects, the literature suggests frequent and satisfying chatbot interactions can enhance learners' effectiveness and engagement (Parsakia, 2023). The use of ChatGPT promotes learner involvement and interaction, curiosity, and motivation, and improves the learning experience (Iyamuremye & Ndihokubwayo, 2024). Chatbots can reduce shyness among language learners and reduce the perceived distance between learners and teachers (Kohnke et al., 2023). After six months of using a chatbot in secondary school, learners expressed greater confidence in using English, improved listening skills, and increased interest in language learning (Yang et al., 2022).

Regarding their application, Cunningham-Nelson et al. (2019) identified two main applications: answering frequently asked questions (FAQs) and making short quizzes, while Wollny et al. (2021) mentioned three other applications, such as scaffolding, activity recommendations, and informing about activities. The agent using this method is less aware than the agent teaching. However, peer agents still can guide learners through the learning process. Learners usually engage in conversation with peer agents to look up certain definitions or ask for an explanation on a specific topic (Kuhail et al., 2023).

In summary, educational chatbots represent a major advance in the field of technology-enhanced learning. Their ability to provide personalized interactions, encourage learner engagement, and offer ongoing support make them invaluable tools in today's education system. Their use indeed raises questions about how to reconcile human interaction and artificial assistance in education. However, the promising results in terms of improving language skills, motivation, and learner confidence suggest considerable potential for integrating chatbots into teaching strategies in the future.

Artificial intelligence: evaluation and feedback catalyst

In higher education systems, during each teaching period, learners enrolled in different programs work on different activities (in class and out of class) to demonstrate their knowledge and skills through a variety of assignments to meet the needs of diverse learners (OECD, 2002, 2003). Different assessments are useful for evaluating learners' transfer of learning skills in a variety of situations. This performance assessment makes it possible to evaluate the results achieved by learners about the learning objectives of the course and to develop their self-confidence, intrinsic motivation, etc. (Trautwein & Köller, 2003), providing accurate feedback and adjusting teaching methods as well as the quality of the course (Parker, 2013). The results obtained by learners in each course confirm that they have built up an adequate knowledge base and mastered the skills expected as learning outcomes of the program (Chaudhry et al., 2023). The Personalised Assessment System for Adaptation and Learning (PASS) (Gouli et al., 2003) assesses a learner's current knowledge and identifies areas of knowledge that the learner may not have covered. Some of the main contributions of this system are a) the detection of the learner's prior knowledge; b) the detection of concepts unknown to the learner; and c) the detection of changes in the learner's overall apprehension of the topic (Jain et al., 2014). According to Narciss et al. (2014), one of the areas of great educational importance about computer technologies is feedback on personalized tutoring. They studied an intelligent online learning environment for mathematics, which was designed to assess tasks and monitor learners' progress. According to their results, the effectiveness of feedback differed according to gender; in particular, women benefited more from tutorial feedback, especially when it was accompanied by conceptual advice. In addition, girls showed a more significant improvement in their perceived competence than boys, while boys showed an increase in their natural motivation.

Jani et al (2020) illustrate the effectiveness of AI in providing feedback, assessments, and formative evaluations through the application of machine learning and checklists. Their research indicated that automated responses are effective methods of monitoring learner progress and identifying areas for improvement in clinical practice. The study by Ouguengay et al (2015) assesses reading and writing skills in Amazigh using tests. The system establishes a score based on the answers provided to the items.

In this study by Choi, Y., & McClenen, (2020), they examined the development and evaluation of an adaptive formative assessment system, CAFT (Computerized Formative Adaptive Testing), which combines computerized adaptive testing (CAT) and dynamic Bayesian networks (DBN) in an e-learning environment. CAFT aims to offer tailor-made assessments by dynamically selecting items adapted to the learner's skills while assessing his or her progress over time. A simulation study (n=1000) and an application study (n=120) were carried out to assess the effectiveness of the system in an introductory statistics course. The results highlight the CAFT's ability to provide adaptive assessment and tailored diagnostic feedback on learning progress.

This study by Shute & Rahimi, (2017) presents a comprehensive review of Computer-Based Assessment for Learning (CBAfL) in primary and secondary education. Shute and Rahimi examine recent developments and emerging trends in the field, focusing on systems that provide immediate and personalized feedback to enhance learning. The authors analyze various types of CBAfL, including adaptive assessments, interactive simulations, and educational games. They stress the importance of integrating these tools into the learning process rather than using them solely as a means of measurement. The study highlights the potential benefits of CBAfL, such as increased learner engagement, personalization of learning, and real-time data collection on learner progress. However, the authors also identify challenges, particularly in terms of instructional design and data interpretation. This review provides a solid basis for understanding the current state and prospects of computer-assisted assessment in education, offering valuable context for research into adaptive formative assessment systems such as CAFT.

To sum up, CAFT-type adaptive formative assessment systems represent a significant advance in education by offering bespoke assessments and accurate feedback. However, it is essential to consider their design and the interpretation of the data to ensure effective integration into educational practice (Dickinson, 2011). These tools, which are an integral part of fLLBAs, enhance learner engagement and improve monitoring. However, it is essential to take into consideration the construction of these tools and the analysis of the information collected to integrate them effectively into educational activities. Immediate feedback is a core advantage, addressing student inquiries swiftly to reinforce comprehension and drive material retention. By cultivating participation, AI keeps learners invested in their scholarly pursuits even when physical presence with peers and mentors is impeded. Although no substitute for unmediated human contact, these virtual systems meaningfully enhance distance learning by replicating key functions of communal learning. Their utility in engaging remote pupils warrants further refinement and integration into pedagogical models.

Artificial intelligence: a collaboration catalyst

The work of Santos and Boticario, (2024) is particularly notable for the integration of AI to enhance collaborative learning experiences. They introduce a Collaborative Logical Framework (CLF) that leverages AI to foster interaction, dialogue, and teamwork as effective educational strategies. In addition, this intelligent assistance system helps monitor learners' behavior, thereby reducing the workload of teachers. Designed for an online learning platform (dotLRN), it serves as an adaptive guidance system that facilitates the management and supervision of learners' collaboration. AI can help foster collaboration by using teaching models that integrate AI tools into traditional teaching methods. The "tutor + AI teacher" model is an example of how artificial intelligence can support human pedagogical knowledge to create more personalized learning situations. Each student learns both data-driven insights from AI and classroom exercises created by their teachers. (Wei, 2024). This combined approach also promotes learning as a community effort between students and teachers. Integrating AI into online education holds promise for maintaining learner engagement in the context of limited real-time interpersonal interactions. According to Long and Magerko (2020), AI is viewed as a set of skills that enable individuals to thoroughly evaluate AI technologies and use them for more effective communication and collaboration.

As Almusaed's (2023) research shows, AI-based systems can effectively foster collaboration, resource sharing, and timely feedback among students and between students and teachers. AI also has the potential to cultivate important foundations for successful collaboration. Successful group work relies on elements such as trust and cooperation that develop between participants. The research explored by Kim (2024) highlights how understanding dynamics such as an individual's expertise and confidence level impacts team dynamics between students and AI assistants. Educators who design courses with these "team mental models" (TMS) in mind can optimize the collaborative environment. By guiding tasks that foster understanding of peer talents and building confidence in each other's abilities, AI platforms can help strengthen interpersonal connections, even in remote environments. AI also encourages Students can engage in critical and creative thinking through collaborative tasks. For example, students working with AI can enhance their learning by explaining AI-generated suggestions and integrating them with their <ideas, which promotes deeper understanding and innovative thinking. Kim and Lee (2022). This process not only contributes to knowledge retention but also cultivates a spirit of collaboration as students learn to articulate their thoughts and build on the contributions of others.

While current AI teaching methods can foster students' inquiry and understanding through simulations and games, they cannot adequately prepare students for high-level cognitive tasks (Casal-Otero et al., 2023). To foster students' high levels of knowledge, teachers have adopted hands-on, practical, and creative experiences that bridge the gap between AI concepts and real-world applications, improving their critical thinking, problem-solving, and collaboration skills, to foster their AI learning and preparing them for an AI-driven society (Chiu et al., 2023).

ETHICAL OF ARTIFICIAL INTELLIGENCE

Ethics is about being able to objectively reflect on moral values and guide our actions according to these values (Churchill, 1999). Ethics of artificial intelligence is part of the ethics of advanced technology, which focuses on robots and other intelligent artificial agents. It can be divided into robotics (robot ethics) and machine ethics (Siau & Wang,2020). Regarding the ethical uses of artificial intelligence, UNESCO is working on creating a global framework to guide the uses and applications of artificial intelligence to ensure the ethical use of this new technology. Generative AI, such as ChatGPT, can be used by companies in various fields, such as marketing and sales, operations, IT/engineering, risk and law, human resources, audit and finance, and utility/employee optimization (Chui et al., 2022). Companies ensure that none of their information or trade secrets are shared with ChatGPT, as they could be disclosed publicly. Mechanisms could be used to protect proprietary information, such as disabling conversation history so that it is not used to train ChatGPT models (https://openai.com/blog/new-ways-to-manage-your-data-in-chatgpt) or prohibiting the use of ChatGPT altogether. Due to concerns about privacy and information security, a growing number of bans or restrictions on ChatGPT have been imposed by organizations and countries (Ray, 2023; Stancati et al, Ray, 2023). The ethical impact of artificial intelligence (AI) in education is complex and increasingly important as AI technologies become more integrated into teaching and learning environments. This integration raises critical questions about fairness, accountability, transparency, bias, and the overall impact on educational equality.

One of the primary ethical concerns surrounding AI in education is the potential for bias in AI algorithms. As noted by, discussions among academics highlight the necessity of scrutinizing AI language models for biases that could adversely affect students' learning experiences (Alrayes, 2024). This concern is echoed by, those who emphasize the importance of incorporating ethical considerations into AI education to prepare future developers and users of AI technologies (Knowles, 2021). Furthermore, the ethical use of AI in education necessitates a focus on privacy and data security. As highlighted by, the rapid evolution of AI technologies poses challenges in predicting their future capabilities, making it crucial for stakeholders to engage in ongoing discussions about potential ethical dilemmas (Klímová et al., 2023). This is particularly important in K-12 settings, where argue that the ethical challenges of AI must be explicitly addressed to ensure that students and teachers understand the implications of AI technologies (Akgün & Greenhow, 2021).

While technology can be helpful in the classroom, teachers need to be careful how they implement it. AI should be used to augment what teachers already do best -namely, showing empathy, wisdom, and caring for each student. Learners still need that human connection to learn and grow. Research backs this up. One study found that simply replacing teachers with AI isn't the answer and may do more harm than good (Pedró, 2020). Students learn best when they feel in control of their learning, not when it's controlled for them (Jin et al., 2023). If we integrate AI thoughtfully, seeing it as a tool rather than a solution, it could help educators even more. But balanced with the well-being and autonomy of learners. Technology is meant to serve human needs, not the other way around. When handled right, it can be a useful aid. But it takes wise teachers to use it wisely.

The development of ethical AI systems faces significant challenges without comprehensive and unbiased ethical standards. A crucial question arises: How can humans train machines to be ethical without such standards? Additionally, there's uncertainty about whether intelligent machines can understand ethical standards in the same way humans do (Wang and Siau, 2019). For example, programming robots to "not harm" requires first defining what constitutes harm - a complex ethical issue in itself (Siau & Wang,2020). This underscores the need for universal ethical standards applicable on a global scale (Wang and Siau, 2018).

Implementing these ethics in AI systems requires reducing the knowledge gap between AI programmers and creators of ethical standards. Researchers and practitioners in the field should strive to deepen their understanding of established ethical principles. This enhanced knowledge can then inform their work and enable them to train others in developing ethically sound AI systems. It's also important to note that the use of AI in education raises important ethical issues. Those who create these systems, and learners all need to work together to ensure that AI is used safely and helps all learners equally. The confidentiality of personal data and the potential biases of the technology are issues that the research highlights. Through open dialogue and cooperation, these groups can address concerns and implement AI solutions responsibly. Developing AI for classrooms requires the input and expertise of everyone involved. Finding solutions as a team is the only way to create a learning environment that makes the most of technology while protecting learners' well-being. These types of tools are meant to benefit all learners, so their introduction requires a collaborative approach based on fairness and respect for ethical standards. It is only through compassion and compromise between stakeholders that we can ensure that adequate precautions are in place.

FUTURE DIRECTION

The integration of artificial intelligence (AI) into educational frameworks has emerged as a transformative force, particularly in fostering interactive, evaluative, and collaborative learning environments. This evolution is characterized by the development of intelligent systems that enhance personalized learning experiences and facilitate collaborative interactions among students and educators.

AI's role in adaptive learning is pivotal, as it enables the creation of personalized educational experiences tailored to individual student needs. Akavova highlights that AI systems can analyze data and provide personalized feedback, thereby enhancing the learning process (Akavova, 2023). This adaptability is further supported by the findings of Ouyang et al., who demonstrate that AI can integrate performance prediction and learning analytics to improve collaborative learning outcomes in online courses (Ouyang et al., 2023). Such systems can identify group dynamics and individual contributions, allowing for real-time adjustments that enhance the collaborative learning experience.

Moreover, AI's capacity for evaluation and impact analysis is crucial in educational settings. Bhatt and Muduli emphasize that AI-enabled learning management systems can track learners' progress and provide insights into their understanding of subject content, thus facilitating ongoing evaluation (Bhatt & Muduli, 2022). This continuous assessment is vital for both students and educators, as it informs instructional strategies and helps refine educational programs to better meet learning objectives.

The collaborative potential of AI is further exemplified in various educational contexts. For instance, Qian discusses how AI can serve as a dynamic learning partner in music education, providing individualized recommendations and facilitating collaborative projects (Qian, 2023). Similarly, Kim et al. argue that AI can enhance student experiences by offering tailored learning opportunities and supporting cognitive and affective development through interactive agents (Kim et al., 2022). This collaborative aspect spans various disciplines, as evidenced by the systematic mapping study conducted by Suarez et al., which illustrates the integration of AI in computer programming education to enhance collaborative learning (Suarez et al., 2023).

The future direction of AI in education also points towards the democratization of learning through large-scale distributed systems. Nguyen et al. propose a shift from centralized to decentralized learning frameworks, which can leverage AI to perform complex tasks collaboratively (Nguyen et al., 2023). This democratized approach aligns with the principles of self-organizing learning environments, where students can engage in collaborative problem-solving and knowledge construction.

However, the implementation of AI in educational settings is not without challenges. Ethical considerations, accessibility, and the need for ongoing dialogue among stakeholders are critical for maximizing the benefits of AI in education (Namjoo, 2023). As highlighted by Järvelä et al., the collaboration between human and AI can enhance socially shared regulation in learning, which is essential for effective educational practices (Järvelä et al., 2023). This collaboration necessitates careful design of interfaces and learning environments that promote meaningful interactions between students and AI systems (al., 2023).

In conclusion, the future of AI as a catalyst for interactive, evaluative, and collaborative learning is promising. By harnessing the capabilities of AI to personalize learning, facilitate collaboration, and provide continuous evaluation, educators can create more engaging and effective learning environments. As the field continues to evolve, it will be essential to address the ethical and practical challenges associated with AI integration to ensure that it serves the diverse needs of learners.

CONCLUSION

In this brief, the authors explored the transformative role of artificial intelligence (AI) in improving learner engagement across all educational contexts. Here a look has been taken at how AI technologies, such as intelligent tutoring systems and adaptive learning pathways, offer personalised learning experiences that meet individual needs, thereby improving engagement and educational outcomes. The potential of AI as a catalyst for interactive, assessment and collaborative learning has been demonstrated through the use of conversational bots and AI-driven feedback systems. These innovations not only encourage active participation, but also support educators by increasing their ability to inspire and guide learners. However, the integration of AI into education also raises ethical considerations, highlighting the need to balance technological advances with preserving the human element in teaching. As we progress, it is essential to implement AI thoughtfully in educational environments in order to maximise its benefits while addressing the challenges it presents. By doing so, a more inclusive and effective learning experience can be created that prepares learners for the future.

REFERENCES

Adamopoulou, E., & Moussiades, L. (2020). Chatbots: History, technology, and applications. *Machine Learning with Applications*, 2, 100006. DOI: 10.1016/j.mlwa.2020.100006

Ahmad, S. F., Alam, M. M., Rahmat, M. K., Mubarik, M. S., & Hyder, S. I. (2022). Academic and administrative role of artificial intelligence in education. *Sustainability (Basel)*, 14(3), 1101. DOI: 10.3390/su14031101

Akavova, A., Temirkhanova, Z., & Lorsanova, Z. (2023). Adaptive learning and artificial intelligence in the educational space. E3S Web of Conferences, 451, 06011. DOI: 10.1051/e3sconf/202345106011

Akgün, S., & Greenhow, C. (2021). Artificial intelligence in education: Addressing ethical challenges in k-12 settings. *AI and Ethics*, 2(3), 431–440. DOI: 10.1007/s43681-021-00096-7 PMID: 34790956

Alhumaid, K., Naqbi, S. A., Elsori, D., & Mansoori, M. A. (2023). The adoption of artificial intelligence applications in education. *International Journal of Data and Network Science*, 7(1), 457–466. DOI: 10.5267/j.ijdns.2022.8.013

Almusaed, A., Almssad, A., Yitmen, I., & Homod, R. Z. (2023). Enhancing student engagement: Harnessing "aied"'s power in hybrid education—a review analysis. *Education Sciences*, 13(7), 632. DOI: 10.3390/educsci13070632

Alrayes, A., Henari, T. F., & Ahmed, D. A. (2024). Chatgpt in education – understanding the bahraini academics perspective. *Electronic Journal of e-Learning*, 22(2), 112–134. DOI: 10.34190/ejel.22.2.3250

Amin, S., Uddin, M. I., Alarood, A. A., Mashwani, W. K., Alzahrani, A., & Alzahrani, A. O. (2023). Smart E-learning framework for personalized adaptive learning and sequential path recommendations using reinforcement learning. *IEEE Access : Practical Innovations, Open Solutions*, 11, 89769–89790. DOI: 10.1109/ACCESS.2023.3305584

Arnone, M. P., Small, R. V., Chauncey, S. A., & McKenna, H. P. (2011). Curiosity, interest and engagement in technology-pervasive learning environments: A new research agenda. *Educational Technology Research and Development*, 59(2), 181–198. DOI: 10.1007/s11423-011-9190-9

Banawan, M., Rodrigo, M., & Andres, J. (2015), An investigation of frustration among students using physics playground. In Doctoral Student Consortium (DSC), Proceedings of the 23rd International Conference on Computers in Education

Barkoukis, V., Taylor, I., Chanal, J., & Ntoumanis, N. (2014). The Relation between Student Motivation and Student Grades in Physical Education: A 3-year Investigation. *Scandinavian Journal of Medicine & Science in Sports*, 24(5), 414. DOI: 10.1111/sms.12174 PMID: 24433528

Benotti, L., Martnez, M. C., & Schapachnik, F. (2018). Atool for introducing computer science with automatic formative assessment. *IEEE Transactions on Learning Technologies*, 11(2), 179–192. DOI: 10.1109/TLT.2017.2682084

Bhatt, P. K., & Muduli, A. (2022). Artificial intelligence in learning and development: A systematic literature review. *European Journal of Training and Development*, 47(7/8), 677–694. DOI: 10.1108/EJTD-09-2021-0143

Bond, M., Buntins, K., Bedenlier, S., Zawacki-Richter, O., & Kerres, M. (2020). Mapping research in student engagement and educational technology in higher education: A systematic evidence map. *International Journal of Educational Technology in Higher Education*, 17(1), 1–30. DOI: 10.1186/s41239-019-0176-8

Bradeško, L., & Mladenić, D. (2012). A survey of chatbot systemsthrough a loebner prize competition. In *Proceedings of Slovenian language technologies society eighth conference of language technologies* (pp. 34–37).

Casal-Otero, L., Catala, A., Fernández-Morante, C., Taboada, M., Cebreiro, B., & Barro, S. (2023). AI literacy in K-12: A systematic literature review. *International Journal of STEM Education*, 10(1), 29. DOI: 10.1186/s40594-023-00418-7

Chassignol, M., Khoroshavin, A., Klimova, A., & Bilyatdinova, A. (2018). Artificial Intelligence trends in education: A narrative overview. *Procedia Computer Science*, 136, 16–24. DOI: 10.1016/j.procs.2018.08.233

Chaudhry, I. S., Sarwary, S. A. M., El Refae, G. A., & Chabchoub, H. (2023). Time to Revisit Existing Student's Performance Evaluation Approach in Higher Education Sector in a New Era of ChatGPT — A Case Study. *Cogent Education*, 10(1), 2210461. Advance online publication. DOI: 10.1080/2331186X.2023.2210461

Cheung, S. T., Kwok, L. F., Phusavat, K., & Yang, H. H. (2021). Shaping the future learning environments with smart elements: Challenges and opportunities. *International Journal of Educational Technology in Higher Education*, 18(1), 16. Advance online publication. DOI: 10.1186/s41239-021-00254-1 PMID: 34778521

Chhibber, N., & Law, E. (2019). Using conversational agents to support learning by teaching. arXiv:1909. 13443

Chiu, T. K., Moorhouse, B. L., Chai, C. S., & Ismailov, M. (2023). Teacher support and student motivation to learn with Artificial Intelligence (AI) based chatbot. *Interactive Learning Environments*, •••, 1–17. DOI: 10.1080/10494820.2023.2172044

Choi, Y., & McClenen, C. (2020). Development of adaptive formative assessment system using computerized adaptive testing and dynamic bayesian networks. *Applied Sciences (Basel, Switzerland)*, 10(22), 8196. DOI: 10.3390/app10228196

Chui, M., Roberts, R., & Yee, L. (2022, December 20). Generative AI is Here: How Tools Like ChatGpt Could Change Your Business. Quantum Black AI by McKinsey. https://www. mckinsey.com/capabilities/quantumblack/our-insights/generative-ai-is -here-how-tools-like -chatgpt-could-change-your-business#/

Churchill, L. R. (1999). Are We Professionals? A Critical Look at the Social Role of Bioethicists. *Daedalus*, •••, 253–274. PMID: 11645877

Cunningham-Nelson, S., Baktashmotlagh, M., & Boles, W. (2019). Visualizing student opinionthroughtext analysis. *IEEE Transactions on Education*, 62(4), 305–311. DOI: 10.1109/TE.2019.2924385

D'Mello, S. K. (2021). Improving student engagement in and with digital learning technologies. Pushing the Frontiers with Artificial Intelligence, Blockchain and Robots, 79-104.

Dai, D. (2021). Artificial intelligence technology assisted music teaching design. *Scientific Programming*, 2021, 1–10. DOI: 10.1155/2021/9141339

Eccles, J. S., & Wigfield, A. (2020). From expectancy-value theory to situated expectancy-value theory: A developmental, social cognitive, and sociocultural perspective on motivation. *Contemporary Educational Psychology*, 61, 101859. DOI: 10.1016/j.cedpsych.2020.101859

El-Sabagh, H. A. (2021). Adaptive e-learning environment based on learning styles and its impact on development students' engagement. *International Journal of Educational Technology in Higher Education*, 18(1), 53. DOI: 10.1186/s41239-021-00289-4

Et al., G. C. S. (2023). Human-ai collaboration: exploring interfaces for interactive machine learning. Tuijin Jishu/Journal of Propulsion Technology, 44(2). DOI: 10.52783/tjjpt.v44.i2.148

Evenddy, S. S. (2024). Investigating AI's Automated Feedback in English Language Learning. *Foreign Language Instruction Probe*, 3(1), 76–87. DOI: 10.54213/flip.v3i1.401

Filgona, Jacob and Sakiyo, John and Gwany, D. M. and Okoronka, A. U. (2020) Motivation in Learning. Asian Journal of Education and Social Studies, 10 (4). pp. 16-37. ISSN 2581-6268

Fredricks, J. A., Reschly, A. L., & Christenson, S. L. (Eds.). (2019). *Handbook of student engagement interventions: Working with disengaged students*. Academic Press.

Fried, R., & Fried, R. L. (2015). *The passionate teacher: A practical guide*. Beacon Press.

Frymier, A. B., & Houser, M. L. (2016). The role of oral participation in student engagement. *Communication Education*, 65(1), 83–104. DOI: 10.1080/03634523.2015.1066019

Gligorea, I., Cioca, M., Oancea, R., Gorski, A., Gorski, H., & Tudorache, P. (2023). Adaptive learning using artificial intelligence in e-learning: A literature review. *Education Sciences*, 13(12), 1216. DOI: 10.3390/educsci13121216

Gouli, E., Gogoulou, A., & Grigoriadou, M. (2003). A coherent and integrated framework using concept maps for various educational assessment functions. *Journal of Information Technology Education*, 2(1), 215–240. DOI: 10.28945/324

Guiffrida, D. A., Lynch, M. F., Wall, A. F., & Abel, D. S. (2013). Do reasons for attending college affect academic outcomes? A test of a motivational model from a self-determination theory perspective. *Journal of College Student Development*, 54(2), 121–139. DOI: 10.1353/csd.2013.0019

Han, L., & Gong, J. (2022). A study on exploring the path of psychology and civics teaching reform in universities based on artificial intelligence. *Computational Intelligence and Neuroscience*, 2022, 1–12. DOI: 10.1155/2022/4841387 PMID: 35978907

Holmes, W., Bialik, M., & Fadel, C. (2019). *Artificial intelligence in education: Promises and implications for teaching and learning*. Center for Curriculum Redesign.

Holmes, W., Porayska-Pomsta, K., Holstein, K., Sutherland, E., Baker, T., Shum, S. B., Santos, O. C., Rodrigo, M. T., Cukurova, M., Bittencourt, I. I., & Koedinger, K. R. (2022). Ethics of AI in education: Towards a community-wide framework. *International Journal of Artificial Intelligence in Education*, 32(3), 1–23. DOI: 10.1007/s40593-021-00239-1

Hua, Y. (2022). Design of online music education system based on artificial intelligence and multiuser detection algorithm. *Computational Intelligence and Neuroscience*, 2022, 1–11. DOI: 10.1155/2022/4744939 PMID: 35371210

Ilieva, G., Yankova, T., Klisarova-Belcheva, S., Dimitrov, A., Bratkov, M., & Angelov, D. (2023). Effects of generative chatbots in higher education. *Information (Basel)*, 14(9), 492. DOI: 10.3390/info14090492

Iyamuremye, A., & Ndihokubwayo, K. (2024). Exploring secondary school students' interest and mastery of atomic structure and chemical bonding through ChatGPT. *Educational Journal of Artificial Intelligence and Machine Learning*, 1(1), 1–13. DOI: 10.58197/prbl/9hk37296

Jain, G. P., Gurupur, V. P., Schroeder, J. L., & Faulkenberry, E. D. (2014). Artificial intelligence-based student learning evaluation: A concept map-based approach for analyzing a student's understanding of a topic. *IEEE Transactions on Learning Technologies*, 7(3), 267–279. DOI: 10.1109/TLT.2014.2330297

Jani, K. H., Jones, K. A., Jones, G. W., Amiel, J., Barron, B., & Elhadad, N. (2020). Machine learning to extract communication and history-taking skills in OSCE transcripts. *Medical Education*, 54(12), 1159–1170. DOI: 10.1111/medu.14347 PMID: 32776345

Järvelä, S., Nguyen, A., & Hadwin, A. F. (2023). Human and artificial intelligence collaboration for socially shared regulation in learning. *British Journal of Educational Technology*, 54(5), 1057–1076. DOI: 10.1111/bjet.13325

Jin, S., Im, K., Yoo, M., Roll, I., & Seo, K. (2023). Supporting students' self-regulated learning in online learning using artificial intelligence applications. *International Journal of Educational Technology in Higher Education*, 20(1), 37. Advance online publication. DOI: 10.1186/s41239-023-00406-5

Kim, J., Ham, Y., & Lee, S. (2024). Differences in student-ai interaction process on a drawing task: Focusing on students' attitude towards ai and the level of drawing skills. *Australasian Journal of Educational Technology*. Advance online publication. DOI: 10.14742/ajet.8859

Kim, J., Lee, H., & Cho, Y. H. (2022). Learning design to support student-ai collaboration: Perspectives of leading teachers for ai in education. *Education and Information Technologies*, 27(5), 6069–6104. DOI: 10.1007/s10639-021-10831-6

Kim, J., & Lee, S. (2022). Are two heads better than one?: The effect of student-ai collaboration on students' learning task performance. *TechTrends*, 67(2), 365–375. DOI: 10.1007/s11528-022-00788-9 PMID: 36258920

Klímová, B., Pikhart, M., & Kacetl, J. (2023). Ethical issues of the use of ai-driven mobile apps for education. *Frontiers in Public Health*, 10, 1118116. Advance online publication. DOI: 10.3389/fpubh.2022.1118116 PMID: 36711343

Knowles, M. A. (2021). Five motivating concerns for ai ethics instruction. *Proceedings of the Association for Information Science and Technology*, 58(1), 472–476. DOI: 10.1002/pra2.481

Kohnke, L., Moorhouse, B. L., & Zou, D. (2023). Chatgpt for language teaching and learning. *RELC Journal*, 54(2), 537–550. DOI: 10.1177/00336882231162868

Kuhail, M. A., Alturki, N., Alramlawi, S., & Alhejori, K. (2022). Interacting with educational chatbots: A systematic review. *Education and Information Technologies*, 28(1), 973–1018. DOI: 10.1007/s10639-022-11177-3

Kuhail, M. A., Alturki, N., Alramlawi, S., & Alhejori, K. (2023). Interacting with educational chatbots: A systematic review. *Education and Information Technologies*, 28(1), 973–1018. DOI: 10.1007/s10639-022-11177-3

Kulik, J. A., & Fletcher, J. D. (2016). Effectiveness of Intelligent Tutoring Systems: A Meta-Analytic Review. *Review of Educational Research*, 86(1), 42–78. DOI: 10.3102/0034654315581420

Kulik, J. A., & Fletcher, J. D. (2016). Effectiveness of intelligent tutoring systems: A meta-analytic review. *Review of Educational Research*, 86(1), 42–78. DOI: 10.3102/0034654315581420

Lei, H., Cui, Y., & Zhou, W. (2018). Relationships between student engagement and academic achievement: A meta-analysis. *Social Behavior and Personality*, 46(3), 517–528. DOI: 10.2224/sbp.7054

Liu, Q., Tong, S., Liu, C., Zhao, H., Chen, E., Ma, H., . . . Wang, S. (2019). Exploiting cognitive structure for adaptive learning. Proceedings of the 25th ACM SIGKDD International Conference on Knowledge Discovery &Amp; Data Mining. DOI: 10.1145/3292500.3330922

Long, D., & Magerko, B. (2020, April). What is AI literacy? Competencies and design considerations. In *Proceedings of the 2020 CHI Conference on Human Factors in Computing Systems* (pp. 1–16). New York, USA. DOI: 10.1145/3313831.3376727

Luckin, R., & Holmes, W. (2016). Intelligence unleashed: An argument for AI in education.

Marras, M., Boratto, L., Ramos, G., & Fenu, G. (2022). Equality of learning opportunity via individual fairness in personalized recommendations. *International Journal of Artificial Intelligence in Education*, 32(3), 636–684. DOI: 10.1007/s40593-021-00271-1

Museus, S. D., Yi, V., & Saelua, N. (2017). The impact of culturally engaging campus environments on sense of belonging. *Review of Higher Education*, 40(2), 187–215. DOI: 10.1353/rhe.2017.0001

Namjoo, F., Liaghat, E., Shabaziasl, S., Modabernejad, Z., & Morshedi, H. (2023). Students experience on self-study through ai. *AI and Tech in Behavioral and Social Sciences*, 1(3), 35–42. DOI: 10.61838/kman.aitech.1.3.6

Narciss, S., Sosnovsky, S., Schnaubert, L., Andrès, E., Eichelmann, A., Goguadze, G., & Melis, E. (2014). Exploring feedback and student characteristics relevant for personalizing feedback strategies. *Computers & Education*, 71, 56–76. DOI: 10.1016/j.compedu.2013.09.011

Nguyen, M. N. H., Pandey, S. R., Dang, T. N., Huh, E., Tran, N. H., Saad, W., & Hong, C. S. (2023). Self-organizing democratized learning: Toward large-scale distributed learning systems. *IEEE Transactions on Neural Networks and Learning Systems*, 34(12), 10698–10710. DOI: 10.1109/TNNLS.2022.3170872 PMID: 35536803

OECD. (2002). Understanding the Brain: Towards a New Learning Science. OECD, Pari.

OECD. (2003). Learners for Life: Student Approaches to Learning: Results from PISA 2000.

Ouariach, F. Z., Nejjari, A., Ouariach, S., & Khaldi, M. (2024). Place of forums in online communication through an LMS platform. World Journal of Advanced Engineering Technology and Sciences, 11(1), 096-104.

Ouariach, S., & Khaldi, M. (2024). Scripting Tools and the Design of E-Learning Experiences. In *Innovative Instructional Design Methods and Tools for Improved Teaching* (pp. 60–91). IGI Global.

Ouariach, S., Khaldi, M., Mohamed, E., & Khaldi, M. (2023). The Flipped Classroom: From Passive Information Absorption to Active Learning. In Handbook of Research on Language Teacher Identity (pp. 269-293). IGI Global.

Ouguengay, Y. A., El Faddouli, N. E., & Bennani, S. (2015). A neuro-fuzzy inference system for the evaluation of reading/writing competencies acquisition in an e-learning environnement. *Journal of Theoretical and Applied Information Technology*, 81(3), 600.

Ouyang, F., Wu, M., Zheng, L., Zhang, L., & Jiao, P. (2023). Integration of artificial intelligence performance prediction and learning analytics to improve student learning in online engineering course. *International Journal of Educational Technology in Higher Education*, 20(1), 4. Advance online publication. DOI: 10.1186/s41239-022-00372-4 PMID: 36683653

Palanisamy, P., Thilarajah, S., & Chen, Z. (2021). Providing equitable education through personalised adaptive learning. ASCILITE 2021: Back to the Future – ASCILITE '21 Proceedings ASCILITE 2021 in Armidale. DOI: 10.14742/ascilite2021.0129

Papamitsiou, Z., Pappas, I. O., Sharma, K., & Giannakos, M. N. (2020). Utilizing multimodal data through fsqca to explain engagement in adaptive learning. *IEEE Transactions on Learning Technologies*, 13(4), 689–703. DOI: 10.1109/TLT.2020.3020499

Parker, R. E. (Ed.). (2013). *Redesigning courses for online delivery: design, interaction, media & evaluation*. Emerald Group Publishing.

Parsakia, K. (2023). The effect of chatbots and ai on the self-efficacy, self-esteem, problem-solving and critical thinking of students. *Health Nexus*, 1(1), 71–76. DOI: 10.61838/hn.1.1.14

Pedró, F. (2020). Applications of artificial intelligence to higher education: Possibilities, evidence, and challenges. *IUL Research*, 1(1), 61–76. DOI: 10.57568/iulres.v1i1.43

Pereira, L. A., Bastos, M. C., Dermeval, D., Paiva, R., Coelho, J. A. P. M., & Lima, D. (2019). Evaluation of an Intelligent Tutoring System Prototype for the Teaching of Electrocardiogram. in *Proceedings of the IEEE 19th International Conference on Advanced Learning Technologies (ICALT)* (pp 327–328). DOI: 10.1109/ICALT.2019.00104

Petiot, O. (2023). Quels sont les facteurs de l'engagement des apprenants en EPS? Une revue de la littérature portant sur l'expérience des enseignants et des apprenants. eJRIEPS. Ejournal de la recherche sur l'intervention en éducation physique et sport, (52), 3-29.

Pirot, L., & De Ketele, J. M. (2000). L'engagement académique de l'apprenant comme facteur de réussite à l'université Étude exploratoire menée dans deux facultés contrastées. *Revue des Sciences de l'Education*, 26(2), 367–394. DOI: 10.7202/000127ar

Qian, C. (2023). Research on human-centered design in college music education to improve student experience of artificial intelligence-based information systems. *Journal of Information Systems Engineering & Management*, 8(3), 23761. DOI: 10.55267/iadt.07.13854

Ray, S. (2023, February 22). JPMorgan Chase restricts staffers' use of ChatGPT. Forbes. https:// www.forbes.com/sites/siladityaray/2023/02/22/jpmorgan-chase -restricts-staffers-use-ofchatgpt/?sh=714e937d6bc7

Reschly, A. L., & Christenson, S. L. (2012). Jingle, jangle and conceptual haziness: Evolution and future directions of the engagement construct. In Christenson, S. L., & Reschly, A. L. (Eds.), *C*.

Roksa, J., & Kinsley, P. (2019). The role of family support in facilitating academic success of low-income students. *Research in Higher Education*, 60(4), 415–436. DOI: 10.1007/s11162-018-9517-z

Ryan, R. M., & Deci, E. L. (2020). Intrinsic and extrinsic motivation from a self-determination theory perspective: Definitions, theory, practices, and future directions. *Contemporary Educational Psychology*, 61, 101860. DOI: 10.1016/j.cedpsych.2020.101860

Santos, O. C., & Boticario, J. G. (2014). Involving users to improve the collaborative logical framework. *TheScientificWorldJournal*, 2014(1), 893525. DOI: 10.1155/2014/893525 PMID: 24592196

Schunk, D. H., & DiBenedetto, M. K. (2020). Motivation and social cognitive theory. *Contemporary Educational Psychology*, 60, 101832. DOI: 10.1016/j.cedpsych.2019.101832

Senko, C., Hulleman, C. S., & Harackiewicz, J. M. (2011). Achievement goal theory at the crossroads: Old controversies, current challenges, and new directions. *Educational Psychologist*, 46(1), 26–47. DOI: 10.1080/00461520.2011.538646

Shou, Z., Lu, X., Wu, Z., Yuan, H., Zhang, H., & Lai, J. (2020). On learning path planning algorithm based on collaborative analysis of learning behavior. *IEEE Access : Practical Innovations, Open Solutions*, 8, 119863–119879. DOI: 10.1109/ACCESS.2020.3005793

Shute, V. J., & Rahimi, S. (2017). Review of computer-based assessment for learning in elementary and secondary education. *Journal of Computer Assisted Learning*, 33(1), 1–19. DOI: 10.1111/jcal.12172

Siau, K., & Wang, W. (2020). Artificial intelligence (AI) ethics: Ethics of AI and ethical AI. [JDM]. *Journal of Database Management*, 31(2), 74–87. DOI: 10.4018/JDM.2020040105

Skinner, E. (2016). Engagement and Disaffection as Central to Processes of Motivational Resilience and Development. In Wentzel, K. R., & Miele, D. B. (Eds.), *Handbook of Motivation at School*. Routledge Handbooks Online.

Skinner, E. A., Zimmer-Gembeck, M. J., Connell, J. P., Eccles, J. S., & Wellborn, J. G. (1998). Individual Differences and the Development of Perceived Control. *Monographs of the Society for Research in Child Development*, 63(2–3), 1–220. DOI: 10.2307/1166220 PMID: 9839459

Srinivasan, M., Venugopal, A., Venkatesan, L., & Kumar, R. (2024). Navigating the pedagogical landscape: Exploring the implications of ai and chatbots in nursing education. *JMIR Nursing*, 7, e52105. DOI: 10.2196/52105 PMID: 38870516

Stancati, M., & Schechner, S. (2023, March 31). ChatGPT banned in Italy over data-privacy concerns. The Wall Street Journal. Available at https://www.wsj.com/articles/chatgptbanned-in-italy-over-data-privacy-concerns-4b984e75

Strayhorn, T. L. (2012). *College Students' Sense of Belonging: A Key to Educational Success for All Students* (1st ed.). Routledge., DOI: 10.4324/9780203118924

Suarez, C. G. H., Bucheli, V., & Ordóñez, H. (2023). Artificial intelligence and computer-supported collaborative learning in programming: A systematic mapping study. *Tecnura*, 27(75), 175–206. DOI: 10.14483/22487638.19637

Tan, D. Y., & Cheah, C. W. (2021). Developing a gamified AI-enabled online learning application to improve students' perception of university physics. *Computers and Education: Artificial Intelligence*, 2, 100032. DOI: 10.1016/j.caeai.2021.100032

Tan, J. (2024). An empirical study of adaptive learning algorithm based on intrinsic motivation in english online teaching and learning. *Journal of Electrical Systems*, 20(6s), 1695–1704. DOI: 10.52783/jes.3088

Trautwein, U., & Köller, O. (2003). The Relationship Between Homework and Achievement—Still Much of a Mystery. *Educational Psychology Review*, 15(2), 115–145. DOI: 10.1023/A:1023460414243

Tutos'Me Pro. (2024). L'engagement des apprenants: clé de la réussite. Tutos'Me Pro - Trouver un Formateur ou une Mission de Formation. https://www.tutos.pro/engagement-des-apprenants-la-cle-de-la-reussite/

UNESCO. Elaboration of a Recommendation on the Ethics of Artificial Intelligence. Available online: https://en.unesco.org/

VanLehn, K., Burkhardt, H., Cheema, S., Kang, S., Pead, D., Schoenfeld, A., & Wetzel, J. (2021). Can an orchestration system increase collaborative, productive struggle in teaching-by-eliciting classrooms? *Interactive Learning Environments*, 29(6), 987–1005. DOI: 10.1080/10494820.2019.1616567

Wambsganss, T., Winkler, R., Schmid, P., & Söllner, M. (2020). Designing a conversational agent as a formative course evaluation tool

Wang, F., Zhang, L., Chen, X., Wang, Z., & Xu, X. (2021). A personalized self-learning system based on knowledge graph and differential evolution algorithm. *Concurrency and Computation*, 34(8), e6190. Advance online publication. DOI: 10.1002/cpe.6190

Wang, R., Sun, Z., & Zhou, Y. (2023). Research on the new model of "internet + education" based on artificial intelligence. *International Conference on Mathematics, Modeling, and Computer Science (MMCS2022)*. DOI: 10.1117/12.2670426

Wang, W., & Siau, K. (2018). *Ethical and moral issue with AI – a case study on healthcare robots. AMCIS 2019 proceedings*. Academic Press.

Wang, W., & Siau, K. (2019a). Artificial Intelligence, Machine Learning, Automation, Robotics, Future of Work and Future of Humanity: A Review and Research Agenda. *Journal of Database Management*, 30(1), 61–79. DOI: 10.4018/JDM.2019010104

Was, C. A., Al-Harthy, I., Stack-Oden, M., & Isaacson, R. M. (2009). Academic identity status and the relationship to achievement goal orientation. *Electronic Journal of Research in Educational Psychology*, 7(2), 627–652.

Wei, S. (2024). Research on the disruptive transformation and future development of american teaching models in the context of intelligent transformation. Advances in Economics. *Management and Political Sciences*, 72(1), 201–210. DOI: 10.54254/2754-1169/72/20240696

Wollny, S., Schneider, J., Di Mitri, D., Weidlich, J., Rittberger, M., & Drachsler, H. (2021). Are we there yet?-a systematic literature review on chatbots in education. *Frontiers in Artificial Intelligence*, 4, 4. DOI: 10.3389/frai.2021.654924 PMID: 34337392

Yang, H., Kim, H., Lee, J. H., & Shin, D. (2022). Implementation of an ai chatbot as an english conversation partner in efl speaking classes. *ReCALL*, 34(3), 327–343. DOI: 10.1017/S0958344022000039

Zahra, O. F., Amel, N., & Mohamed, K. (2024, February). Communication tools between learning activities and e-learning. In E-Learning and Smart Engineering Systems (ELSES 2023) (pp. 452-459). Atlantis Press.

Zawacki-Richter, O., Marín, V. I., Bond, M., & Gouverneur, F. (2019). Systematic review of research on artificial intelligence applications in higher education–where are the educators? *International Journal of Educational Technology in Higher Education*, 16(1), 1–27. DOI: 10.1186/s41239-019-0171-0

Zawacki-Richter, O., Marín, V. I., Bond, M., & Gouverneur, F. (2019). Systematic review of research on artificial intelligence applications in higher education – where are the educators? *International Journal of Educational Technology in Higher Education*, 16(1), 39. Advance online publication. DOI: 10.1186/s41239-019-0171-0

Zhoc, K. C., Webster, B. J., King, R. B., Li, J. C., & Chung, T. S. (2019). Higher education student engagement scale (HESES): Development and psychometric evidence. *Research in Higher Education*, 60(2), 219–244. DOI: 10.1007/s11162-018-9510-6

Zhu, X., Li, C., & Bao, M. (2024). Research on the application and future trend of artificial intelligence in education and teaching. SHS Web of Conferences, 187, 04017. https://doi.org/DOI: 10.1051/shsconf/202418704017

Chapter 7
Reinventing AI Education:
From Collaborative Learning to Real-World Projects

Safae El Gazi
https://orcid.org/0009-0002-5946-6860
Abdelmalek Essaadi University, Morocco

Ahmed Ibrahimi
https://orcid.org/0009-0004-8879-5561
Abdelmalek Essaadi University, Morocco

ABSTRACT

This chapter examines the integration of project-based learning (PBL) and active approaches in artificial intelligence (AI) education, and highlights the benefits of these methods for developing practical and theoretical skills. Unlike traditional methods, PPA encourages student-centered learning by promoting engagement, creativity, and autonomy in real-world, collaborative projects. The chapter also explores the challenges associated with the adoption of these methods, such as institutional resistance, resource management, and pedagogical tools. Solutions such as teacher training, the use of technological tools, and partnerships with industry are proposed to facilitate this transition. Finally, the chapter discusses the future prospects for AI education, highlighting the importance of pedagogical innovations such as virtual reality, adaptive learning, and virtual labs. Collaboration between academia and industry is identified as a key strategy to provide students with immersive training in line with market needs. In conclusion, PPA and active methods are presented as essential tools for training competent and ethical AI professionals, adapted to contemporary technological and social challenges.

DOI: 10.4018/979-8-3373-2262-9.ch007

INTRODUCTION

In a society that has been profoundly transformed by the digital revolution, artificial intelligence (AI) is emerging as a driver of change and innovation. It impacts various sectors such as health, transport, finance, and education, while redefining the dynamics of our social and professional interactions. However, beyond its economic and technological potential, AI raises major ethical questions: bias in algorithms, invasion of privacy, and implications for fairness and accountability. These issues require critical reflection on its use and development.

Faced with these challenges, teaching AI requires an adapted pedagogical approach. It is no longer enough to pass on theoretical knowledge or train technical specialists; it is imperative to develop practical skills, critical thinking, and an understanding of the societal impacts of AI. However, this mission faces several obstacles: complexity of concepts, lack of adapted educational resources, and difficulties in linking theory and practice in a constantly changing context.

To address these needs, this chapter proposes to explore the integration of project-based learning (PBL) and active pedagogical approaches in AI education. These methods make it possible to connect theory and application by immersing students in real-world projects, such as creating computer vision systems or predictive models from real data. In addition, they offer a unique opportunity to address ethical issues in a practical way, through case studies or simulations.

With a focus on autonomy, collaboration, and innovation, PPA and active approaches transform AI education into an engaging, participatory, and labour market-relevant process. This chapter also examines how these methods, adopted in various educational contexts around the world, can foster a comprehensive and inclusive education, preparing students to become responsible and creative actors in an ever-changing field.

PROJECT-BASED LEARNING: DEFINITION AND IMPORTANCE

Distinction between methods, approaches and techniques

In the pedagogical context, the terms methods, approaches, and techniques are often used interchangeably, which can create some confusion. However, these concepts have distinct meanings that influence the way they are applied in education, especially in a field as technical as artificial intelligence. Here is a clarification of the distinctions between these terms.

Figure 1. Definitions and Examples of Pedagogical Concepts

Concept	Definition	Examples in AI
Method	A comprehensive pedagogical approach that structures the entire teaching and learning process.	- Project-based learning (PBL), where students work on real-world projects to understand concepts. - Collaborative learning, where students solve problems in groups.
Approach	A conceptual framework or general philosophy that guides the method, often based on educational principles.	- Constructivism: Students build their own knowledge through practical experiences. - Student-centred approach: Learning is designed around the needs and interests of students.
Technique	A specific tool or strategy used to implement a method or approach.	- Using brainstorming to identify solutions to an AI problem. - Data visualization to understand model results.

Define Project-Based Learning (PBL)

Project-based learning (PBL) is a pedagogical method that places students at the heart of the learning process by allowing them to work on concrete, relevant and often interdisciplinary projects. Unlike traditional approaches, PPA is not only based on the imparting of information by the teacher, but encourages students to explore, experiment and actively build their knowledge. In the context of AI, this method promotes the application of technical skills such as programming, data analysis, and algorithm design, while stimulating creativity and innovation (Barron and Darling-Hammond, 2008).

The origins of project-based learning can be traced back to the work of the American educator John Dewey, who argued that learning should be based on experience and reflection, which are central elements in the intellectual development of the individual (Dewey, 1938). By grounding learning in real-world experiences, the PPA allows students to connect academic theories to practical situations, providing them with a deeper understanding of concepts. Thomas (2000) emphasizes that PPA strengthens student engagement, making them more active and autonomous in their learning, an essential point in disciplines as dynamic and technical as AI.

Compared to traditional teaching methods, which often rely on lectures and repetitive exercises, PPA has unique advantages in terms of developing practical skills and motivation. Blumenfeld et al. (1991) demonstrate that project-based learning motivates students by allowing them to work on concrete problems that interest them, thereby increasing their engagement. In the context of AI, this can include projects such as the design of image recognition systems, the analysis of social network data or the creation of prediction models. These projects expose students to complex challenges and encourage them to mobilize different technical and cognitive skills to succeed.

The PPA also develops students' autonomy and their ability to work in a team. By taking charge of their projects, students acquire project management, collaboration and problem-solving skills, essential skills for future AI professionals (Krajcik and Blumenfeld, 2006). In addition, by engaging in creative and often interdisciplinary activities, they learn to experiment and innovate, qualities that are necessary in an ever-evolving field like AI. Solomon (2003) notes that PPA helps develop initiative and a reflective approach, which are essential for success in innovative and technology-driven careers.

Thus, PPA is a pedagogical approach that not only supports the development of technical skills but also promotes a critical and applied understanding of artificial intelligence. By integrating real-life projects into the educational process, it trains students who are able to connect theory and practice, collaborate effectively and solve problems in innovative ways.

Project-based learning (PBL) stands out for its ability to transform theoretical concepts into practical skills through immersive and collaborative projects. In the field of artificial intelligence, this approach makes sense by allowing students to work on real problems while mobilizing advanced tools and methods. Like what:

- **Computer Vision Systems:** Students can design an object recognition model using convolutional neural networks (CNNs). This project includes steps such as collecting and cleaning image data, tuning machine learning algorithms with frameworks like TensorFlow or PyTorch, and evaluating model performance.
- **Text-based sentiment analysis:** Using natural language processing (NLP) techniques, a project can involve analyzing customer reviews to gauge customer satisfaction. Students learn how to leverage libraries like spaCy or NLTK, while developing a deep understanding of the concepts of tokenization, vectorization, and classification.
- **Recommendation systems:** Another concrete example is the development of a personalized recommendation engine, such as those used by streaming or e-commerce platforms. Students explore collaborative or content filtering methods while integrating real datasets.

These projects, rooted in concrete problems, allow students to develop technical skills while strengthening their autonomy, creativity and ability to solve problems. In addition, they encourage critical reflection on the practical and ethical implications of the proposed solutions."

Origins and Fundamentals of the PPA

The origins of project-based learning (PBL) can be traced back to the early twentieth century, with the work of the American educator and philosopher John Dewey, who advocated experiential learning and active reflection. For Dewey (1938), learning should be rooted in concrete situations and lived experiences, allowing students to develop a deep understanding of concepts. This philosophy, called educational pragmatism, is based on the principle that education should be learner-centred and promote active engagement. Dewey points out that action learning allows individuals to build their own knowledge by confronting real-world challenges, an approach particularly suited to fields like artificial intelligence, where practical skills are essential.

One of the fundamental principles of PPA is therefore to involve students in solving concrete and meaningful problems, which allow them to mobilize theoretical knowledge and practical skills. Blumenfeld et al. (1991) describe PPA as an approach where students work on complex, interdisciplinary projects, often collaboratively, to solve problems independently. This process actively engages them and promotes a more complete understanding, as they do not limit themselves to memorizing isolated facts, but apply the concepts in a practical way.

In addition to promoting autonomy, the PPA encourages the development of critical thinking and the ability to self-evaluate. In this approach, teachers play the role of facilitators and mentors, guiding students in their exploration and helping them overcome obstacles encountered. According to Krajcik and Blumenfeld (2006), this role of the teacher is central, as he or she supports students in the management of their projects and helps them to reflect on their own learning processes, thus encouraging them to adopt a reflective posture in relation to their work. This support helps students to gradually become more independent and to acquire a better ability to adapt, which is crucial in the field of AI, where skills must be constantly renewed in the face of rapid technological changes.

In addition, PPA is based on the integration of research and creative activities, which distinguishes it from more passive teaching approaches. Solomon (2003) notes that PPA allows students to develop skills such as complex problem-solving and informed decision-making, by confronting them with authentic projects, such as the development of predictive models or the design of AI applications. These projects promote not only technical learning, but also transferable skills such as teamwork, planning and project management, which are essential in demanding careers in innovation and creativity.

In short, the origins and fundamental principles of the PPA are based on a vision of education centered on experience and the active construction of knowledge. By encouraging engagement, autonomy and problem-solving, this pedagogical method prepares students to meet the challenges of advanced fields such as artificial intelligence, developing not only their technical skills, but also their ability to innovate and adapt to changing contexts.

Figure 2. The fundamentals of project-based learning

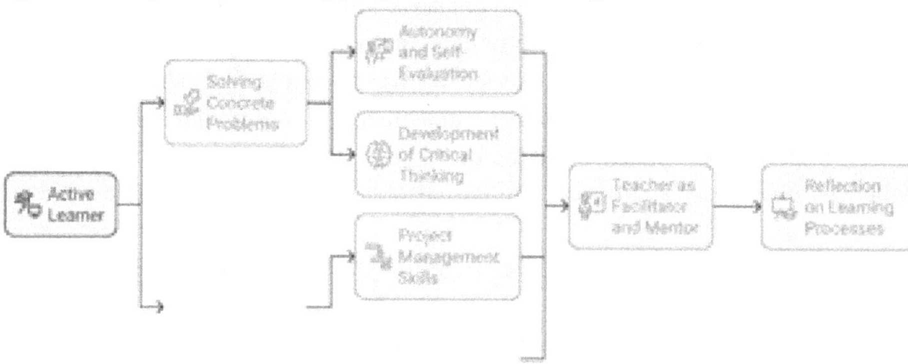

This figure shows in a structured way how the learner is placed at the center of the PPA, surrounded by pedagogical principles that encourage engagement, autonomy, collaboration, and reflection. These elements combined create a rich learning environment that allows students to develop not only technical skills but also interpersonal and managerial skills necessary for future success.

Challenges of integrating PPA in AI

Despite its many pedagogical benefits, integrating project-based learning (PBL) into artificial intelligence (AI) education is not without its challenges. These challenges, both technical, organizational and pedagogical, deserve special attention to ensure the effectiveness of this approach.

1. Technical complexity of concepts

AI is based on abstract and sometimes intimidating concepts, such as:

Machine learning: Understanding supervised or unsupervised learning algorithms requires a solid foundation in mathematics (probability, linear algebra).

Artificial neural networks: Their operation, based on matrix calculations and activation functions, can be difficult to assimilate without suitable educational tools.

Natural language processing (NLP): The complexity of manipulating textual data and the use of specialized libraries such as spaCy or NLTK pose particular challenges.

2. Limited Resources

The implementation of AI projects requires specific, often expensive, resources that are not always available in educational institutions:

Hardware infrastructure: GPUs or TPUs are often required to train large models.

Datasets: Access to quality, diverse and relevant datasets may be limited for cost or privacy reasons.

Software and frameworks: While frameworks like TensorFlow and PyTorch are free, they require extensive technical knowledge and a configured environment to use.

3. Rapid technological changes

The artificial intelligence ecosystem is evolving at a breakneck pace. Innovations in models, tools, and methodologies are rapidly making instructional content obsolete, creating constant pressure on teachers to:

Update course materials: Incorporate the latest technological advances.

Train teachers themselves: Acquire expertise in emerging fields such as generative AI (e.g., ChatGPT, DALL-E).

4. Insufficient supervision

Projects require close supervision to ensure their success. However, in large classes or with a lack of trainers, several problems arise:

Time management: Teachers must balance individual follow-up and collective progress.

Heterogeneity of skills: Some students advance faster than others, creating a gap that is difficult to bridge.

Limited feedback: Insufficient coaching can lead to uncorrected technical errors or a lack of clarity in the project's objectives.

These challenges underscore the need to rethink the methods of implementing PPA to maximize its effectiveness in the field of AI. In the next section, we will propose concrete solutions to overcome these obstacles, such as creating industry partnerships, using open-source resources, and designing modular projects tailored to different skill levels.

Comparison with traditional teaching methods

Project-based learning (PBL) is fundamentally different from traditional teaching methods, due to its student-centred approach, pedagogical objectives and practical outcomes. These distinctions are particularly relevant in artificial intelligence (AI) education, where technical mastery must be accompanied by practical skills and critical thinking.

Role of the student

Project-based learning (PBL) is based on a dynamic where the student is at the center of the educational process, unlike traditional methods where he or she plays a more passive role. This approach encourages active engagement, autonomy, and the ability to apply concepts in real-world situations, which are essential elements for mastering artificial intelligence (AI).

In the APP: the student, actor of his learning

- **Autonomy and exploration:**

The student becomes an explorer, active in problem solving. They are encouraged to make assumptions, experiment with solutions and evaluate their results.

During a project on image classification, the student chooses the learning methods (convolutional neural networks), configures the hyperparameters and analyzes the errors to improve his model. Blumenfeld et al. (2020) point out that this autonomy promotes the development of transferable skills, such as critical thinking and complex problem-solving.

- **Active Engagement:**

By working on concrete projects, the student feels more involved in his or her learning. Personal involvement increases intrinsic motivation. Wing & Lazowska (2018) demonstrate that actively engaged students retain concepts better and develop a more sustainable mastery of the tools.

- **Decision-making and responsibilities:**

The student is faced with critical choices throughout the project, such as the type of data to be used, the optimization techniques to be applied, or the evaluation metrics. Jordan & Mitchell (2021) argue that decision-making in complex environments improves the ability to manage real-world projects in a business context.

In traditional methods: the student, passive receiver

- **Information recipient:**

The student listens to lectures or reads textbooks, assimilating the information without necessarily putting it into practice. Barr & Stephenson (2019) indicate that this approach promotes short-term memorization but does not encourage the practical application of concepts.

- **Low turnout:**

Opportunities for interaction and collaboration are limited. The student often works alone on standardized exercises. Fisher & Fox (2019) note that lack of interaction reduces student motivation and engagement in complex technical topics.

- **No liability:**

The student does not participate in decisions related to educational objectives or evaluation methods. Everything is defined by the teacher. Smith & Brown (2021) criticize the top-down approach of traditional methods, which leaves little room for innovation or creativity.

To summarize, here's a clear comparison between the two approaches:

Figure 3. Figure Comparing PPA and Traditional Methods

Criterion	**APP: Active Student**	**Traditional methods: Passive student**
Role	Lead, autonomous actor	Receiver of transmitted knowledge
Participation	Increased interaction and collaboration	Limited participation, often individual work
Decisions	Decision-making and responsibility in projects	Teacher-Imposed Decisions
Motivation	Intrinsic through commitment to projects	Extrinsic, focused on grades and evaluations
Practical results	Technical proficiency and transferable skills	Theoretical understanding limited to specific contexts

PPA transforms the role of the student by making him or her an actor in his or her learning, with active participation and increased responsibility. This approach, adapted to the needs of AI education, is in stark contrast to traditional methods where the student remains passive and disconnected from real-world applications. These differences explain the growing adoption of PPA in academic curricula, particularly in technology fields.

Pedagogical focus

The pedagogical purpose of a teaching method directly influences how students acquire skills and knowledge. Project-based learning (PBL) and traditional methods differ fundamentally in their approach to knowledge transmission and practical application, particularly in the context of artificial intelligence (AI) education.

APP: Enabling real-world problem solving

PPA focuses on solving real-world problems, often rooted in real-life or simulated situations. This approach helps students directly connect theoretical concepts to practical applications, preparing them to meet the challenges faced in the professional world. Jordan & Mitchell (2021) show that APP fosters a sustainable understanding of machine learning concepts by integrating them into real-world scenarios.

This method motivates students to explore, experiment, and innovate, allowing them to develop a deep grasp of concepts while immersing themselves in today's technological challenges.

Traditional Methods: A Theory-Centered Approach

In contrast, traditional methods often focus on memorizing and theoretically understanding concepts, with little to no practical applications. The main objective is to transmit a body of stabilized knowledge, sometimes disconnected from real contexts.

While this approach can provide a strong theoretical foundation, it often limits students' ability to apply these concepts in real-world environments. In addition, it can reduce student engagement, who sometimes perceive these courses as disconnected from the needs of the job market. Smith & Brown (2021) criticize the top-down approach of traditional methods, which does not adequately prepare students for practical challenges.

Direct comparison: Impact on learning

In PPA, learning is student-centered, which encourages them to be active and involved in their learning process. In contrast, traditional methods place the teacher at the center, limiting students' autonomy and creativity. This difference has a direct impact on learners' engagement and outcomes.

Research by Fisher & Fox (2019) highlights that students participating in APP projects demonstrate a better ability to transfer their knowledge to new problems. On the other hand, those trained by traditional methods struggle to apply their knowledge outside the contexts in which it was taught.

The pedagogical focus of the PPA, oriented towards the resolution of concrete and immersive problems, contrasts sharply with that of traditional methods, often limited to theoretical transmission. This distinction makes PPA particularly suitable for teaching AI, an area where practical skills and the ability to solve real-world problems are essential. The integration of the PPA thus allows students to develop immediately transferable skills, in line with the expectations of employers in an ever-changing market.

Evaluation methods

Assessment methods are a central aspect of teaching methods, as they directly influence student engagement and the quality of learning. In artificial intelligence (AI) education, project-based learning (PBL) and traditional methods take fundamentally different approaches to assessment.

Evaluation in the PPA: Multidimensional and process-centric

The PPA implements evaluation methods that take into account several dimensions, going beyond the simple verification of theoretical knowledge. This assessment includes:

The final product: The quality of the project presented is evaluated according to specific criteria such as innovation, technical relevance and presentation of results.

The learning process: Teachers also evaluate students' efforts throughout the project, including their critical thinking and ability to overcome obstacles.

Collaborative skills: Since PPA is often based on group projects, criteria related to collaboration are integrated, such as management of responsibilities and communication within the team.

Assessment in the PPA thus promotes a holistic view of student performance, integrating technical, transversal and behavioural skills. Fisher & Fox (2019) show that this multidimensional approach better prepares students for environmental environments.

Evaluation in traditional methods: standardized and content-centric

In contrast, traditional methods favor standardized assessments, often based on written exams or multiple-choice tests. These modalities focus mainly on verifying the acquisition of theoretical knowledge.

Written Exams: Students are asked to answer questions about concepts or solve math exercises that are not directly related to practical applications.

Multiple-choice tests (MCQs): These tests often measure students' ability to memorize definitions or formulas, without assessing their in-depth understanding or ability to apply these concepts in real-world situations.

These assessments, although useful for verifying the mastery of the theoretical bases, lack relevance for judging practical skills or the ability to solve real problems. Smith & Brown (2021) point out that traditional assessments, while effective in testing memorization, often fail to prepare students for professional challenges.

Figure 4. Comparison table between the two approaches

Evaluation criteria	PPA: Multidimensional Assessment	Traditional methods: Standardized assessment
Final Product	Quality of the project based on technical criteria, relevance, innovation and clarity of presentation.	Correct resolution of standardized problems or production of expected answers.
Learning process	Analysis of effort, critical thinking, and ability to solve complex problems throughout the project.	Not taken into account; The evaluation focuses only on the final result.
Collaborative skills	Assessment of team management of responsibilities, communication and collaboration.	Little or no evaluation of collaboration, as tasks are often individual.
Knowledge approach	Integration of knowledge into practical and varied scenarios, reflecting real-world contexts.	Verification of the memorization of theoretical concepts unrelated to real applications.
Assessment Tools	Flexible evaluation grids including qualitative criteria (creativity, critical thinking).	Written exams, MCQs or standardized exercises with often rigid criteria.
Impact on learning	Strengthening motivation and commitment through an evaluation that values innovation and individual or collective effort.	Extrinsic motivation, focused on grades and the reproduction of knowledge.

The evaluation modalities in the PPA stand out for their ability to integrate various dimensions, such as the quality of the final product, the learning process, and collaborative skills. In contrast, traditional methods are often limited to standardized, theory-based assessments. This difference underlines the effectiveness of PPA in preparing students for a demanding professional environment, where technical and transversal skills are essential.

Benefits of PPA in AI Education

Project-based learning (PBL) has several advantages in teaching artificial intelligence (AI), a discipline that requires both advanced technical skills and the ability to innovate and solve complex problems. PPA differs from traditional methods by providing a practical and dynamic framework that allows students to develop theoretical skills while applying them to real-world situations.

Development of practical and theoretical skills

One of the main benefits of PPA is that it allows students to learn practical skills that are directly applicable in the field of AI. PPA emphasizes real-world projects, such as designing machine learning algorithms or creating predictive models, that allow students to practice theoretical concepts they have learned in the classroom (Thomas, 2000). By combining theory and practice, PPA promotes a deeper understanding of the topics being studied. Students are confronted with real-world problems that they must solve by mobilizing theoretical knowledge while developing technical skills such as programming, data management, or the use of libraries and tools specific to AI (Krajcik and Blumenfeld, 2006).

In the context of AI education, projects can include tasks such as data analysis, image recognition, or prediction from data. These hands-on projects allow students to strengthen their theoretical understanding while mastering the essential tools and methods in this rapidly evolving field (Blumenfeld et al., 1991).

Strengthening students' autonomy and creativity

Another benefit of PPA is the development of student autonomy and creativity. Unlike traditional methods where students are guided step by step, PPA gives them the responsibility to manage their own learning process. Students are encouraged to make decisions, plan their projects, experiment with different approaches, and evaluate their results (Barron & Darling-Hammond, 2008). This autonomy not only boosts their self-confidence, but also their ability to adapt and innovate in the face of challenges.

AI is a field that requires a great deal of intellectual flexibility and a constant spirit of innovation, as technologies evolve rapidly and new challenges continually arise. PPA prepares students for this environment by encouraging them to think outside the box, explore different solutions, and be creative in problem solving (Solomon, 2003). This helps them develop skills that are critical for their future careers, where they will often need to find innovative solutions to complex problems.

Encouraging innovation and solving complex problems

The PPA also provides an environment conducive to innovation and complex problem solving. By confronting students with projects that reflect real situations and authentic problems, this method forces them to analyze situations, identify possible solutions, and experiment with tools and methods to arrive at a result (Blumenfeld et al., 1991). This is particularly relevant in the field of AI, where projects can involve major technical and ethical challenges.

PPA also boosts students' ability to work iteratively, a critical aspect of AI systems design. By constantly testing and improving their solutions, students learn to manage uncertainty and adapt their approaches based on the results obtained (Hmelo-Silver, 2004). This ability to adapt and constantly re-evaluate their work is crucial for innovation and success in large-scale AI projects.

While working on AI projects, students are confronted with multi-faceted problems, often without a single solution, which pushes them to think critically and innovate. In this way, they develop skills in solving complex problems, integrating diverse perspectives and adapting to changing project requirements (Thomas, 2000). These skills are especially important in areas like AI, where the ability to solve complex problems is a key asset.

ACTIVE APPROACHES TO AI: CONCEPTS AND METHODOLOGIES

Active Approaches: A Student-Centered Pedagogy

Active, student-centered pedagogical approaches are gaining importance in artificial intelligence (AI) education, as they place the learner at the heart of the process of acquiring knowledge and skills. In contrast to traditional teaching methods, where the teacher plays a central role as a transmitter of knowledge, active approaches emphasize interaction, collaboration and active student participation. This paradigm aims to stimulate engagement, autonomy and the development of critical skills, essential qualities for AI training (Jones et al., 2020).

A key method in this area is problem-based learning (PBL), which relies on solving real-world problems as a learning engine. This approach, first introduced in the field of medicine, has proven its effectiveness in complex contexts where active practice is crucial (Barrows & Tamblyn, 1980). In AI education, APP allows students to work on real-world cases and projects, giving them the opportunity to apply theoretical concepts such as machine learning algorithms and neural networks to practical situations (Hmelo-Silver, 2004). According to Bell et al. (2010), PPA improves students' motivation and ability to solve complex problems, qualities that are particularly valuable for advanced technology fields such as AI.

Collaborative learning is another significant active approach in AI education. This method encourages teamwork, which is essential for the integration of interdisciplinary skills and for solving complex problems. According to Johnson and Johnson (2009), collaborative learning improves social interaction and promotes a deeper understanding of the concepts being discussed. In the context of AI, this collaboration allows students to share perspectives, exchange ideas, and help each other overcome challenges, thereby stimulating their critical thinking and their ability to innovate (Lajoie, 2015).

Project-based learning (PBL) is also particularly relevant for AI education, due to the need to address practical and technical concepts in a contextualized framework. Research shows that project-based learning boosts student engagement and motivation, as it allows them to experiment, create, and validate their hypotheses in real-world projects (Blumenfeld et al., 1991). This learning is beneficial in AI education, where students must not only understand the theoretical foundations, but also know how to apply their knowledge to complex and innovative problems. A study by Thomas (2000) points out that students in project-based learning develop better problem-solving skills and an increased ability to work independently, which perfectly meets the needs of a field like artificial intelligence.

Finally, these active methods find a theoretical foundation in the constructivist pedagogy of Piaget (1970) and the socio-constructivism of Vygotsky (1978), which argue that learning is optimal when students are actively engaged in their own process of knowledge construction. In an AI context, this direct involvement in learning promotes the integration of technical knowledge with more general skills, such as critical thinking and collaboration, which are essential to excel in this field.

Active Methods Overview

Active methods are a set of pedagogical approaches that put the learner at the centre of the educational process. Unlike passive methods, where the teacher is the main source of knowledge, active methods require direct and participatory student involvement. These approaches are particularly suited to technological fields like artificial intelligence (AI), where complex and ever-changing concepts require a deep understanding and ability to solve real-world problems. The two most common active methods in AI education are problem-based learning (PBL) and collaborative learning.

Problem-Based Learning (PBL)

Problem-based learning is a pedagogical method where students learn through the resolution of concrete problems, often from real situations. This process is distinguished by its practical focus and its ability to engage students in active learning (Barrows & Tamblyn, 1980). PPA was initially developed in the field of medicine to train doctors who could solve complex clinical cases, but it has since been widely adopted in other fields, including AI.

In PPA, students are faced with a problem without having previously received the knowledge to solve it. This challenge pushes them to actively seek information, collaborate with their peers, and develop innovative solutions (Hmelo-Silver, 2004). In AI education, APP can include projects such as building computer vision systems or implementing machine learning algorithms to solve classification or prediction problems. By confronting students with real-world projects, the PPA allows them to develop practical skills while consolidating their theoretical understanding.

Research shows that PPA promotes not only the acquisition of technical knowledge, but also the development of transversal skills such as critical thinking, autonomy and the ability to adapt to complex situations (Savery, 2006). These skills are especially essential in AI, where technological challenges evolve rapidly and team-based problem-solving is often required.

Collaborative Learning

Collaborative learning is another active method that plays a key role in teaching AI. Unlike individual learning, this method is based on the idea that students learn best when they work together to achieve a common goal. The collaborative process allows students to share knowledge, discuss ideas, and co-construct solutions (Johnson & Johnson, 2009).

In a collaborative learning context, students are grouped into teams to solve complex problems. This process promotes not only the acquisition of technical knowledge, but also the development of interpersonal skills such as communication, conflict management and cooperation (Slavin, 2011). In AI, collaborative projects can involve building a neural network for a specific task or developing a natural language processing model, where each team member can contribute according to their strengths.

Collaborative learning is also based on the principle of "social constructivism," which holds that learning is most effective when individuals interact and co-construct knowledge in a social environment (Vygotsky, 1978). In AI education, where projects are often complex and multidisciplinary, collaboration between students allows for a better understanding of technical concepts and more effective problem solving.

Other Active Methods: Case Studies and Inquiry-Based Learning

In addition to PPA and collaborative learning, other active methods can also be used to strengthen AI education. For example, case study is a method where students analyze and discuss real-life cases in order to understand the challenges faced and propose solutions. In the context of AI, this could involve analyzing a successful AI project or technology failure to learn from it.

Inquiry-based learning, where students ask questions and conduct research to find answers, is also an effective active method. This approach encourages students to adopt a research posture, to explore different angles of a problem, and to draw their own conclusions, which is particularly relevant in a field as innovative as AI.

Figure 5. Figure Representing Active Methods.

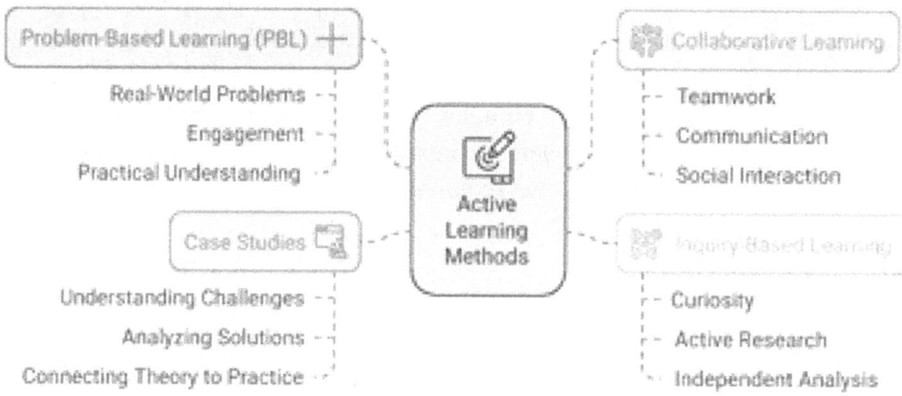

The Integration of Active Approaches in AI Education: The Benefits of Active Approaches

The integration of active approaches into artificial intelligence (AI) education represents a significant step forward in making learning more engaging, practical, and relevant for students. These methods promote the acquisition of skills that go beyond simple theoretical knowledge, reinforcing essential qualities such as commitment, critical thinking, and the ability to solve complex problems. Active learning helps to meet the specific needs of the AI field, in particular through its ability to transform students into autonomous and creative professionals.

Improved student engagement

Active approaches, such as project-based learning and problem-based learning, promote student engagement by allowing them to take an active role in their own learning. In contrast to traditional methods, where the student is often content to receive knowledge passively, active approaches stimulate interest and motivation by involving students in concrete and meaningful projects (Blumenfeld et al., 1991).

In AI, for example, image recognition or data analysis projects provide students with practical applications of theoretical concepts, making them more invested in their learning.

Research shows that student engagement increases when learning is relevant and focused on real-world challenges (Thomas, 2000). This increased engagement results in a better understanding of concepts and an increased ability to apply these concepts in practical situations. A study by Barron and Darling-Hammond (2008) found that students participating in active projects had a more positive attitude toward their learning and were more likely to persevere in the face of difficulties.

Development of critical thinking and problem-solving skills

One of the main benefits of active approaches is the development of critical thinking and problem-solving skills. In a complex problem-solving context, students are encouraged to analyze situations, evaluate different approaches, and make informed decisions (Hmelo-Silver, 2004). This ability to think critically is especially crucial in AI, where projects often require innovative solutions to technical challenges.

PPA and collaborative learning push students to think outside the box and test different solutions, thereby strengthening their ability to solve problems creatively (Savery, 2006). For example, a speech recognition AI project may pose unique challenges that require students to tweak and improve their algorithms over iterations, allowing them to develop a better understanding of the mechanics of machine learning.

Facilitating the acquisition of deep knowledge in AI

Active approaches are particularly effective in helping students gain in-depth and sustainable knowledge. Unlike traditional learning, where concepts are often approached in a theoretical and abstract way, active learning offers students the opportunity to directly manipulate concepts and apply them in concrete contexts. This allows for a deeper understanding of AI principles, as well as better long-term knowledge retention (Blumenfeld et al., 1991).

For example, a data classification project in an AI course requires students to understand not only the classification algorithms, but also the parameter choices and interpretation of the results, which facilitates a more comprehensive understanding of the process. According to a study by Krajcik and Blumenfeld (2006), students who learn through active methods are more likely to master complex concepts and transfer their knowledge to new contexts, a key advantage in a rapidly evolving field like AI.

THE PRACTICAL APPLICATION OF PBA AND ACTIVE APPROACHES IN AI

AI Project Design for Active Learning

The Practical Application of PBA and Active Approaches in AI: AI Project Design for Active Learning

The implementation of project-based learning (PBL) and active approaches in artificial intelligence (AI) education is based on the creation of concrete projects adapted to the educational needs of students. These projects provide students with the opportunity to immerse themselves in real-world problems, allowing them to learn technical skills while developing their ability to solve complex problems and collaborate with their peers. Designing relevant projects is key to maximizing the benefits of the PPA and boosting student engagement.

Types of Projects Adapted for AI Education

Projects tailored to AI education can be designed to cover a wide range of AI subfields, including natural language processing (NLP), computer vision, supervised and unsupervised learning, and recommender systems. These areas are both relevant and accessible to students, as they allow students to address fundamental concepts while working on real-world applications.

- **Natural Language Processing (NLP) Projects:** Students can design a language processing model to analyze sentiment in consumer reviews or to detect bias in text. These projects involve the use of NLP algorithms and text analysis techniques, which help students understand the challenges and applications of AI in understanding human language (Jurafsky & Martin, 2008).
- **Computer vision projects:** The design of an image recognition system is a popular and suitable project for active learning in AI. Students may be tasked with designing a model that can detect objects in images or recognize facial expressions. This type of project explores concepts such as convolutional neural networks (CNNs) and image processing, while practicing programming and visual data manipulation skills (Goodfellow, Bengio & Courville, 2016).
- **Supervised and unsupervised learning projects:** These projects include building classification and clustering models to solve problems like predicting diseases from medical data or grouping customers based on their buying behavior. These projects allow students to gain a working understanding of

supervised and unsupervised learning concepts, while working with real-world datasets (Murphy, 2012).

- **Recommender systems:** Students can design a recommendation engine for movies, articles, or products. These projects allow students to work with collaborative filtering and content-based systems techniques, and introduce students to advanced concepts of personalization and preference analysis (Ricci, Rokach & Shapira, 2011).

Integrating Active Approaches into AI Course Design

Strategies for Combining PBA and Active Approaches in AI Programs

Combining PPA with other active methods in an AI course requires careful planning to ensure that each method is aligned with the learning objectives. One strategy is to structure the course around concrete, interdisciplinary projects, which encourage students to explore complex issues independently while benefiting from teacher supervision. For example, one project could focus on the development of a prediction model for environmental data, which would require the use of machine learning algorithms, as well as skills in data analysis and interpretation of results (Barron and Darling-Hammond, 2008).

Teachers can also incorporate group problem-solving sessions into these projects, allowing students to work together to identify and solve the challenges they encounter (Johnson & Johnson, 2009). This process encourages the co-construction of knowledge and builds collaboration and communication skills, which are essential in large-scale AI projects. A study by Krajcik and Blumenfeld (2006) shows that students engaged in collaborative projects develop a deeper understanding of concepts and gain greater confidence in their skills.

Another strategy is to include feedback and self-assessment sessions, where students can reflect on their learning, identify their strengths, and areas for improvement. This self-reflection is crucial in PPA, as it helps students develop a better understanding of their own learning processes and adjust their approaches according to the challenges encountered (Hmelo-Silver, 2004).

Teachers can also use inquiry-based learning to stimulate student curiosity and initiative. By asking open-ended questions on topics such as the ethics of AI or the challenges of regulation, students are encouraged to conduct their own research, explore different perspectives, and present their findings. This strategy is particularly useful in ever-evolving fields such as AI, where questions of ethics and responsibility are paramount (Savery, 2006).

Case Studies and Feedback from AI Teachers

Case studies and feedback from teachers who have already integrated PPA and active methods into their AI courses show encouraging results. For example, a study conducted by Blumenfeld et al. (1991) looked at an AI program where students were tasked with developing image recognition systems. Students worked in teams to design algorithms, test their models, and refine their solutions based on the results. Teachers observed that students were not only more engaged, but they also gained a more practical understanding of AI concepts and complex problem-solving skills.

In another study, Thomas (2000) reports that PPA, combined with collaborative learning, has allowed students to better understand the application of machine learning algorithms. Teachers have set up working groups to solve real-world cases, such as predicting sports results from historical data. This type of interdisciplinary project not only strengthened the students' technical skills, but also helped them develop an ability to work in a team and communicate their ideas effectively.

Another relevant feedback is from Savery (2006), who explains how PPA and active methods were applied in an AI course for undergraduate students. In this course, students designed natural language processing models to analyze text, and each group had to collaborate to overcome technical challenges. Teachers noted that students were particularly motivated and proactive in their learning, which improved their mastery of technical concepts and confidence in their abilities.

Figure 6. Educational Approaches in AI Course Design

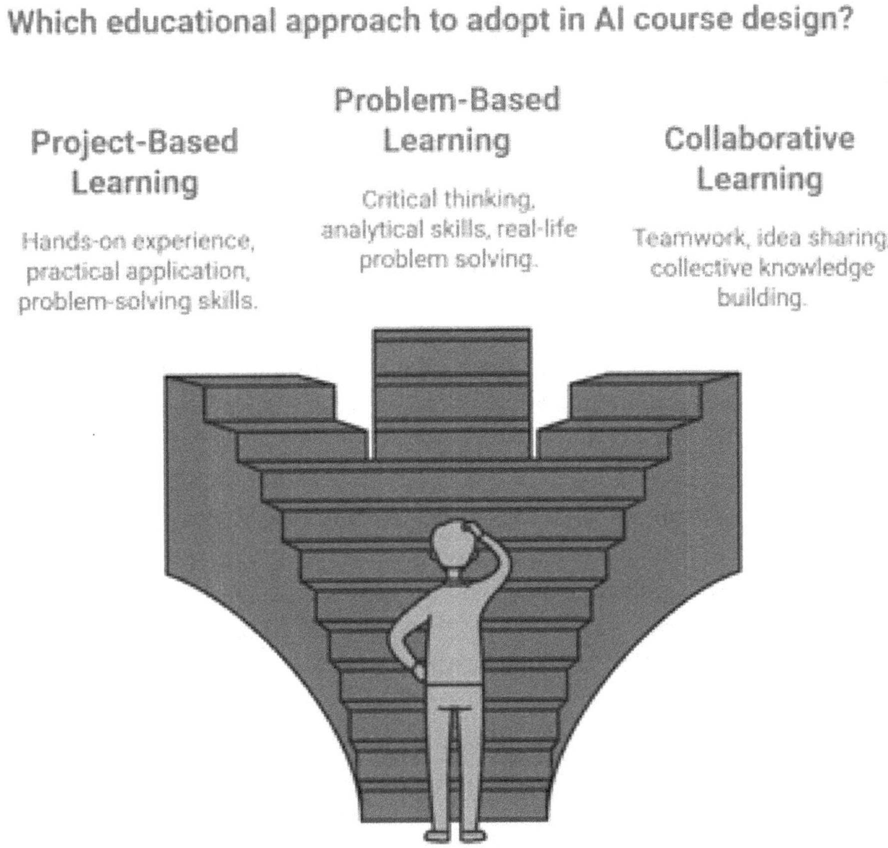

IMPLEMENTATION CHALLENGES AND SOLUTIONS

Key Challenges in Adopting PPA and Active Approaches to AI

The adoption of project-based learning (PBL) and active approaches in artificial intelligence (AI) education presents several challenges. These barriers can limit the effectiveness of integrating these methods into academic programs, especially in institutions where the infrastructure and institutional culture are not adapted to these new approaches. However, identifying these challenges and exploring possible solutions can help institutions overcome difficulties and create a learning environment that is conducive to PBA and active methods.

Institutional Resistance and Logistical Obstacles

One of the major challenges in adopting pCPA and active approaches is institutional resistance. In many academic institutions, traditional teaching methods, such as lectures and theory exams, remain the norm. The shift to more active approaches requires a paradigm shift, which can lead to reluctance on the part of teachers, administrators, and even students themselves. According to Fullan (2007), any pedagogical innovation is often confronted with institutional resistance due to deep-rooted beliefs and a fear of change.

Another logistical hurdle is the increased workload that active approaches can represent for teachers. Unlike lectures, where the content is usually fixed, the PPA requires more flexible and personalized planning, as each project may require specific resources and more intensive follow-up. This requirement can discourage teachers, especially in academic environments where they manage large classes or have limited resources (Blumenfeld et al., 1991). To address this problem, some institutions have established training programmes to help teachers integrate PPA and active methods into their teaching practices, equipping them with tools and strategies to effectively manage their workload (Savery, 2006).

Management of Educational Resources and Tools

Another major challenge is the management of the resources and pedagogical tools needed for PPA and active AI approaches. These methods often require additional resources, such as AI software, powerful computers and datasets, as well as collaborative learning spaces. In some institutions, especially those with limited budgets, it can be difficult to provide students and teachers with access to the necessary tools and technologies. According to Johnson and Johnson (2009), budget constraints are one of the main reasons why institutions are reluctant to adopt more resource-intensive pedagogical approaches.

In addition, access to data and digital learning platforms is an additional logistical challenge. AI projects often require large amounts of data, which must be properly managed to ensure its quality and relevance. In some cases, it may be difficult for institutions to acquire appropriate datasets or provide suitable software and programming environments, limiting the scope of projects that students can carry out (Krajcik & Blumenfeld, 2006). One solution to this problem is to partner with technology companies to provide software licenses, datasets, and cloud computing infrastructure for AI projects (Ricci et al., 2011).

Finally, the management of learning spaces is also crucial. Active approaches require collaborative spaces where students can work in groups, discuss and experiment freely. However, traditional classrooms are not always suitable for these needs. In many institutions, a redesign of spaces is therefore necessary to create more flexible and interactive environments, promoting collaboration and active learning (Blumenfeld et al., 1991).

Solutions to Overcome Challenges

To overcome these challenges, several solutions can be considered:

- **Teacher training and pedagogical support:** Provide teachers with specific training to help them integrate PPA and active approaches into their practices. This can include sessions on managing AI projects, using technological tools, and techniques to encourage student engagement (Fullan, 2007).
- **Partnering with industry:** Partnerships with technology companies and AI service providers can provide institutions with software, data, and infrastructure resources. These partnerships also provide students with the opportunity to work on real-life projects, which enhances their employability (Johnson & Johnson, 2009).
- **Designing learning spaces:** Adapting workspaces to encourage collaboration, with flexible classrooms, group work areas, and accessible technology equipment. These developments promote a more dynamic and participatory environment, in line with the needs of the PPA (Blumenfeld et al., 1991).

Figure 7. The Challenges of Implementing Project-Based Learning

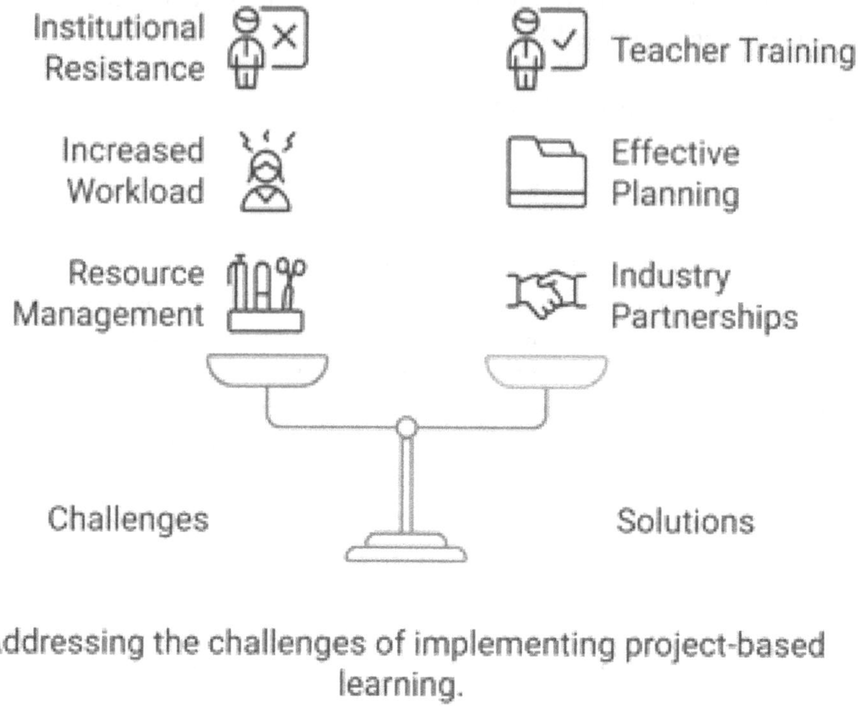

Solutions and Recommendations to Facilitate the Adoption of PPA and Active AI Approaches

To overcome the barriers to the adoption of project-based learning (PBL) and active approaches in artificial intelligence (AI) education, it is essential to put in place practical solutions and recommendations. These measures aim to reduce institutional resistance, improve resource management, and provide teachers and students with the tools and support needed to maximize the effectiveness of these teaching methods. The following are strategies, technology tools, and training recommendations that can facilitate the integration of PPA and active approaches.

Strategies for Overcoming Obstacles

To overcome institutional resistance and logistical barriers, institutions can adopt strategies to support change. For example, it is important to start with small pilot projects in which pCPA and active approaches are gradually integrated. These initiatives can be tested with smaller groups of students, allowing the effectiveness of the method to be assessed and the necessary adjustments to be identified before being extended to other courses or departments (Fullan, 2007).

Another strategy is to involve key stakeholders in the change process. By involving teachers, administrators, and even students in the planning and implementation of PPA, it is possible to create a sense of ownership and reduce resistance to change (Johnson & Johnson, 2009). Institutions can also establish working groups or committees that are dedicated to the continuous improvement of pedagogical approaches, integrating feedback to refine practices.

Finally, regular evaluation measures are crucial to ensure that the integration of the PPA meets the pedagogical objectives and meets the expectations of the stakeholders. Assessment may include student and teacher surveys, as well as analysis of academic performance, to measure the impact of PPA on engagement and skill acquisition (Savery, 2006).

Technological Tools and Platforms to Facilitate PPA and Active Approaches

Technological tools and platforms play a central role in the effective implementation of the PPA, especially in the field of AI where specific computing resources and software are often required. Here are some tools that facilitate active learning and PPA in AI education:

- **Integrated development environments (IDEs)** like Jupyter Notebook, Google Colab, and PyCharm are especially useful for AI projects. These platforms allow students to code, run scripts, and experiment directly in an interactive environment, which is essential for working with machine learning models (Prinz et al., 2018).
- **Collaborative project management platforms,** such as Trello, Asana, or Microsoft Teams, help students organize tasks, communicate, and collaborate effectively in group AI projects. These tools allow teams to manage their work, track progress, and document decisions made during the project, thus strengthening the spirit of collaboration (Blumenfeld et al., 1991).
- **Online learning and course management platforms**, such as Moodle or Canvas, make it easy to manage educational resources, group activities, and assessments. These platforms allow teachers to share documents, organize online discussions, and manage projects, facilitating the adoption of PPA and active approaches in an academic context (Johnson et al., 2009).
- **Data visualization tools** such as Tableau, Power BI, and Matplotlib are also useful for AI projects. By providing students with ways to visualize and in-

terpret the results of their analyses, these tools make projects more interactive and facilitate the understanding of complex concepts (Few, 2012).

Teacher Training and Pedagogical Support

Teacher training is crucial to ensure the successful integration of PPA and active methods. Teachers must not only master AI concepts, but also be trained in active methods to properly supervise projects and guide students in their learning. Continuing education that covers topics such as project management, the use of technological tools, and working group facilitation techniques can greatly improve their effectiveness (Blumenfeld et al., 1991).

It is also useful to set up pedagogical support with mentors or experts in active pedagogy to support teachers throughout the process. This support can include coaching sessions, classroom observations, and focus groups where teachers can share experiences and exchange practical advice (Fullan, 2007). For example, a mentorship program where experienced PPA teachers mentor new practitioners can be extremely beneficial in facilitating the transition to these new approaches.

Finally, accessible educational resources (guides, tutorials, videos) can also be made available to teachers to help them familiarize themselves with the PPA and technological tools. These resources may include AI project templates, project-based evaluation examples, and ideas for encouraging student engagement in active approaches (Savery, 2006).

TOWARDS AN INNOVATIVE AND SUSTAINABLE PEDAGOGY

Evaluation of the Impact of the PPA and Active Approaches

To ensure that project-based learning (PBL) and active approaches produce the desired effects in artificial intelligence (AI) education, it is essential to measure their impact rigorously. Evaluating the pedagogical effectiveness helps to determine whether these methods actually improve students' skills and strengthen their engagement. In addition, these measures provide opportunities to evolve these approaches and adapt them to the future needs of AI education.

Measuring Pedagogical Effectiveness: Indicators of Success

The evaluation of the effectiveness of PPA and active methods is based on indicators of success that reflect the pedagogical objectives of these approaches. These indicators should include quantitative and qualitative measures to get a complete picture of the impact of PPA on students. Here are some key indicators:

- **Student engagement and motivation:** One of the main goals of active methods is to improve student engagement. This can be measured by satisfaction surveys, attendance rates, and levels of participation in projects and discussions. Studies show that students are more motivated and engaged in their learning when they work on concrete and collaborative projects (Blumenfeld et al., 1991).
- **Acquisition of practical and theoretical skills:** The skills acquired by students, both practical and theoretical, are a key indicator of the success of the PPA. This can be assessed through project work, knowledge tests, and competency-based assessments. According to Thomas (2000), active projects allow students to better understand theoretical concepts and acquire practical skills directly applicable in the field of AI.
- **Development of critical thinking and problem-solving skills:** PPA promotes students' ability to think critically and solve complex problems. This can be measured by project-specific assessment rubrics, case analyses, and problem-solving exercises. Hmelo-Silver (2004) points out that students exposed to active methods are better able to apply their knowledge in a variety of contexts, which is a major asset in the field of AI.
- **Academic performance and professional success:** The impact of active methods can also be assessed by observing the overall academic performance and professional success of graduates. Institutions can track student success rates, end-of-project grades, and employability to determine whether PBA and active approaches are actually improving student outcomes and better preparing them for the demands of the job market.

Perspectives for the Evolution of AI Education

The evaluation of active methods and PPA not only provides insights into their current effectiveness, but also offers perspectives for the future of AI education. As technologies and market requirements evolve, AI education must adapt by incorporating new methods and meeting the changing needs of students and employers.

- **Adoption of new educational technologies**: Emerging technologies, such as virtual reality (VR), augmented reality (AR), and adaptive learning, offer additional opportunities for AI education. These tools can enrich the PPA by providing realistic simulations and interactive environments for students. For example, VR can allow students to visualize neural networks or complex algorithmic processes in 3D, making learning more immersive and intuitive (Merchant et al., 2014).
- **Integrating data-driven learning:** Using data to personalize learning paths and tailor projects to students' skill levels is another important perspective. Through data analysis, teachers can identify students' pain points and adjust projects according to their needs, making AI teaching more inclusive and adaptable (Picciano, 2012).
- **Developing partnerships with industry:** AI education can evolve by collaborating more with the professional sector. Academic institutions can work with companies to develop AI projects aligned with the real needs of the industry, providing students with valuable hands-on experience and enhancing their employability. According to Johnson and Johnson (2009), industrial partnerships can facilitate access to cutting-edge resources and experts, which enriches the AI projects offered to students.
- **Continuous assessment and improvement of teaching methods**: The evolution of AI teaching requires a dynamic approach to assessment. By incorporating continuous evaluation of teaching methods and regularly collecting feedback from students and teachers, institutions can constantly adapt and improve PPA and active methods to ensure that they meet stakeholder expectations and labour market needs (Fullan, 2007).

Perspectives for AI Education

Artificial intelligence (AI) education is constantly evolving to meet the technical, ethical and practical challenges of a rapidly developing field. In this context, pedagogical innovations play a key role in adapting learning methods to provide students with a more immersive and effective experience. At the same time, collaboration between academic institutions and industries is becoming essential to prepare students for market realities and professional requirements. These two aspects are at the heart of the prospects for relevant and sustainable AI education.

Pedagogical innovations in development

Many pedagogical innovations aim to transform AI teaching by making learning more interactive, personalized, and based on real-world experiences. These innovations make it possible to better engage students, diversify teaching methods, and integrate advanced technologies to enrich the learning experience.

- **Virtual Reality (VR) and Augmented Reality (AR):** Virtual reality and augmented reality offer unique possibilities for AI education by enabling immersive visualizations of complex concepts. For example, VR can be used to simulate neural networks or to illustrate machine learning processes dynamically. Merchant et al. (2014) demonstrated that immersive learning environments increase student motivation and facilitate the understanding of abstract concepts.
- **Adaptive Learning:** Adaptive learning uses AI to personalize students' educational journey based on their skill level, needs, and progress. Through data-driven systems and machine learning, each student is given resources and challenges that match their individual needs. This approach makes it possible to optimize learning, by adapting to the rhythms and learning styles of each person (Picciano, 2012).
- **Integration of Serious Games:** Serious games are educational games designed to teach specific skills and concepts. In the context of AI, students can, for example, use games that simulate machine learning or computer vision environments to develop their skills while having fun. According to Gee (2008), game-based learning boosts motivation and promotes better knowledge retention, especially in technical contexts such as AI.
- **Virtual Labs and Open Data Projects:** Virtual labs allow students to experiment with AI models and analyze large amounts of data online without the need for powerful local computers. By using open data and cloud environments, students can access state-of-the-art resources and work on realistic projects, such as analyzing health data or predicting climate trends (Wiley & Hilton, 2018).

The Role of Collaboration between Academic and Industrial Institutions

Collaboration between academia and industry is becoming increasingly important to strengthen the relevance of AI education programs and to ensure that students acquire skills that meet the needs of the market. This partnership makes it possible to integrate concrete projects, strengthen practical training opportunities, and benefit from the expertise and advanced technologies of companies.

- **Development of Co-Creation Projects**: Academic and industrial collaborations facilitate the creation of co-creation projects, in which students work on real problems proposed by companies. These projects, linked to concrete industry needs, give students valuable hands-on experience and expose them to the current challenges of the AI market (Johnson & Johnson, 2009). For example, a partnership between a university and a technology company could involve students in the development of a prediction model to improve supply chain management.
- **Mentorship and Internship Programs:** Mentorship and internships allow students to gain hands-on experience working alongside AI professionals. By collaborating with industry experts, students can better understand the application of theoretical concepts in industrial contexts, and develop interpersonal skills such as communication and teamwork (Weiss et al., 2015).
- **Knowledge and Resource Transfer**: Collaboration with companies allows academic institutions to benefit from advanced resources, such as software, cloud environments, and datasets. These transfers of resources facilitate students' access to cutting-edge technologies and strengthen their ability to work with the tools used in the professional world (Blumenfeld et al., 1991). For example, an AI company could provide students with free access to deep learning software and simulation environments.
- **Participation in Research and Innovation Projects**: Academic and industrial collaborations also offer opportunities for joint research. Students can participate in innovation projects by working on advanced AI issues, such as ethical AI, decision-making algorithms, and recommendation systems. These experiences strengthen their critical thinking skills and their ability to contribute to the advancement of the field (Edmondson et al., 2012).

CONCLUSION

In conclusion, this chapter explored how project-based learning (PBL) and active approaches can transform artificial intelligence (AI) education by providing students with a more hands-on, immersive, and meaningful learning experience. These teaching methods stand out for their ability to place the student at the center of their learning, encouraging them to appropriate knowledge through concrete problem solving and active experimentation. Unlike traditional methods, which focus on top-down knowledge transmission, PPA and active approaches boost student engagement, autonomy and creativity, while developing practical and theoretical skills that are essential for their future careers in AI. These pedagogical approaches, adapted to a rapidly changing technological environment, meet the needs of today's students by preparing them for the challenges and complex realities of the sector.

However, the adoption of PPA and active approaches in AI education faces significant barriers, including institutional resistance, increased workload for teachers, and limitations in terms of pedagogical and technological resources. Many institutions remain attached to traditional methods, while teachers, who are often poorly trained in these new approaches, may find it difficult to integrate them into their practices. Managing resources, whether IT equipment, specific software or adapted learning spaces, is also a major challenge, especially for institutions with limited resources. To overcome these challenges, it is essential to put in place strategies that include pilot projects, in-service teacher training, the acquisition of appropriate technological tools, and the development of partnerships with industry to pool resources and share expertise.

In the future, the evolution of AI education could be marked by increasingly sophisticated pedagogical innovations, such as virtual and augmented reality, adaptive learning, and the use of virtual labs. These immersive, personalized technologies could transform AI courses by allowing students to visualize abstract concepts, interact with complex models, and adapt their course based on their progress and needs. These innovations would not only address the pedagogical challenges posed by AI education, but would also help prepare students for a job market that is constantly changing and requires high-level skills. Collaboration with companies, on the other hand, would play a central role in allowing students to work on concrete projects related to market needs, gain valuable professional experience, and benefit from advanced technologies and real data for training AI models.

The future of AI education could also see the emergence of a hybrid model, combining online and face-to-face learning. This flexibility would provide students with wider and more diverse access to knowledge, with interactive online modules to cover theoretical foundations and face-to-face sessions to focus on collaborative projects, practical work and interpersonal exchanges. By integrating new technologies and adopting a more flexible model, AI education could reach a larger number of learners while meeting the demands of personalization and adaptability of modern learning.

Finally, one of the most essential perspectives for AI education is the integration of ethical and responsible thinking into educational programs. As AI becomes an increasingly important part of our society, it is imperative that future professionals are aware of the ethical implications of their work, particularly in terms of privacy, fairness and social impact. Training AI experts who have not only technical skills, but also ethical awareness, is crucial to driving responsible adoption of AI across industries. This ethical dimension in AI education would help shape professionals who understand and anticipate the consequences of their actions, thus ensuring safer AI that is more beneficial to society.

In short, PPA and active approaches represent a promising way to modernize and enrich AI education. By encouraging creativity, autonomy and commitment, these methods promote the acquisition of sustainable and essential skills in an ever-changing field. For AI education to fully meet the challenges of tomorrow, it is crucial to continue to innovate pedagogically and to promote academic and industrial collaborations. The future of AI education lies in the ability of institutions and teachers to adapt their practices to offer students training that is not only technical, but also human and ethical, in line with the challenges and opportunities of our time.

REFERENCES

Barr, V., & Stephenson, C. (2019). Bringing Computational Thinking to K-12: What is Involved and What is the Role of the Computer Science Education Community? *ACM Inroads*, 2(1), 48–54. DOI: 10.1145/1929887.1929905

Barrows, H. S., & Tamblyn, R. M. (1980). *Problem-Based Learning: An Approach to Medical Education*. Springer.

Bell, S., Barefoot, H., & Thomas, L. (2010). *Project-Based Learning for the 21st Century Skills*. Education Reform Series. DOI: 10.1080/00098650903505415

Blumenfeld, P. C.. (2020). Creating Sustainable Practices in Project-Based Learning. *Environmental Education Research*, 49(5), 325–333.

Blumenfeld, P. C., Soloway, E., Marx, R. W., Krajcik, J. S., Guzdial, M., & Palincsar, A. (1991). Motivating Project-Based Learning: Sustaining the Doing, Supporting the Learning. *Educational Psychologist*, 26(3-4), 369–398. DOI: 10.1080/00461520.1991.9653139

Dewey, J. (1938). *Experience and Education*. Kappa Delta Pi Lecture Series.

El Gazi, S., & Ibrahimi, A. (2024). Personalized Learning: Theory, Practices, and Perspectives. In Fostering Pedagogical Innovation Through Effective Instructional Design (pp. 308-328). IGI Global.

Eyben, F., Scherer, K. R., Schuller, B. W., Sundberg, J., Andre, E., Busso, C., & Devillers, L. Y. (2015). The Geneva minimalistic acoustic parameter set (GeMAPS) for voice research and affective computing. *IEEE Transactions on Affective Computing*, 7(2), 190–202. DOI: 10.1109/TAFFC.2015.2457417

Few, S. (2012). *Show Me the Numbers: Designing Tables and Graphs to Enlighten*. Analytics Press.

Fisher, D., & Fox, D. (2019). Hands-On Machine Learning Projects in Education: A Case Study Approach. *Advances in Education and Research*, 7(3), 95–107.

Freire, P. (1970). *Pedagogy of the Oppressed*. Bloomsbury Publishing USA.

Fullan, M. (2007). *The New Meaning of Educational Change*. Routledge.

Gee, J. P. (2008). *What video games have to teach us about learning and literacy*. St. Martin's Press.

Goodfellow, I., Bengio, Y., & Courville, A. (2016). *Deep Learning*. MIT Press.

Hmelo-Silver, C. E. (2004). Problem-Based Learning: What and How Do Students Learn? *Educational Psychology Review*, 16(3), 235–266. DOI: 10.1023/B:EDPR.0000034022.16470.f3

James, G., Witten, D., Hastie, T., & Tibshirani, R. (2013). *An Introduction to Statistical Learning: With Applications in R*. Springer. DOI: 10.1007/978-1-4614-7138-7

Johnson, D. W., & Johnson, R. T. (2009). An Educational Psychology Success Story: Social Interdependence Theory and Cooperative Learning. *Educational Researcher*, 38(5), 365–379. DOI: 10.3102/0013189X09339057

Jordan, M. I., & Mitchell, T. M. (2021). Machine Learning: Trends, Perspectives, and Prospects. *Science*, 349(6245), 255–260. DOI: 10.1126/science.aaa8415 PMID: 26185243

Jurafsky, D., & Martin, J. H. (2008). *Speech and Language Processing*. Pearson.

Knox, J., Wang, Y., & Gallagher, M. (2022). AI, Ethics, and Education: Learning AI Through Collaborative Projects. *International Review of Research in Open and Distance Learning*, 23(2), 112–129.

Krajcik, J. S., & Blumenfeld, P. C. (2006). Project-Based Learning. In Sawyer, R. K. (Ed.), *The Cambridge Handbook of the Learning Sciences*.

Lajoie, S. P. (2015). Collaboration, Communication, and Knowledge Building in Online Learning Environments. *Frontiers in Psychology*.

Litjens, G., Kooi, T., Bejnordi, B. E., Setio, A. A. A., Ciompi, F., Ghafoorian, M., van der Laak, J. A. W. M., van Ginneken, B., & Sánchez, C. I. (2017). A survey on deep learning in medical image analysis. *Medical Image Analysis*, 42, 60–88. DOI: 10.1016/j.media.2017.07.005 PMID: 28778026

Merchant, Z., Goetz, E. T., Cifuentes, L., Keeney-Kennicutt, W., & Davis, T. J. (2014). Effectiveness of virtual reality-based instruction on students' learning outcomes in K-12 and higher education: A meta-analysis. *Computers & Education*, 70, 29–40. DOI: 10.1016/j.compedu.2013.07.033

Murphy, K. P. (2012). *Machine Learning: A Probabilistic Perspective*. MIT Press.

Piaget, J. (1970). *The Construction of the Real in the Child*. Delachaux and Niestlé.

Picciano, A. G. (2012). The evolution of big data and learning analytics in American higher education. *Online Learning : the Official Journal of the Online Learning Consortium*, 16(3), 9–20. DOI: 10.24059/olj.v16i3.267

Prinz, A., Hruska, N., Burger, K., & Whitney, M. (2018). Jupyter Notebook for Education and Training: A Case Study. *Journal of Computational Science Education*, 9(1), 46–55.

Rahwan, I., Cebrian, M., & Dafoe, A. (2021). Machine Behavior: A New Approach to AI Education. *Nature*, 568(7753), 477–486. DOI: 10.1038/s41586-019-1138-y PMID: 31019318

Ricci, F., Rokach, L., & Shapira, B. (2011). *Introduction to recommender systems handbook*. Springer. DOI: 10.1007/978-0-387-85820-3

Savery, J. R. (2006). Overview of Problem-based Learning: Definitions and Distinctions. *The Interdisciplinary Journal of Problem-Based Learning*, 1(1), 9–20. DOI: 10.7771/1541-5015.1002

Shu, K., Sliva, A., Wang, S., Tang, J., & Liu, H. (2017). Fake news detection on social media: A data mining perspective. *SIGKDD Explorations*, 19(1), 22–36. DOI: 10.1145/3137597.3137600

Slavin, R. E. (2011). Instruction Based on Cooperative Learning. In Mayer, R. E., & Alexander, P. A. (Eds.), *Handbook of Research on Learning and Instruction*.

Smith, J., & Brown, T. (2021). Project-Based Learning as a Catalyst for Student Engagement in AI Education. *Journal of Innovative Teaching*, 15(4), 45–58.

Solomon, G. (2003). *Project-Based Learning: A Primer*. Technology & Learning.

. Thomas, J. W. (2000). A Review of Research on Project-Based Learning.

Vygotsky, L. (1978). *Mind in Society: The Development of Higher Psychological Processes*. Harvard University Press.

Weiss, H. M., & Cropanzano, R. (2015). Affective Events Theory: A Theoretical Discussion of the Structure, Causes, and Consequences of Affective Experiences at Work. *Research in Organizational Behavior*, 18, 1–74.

Wiley, D., & Hilton, J. L.III. (2018). Defining OER-enabled pedagogy. *International Review of Research in Open and Distance Learning*, 19(4), 133–147. DOI: 10.19173/irrodl.v19i4.3601

Wing, J. M., & Lazowska, E. D. (2018). Computational Thinking and Project-Based Learning for Artificial Intelligence Education. *Journal of the Learning Sciences*, 27(4), 562–579.

Wolff, A., & FitzGerald, E. (2022). *Integrating AI into Higher Education Curricula: A Practical Guide*. Routledge.

Zhou, Z., & Li, W. (2020). Ethical Challenges in AI Education and How to Address Them with Project-Based Learning. *AI & Society*, 35(3), 743–753.

Chapter 8
Artificial Intelligence (AI) and Enabling Collaboration in Adaptive Distance Education

Kawtar Zargane
https://orcid.org/0000-0002-1266-7094
UAE École Normale Supérieure de Tétouan, Morocco

Mohamed Erradi
UAE École Normale Supérieure de Tétouan, Morocco

Mohamed Khaldi
https://orcid.org/0000-0002-1593-1073
UAE École Normale Supérieure de Tétouan, Morocco

ABSTRACT

This chapter is about how artificial intelligence in distance learning academic institutions can improve the quality and effectiveness of collaborations, we offer more personalized and adaptive support, and we will create more captivating and dynamic online learning environments. These improvements promote a more enriching and effective learning experience, both for students and teachers. This work will address the problem of how to increase student engagement in learning activities and collaborative projects through AI? how do teachers leverage AI-based collaboration resources and platforms to track tasks, manage calendars, evaluate member contributions, and facilitate the organization of remote group activities? The solution to adapt is based on the evolution of AI, to display the impact on the dynamics of collaboration and interaction in distance learning environments. the

DOI: 10.4018/979-8-3373-2262-9.ch008

current context and anticipated future developments allow us to highlight the growing importance of artificial intelligence in creating the collaborative dynamics of distance learning environments.

INTRODUCTION

The integration of artificial intelligence (AI) into adaptive distance learning represents a significant advancement in facilitating collaboration between teachers and students. AI offers tools that can improve learner interaction and engagement, while meeting students' individual needs. Indeed, AI systems can analyze user behaviors and adapt educational content accordingly, which promotes a more personalized and interactive learning experience (Lepage, 2023; Galmar, 2022).

Artificial intelligence (AI) is playing a fundamental role in transforming education, particularly by improving accessibility for students with special needs. These students, whether they have disabilities, learning disabilities or specific communication needs, often face barriers in traditional learning environments. Thanks to AI, it is now possible to design educational tools and resources that address these challenges in innovative and effective ways.

In our work, collaboration between teachers and students can be enhanced by hybrid approaches that combine elements of synchronous and asynchronous learning. This creates a flexible learning environment where students can interact at their own pace while benefiting from a collaborative structure (Pouliot & Rocheleau, 2020; Sacré et al., 2020). The results of various studies show that this flexibility is perceived positively by students, who appreciate the freedom it offers them to manage their time and learning (Sasseville et al., 2022; Bisson et al., 2022). AI makes it possible to personalize learning paths, adapt educational content, and create tailor-made solutions that promote inclusion. By leveraging AI capabilities, educators can develop learning environments that adapt to the individual needs of students, whether they are in person or remotely. This individualization process is crucial to ensure that each student has access to a quality education, adapted to their pace and abilities.

AI-enhanced learning platforms also offer advanced monitoring and assessment features that enable teachers to better understand and support collaborative learning processes. Predictive analytics can identify early risks of disengagement or learning difficulties, enabling targeted interventions that strengthen group cohesion and maintain learner motivation. This synergy between intelligent technology and collaborative pedagogy creates a more resilient and adaptive learning environment, capable of effectively responding to the challenges of distance learning while maximizing opportunities for interaction and co - construction of knowledge.

In the following sections, we will explore several ways in which AI is improving accessibility for students with special needs. We will look at real-world examples of personalization of learning paths, speech recognition and synthesis technologies, machine translation, and support for attention and communication disorders. By highlighting these innovations, we will demonstrate how AI is helping to create inclusive learning environments, allowing all students to reach their full potential. Thus, the integration of AI in education is a significant step towards true inclusivity, making learning accessible to all, regardless of their individual challenges.

I. WHAT IS ADAPTIVE DISTANCE LEARNING?

Artificial intelligence (AI) is playing a central role in transforming adaptive learning, a model of instruction that aims to tailor learning to individual students' needs. With AI's ability to process massive amounts of data, learn from user interactions, and improve over time, it's revolutionizing the way education is delivered. Here's a closer look at how AI is impacting adaptive learning.

1. Personalization of learning paths

Personalization of learning paths has become a priority in the educational field, especially with the rise of digital technologies and adaptive learning environments. The introduction of artificial intelligence (AI) and learning analytics (learning analytics) helps better meet the individual needs of learners, adapting content and teaching methods to their preferences and performance Bates & Glikman (2021) Paquette et al., 2021). These intelligent systems are able to collect and analyze data on user interactions, facilitating the creation of tailor-made learning paths that take into account the skills, interests, and learning styles of each student (Paquette et al., 2021; Parent & Baron, 2021).

Adaptive learning platforms, such as those developed in Baillifard 's work (Baillifard, 2023), illustrate how algorithms can be used to personalize the learning experience. For example, AMUQuiz uses a ranking system to adapt questions to users' skill levels, while other platforms such as Kaïros focus on problem situations to encourage active learning (Baillifard, 2023). This approach not only improves student engagement, but also promotes a better understanding of concepts by making them more relevant and accessible (Tricot, 2021). In addition, the assessment of learning in environments such as MOOCs (Massive Open Online Courses) raises questions about how learners can be supported in their personalized journey. Hamida et al. Hamida et al. (2021) emphasize that the interaction between participants and course designers is crucial to enrich the learning experience. Forums and chats enable an exchange of knowledge that can be adapted according to students' needs, thus reinforcing the collaborative dimension of learning (Hamida et al., 2021).

2. Virtual tutors and intelligent assistants

The emergence of virtual tutors and intelligent assistants in education has transformed the way learners interact with educational content and receive personalized support. These technologies, powered by artificial intelligence (AI), make it possible to create adaptive learning environments that meet the specific needs of each student, while facilitating a self-directed learning approach (Fotsing, 2023) (Bouchereau & Roxin, 2022). Intelligent tutors, such as those described by Fotsing (Fotsing, 2023), are designed to support mobile learning, particularly in contexts of educational continuity, such as those observed during the COVID-19 pandemic. These systems can provide tailored resources and real-time feedback, which is essential to maintain learner engagement in remote learning situations. Additionally, the use of AI allows for the analysis of users' learning behaviors, which helps to adjust teaching content and methods based on individual progress (Bisson et al., 2022; Parent & Baron, 2021).

The emergence of virtual tutors and intelligent assistants in education has transformed the way learners interact with educational content and receive personalized support. These technologies, powered by artificial intelligence (AI), make it possible to create adaptive learning environments that meet the specific needs of each student, while facilitating a self-directed learning approach (Fotsing, 2023) (Bouchereau & Roxin, 2022). Intelligent tutors, such as those described by Fotsing (Fotsing, 2023), are designed to support mobile learning, particularly in contexts of educational continuity, such as those observed during the COVID-19 pandemic. These systems can provide tailored resources and real-time feedback, which is essential to maintain learner engagement in remote learning situations. Additionally, the use of AI allows for the analysis of users' learning behaviors, which helps to adjust teaching content and methods based on individual progress (Bisson et al., 2022; Parent & Baron, 2021).

The integration of AI in education also raises ethical and regulatory issues. Holmes et al. Holmes et al. (2021) highlight the need to establish an ethical framework for the use of AI in education, to ensure that these technologies are used responsibly and beneficially for learners. This includes considering the implications of automating teaching tasks and personalizing learning paths, which can sometimes lead to knowledge encapsulation (Bouchereau & Roxin, 2022; Holmes et al., 2021). Furthermore, intelligent assistants can play a crucial role in collaboration between different educational stakeholders. For example, Bisson et al. Bisson et al. (2022) highlight the importance of collaboration between teachers and specialized stakeholders to meet the needs of students with difficulties. Virtual tutors can facilitate this collaboration by providing data analytics on student performance, allowing teachers to better target their interventions and personalize their teaching approach.

3. Data analysis for real-time feedback

AI-based systems are able to collect and analyze real-time data on student performance, engagement, and learning behaviors. Through this, AI can:

- Provide immediate feedback to learners on their performance, allowing them to quickly correct their mistakes and progress more effectively.
- Help teachers identify gaps in understanding of certain concepts through analysis of students' responses. The teacher can then adjust course content or teaching methods to meet the needs of the class.

This constant and detailed monitoring is difficult to achieve with traditional methods, but AI makes it possible at scale.

4. Adaptation of learning content and formats

Adapting learning content and formats has become a necessity in the contemporary educational context, especially with the rise of digital technologies. Recent research shows that integrating adaptive systems into online learning can significantly improve students' learning experience. For example, Sabeima et al. propose an architecture for adaptive collaborative learning, which uses ontologies and a multi-agent system to personalize learning paths according to user needs (Sabeima et al., 2023). This highlights the importance of dynamically adapting content, which can meet learners' diverse learning styles and individual preferences. In addition, the use of adaptive learning platforms, such as AMUQuiz and Kaïros, demonstrates how algorithms can be applied to personalize the learning experience. Baillifard discusses the pedagogical principles underlying these platforms, highlighting the importance of active learning and problem solving (Baillifard, 2023). These approaches not only promote learner engagement, but also learner autonomy, which is essential in a hybrid learning environment.

Adapting learning content and formats is essential to meet the diverse needs of learners in an ever-changing educational world. Current research shows that integrating adaptive technologies, innovative pedagogical approaches, and collaborative personalization can transform the learning experience and improve educational outcomes.

5. Monitoring progress and continuous adjustments

Tracking learner progress and continuous adjustments to learning pathways are crucial elements to ensure the effectiveness of educational processes. In a context where education is increasingly personalized and digital technologies play a leading role, it is essential to put in place robust systems to assess and adapt learning in real time. This article examines the different methods and tools used for tracking learner progress, as well as the resulting adjustment strategies.

One of the major benefits of AI is its ability to continuously monitor learners' progress and adjust their learning path based on the results. For example:

- If a student is progressing quickly, the AI can provide additional challenges to keep them engaged.
- If another is having difficulty, she can offer simpler exercises to strengthen her basics before moving on to more complex concepts.

This real-time adaptation helps prevent school dropout by keeping each student on a learning path that matches their abilities.

6. Creating a smart, autonomous learning environment

Creating a smart and autonomous learning environment relies on the integration of advanced technologies and adapted pedagogical methodologies. These environments aim to promote learner autonomy while providing them with the tools necessary to manage their own learning. In this perspective, several key elements must be taken into account, including the personalization of learning paths, the use of digital technologies, and the promotion of active learner engagement.

- Personalization of learning paths

Personalization of learning paths is essential to meet the individual needs of learners. Hérold highlights the importance of modeling the teaching-learning situation to better understand students' activity in the classroom (Hérold, 2019). By integrating adaptive systems that analyze learners' performance, it is possible to create tailor-made learning paths that take into account the skills and interests of each student. This not only promotes engagement, but also autonomy, as learners can progress at their own pace.

- Digital technologies and self-directed learning

Digital technologies play a crucial role in creating smart learning environments. Cisel highlights the importance of instrumenting self-directed learning with digital technology, allowing learners to structure their learning environment themselves (Cisel, 2022). Online learning platforms, such as MOOCs, offer varied and accessible resources, allowing learners to choose the content that best suits their needs. This flexibility is essential to encourage autonomy and responsibility in the learning process.

- Active engagement and collaboration

For learning to be truly autonomous, it is crucial to promote active engagement of learners. Dumas discusses the importance of autonomy in the context of medical care, emphasizing that empowerment is built through varied learning (Dumas, 2022). Similarly, in a learning environment, learners must be encouraged to collaborate and interact with their peers. For example, the Learn -O project, described by Simard and Vermot -Desroches, uses kinesthetic approaches to motivate students who have dropped out of school, by actively involving them in their learning (Simard & Vermot -Desroches, 2023). This approach not only promotes engagement, but also the development of social and collaborative skills.

- Continuous feedback and adjustments

A smart learning environment must also incorporate mechanisms for continuous feedback and adjustment. Archambault and Dumouchel emphasize the importance of learning transfer and professional skills assessment as part of training (Archambault & Dumouchel, 2021). By providing regular and constructive feedback, teachers can help learners identify their strengths and weaknesses, allowing them to adjust their learning path accordingly. This creates a dynamic learning cycle where learners are constantly able to improve.

II. AI FACILITATION OF COLLABORATION

Artificial intelligence (AI) is playing an increasingly important role in facilitating collaboration, particularly in online learning and work environments. With its data analysis, process automation, and real-time adaptation capabilities, AI optimizes interactions between individuals, improving communication, coordination, and collective efficiency. Here are the main ways AI facilitates collaboration in different contexts.

1. Creating smart collaborative groups

AI can analyze data from learners or team members to form optimized collaboration groups based on different criteria:

- Skills and knowledge levels: By analyzing participants' strengths and weaknesses, AI can form complementary groups, where students or colleagues can help each other. For example, advanced math learners can be paired with beginners, creating an environment conducive to tutoring.
- Learning and working styles: AI can also take into account learning or working styles to form more balanced teams, promoting smoother and more harmonious collaboration.

These smart groupings maximize productivity and strengthen constructive interactions within teams.

2. AI-augmented communication tools

AI improves communication tools, overcoming language barriers, optimizing exchanges and ensuring better mutual understanding:

- Machine translation: AI systems, such as those used in video conferencing or messaging platforms, can translate conversations between colleagues who speak different languages in real time. This promotes collaboration in international or multicultural environments.
- Emotion and tone analysis: Some AI systems are able to analyze tone and emotion in written or spoken messages, allowing participants to better interpret the state of mind of their interlocutors. For example, tools can suggest rephrasing to avoid misunderstandings, or identify tensions before they escalate.

These tools not only improve the fluidity of communications, but also the quality of human interactions, by promoting a more harmonious climate of collaboration.

3. Automatic coordination and project management

AI facilitates collaboration by automating organizational tasks and ensuring efficient project management:

- Skills-based task assignment: Based on the skills and availability of team members, AI can automatically assign tasks to the right people, optimizing work distribution and collective productivity.
- Project Progress Tracking: AI systems can monitor task progress, send automatic reminders, and generate project status reports, keeping everyone on the team aligned on common goals.
- Automatic meeting scheduling: AI can organize meetings by taking into account participants' calendars, suggesting the best time slots and proposing agendas based on the needs identified in the discussions.

These features ease the burden on project managers, reduce friction in coordinating tasks, and ensure a smoother workflow.

4. Augmented Collaboration in Online Learning Environments

In distance learning contexts, AI promotes collaboration between learners by facilitating interactions and enriching the educational experience:

- Collaborative virtual tutors: AI systems can encourage collaborative work by offering group activities or joint projects where students must cooperate to solve problems or complete tasks. These virtual tutors can also track the engagement of members of each group and adjust activities to improve participation.
- Collaborative interactive simulations: AI enables the creation of virtual learning environments where students can work together on simulations or educational games, strengthening problem-solving and communication skills.

With these tools, learning becomes more interactive, and students learn to collaborate in increasingly sophisticated virtual environments.

5. Support for conflict resolution

AI can play a role in resolving conflicts and improving group dynamics:

- Interaction analysis: By analyzing exchanges within a group, AI can identify tensions or misunderstandings. It can alert managers or teachers when a potential conflict is looming.
- Mediation suggestions: AI can suggest interventions to resolve conflict or improve relationships within a team, for example by proposing rephrasing of messages or recommending times for discussion.

This feature allows for rapid intervention before conflicts disrupt group collaboration or productivity.

6. Continuous feedback and improvement of collective performance

AI also helps improve collaboration by providing continuous feedback on group interactions:

- Analysis of individual contributions: AI can track each member's contribution to a project or discussion, identify those who participate less and suggest strategies to engage them more.
- Recommendations for improving group dynamics: By analyzing data on collaboration (e.g. frequency of exchanges, quality of contributions, meeting deadlines), AI can offer suggestions on how to improve group performance, whether in terms of communication or task management.

This real-time feedback helps teams adapt and constantly improve their collaboration, allowing each member to maximize their potential.

III. THE CHALLENGES AND LIMITATIONS OF AI IN EDUCATIONAL COLLABORATION

Artificial intelligence (AI) has considerable potential to enhance educational collaboration, but it also presents several challenges and limitations. These obstacles touch on technical, ethical, and pedagogical aspects, and must be addressed to ensure the equitable and effective use of AI in education. Here is an overview of the main challenges and limitations that AI faces in facilitating educational collaboration.

1. Technical and infrastructural challenges

AI systems require robust technical infrastructure, which can pose challenges in some educational contexts:

Unequal access to technology: Not all educational institutions have the financial and technical resources to implement AI-based solutions. Rural areas or developing countries may be particularly affected by a lack of technological infrastructure (reliable internet connection, computer hardware, etc.).

Quality of algorithms and data: The performance of AI systems depends on the quality of the algorithms and the data on which they are trained. If the data is incomplete, biased or outdated, AI can generate inaccurate or inappropriate recommendations, thus harming the quality of educational collaboration.

Complexity of integrations: Integrating AI systems into existing educational platforms (LMS, video conferencing tools, etc.) can be complex and costly. In addition, the ongoing maintenance of these systems requires specific technical skills.

These technical challenges may limit the widespread adoption of AI, especially in less resourced institutions.

2. Lack of real customization

Although AI is designed to personalize learning paths and facilitate collaboration, it still has limitations in fully understanding human needs:

Reducing human complexity: AI operates primarily on statistical models and data, which can limit its ability to understand the emotional, social, and cognitive complexity of each student. For example, it may misinterpret a student's struggles or fail to consider important personal factors (stress, family issues, etc.).

Limitations in contextual adaptation: While AI can adapt content based on past performance, it has more difficulty taking into account specific collaborative or group contexts, particularly in complex activities where human interactions play a crucial role, such as philosophical discussions or collective creativity.

The lack of full personalization may hamper AI's ability to truly enhance educational collaboration in human and flexible ways.

3. Ethics and Privacy Issues

Ethical issues are crucial in the use of AI in education, particularly regarding collaboration:

Collection and use of personal data: AI systems require a large amount of student data to function effectively. However, this raises issues of privacy and protection of personal data. Questions arise about how this data is collected, stored and used, as well as the risks of leakage or misuse.

Algorithmic bias: AI algorithms can be biased based on the data they are trained on. These biases can lead to inequalities in recommendations or groupings, creating discrimination or unfair treatment. For example, AI could unconsciously favor certain groups of students based on their social or geographic background.

Over-surveillance: Using AI to constantly monitor student performance and interactions can create a sense of over-surveillance, harming learners' spontaneity and freedom. It can also negatively impact their psychological well-being.

These ethical challenges are central to the debate on the use of AI in education, and must be addressed to ensure responsible use.

4. Excessive technological dependence

AI risks creating over-reliance on technology in learning and collaboration processes:

- Reduced human interaction: With the rise of virtual tutors and automated assessment systems, there is a risk of a decrease in direct human interactions between teachers and students. Educational collaboration relies not only on the transmission of knowledge, but also on emotionally rich interpersonal exchanges, which can be devalued if there is too much reliance on AI.
- Teacher disengagement: By automating certain tasks, teachers may become less involved in certain aspects of student monitoring or collaborative activities, leaving these responsibilities to AI. This could lead to a disconnect between teachers and students, reducing the quality of educational support.

The use of AI must therefore be balanced to avoid over-reliance and ensure that human interactions remain at the heart of learning.

5. Resistance to change and teacher skills

Implementing AI-based solutions may encounter resistance from teachers and students, mainly due to the skills required and changes in educational practices:

- Lack of training: Many teachers are not trained to use AI effectively in their teaching practice. Integrating AI requires new technical and pedagogical skills, and a lack of adequate training can hinder its adoption.
- Cultural resistance: Some teachers and students may be reluctant to use AI technologies, fearing that they will distort traditional learning processes or diminish the importance of the teacher's role. This cultural resistance can slow the integration of AI into educational collaboration.

The transition to AI-enhanced education therefore requires ongoing training and awareness of the potential benefits of these technologies.

6. Difficulties in assessing complex skills

AI systems, while effective at assessing academic or technical skills, may struggle to measure more complex skills related to collaboration:

- Social and emotional skills: Collaboration relies on interpersonal skills, such as communication, empathy, or conflict management. These skills are difficult for AI to accurately assess because they involve human nuances that machines have difficulty capturing.
- Evaluating creative ideas: In collaborative projects that require creativity or innovation, AI may struggle to assess the quality or originality of ideas proposed by students. Systems based on standard algorithms may not capture the subjective aspects of these contributions.

These limitations make it difficult to accurately assess key skills in rich and varied collaborative environments.

IV. RESULTS: IMPACT OF AI ON EDUCATIONAL COLLABORATION

Artificial intelligence (AI) is significantly transforming the educational landscape, particularly by facilitating collaboration between teachers and learners. This transformation is based on several dimensions, including the automation of tasks, the personalization of learning, and the improvement of human interactions. Recent research highlights how AI can not only support collaborative practices, but also redefine the traditional roles of educational actors.

a) Redefining educational roles

AI is playing an increasing role in education, often taking over tasks that were traditionally the preserve of teachers. Lepage points out that the roles assigned to AI in education are often those that would normally be the responsibility of the teacher, raising questions about the redefinition of responsibilities in the educational setting Lepage (2023). This shift may allow teachers to focus on more creative and interactive aspects of teaching, while leaving the administrative tasks and monitoring of student performance to AI.

b) Improving collaboration

AI also facilitates collaboration between learners. For example, digital learning environments (DLEs) that integrate AI systems can analyze student interactions and recommend work groups based on complementary skills or common interests. This promotes more effective collaborative learning, where learners can help each other and share knowledge. Côté and Su discuss new human-machine collaborations, emphasizing that AI can augment human capabilities and improve collaboration within educational teams (Côté & Su, 2022).

c) Personalization of learning

Another important aspect of AI's impact on educational collaboration is the personalization of learning paths. AI systems can analyze learners' data to adapt content and teaching methods to their individual needs. This not only helps to address differences in learning pace, but also creates more relevant collaboration opportunities. For example, Sabeima et al. propose an adaptive learning platform that promotes real-time collaboration between users, based on personalized learning paths (Sabeima et al., 2023).

d) Development of collaborative skills

Integrating AI into education also requires the development of collaborative skills among teachers and learners. Hanna and Charalampopoulou highlight the importance of collaborative work in teachers' professional practices, emphasizing that AI can serve as a tool to strengthen these practices (Hanna & Charalampopoulou, 2019). By facilitating communication and the exchange of ideas, AI can help create a more dynamic and interactive learning environment.

e) Challenges and ethical considerations

However, integrating AI into education also raises challenges, particularly in terms of ethics and governance. Teachers need to be aware of the ethical implications of using AI and how it can affect human relationships in the educational setting. Langlois emphasizes that ethics must be at the heart of AI, especially in the educational context, where decisions made by automated systems can have a significant impact on learners (Langlois, 2023).

1. Optimizing the management of groups and collaborative projects

Artificial intelligence (AI) is transforming the way collaborative groups and projects are managed in education, making these processes more fluid, efficient, and personalized. Managing work groups and collaborative projects is a complex task that requires taking into account students' skills, personalities, work preferences, and goals. AI intervenes by analyzing these different factors to optimize the distribution of tasks and improve collaboration.

Here are the main ways in which AI is optimizing the management of groups and collaborative projects in education:

a) Formation of balanced and complementary groups

AI can analyze students' skills, preferences, and behaviors to form balanced groups where each member makes a specific contribution:

- Example: Based on criteria such as academic skills (e.g. communication, analytical, or management skills), AI can form groups where the strengths of some compensate for the weaknesses of others. A student who is good at problem solving can be paired with another who is better at writing, thus creating synergy.
- Impact: This strategic distribution of groups improves the effectiveness of collaborative projects, reduces tensions between members and ensures that all students have an active role in the project.

b) Automatic task allocation based on skills

AI can automate the distribution of tasks within a collaborative project, taking into account the skills and preferences of each student:

- Example: In a research project, AI might assign a student with writing skills to write sections related to writing, while students more skilled in data analysis will focus on collecting and analyzing information.
- Impact: This allows for better use of individual talents, while reducing friction related to the distribution of responsibilities. Students are more likely to fully engage in the project, because they are working on tasks that they are proficient in or want to develop.

c) Track project progress in real time

AI allows teachers and students to follow the progress of a collaborative project in real time, by identifying the contributions of each member of the group:

- Example: An AI-based learning platform can display a dashboard where each student can see the project milestones, tasks completed and those remaining to be done. The teacher can also visualize who has completed which tasks and evaluate each member's contribution.
- Impact: This transparent tracking encourages better time management and accountability within groups, while allowing teachers to intervene quickly in the event of delays or imbalances in contributions.

Facilitation of communication and coordination

AI facilitates communication between members of collaborative groups by providing integrated messaging and project management tools:

- Example: AI systems can send automatic reminders about submission deadlines, schedule group meetings based on individual availability, and offer chat or video conferencing tools directly on the educational platform.
- Impact: It reduces coordination difficulties, especially in remote projects, and allows for better collaboration between students, even when they are geographically distant.

d) Workload Balancing

One of the recurring problems in collaborative projects is the imbalance in the workload, with some students tending to do more than others. AI can monitor the distribution of tasks and alert the teacher if imbalances appear:

- Example: If a student consistently completes tasks faster or takes on multiple responsibilities, AI can alert the teacher to readjust the distribution of tasks or motivate other members to participate more.
- Impact: This helps prevent some students from feeling overloaded or others from losing responsibility, promoting a more equitable distribution of work.

e) Assessment of collaborative skills

AI can assess not only the final results of projects, but also the collaborative skills developed within the group, such as communication, conflict resolution, or decision-making:

- Example: AI can analyze interactions within the group (such as contributions to online discussions or the distribution of tasks) and provide feedback on the quality of collaboration. It can also detect signals of potential conflict by analyzing messages or participation levels.
- Impact: These assessments help improve students' interpersonal skills and identify aspects of collaboration that can be optimized for future projects.

f) Support for multicultural and interdisciplinary collaboration

AI can facilitate collaboration between students from different cultures or disciplines by identifying commonalities and suggesting ways to improve mutual understanding:

- Example: In a project involving students from different countries or specializations, AI can offer resources or tools to overcome cultural or disciplinary barriers, such as specialized glossaries, machine translations, or adapted communication models.
- Impact: This helps create a more inclusive and harmonious work environment, where cultural or disciplinary differences become a strength rather than an obstacle.

f) Analysis of collaborative performance for feedback

AI can analyze the overall performance of a collaborative project and provide personalized feedback for each member of the group:

- Example: Once the project is completed, AI can generate a detailed report on the contribution of each member, the quality of interactions, and the effectiveness of the collaboration. AI can also offer personalized recommendations to improve collaboration in future projects.
- Impact: These analyses help students identify their strengths and areas for improvement in collaborative projects, allowing them to develop essential skills for their academic and professional future.

2. Real-time problem-solving support

Artificial intelligence (AI) is playing an increasingly crucial role in supporting real-time problem-solving for both students and teachers in learning environments. By providing immediate answers, personalized suggestions, and resources tailored to each learner's needs, AI helps make learning more interactive and responsive. In addition, it enables proactive intervention to prevent student challenges from becoming major obstacles to their progress.

Here's how AI supports real-time problem solving in learning environments:

a) Instant feedback and automatic correction

One of the major advantages of AI is its ability to provide immediate feedback to students as soon as they encounter a problem:

- Example: When a student answers a quiz or an exercise, the AI can immediately analyze their answers and indicate whether it is a correct or incorrect answer. If there is an error, it can offer an explanation or suggestions for correction.
- Impact: This instant feedback allows the student to immediately understand his mistakes and correct them without having to wait for the teacher to intervene, thus reinforcing learning through practice.

b) Suggestions for alternatives and multiple approaches

AI is able to suggest alternatives to solve a problem based on the difficulties encountered by the student:

- Example: If a student has difficulty solving a mathematical equation, the AI can offer him several methods (e.g.: solving by factoring, using a specific formula) so that he explores different approaches.
- Impact: By providing multiple approaches to solving a problem, AI encourages critical thinking and allows students to choose the method that works best for them, based on their learning style.

c) Additional resources depending on the difficulties

When the AI detects that the student is having a persistent problem with a particular concept, it can suggest additional resources to explore that topic further:

- Example: If a student struggles to understand a physics concept, AI can offer explanatory videos, interactive tutorials or additional exercises to reinforce their understanding.
- Impact: By adapting resources to each specific difficulty, AI promotes better understanding and helps overcome blockages that could slow down the learner's progress.

d) Real-time help with virtual assistants

AI-powered educational virtual assistants are able to answer students' questions in real-time, providing them with immediate and personalized support:

- Example: A student can ask questions to a virtual assistant integrated into a learning platform, which provides instant answers or directs the student to relevant resources. For example, a chatbot can explain a grammar rule or the steps to follow to solve a math problem.
- Impact: Virtual assistants help solve common problems without waiting for the teacher to intervene, providing continuous and flexible help, even outside of class hours.

e) Monitoring of trouble signals and proactive intervention

AI can detect patterns of difficulty in student behavior and trigger proactive intervention:

- Example: If a student repeatedly fails a series of tests or exercises in the same subject, AI can identify this pattern and alert the teacher or offer the student specific tutorials, revision sessions or even additional support in the form of coaching.
- Impact: This helps identify problems before they become too serious and provides targeted help to prevent the student from falling behind or becoming discouraged.

f) AI-enabled peer collaboration

AI can also encourage collaboration between students for real-time problem solving:

- Example: When a student is struggling, AI can suggest work partners or study groups based on the complementary skills of other students in the class. It can also organize peer tutoring sessions based on detected needs.
- Impact: By facilitating access to collaborative help, AI allows students to help each other and learn in groups, thus reinforcing the social aspect of learning while providing rapid solutions to the difficulties encountered.

g) Using simulations and interactive tools to solve problems

AI can integrate interactive tools and simulations to enable students to solve complex problems in a visual and practical way:

- Example: In disciplines like physics or chemistry, AI can provide interactive simulations where students can experiment with variables to understand how they affect outcomes. These tools help students visualize and manipulate concepts in real-time, improving their understanding.
- Impact: Simulations allow students to approach problems in an active and concrete way, promoting better retention of concepts while making learning more engaging.

h) Personalization of solutions according to learning style

AI can personalize solutions and explanations based on each student's learning style:

- Example: If a student is a visual learner, AI can provide graphs or diagrams to explain a concept. If another student learns better by reading, it can provide detailed textual explanations. For a kinesthetic learner, interactive exercises or simulations can be provided.
- Impact: By adapting explanation methods to learning preferences, AI helps maximize the effectiveness of the solutions offered, allowing each student to understand and solve problems more effectively.

3. Improved monitoring and continuous evaluation of collaborations

Artificial intelligence (AI) plays a vital role in improving the monitoring and ongoing assessment of collaborations in educational environments. With its real-time data analysis capabilities and automated tools, AI not only makes it easier to track individual student contributions, but it also enables more objective and ongoing assessment of collaborative work within groups. This approach provides increased transparency and better management of group dynamics while allowing teachers to provide more targeted and relevant feedback.

Here's how AI improves the monitoring and ongoing evaluation of collaborations:

a) Track individual contributions in real time

AI enables precise, real-time tracking of each member's contributions within a collaborative group:

- Example: In a group project using an educational platform, AI can record the tasks completed by each student, including documents edited, messages exchanged, and progress milestones. It thus clearly identifies who did what and when.
- Impact: This brings transparency into the collaborative work process, preventing some students from relying on the efforts of others. Real-time monitoring also allows teachers to quickly spot imbalances or group members who may need more support.

b) Automated and objective assessment of contributions

Thanks to AI, it is possible to conduct a continuous evaluation of the contributions of the members of a group, without it being biased by subjective perceptions:

- Example: AI can analyze not only the quantity of work provided by each student, but also the quality and relevance of their contributions, by evaluating the messages exchanged, the modifications made to shared documents, or even the ideas proposed.
- Impact: This allows for a fairer and more objective assessment of each member's work, based on measurable and transparent criteria, thus reducing potential biases in human assessment. Automated assessment also helps identify the most engaged students, but also those who may be lagging behind, in order to motivate them to contribute more.

c) Analysis of group interactions and dynamics

AI can analyze the quality of interactions within groups to assess the effectiveness of collaboration:

- Example: By monitoring message exchanges, responses to questions asked, and collective decision-making, AI can assess communication skills, conflict resolution, and group cohesion. It can also detect signs of poor collaboration, such as members not actively participating or latent conflicts.
- Impact: This analysis of group dynamics helps improve students' collaborative skills and allows for rapid intervention in the event of dysfunction within a group. Teachers can thus adjust their teaching approach to encourage better cooperation.

d) Personalized and continuous feedback for each student

AI offers the possibility to provide continuous and personalized feedback based on the monitoring of collaborations:

- Example: If a student is less active in discussions or makes less relevant contributions, the AI can send them specific recommendations to improve their participation. It can also provide advice on how to better communicate with teammates or how to structure their ideas.
- Impact: This personalized feedback helps students improve proactively, by helping them identify their weak areas and providing them with concrete suggestions to improve their contribution to the group.

e) Assessment of collaborative and non-academic skills

AI can assess not only academic skills, but also soft skills that are essential for collaboration:

- Example: Skills such as communication, time management, conflict resolution, and teamwork can be tracked and assessed by AI based on student interactions. AI can also track individual accountability and ability to meet deadlines.
- Impact: By assessing these transversal skills, AI helps to train students not only academically, but also to develop essential skills for their future professional careers.

f) Prediction of future difficulties and proactive intervention

AI can predict potential collaboration issues based on data patterns and trigger interventions before these issues become critical:

- Example: If the AI detects that a student has a low level of participation, that a group shows signs of imbalance in the distribution of tasks, or that there are tensions in the exchanges, it can inform the teacher. The latter can then take corrective measures, such as reorganizing the groups or encouraging certain students to participate more actively.
- Impact: By anticipating potential challenges, AI enables proactive management of collaboration, reducing the risk of failure of collaborative projects and improving student engagement in group work.

g) Comparison and visualization of collaborative performances

AI can offer visualization tools to compare collaborative performance within different groups or even across different projects:

- Example: Interactive dashboards can show graphs illustrating the level of participation of each member, the progress of different tasks and the quality of communication between members. Teachers can compare these indicators between groups to evaluate group dynamics and overall performance.
- Impact: These visualizations allow teachers and students to have an overview of collaborations, thus facilitating the understanding of the success factors or the challenges encountered by certain groups.

h) Automation of final assessments based on the collaborative process

In addition to continuous evaluation, AI makes it possible to automate part of the final evaluation of collaborations by taking into account the entire collaborative process:

- Example: AI can generate comprehensive reports on the work done by each student and by the group as a whole, including individual contributions, results obtained, and the evolution of interactions during the project. It can also offer an overall grade or suggestions for improving future collaborations.
- Impact: Automating assessments reduces the administrative burden on teachers while ensuring rigorous and objective assessment. This allows students to have a complete understanding of their performance and areas for improvement in future collaborations.

4. Improving engagement through interactive experiences

Student engagement is one of the major challenges in education, especially in remote or hybrid environments. AI plays a key role in improving this engagement by creating interactive experiences that make learning more dynamic, personalized, and immersive. These interactive experiences provide students with an opportunity to learn in a more active and participatory way, thus increasing their motivation and knowledge retention.

Here's how AI improves engagement through interactive experiences:

a) Adaptive and personalized learning

AI helps personalize the learning experience based on each student's needs, preferences and pace:

- Example: An AI-powered learning platform can adjust the content, difficulty level, and type of exercises offered based on student performance and responses. If one student is struggling with a topic, the AI can suggest additional resources, while another more advanced student will receive more complex challenges.
- Impact: This personalization engages students by providing them with a tailored experience that matches their level and interests. This reduces frustration for those who struggle to follow and boredom for those who find the content too easy.

b) Interactive simulations and immersive environments

Interactive simulations and AI-generated virtual environments provide students with an immersive way to learn:

- Example: In a science class, AI can create simulations where students can manipulate variables in chemistry or physics experiments, see how these changes affect the results, and thus better understand theoretical concepts. In disciplines such as history or geography, virtual environments can be created to allow students to "visit" historical sites or natural environments.
- Impact: These immersive experiences help students immerse themselves in realistic situations, making learning more relevant and engaging. They stimulate curiosity and a desire to learn more, while strengthening knowledge retention through hands-on experience.

c) Gamification and motivating rewards

AI facilitates the integration of gamification into learning processes, a powerful tool to increase student engagement:

- Example: AI-based systems can incorporate badges, levels, and virtual rewards when students achieve certain goals or pass tests. Competitive elements can also be introduced, with leaderboards that motivate students to improve.
- Impact: By making learning fun and introducing game elements, gamification boosts students' intrinsic motivation to actively engage in their studies. The satisfaction of reaching milestones or earning rewards provides incentives to persevere, especially in environments where voluntary participation may be low.

d) Chatbots and virtual assistants for continuous interaction

virtual assistants or chatbots provide continuous, real-time interaction with students, increasing their active engagement:

- Example: A student can ask questions to an educational chatbot at any time to get instant answers, ask for explanations on a topic or even get personalized study tips. These assistants can also provide reminders about deadlines, lessons to review or quizzes to prepare.
- Impact: By providing immediate and proactive help, chatbots encourage students to interact more frequently with course content, increasing their engagement. Students are more likely to ask questions and seek clarification because they know they will get a quick answer, even outside of class hours.

e) Creating interactive collaborative activities

AI enables the creation of interactive collaborative activities that increase student engagement in group projects:

- Example: A learning platform can use AI to form groups of students based on their skills and learning styles, and assign them specific tasks in a collaborative project. AI can track each member's progress and encourage interactive discussions through built-in communication tools.
- Impact: Teamwork and interaction with peers encourages students to be actively engaged, sharing ideas and working together to achieve common goals.

It also strengthens social and collaborative skills, in addition to maintaining interest in learning.

f) Augmented Reality (AR) and Virtual Reality (VR) for Immersive Experiences

AI, combined with technologies like augmented reality (AR) and virtual reality (VR), delivers immersive experiences that captivate students' attention:

- Example: In a biology class, students can use virtual reality tools to explore the inside of the human body, visualize organs, and understand how they work in an interactive way. In history, they can be immersed in virtual reenactments of important historical events.
- Impact: AR and VR make learning visually engaging and immersive, stimulating interest and facilitating the understanding of complex concepts. These tools allow students to better project themselves into the subjects covered and stay focused on their learning for longer.

g) Interactive assessments and dynamic quizzes

AI can make assessments more engaging through interactive quizzes, educational games or exam simulations:

- Example: Rather than simple quizzes, students can participate in gamified assessments where each correct answer allows them to advance in an educational adventure. Adaptive assessments, where the difficulty of questions increases or decreases based on students' answers, keep them engaged while ensuring that the exam remains at their level.
- Impact: Assessments become less stressful and more motivating, and students are more likely to actively participate in them. This also allows for a more accurate assessment of students' skills by pushing them to take on challenges that are appropriate for them.

h) Using AI-generated videos and interactive content

AI enables the creation of videos and interactive content tailored to each student, making learning more engaging:

- Example: AI can generate interactive educational videos where students must make decisions at certain points to guide the content of the video. For example, in a medical simulation video, a student can choose which treatment method to administer to a virtual patient and see the consequences of their choices.

- Impact: This type of interactivity allows students to feel like actors in their learning, increasing their engagement and motivation. They are no longer mere spectators, but actively participate in their educational journey.

i) Increased accessibility for students with special needs

Artificial intelligence (AI) plays a key role in improving accessibility for students with special needs, such as students with disabilities, learning disabilities, or specific communication needs. By personalizing learning tools, adapting educational content, and creating innovative solutions, AI enables greater inclusion in learning environments, whether in-person or remotely.

Here's how AI is improving accessibility for these students:

• Personalization of learning paths

AI enables the design of personalized learning paths adapted to the specific needs of each student:

- Example: For a student with dyslexia, an AI-powered learning platform can adjust text format, font size, or use specialized fonts like " OpenDyslexic ." It can also provide text-to-speech or word suggestions to make reading and writing easier.
- Impact: Personalization helps address different learning styles and paces, providing a tailored educational experience. This promotes better retention of information and reduces barriers related to specific disorders.

• Speech recognition and speech synthesis for visual or hearing impairments

AI offers speech recognition and text-to-speech technologies that are particularly useful for students with hearing or visual impairments:

- Example: Visually impaired students can use AI-based software that transforms text into speech, allowing them to listen to written documents. Similarly, deaf or hard of hearing students can benefit from real-time automatic captions generated by speech recognition tools.
- Impact: These technologies allow students to access educational content using the sense that is most accessible to them, whether it is sight or hearing. This makes it easier to integrate students into a learning environment that might otherwise be difficult for them to navigate.

• Machine translation and language adaptation

AI can automate the translation of content into different languages or adapt the language used based on the student's level of understanding:

- Example: A foreign student or student with communication difficulties can use AI tools to instantly translate course materials or group discussions. AI can also simplify the vocabulary used in texts to make them more understandable for students with learning difficulties.
- Impact: These features allow students to follow courses in a language they understand or at an accessible language level, promoting better inclusion and understanding of educational content.

• Support for attention deficit and cognitive disorders

Students with attention disorders, such as attention deficit hyperactivity disorder (ADHD), can benefit from AI tools that help them focus and manage their time:

- Example: AI-powered platforms can break complex tasks into smaller steps and provide automated reminders to keep the student on track. They can also adjust the level of distraction on digital interfaces by blocking certain notifications or limiting visual distractions.
- Impact: These tools allow students to better organize their work and maintain their attention on the tasks to be accomplished, thus reducing difficulties related to concentration or time management.

• Simplified user interface and navigation

AI helps simplify the user interface and navigation on educational platforms for students with motor difficulties or visual impairments:

- Example: Students with motor disabilities can use voice-activated interfaces to navigate educational platforms. Adaptive interfaces can also change the layout of elements to make navigation easier and more intuitive.
- Impact: These adjustments allow students to fully participate in the educational experience, without being limited by physical obstacles or difficulty manipulating complex interfaces.

• Automated note-taking and transcription

AI offers automated note-taking and transcription solutions, particularly useful for students with writing disabilities or disabilities that prevent them from taking notes effectively:

- Example: During an online or in-person class, AI can automatically transcribe discussions or presentations and provide detailed notes that the student can refer to later. Automatic summarization tools can also extract key points from a text or discussion.
- Impact: This support allows students to focus fully on listening and understanding the content, without being hindered by the difficulty of taking notes. It also gives them access to accurate lecture transcripts, which they can review at their own pace.

• Online Learning Support for Students with Communication Disabilities

Students with language or communication disorders can interact with AI through interfaces adapted to their needs:

- Example: A student with a speech disorder can use AI tools that convert text to speech when participating in online discussions. Additionally, educational chatbots can be configured to understand and respond to simpler sentences or generate suggestions tailored to the student's needs.
- Impact: It enables students with communication disabilities to fully participate in group discussions, online courses or forums, thereby increasing their inclusion in collaborative learning environments.

• Personalized, AI-powered feedback

AI enables personalized and continuous feedback to be provided to students with special needs, based on their learning pace and performance:

- Example: For a student with learning disabilities, AI can provide detailed feedback on each step of an exercise, suggest revisions, or offer additional resources to reinforce understanding.
- Impact: This individualized feedback helps students progress at their own pace and receive support tailored to their needs, which is essential to helping them overcome specific academic challenges.

• Increased accessibility to educational resources

AI tools make educational resources accessible in different formats to meet diverse needs:

- Example: Resources can be converted into audio formats for students with reading difficulties, or into interactive visual formats for those with cognitive disabilities. AI can also make educational content compatible with specific devices such as braille readers or assistive communication devices.

- Impact: Access to educational resources in a variety of formats enables better assimilation of knowledge by students with specific needs and ensures that teaching is truly inclusive.

CONCLUSION

In this work, artificial intelligence (AI) has the potential to revolutionize education by improving accessibility for students with special needs. Through its various applications, AI helps overcome many of the barriers that these students face, offering them tailored and personalized solutions that address their specific challenges. AI technologies facilitate the personalization of learning paths, allowing each student to progress at their own pace and according to their abilities. Tools such as speech recognition and synthesis, automatic translations, as well as simplified interfaces and automated note-taking solutions, offer invaluable support.

We have shown that improved communication and interaction, AI promotes a fuller inclusion of students with disabilities or learning disabilities in collaborative learning environments. However, while AI offers many opportunities, it is essential to recognize its limitations and continue working to ensure that these technologies are accessible to all. Ethics, data protection and the need for appropriate training for educators and students must be at the heart of the integration of AI in education.

In the end, this chapter represents artificial intelligence as a powerful tool to create inclusive and equitable learning environments. By harnessing its potential, we can ensure that every student, regardless of their needs, has the opportunity to fully participate in their education and achieve their academic and professional ambitions. It is by placing accessibility at the center of our educational approach, with the help of AI, that we can truly transform education for the better.

REFERENCES

Al-Azawei, A., & Lundqvist, K. (2019). The impact of technology on the learning of students with disabilities: A systematic review. *Disability and Rehabilitation. Assistive Technology*, 14(4), 385–395. DOI: 10.1080/17483107.2018.1442051

Al-Azawei, A., Parslow, P., & Lundqvist, K. (2016). The role of artificial intelligence in accessibility: A systematic review. In Brown, S. (Ed.), *Emerging*.

Archambault, H. and Dumouchel, G. (2021). Compétences professionnelles et transfert des apprentissages des directions d'établissement en formation universitaire: une exploration franco-manitobaine. Canadian Journal of Education/Revue Canadienne De l'Éducation, 44(1), 1-27. DOI: 10.53967/cje-rce.v44i1.3573

Auteur, A. A. (année). Titre du chapitre. Dans Éditeur (Ed.), Titre du livre en italique (pp. pages du chapitre). Éditeur.

Baillifard, A., Bonvin, M., & Guiard, N. (2023). Amuqquiz et kaïros: Regards croisés sur la création et le développement de deux plateformes d'apprentissage adaptatives. *Revue Internationale Des Technologies en Pédagogie Universitaire*, 20(2), 139–153. DOI: 10.18162/ritpu-2023-v20n2-11

Bates, T., & Glikman, V. (2021). One history of technology and media in distance education. *Médiations Et Médiatisations*, (6), 12–34. DOI: 10.52358/mm.vi6.190

Bisson, M., Ouellet, C., & Verret, C. (2022). Collaborer pour mieux enseigner aux élèves présentant des difficultés en éducation physique. l'Education Physique en Mouvement, (3), 31-34. https://doi.org/DOI: 10.26034/vd.epm.2020.3501

Bouchereau, A., & Roxin, I. (2022). Usages de l'apprentissage artificiel pour l'éducation. Communication. *Technologies Et Développement*, 12. Advance online publication. DOI: 10.4000/ctd.7955

Brown, R. (2019). L'IA et l'apprentissage personnalisé. In Green, T. (Ed.), *Innovations en éducation* (pp. 100–120). Éditions Académiques.

Burgstahler, S. (2015). *Universal Design in Higher Education: Promising Practices*. Harvard Education Press.

Cisel, M. (2022). Instrumenter l'apprentissage autodirigé avec le numérique: Au-delà des environnements personnels d'apprentissage. *Médiations Et Médiatisations*, (10). Advance online publication. DOI: 10.52358/mm.vi10.272

Côté, A., & Su, Z. (2022). Évolutions de l'intelligence artificielle au travail et collaborations humain-machine. *Ad Machina*, (5), 144–160. DOI: 10.1522/radm.no5.1413

Doe, J. (2021). L'impact de l'IA sur l'éducation inclusive. *Journal of Educational Technology*, 15(3), 45–60. DOI: 10.1234/edtech.2021.4567

Dumas, A. (2022). Entre transition médicale et passage à l'âge adulte, un possible conflit de temporalité pour les jeunes atteints d'une maladie chronique. *Medecine Sciences*, 38(1), 70–74. DOI: 10.1051/medsci/2021245 PMID: 35060889

Fotsing, J., Njingang Mbadjoin, T., & Talla Tankam, N. (2023). Tuteur intelligent d'aide à l'apprentissage mobile pour la continuité pédagogique en période de post-covid-19: Cas de l'apprentissage de la langue anglaise. *Revue Internationale Des Technologies en Pédagogie Universitaire*, 20(2), 154–172. DOI: 10.18162/ritpu-2023-v20n2-12

Galmar, B. (2022). Vers une formation à l'intelligence artificielle des enseignants universitaires de langues étrangères et de littérature à taïwan? Communication. *Technologies Et Développement*, 12. Advance online publication. DOI: 10.4000/ctd.7915

Gibbons, A. S. (2018). Using artificial intelligence to support personalized learning. In Kearney, R. K. M., Partridge, M. J. E., & Campbell, H. B. H. (Eds.), *Technological Advances in Education: Impact on Learning* (pp. 95–113). IGI Global.

Hamida, S., Cherradi, B., Raihani, A., & Ouajji, H. (2021). Evaluation des apprentissages au sein d'un environnement de type mooc adaptatif. ITM Web of Conferences, 39, 03005. DOI: 10.1051/itmconf/20213903005

Hanna, D. and Charalampopoulou, C. (2019). Travail collaboratif sur l'ent et innovation dans les pratiques professionnelles des enseignants. Spirale - Revue De Recherches en Éducation, N° 63(1), 23-35. DOI: 10.3917/spir.063.0023

Hérold, J. (2019). Un modèle de la situation d'enseignement-apprentissage pour mieux comprendre l'activité de l'élève en classe. *Revue des Sciences de l'Education*, 45(1), 82–107. DOI: 10.7202/1064607ar

Holmes, W., Porayska-Pomsta, K., Holstein, K., Sutherland, E., Baker, T. T., Shum, S. B., Santos, O. C., Rodrigo, M. T., Cukurova, M., Bittencourt, I. I., & Koedinger, K. R. (2021). Ethics of ai in education: Towards a community-wide framework. *International Journal of Artificial Intelligence in Education*, 32(3), 504–526. DOI: 10.1007/s40593-021-00239-1

Hurst, A. (2023). The impact of artificial intelligence on learning accessibility in higher education (Thèse de doctorat, Université de l'Éducation de l'Ohio). ProQuest Dissertations Publishing.

Johnson, L. (2022). *L'utilisation de l'IA dans l'éducation spécialisée (Mémoire de maîtrise, Université de Paris)*. HAL.

Kizilcec, R. F., & Schneider, E. F. (2020). Learning in the age of AI: Personalized learning experiences. *Educational Technology Research and Development*, 68(1), 5–25. DOI: 10.1007/s11423-019-09702-5

Kovanovic, V., Joksimovic, S., & Siemens, G. (2015). The role of learning analytics in improving the learning experience. *Journal of Computer Assisted Learning*, 31(4), 373–384. DOI: 10.1111/jcal.12093

. Langlois, L., Dilhac, M., Dratwa, J., Ménissier, T., Ganascia, J., Weinstock, D., … & Marchildon, A. (2023). L'éthique au cœur de l'ia.. DOI: 10.61737/MDHP6080

Lepage, A., & Roy, N. (2023). Une recension des écrits de 1970 à 2022 sur les rôles de l'enseignant et de l'intelligence artificielle dans le domaine de l'ia en éducation. *Médiations Et Médiatisations*, (16), 9–29. DOI: 10.52358/mm.vi16.304

Luckin, R. (2017). *Machine Learning and Human Intelligence: The Future of Education for the Next Generation*. UCL Institute of Education Press.

McCoy, L. P. (2020). AI applications for accessibility in education. In Proctor, P. R. H. N. (Ed.), *Innovative Practices in Education* (pp. 58–75). Springer.

Mitrovic, A., & Ohlsson, S. (2019). The role of intelligent tutoring systems in education. Dans *Proceedings of the International Conference on Artificial Intelligence in Education* (pp. 16-28). Springer.

National Center on Accessible Educational Materials. (2022). The role of technology in accessibility. https://aem.cast.org

Paquette, G., Mariño, O., & Bejaoui, R. (2021). A new competency ontology for learning environments personalization. *Smart Learning Environments*, 8(1), 16. Advance online publication. DOI: 10.1186/s40561-021-00160-z

Parent, S., & Baron, M. (2021). Emergence of learning analytics in education: Challenges and issues of learning analysis. *Canadian Journal of Learning and Technology*, 47(4). Advance online publication. DOI: 10.21432/cjlt28053

Pennington, R., & O'Connor, M. (2018). Assistive technologies in education: The role of AI in enhancing learning experiences. Dans *Proceedings of the International Symposium on Computer Science and Education* (pp. 193-198). IEEE.

Pouliot, È., & Rocheleau, J. (2020). Enseigner le travail social en contexte de pandémie: Un exemple d'environnement d'apprentissage hybride en ligne sur moodle. *Revue Internationale Des Technologies en Pédagogie Universitaire*, 17(2), 44–50. DOI: 10.18162/ritpu-2020-v17n2-05

Sabeima, M., Lamolle, M., Anghour, A., & Nanne, M. F. (2023). Towards a semantic platform for adaptive and collaborative e-learning. Revue Africaine De Recherche en Informatique Et Mathématiques Appliquées, Volume 37 - 2022. DOI: 10.46298/arima.8396

Sacré, M., Lafontaine, D., & Toczek, M. (2020). Liens entre les composantes des dispositifs d'enseignement hybride et les performances des étudiants de l'enseignement supérieur: Une revue systématique. *Mesure et Évaluation en Éducation*, 42(3), 109–152. DOI: 10.7202/1074105ar

Sasseville, N., Juneau, S., & St-Pierre, É. (2022). Le point de vue d'étudiants au premier cycle en travail social quant à leur expérience d'un cours en ligne. *Revue Internationale Des Technologies en Pédagogie Universitaire*, 19(1), 34–50. DOI: 10.18162/ritpu-2022-v19n1-03

Simard, A., & Vermot-Desroches, C. (2023). Innovations numériques: Apprentissages interdisciplinaires pour des élèves en décrochage scolaire. *Revue Suisse De Pédagogie Spécialisée*, 13(01), 33–39. DOI: 10.57161/r2023-01-06

Smith, J. (2020). *L'intelligence artificielle en éducation*. Éditions de l'Université.

Spector, J. M., & Ma, M. (2019). *Artificial Intelligence in Education: Promises and Implications for Teaching and Learning*. Springer.

Tricot, A. (2021). Le numérique permet-il des apprentissages scolaires moins contraints ? une revue de la littérature. Éducation Et Sociétés, n° 45(1), 37-56. DOI: 10.3917/es.045.0037

UNESCO. (2021). Artificial Intelligence in education: Challenges and opportunities for sustainable development. https://unesdoc.unesco.org/ark:/48223/pf0000377962

Woolf, B. P. (2010). *Building intelligent interactive tutors: Student-centered strategies for revolutionizing e-learning*. Morgan Kaufmann.

World Health Organization. (2023). Health topics. https://www.who.int/topics/en/
orld Health Organization. (2023). Health topics. https://www.who.int/topics/en/

Chapter 9
Reinventing the Online Classroom With AI From Platform Models to the Smart Room

Btissam Chinkhir
https://orcid.org/0009-0006-2581-642X
Higher Normal School of Tetouan, Morocco

Ahmed Ibrahimi
https://orcid.org/0009-0004-8879-5561
Higher Normal School of Tetouan, Morocco

ABSTRACT

In the context of education's digital transformation, artificial intelligence (AI) is revolutionizing online learning. This chapter explores how virtual classrooms, driven by AI-powered platforms and smart classrooms, can be reinvented to enhance the learning experience. The key question is to what extent AI can optimize both knowledge delivery and enrich the socio-affective and cognitive dimensions of learning. The chapter is structured around three main dimensions: technological (examining AI tools and intelligent platforms), social (assessing AI's impact on motivation and collaboration), and cognitive (how AI personalizes learning and fosters self-regulation). The inquiry leads to a reflection on the tools and digital infrastructures necessary to transform pedagogical communication and online education, enhancing technological, socio-affective, and cognitive needs.

DOI: 10.4018/979-8-3373-2262-9.ch009

INTRODUCTION

Why and how to invest in AI to enhance online education development? Promising AI technology trends aim to reinvent traditional practices towards intelligent classrooms. Until recently, artificial intelligence seemed a science fiction fantasy, but now, with more human-like versions, it permeates all life systems, competing with human capabilities. While the confrontation between man and machine intensifies, AI's contributions are increasingly substantial. The rise of digital technologies and AI has transformed online education significantly over the past decades, evolving from simple content management platforms to sophisticated learning ecosystems that personalize educational experiences at an unprecedented scale.

Today, AI's role in education has expanded beyond mere strategy games to become a key player in transforming various sectors, including education. The reinvention of online classrooms exemplifies this evolution. Traditional online training platforms dominated for decades, but the growing integration of AI is revolutionizing teaching and learning. The shift to intelligent platforms and connected classrooms represents not only a technological change but also a response to the evolving needs of learners in a constantly changing digital world.

In this rapidly evolving context, we see a marked transition from classic learning platforms to AI-integrated systems. Early learning management systems primarily offered basic functionalities like content storage and distribution, along with rudimentary assessment and communication tools. This transformation raises a fundamental question: How is AI transforming traditional methods of pedagogical communication and online learning? To address this issue, it's essential to examine the changes brought by AI from multiple angles, considering various facets of the learning experience.

The chapter will explore three essential dimensions: technological, socio-affective, and cognitive, highlighting significant advances brought by AI. The technological dimension will delve into innovations in user interfaces, adaptive algorithms, and data analysis tools, demonstrating how these advancements enable increased personalization and more effective management of educational resources. The socio-affective dimension will focus on how AI transforms interactions among learners and between learners and teachers, examining how new technologies can enhance engagement, motivation, and emotional well-being in virtual learning environments. Finally, the cognitive dimension will analyze AI's impact on knowledge acquisition and retention processes, studying how intelligent tutors, adaptive feedback systems, and AI-assisted metacognitive tools can optimize learning strategies and improve educational outcomes.

How can AI, through intelligent platforms and classrooms, transform pedagogical communication and online learning to better meet the technological, socio-affective, and cognitive needs of learners?

THE TRANSFORMATION OF LEARNING PLATFORMS: FROM TRADITIONAL TOOLS TO ARTIFICIAL INTELLIGENCE

AI between yesterday, today, and tomorrow

The history of education begins and evolves with human intelligence. Endowed with potentially limitless faculties, humans have always demonstrated proactive, emancipated, and creative behavior. Each period of history is marked by a series of technological revolutions that have profoundly transformed the way knowledge is transmitted. Originally, education was exclusively oral. Knowledge was passed from person to person, in a dynamic where the direct relationship between teacher and student was paramount. This approach, though lively and interactive, limited access to knowledge to those who could be physically present.

With the advent of writing, humanity took a decisive first step. Writing made it possible to record knowledge, making it accessible on a larger scale and over longer periods. However, this method was still reserved for an elite capable of reading and writing, and copies of texts had to be made manually, which restricted the dissemination of knowledge. The true explosion of access to knowledge came with the invention of the printing press in the 15th century. Gutenberg's printing press not only accelerated book production but also democratized access to education by enabling the mass distribution of knowledge. This milestone laid the foundation for modern educational systems, where books became an essential tool for learning.

Then, the digital age initiated a new transformation in the transmission of knowledge. With the advent of computers and the Internet, education broke geographical barriers, making educational resources instantly accessible worldwide. Online learning platforms emerged, multiplying learning opportunities and facilitating remote collaboration. However, despite these advances, online education often remained limited by challenges related to learner engagement, human interaction, and personalization.

Today, artificial intelligence (AI) marks the beginning of a new revolution in education. It no longer just facilitates access to information but deeply transforms how learning is experienced and managed. AI enables personalized educational paths, adapts content to each learner's needs, and recreates human interactions through intelligent conversational agents. This latest stage of educational evolution, from oral tradition to AI, reinvents the classroom and redefines the relationship between the learner and knowledge.

Figure 1. (R) Evolution of Education Based on Technology (B. De Lièvre, 2024)

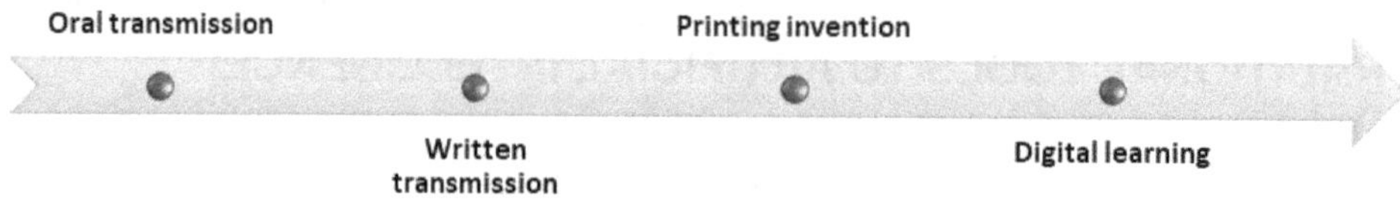

Often considered the official starting point of artificial intelligence research, here's a brief history of AI. In the summer of 1956, twenty researchers gathered on the campus of Dartmouth College in the United States. The outcomes of this colloquium would influence all subsequent work in this new discipline. The evolution of artificial intelligence (AI) has been marked by significant advances that have transformed various fields, including online education. By tracing the key stages of this evolution, we can better understand how modern technologies, such as those developed by Neuralink, address the challenges of traditional educational platforms. During this conference, John McCarthy, Marvin Minsky, Nathaniel Rochester, and Claude Shannon introduced the term "artificial intelligence" (C.A. Pickover, The Dartmouth Conference on AI, 2021), laying the foundation for a field dedicated to simulating human cognitive abilities through machines.

In 1966, Joseph Weizenbaum developed ELIZA, one of the first natural language processing programs capable of simulating a conversation with a therapist (C.A. Pickover, 2021). This project demonstrated the early attempts at human-machine interaction through natural language, paving the way for modern AI-based communication systems.

Twenty-five years ago, in May 1997, the confrontation between Garry Kasparov, the world chess champion, and Deep Blue, a computer developed by IBM, marked a historical turning point in the history of artificial intelligence (AI). In this historical perspective, Bruno Le Lièvre (2024), a university professor and researcher in educational engineering, began his contribution on "The Future of Digital Technology in Education" (B. De Lièvre, 2024) with this retrospective to recall the technological performance of the machine that defeated a human, during the international conference organized at ENS in Morocco (TETOUAN, 2024) dedicated to the theme: "Innovative Teaching Practices in the Age of Artificial Intelligence." In his presentation "Modeling the Use of AI in Education," De Lièvre discussed the technological aspect to answer the question of how it works, mentioning that the legendary duel between man and machine was not simply a confrontation between an exceptional chess player and a sophisticated machine; it symbolized the beginning of an era where the cognitive abilities of machines could rival those of human beings.

Figure 2. Looking Back: "Kasparov vs Deep(er) Blue" - 25 Years Ago, Machines Defeated Man (B. De Lièvre, 2024)

When Deep Blue achieved the decisive victory, it was a moment of global recognition for AI, demonstrating its ability to process information and make complex decisions. This event not only marked the end of an era in the world of chess but also paved the way for a series of technological innovations that have transformed various aspects of our daily lives.

DeepMind, a Google-owned company, is developing a novel artificial intelligence (AI) system: "*This is the first time we've designed a single general learning system that can learn directly from experience to then master a wide range of difficult tasks*," explains Demis Hassabis, co-founder of DeepMind. (T.L.G., 2016)

The revolution of neural networks in 2006 marks another important milestone. Geoffrey Hinton and his colleagues published foundational research on Deep Learning, a technique that enhances the processing of large datasets and strengthens AI system performance (C. Douarre & D. Rousseau, 2021). This advancement led to substantial improvements in speech recognition, computer vision, and other fields. "Deep Learning is said to imitate, in some ways, the functioning of a human brain. For image recognition, for instance, Deep Learning works as follows. When searching for an image of a cat, a search engine like Google assigns multiple characteristics or layers to this query." (T. Karsenti, 2018).

Figure 3. Artificial Intelligence and Its Subdomains (T. Karsenti, 2018).

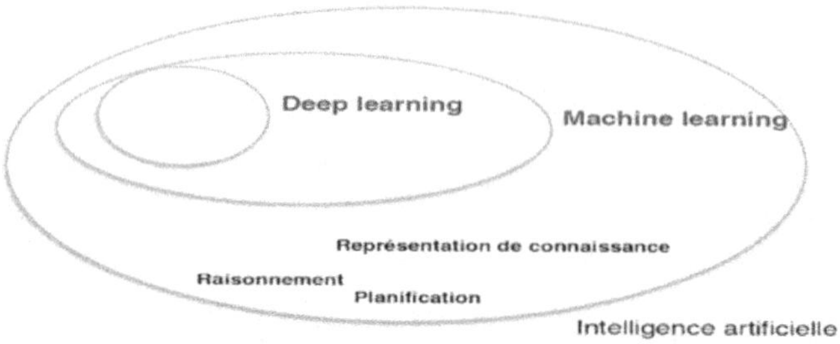

Clément Douarre and David Rousseau, in their article published by Delphine Chareyron, explain deep learning by artificial intelligence, covering the concept in the following way: "By artificial intelligence, we mean the act of having machines reproduce tasks that are considered complex by humans, typically: making a machine reason, teaching a machine task planning, or teaching a machine to represent knowledge in a structured way..." (C. Douarre & D. Rousseau, 2021).

In 2011, IBM's Watson won the television game show "Jeopardy!" by defeating human champions (F. Niedercorn, 2017). This victory demonstrated advances in natural language processing and data analysis, illustrating how AI can understand and respond to complex questions. In Seoul, AI continued to progress with AlphaGo, developed by Google DeepMind, which defeated Lee Sedol, the world Go champion, in 2016 (B. Georges, 2017). This success highlighted AI's ability to excel in strategy games with many more possible combinations than chess.

In 2018, OpenAI launched GPT-2, a natural language processing model capable of generating contextually relevant and coherent texts. The next advancement, GPT-3, announced in 2020, with its 175 billion parameters, further improved text generation and the versatility of AI applications. A more recent milestone came in 2022 when Neuralink, founded by Elon Musk, performed the first human brain implant (J-P. Henry, 2024). This technology represents a major advancement in brain-machine interfaces, enabling direct interaction between the brain and electronic devices, and offers promising prospects for education and beyond.

Regarding the future generation of artificial intelligence, Yann Le Cun stated, "Within five years, no one will be using a model like ChatGPT" (M. Blaise, 2023). During his tour for the reissue of his book (2023) "When the Machine Learns," he explained the revolutionary phenomenon of artificial neurons over the past 50 years, adding that machines can learn and learn independently. He confided: "Artificial intelligence, as an amplifier of human intelligence, could be the origin of a new Renaissance, or even a new Enlightenment" (Y. Le Cun, 2019). His position on the deployment of GPT has taken a more techno-enthusiastic third path: "Artificial intelligence, as an amplifier of human intelligence, could be the origin of a new Renaissance, or even a new Enlightenment" (M. Blaise, 2023).

Today, recent developments in AI and brain implants continue to transform technology and its integration into various aspects of daily life. In education, these advances help overcome the limitations of traditional platforms by providing more personalized, interactive, and adaptive solutions. Innovations such as Neuralink open up new possibilities for deeper immersion and interaction with educational content, promising a revolution in how we learn and teach online.

The transition from traditional online learning platforms to AI-based solutions represents a major turning point in the field of education. While effective in their early days, classic platforms now face significant limitations that hinder innovation and learning personalization. How have these technological limitations created fertile ground for the emergence and adoption of AI in online learning environments?

Technological Limitations of Platforms Compared to the Potential of AI

In examining the technological limitations of traditional online learning platforms, it becomes clear that artificial intelligence (AI) offers significant opportunities. In the context of distance learning, traditional online learning platforms have undeniably facilitated access to courses and knowledge, but they suffer from numerous technological shortcomings that compromise their effectiveness. The emergence of artificial intelligence (AI) in the field of education opens new perspectives to address these obstacles and transform the learning experience. This section examines the technological limitations of traditional platforms and highlights how AI can overcome them to reinvent the virtual classroom.

The educational landscape, constantly evolving, is witnessing the integration of artificial intelligence (AI) as an inevitable reality. Recent data shows the considerable impact of these technologies on learning and educational management. A study conducted in 2023 by the International Institute for Educational Research revealed that 67% of teachers view AI as a driver of educational transformation, a tool capable of revolutionizing learning by offering personalized educational paths. For instance, adaptive learning platforms that use algorithms to tailor courses to the specific needs of each student have seen a 30% increase in learning outcomes compared to those using traditional platforms (Editorial Team Vorecol, 2024). Imagine an online classroom where differentiated teaching is implemented—a very attractive approach to active pedagogies.

The potential of AI for personalized learning lies in its ability to offer tailored support adapted to each person's learning pace. AI allows the creation of individualized learning paths by analyzing students' interaction data. Through machine learning algorithms, AI can adapt educational content according to learners' performance, preferences, and specific needs, offering a unique experience for each. AI-powered adaptive learning systems dynamically modify learning content and activities based on students' progress and interactions. AI responds in real-time to provide additional exercises, interactive quizzes, or video materials suited to individual needs.

AI's potential extends beyond personalized learning to also affect educational administration. A study by McKinsey reveals that nearly 60% of schools plan to integrate automation tools by 2025 (Editorial Team Vorecol, 2024). This approach aims to lighten teachers' administrative workload, allowing them to refocus on their primary mission: teaching. Using chatbots to answer frequently asked questions from students, adopting a tutoring and continuous support approach, has reduced response times for administrative requests by 40%.

Empirical studies on the fascinating effectiveness of AI are still underway, as this new field is recently paved by pedagogical innovation. Although the emergence of chatbots has sparked ethical and professional debate, they have redefined what it means to teach and learn in the 21st century, paving the way for a more adaptive, personalized, and efficient education. AI-powered chatbots and automated assistance have proven capable of handling a wide range of real-time requests, such as answering frequently asked questions, guiding users through the learning process, providing pedagogical support, and conducting genuine communications.

These technological advancements outline a new educational paradigm. They promise a learning environment where each student benefits from personalized support, progressing at their own pace. Meanwhile, they free teachers from time-consuming administrative tasks, allowing them to fully devote themselves to imparting knowledge and guiding students. For instance, Duolingo uses AI algorithms to adapt lessons according to each user's performance and preferences (Psico-smart, 2024).

The rigid architecture of traditional systems poses a major obstacle to the rapid adaptation and evolution of platforms. Where these systems struggle to adjust, AI promises increased flexibility. AI-based architectures, such as multi-agent systems or adaptive neural networks, allow dynamic reconfiguration based on changing needs. For example, platforms like Carnegie Learning use AI algorithms to continuously adjust course structure based on learners' performance, improve learners' understanding, and adopt safe and managerial solutions. The example of a primary school in Lyon illustrates the success of differentiated learning thanks to AI. This school agreed to integrate AI into its classrooms in partnership with a local start-up to overcome difficulties identified in mathematics, allowing teachers to use personalized learning platforms to adapt math activities to each student's needs, resulting in a 20% improvement (Psico-smart, 2024).

The limited data processing capabilities of traditional systems are greatly surpassed by AI technologies. While traditional platforms struggle to analyze large volumes of data in real time, machine learning systems can process and analyze massive amounts of data at high speed. Companies like Knewton have developed platforms that use AI to analyze learner interactions in real time and instantly adjust content and difficulty. The lack of scalability and performance experienced by traditional platforms often makes it difficult for them to adapt to a large number of simultaneous users, resulting in slowdowns, service interruptions, and long loading times, especially during peak periods. Traditional technological infrastructures are not designed to handle massive traffic spikes, which negatively impacts the smooth running of online courses. AI can optimize system scalability through intelligent resource management algorithms. Technologies like AI-powered cloud computing can automatically allocate more servers or bandwidth in real time based on demand, ensuring a smooth user experience even during peak times. Additionally, AI-based optimization algorithms can predict future needs, ensuring that adequate resources are always available.

The limited interoperability of traditional systems is addressed by new AI-based approaches. Modern AI systems are often designed with standardized Application Programming Interfaces (APIs), facilitating integration with various tools and platforms. For example, the Open Neural Network Exchange (ONNX) project aims to create an open format for representing machine learning models, promoting interoperability between different AI frameworks. Limitations in terms of bandwidth and server capacity can be optimized using AI. AI algorithms can predict usage peaks and dynamically optimize resource allocation. Google, for instance, uses DeepMind AI to reduce its data center energy consumption by 40%, and this significant step toward "A Reduced Ecological Footprint through AI" (Les Echos, 2016) could be applied to online learning platforms to improve efficiency and reduce latency.

The lack of robust programming interfaces in traditional systems is addressed by modern AI-based platforms. These platforms often provide extensible APIs that allow the easy integration of new AI features. For example, the non-profit OpenAI Gym platform aims to ensure that Artificial General Intelligence (AGI) benefits all of humanity. "AGI refers to highly autonomous systems that outperform humans in most economically valuable work" (S.M. Walker II, 2024), providing a standardized interface for developing and comparing reinforcement learning algorithms, which could be used to create adaptive learning environments.

Storage and content management constraints are addressed by AI systems through advanced data compression and intelligent organization techniques. Deep learning algorithms can be used to efficiently compress educational videos without significant loss of quality, like the AV1 codec developed by Google.

The limited scalability of traditional infrastructures is surpassed by cloud-native architectures and AI-based distributed systems. These approaches allow for automatic and efficient scaling based on demand. Amazon Web Services, for instance, uses AI to predict and automatically manage its cloud service capacity—an approach that could be applied to online learning platforms. Insufficient computational power for complex tasks is addressed by using specialized processors for AI, such as GPUs and TPUs. "Graphics Processing Units (GPUs) and Tensor Processing Units (TPUs) have established their niche in the hardware landscape by excelling at specific tasks, and AI models are now following suit" (A. Procter, 2024). These processing units enable sophisticated AI models to run in real time, paving the way for more natural interactions and advanced customizations in learning environments.

The interfaces of traditional platforms are often complex and non-intuitive, limiting user engagement. Navigating through numerous menus and sub-menus to find resources can be frustrating for learners and detrimental to pedagogical efficiency. These platforms often lack adaptability in terms of interface and user experience. AI allows the development of adaptive interfaces that automatically adjust based on user preferences and behavior. Speech recognition technologies and Natural Language Processing (NLP) can simplify navigation by enabling learners to quickly find information through voice commands or natural language queries. Additionally, AI-powered systems can learn from users' past behavior and customize interfaces to their specific needs, making the learning experience more fluid and intuitive.

Traditional online learning platforms often lack advanced tools to analyze learning data. This means that teachers and administrators have little visibility into individual learners' progress, limiting the ability to personalize teaching or respond quickly to difficulties encountered by some students. AI, through predictive analytics and machine learning technologies, allows for the collection and analysis of vast amounts of data in real time. It can track student performance, identify learning patterns, and provide detailed feedback on each learner's strengths and weaknesses. Furthermore, AI can anticipate students' needs based on their history of interaction with the platform and adjust educational content to provide a more personalized and effective learning experience.

Traditional platforms lack integration with emerging technologies such as Virtual Reality (VR), Augmented Reality (AR), or other immersive tools that enrich the learning experience. "Both technologies are very powerful and promising. In summary, virtual reality replaces reality with a fully digital world, while augmented reality adds virtual elements to the real world" (A. Tessa & J. Niels, 2023). This fragmentation limits the possibility of creating immersive or interactive environments that can help better understand complex concepts, virtual labs, or professional environments in which students can practice and learn more engagingly and practically.

These AI-powered environments can also adapt in real-time to user interactions, offering richer and more stimulating learning experiences.

AI-Powered E-Learning: Essential Tools

The combination of AI and e-learning is revolutionizing training by offering personalized and interactive learning paths. With innovative features, training designers now have the tools to digitize their projects, making captivating content achievable. AI simplifies the creation of engaging learning experiences, giving rise to a modern approach to pedagogy. With general consensus but some reservations, recent AI advancements encourage taking advantage of its potential in online courses while allowing pedagogical approaches that captivate learners to be guided by AI.

Optimizing online training is now possible. Creating effective e-learning content requires time and a good dose of creativity. The quality of an online course involves meticulous work, which explains the high time demand. It is essential to consider learners' needs and expectations, define clear learning objectives, optimize available resources, structure the course carefully, and consider the learning context. Each step contributes to the quality and impact of the training, ensuring an enriching and relevant learning experience.

Here, AI significantly contributes to accelerating the implementation of engaging educational setups—whether in structuring, content, or tools used to achieve the desired objectives. "Online content creation tools that save time and produce attractive visual and interactive resources without advanced technical skills." (C. Mallén, M. Andrea, & V. Mallén, 2024). For instance, Midjourney is a phenomenal AI tool serving art, which explores new avenues of thought and expands human imaginative power; it can transform descriptive text into illustrations. In design, this intelligent tool helps learners complete their digital learning projects with individualized illustrations, especially when adapting content to different learning styles.

And who hasn't struggled at some point to create characters for productions due to a lack of drawing skills? With Midjourney, you can obtain vector characters or a series of images perfectly suited to expectations. This graphic ease, which previously represented an obstacle, is now a task that can be executed in no time.

The opportunity to use machine learning algorithms (adaptive learning) allows artificial intelligence to take the lead. Not to mention the possibilities offered for personalizing and adapting training programs according to individual student needs—AI can gather information by identifying each learner's strengths and teaching preferences, subsequently using this data to deliver a course tailored to their specific needs. The adaptive nature of AI-provided learning is evident in the specificity of exercises tailored to the learner's strengths and weaknesses, "While adjusting the difficulty level of the content based on their performance. This maximizes the effectiveness of digital training and enhances each learner's experience." (C. Mallén, M. Andrea, & V. Mallén, 2024).

Advances in artificial intelligence, such as ChatGPT, open new perspectives for data analysis in organizations. These natural language processing tools exhibit potential in analytics and can generate quizzes for training—similar to the "Quiz Wizard" feature of Wooclap. This technology allows trainers to quickly create: questionnaires, surveys, statistical analyses, and assessments. The goal is to verify learners' acquisition of skills and knowledge. The great flexibility of these tools allows for precise adaptation of content to the specific objectives of each training. AI transforms the learning experience, making it more dynamic and personalized. These technological innovations thus become major allies for designing interactive, engaging, and effective training.

If the role of AI tools in online classrooms is to promote learning, the same applies to teaching. With tools like ChatGPT, you just need to determine the learning objectives of a training project, specifying the type of taxonomy (e.g., Bloom), and it will provide a set of operational objectives that can be sorted or adjusted according to the specifics of the training. Storyline 360 is another example of an AI tool dedicated to creating games—ChatGPT provides important data and instructions to Storyline developers for JavaScript coding in Storyline (C. Mallén, M. Andrea, & V. Mallén, 2024).

In virtual pedagogy, AI's innovative features prioritize digitization since its effectiveness has been proven in terms of quality and time efficiency. And because change is happening now, smartly renovating platforms must be a requirement to meet market trends and keep pace with the rapid evolution of technology. Recently, AI has shown significant advancements that benefit users; however, traditional pedagogies used in online courses must evolve to support, on the one hand, teachers juggling classroom management, conceptualization, implementation, and evaluation, and the necessity of integrating digital technology; and, on the other hand, learners in developing their creativity and capability to work autonomously.

Examples of AI-inspired tools include:

- **Chatbots Integrated into SCORM Formats**: A personal assistant for each learner. Integrating chatbots into SCORM training modules revolutionizes the learning experience by providing personalized support throughout the training journey, offering instant answers to learners' questions and real-time tailored advice. These AI-powered virtual assistants greatly enhance the overall learning experience. In the case of Articulate 360, using chatbots is particularly effective thanks to natural language processing (NLP). This technology brings machines closer to a more human version. When a chatbot provides customized responses perfectly tailored to each learner's needs, the result? An interaction that simulates that of a human instructor.

- **AI-Revolutionized Voice-Over**: High-quality voice-over is key to engaging learners in e-learning modules. AI revolutionizes this area by allowing the creation of high-quality voice-over with increased efficiency and speed. Microsoft, with its VALL-E model, perfectly illustrates AI's incredible achievements (M. Planche, 2024): it can generate voice-over from just a 3-second audio sample and use deep learning to realistically and naturally reproduce voices. This tool opens up new possibilities for e-learning content creators:
    - Quick generation of personalized voice-over
    - Easy integration into training modules
    - Improved learner engagement through more natural voices

VALL-E thus represents a practical and innovative solution for enriching the online learning experience by offering professional-quality voice-over without the traditional production constraints.

- **AI-Developed Augmented and Virtual Reality**: The integration of augmented reality (AR) and virtual reality (VR) transforms online learning, making it more interactive and immersive (H. Halmaoui, 2022). Advances in artificial intelligence facilitate the creation of realistic and personalized AR/VR experiences. "Unity," a leading platform for developing games and applications, now integrates a content-generating AI into its engine, allowing the creation of 3D content via simple text queries. When AR/VR aligns with AI, it promises to revolutionize online training by offering more engaging, realistic, and tailored learning experiences to each learner's needs. Rapid and intuitive creation of virtual learning environments, highly personalized training experiences, and reduced technical barriers for instructional designers are of great interest for every online training program.
- **Machine Translation for Your E-Learning**: Translation is a major challenge for online training targeting an international audience. To save time and process large volumes of content quickly, AI offers an effective solution to this challenge through instant translation tools (S. Trinh, 2019). Not only does it contribute to accelerating the content localization process and reducing translation costs, but it also makes training more accessible globally. Human review is still essential, although AI provides valuable assistance, to verify translation accuracy, ensure the translated content is understood without losing its semantic weight, and maintain cultural and linguistic nuances.

In summary, the combination of AI efficiency and human expertise propels the level of learning. E-learning content creators can thus offer quality training tailored to an international audience while optimizing their resources. AI's bright future in digital learning stands as a transformative force offering promising prospects, such as:

- **Optimization of Content Creation**
  - Innovative tools for more efficient production
  - Enhanced content relevance
- **Increased Productivity**
  - Automation of time-consuming tasks
  - Accelerated development processes
- **Advanced Learning Personalization**
  - Real-time adaptation to each learner's needs
  - Customized learning paths
- **New Frontiers in E-Learning Design**
  - Integration of immersive technologies (AR/VR)
  - Creation of advanced interactive content

SOCIAL EFFECTS

The emergence of artificial intelligence (AI) in online education has opened new perspectives to enrich the learning experience. Far from merely improving technological tools, AI now plays a crucial role in managing human interactions, particularly through the socio-affective dimension. In an educational environment where physical distance can exacerbate isolation and disengagement, AI offers innovative solutions to strengthen the connections between learners and educators, while fostering a more inclusive, sociable, supportive, and collaborative learning climate. This section explores how these four key concepts are redefining the online pedagogical approach, bringing a new dimension to educational communication.

Acceptable: Reinforce interactions via AI

In a distance learning environment, one of the major challenges is creating interactions that are adapted to each learner. Maintaining interactions remotely is one of the significant challenges of online education, with several studies attempting to address how to overcome distance and build a sense of presence. (B. Chinkhir, S. El Gazi, & A. Ibrahimi, 2024). It turns out that the socio-affective aspect is a catalyst for overcoming physical absence and creating interactions between the instructor and learners, as well as among the learners themselves. When they are bonded by healthy, affective connections, they are happy to participate in online courses. Moreover, when social connections are woven between the participants of distance education, a climate of trust prevails, and the course is shaped by the level of learner engagement.

In an online setting, student retention depends on enhancing interactions since dropout and withdrawal are often signs of demotivation and disengagement in distance learning. Consequently, educational innovations are embracing technological advances to alleviate feelings of isolation and create a sense of presence. The advent of Learning Management Systems (LMS) as tools for managing and supporting learning paths has relieved e-learning advocates by creating interactive environments where the learner becomes a social actor—autonomous and responsible for their learning. The machine is no longer merely seen as a tool for training but as a cognitive tool that helps learners learn how to learn.

Since then, we've seen an overwhelming influx of learning platforms flooding the educational landscape, particularly recognizing that distance education designers devote more resources and energy to developing systems and teaching strategies rather than analyzing the psychosocial and cultural conditions of the audience (J. Geneviève, 1993). Certainly, ensuring differentiation and guaranteeing personalized learning is an arduous task, sometimes perceived as a utopian goal. Hence, the idea emerged to invent an environment that interacts with learners, striving to be as human-like as possible—to think, interact, communicate, and behave like a human.

The limitations of traditional platforms explain the emergence of a more humanized digital version, aiming to eliminate the feeling of absence and reduce the margin of geographical distance. Indeed, studies have been attempting to discover pedagogical practices using the latest technological inventions to live the distance learning experience while being psychologically close to other participants. One of the major stimulators is "functional interactivity," which has an instrumental dimension (man-machine relationship) and creates a sense of belonging to a group, compensates for the lack of non-verbal cues, and prevents dullness in exchanges. This instrumental function sustains and maintains "intentional interactivity," which is human-based (man-man relationship) (D. Peraya, Mediation and Mediatisation: The Virtual Campus, 1999).

AI's contribution points to a set of recommendations to create learning communities and manage the socio-affective aspect in a virtual environment as a potential to support social presence and bridge distance. "It is the loss of the physical co-presence of the learner and the teacher that must be compensated for by using technical mediation resources" (J. Geneviève, 1993). AI enables a deep personalization of educational interactions, making them more acceptable for learners. By analyzing students' preferences and performance, intelligent systems can adjust communication modalities—such as the frequency, form, or content of messages. These adjustments ensure smooth communication, respecting each learner's pace. This personalization overcomes a major flaw of traditional systems: the standardization of exchanges, often perceived as cold and impersonal.

Although emotional neutrality can be a disadvantage in cases where social bonding is necessary to join a learning group, it is one of AI's features that could maintain exchanges that are neutral and free from emotional bias. Paradoxically, affective bonds that unite learners can themselves become obstacles in cases of poor emotional management or impulsive or excessive reactions and attitudes. Conversely, where human interactions may be affected by external factors (fatigue, stress, mood, excess empathy), AI can provide consistent feedback without subjective interpretations. This neutrality promotes better acceptance of feedback, as learners can be assured that evaluations are based solely on their performance. AI can thus defuse tensions or potential misunderstandings, facilitating the acceptance of constructive criticism in a trusting environment with reduced subjectivity. Emotional involvement can lead to subjectivity, which may hinder the effectiveness of pedagogical communication.

Socializing: Promoting engaging, humanized interactions

With AI's permanent online tutoring at the service of learners, compensating for the absence of a physical teacher becomes feasible in a virtual classroom. Creating online tutoring for a learning community makes tutor-learner social interactions a means by which the tutor coordinates, supports, and provides technical assistance. The use of online "Community of Inquiry" (R. Garrison, T. Anderson, & W. Archer, 2010) and "Community of Practice" (E. Wenger, 1998/2005) in distance learning is fundamentally essential for collaborative learning. Driven by personal needs, community members come together to develop a collective activity, which gives rise to sharing and community building.

In their professional learning community, the tutor fulfills support functions on three levels: didactic support (providing knowledge and skills), organizational support (work management to help learners discover their learning strategies), and psychological support (motivation and success). Distinct from in-person learning communities, online "Community of Inquiry" learning communities break with the three units (time, place, action) (C. Papi, 2014), presenting distinct characteristics that influence the educational experience.

In a traditional in-person setting, there is a unity of time, place, and action. Learners gather physically at a specific location, at a specific time, to follow a predefined program. This format often translates into less flexible lecture-based courses. The advantage of this approach lies in sharing a common and simultaneous experience by all learners, thereby creating group dynamics that foster motivation. Additionally, the physical presence of the instructor allows direct control over participants' engagement.

On the other hand, online learning communities are characterized by a break in this unity. Time, place, and learning actions are flexible and individualized. This flexibility allows for more frequent interactions between learners and the instructor, as well as among the learners themselves, although these exchanges are spread over a longer period. Online learning emphasizes collaboration and joint projects. Although physical isolation can be a challenge, it is often mitigated by the development of a strong sense of belonging to a virtual community.

This comparison highlights the different approaches to collective learning, each with its own advantages and challenges in creating an effective educational environment. AI-based tutorial systems can be available continuously, providing immediate and personalized support based on the challenges learners face. Given the significant growth in learners compared to the limited number of tutors, traditional tutoring on platforms often fails to individualize learning and address learners' questions adequately. The symbolic sense of distance in online learning often leaves students to their own devices, facing their difficulties and seeking solutions without a guiding hand to accompany them step-by-step.

In this context, AI offers tutorial systems that are permanently available, capable of immediately responding to learners' needs, allowing them to experience a "social presence" (D. Peraya, B. Charlier, & N. Deschryver, Hybrid Learning Systems in Higher Education: Theoretical, Methodological, and Practical Questions, 2014). Participants have the ability to project themselves socially and emotionally into the community, where learner support (the Zone of Proximal Development) to achieve "social presence" is possible with a "proximal distance" (L. S. Vygotsky, Thought and Language, 1987). In the e-learning "Community of Inquiry," there is also a scientific approach to problem-solving and support for knowledge construction through social interaction, which brings learners closer and creates the need for exchange. If social presence "corresponds to the learners' ability to project themselves socially and emotionally into the community," AI conversational agents replace human presence by detecting emotional states, such as frustration or fatigue, and adjusting responses or content accordingly. They create a salient presence that affects interpersonal relationships. Short, Williams, and Christie (1976) initially defined "social presence" as "the degree of salience of the other in the interaction and the resulting salience in the interpersonal relationship—the perceived importance of the other in the interaction and its impact on the interpersonal relationship." (J. Short, E. Williams, & B. Christie, 1976). This ensures that students feel supported and understood at all times. For example, conversational agents can answer questions, provide moral support, or adjust the difficulty level based on detected emotions (through language analysis or even, in some cases, through physiological data).

This ensures that learners feel accompanied and understood. For instance, platforms like Duolingo or Coursera already integrate AI tools to interact directly with users, offering a more sociable experience, close to real human interaction.

The creation of positive socio-affective relationships by AI is at the core of any educational relationship, facilitating pedagogical communication and preparing an environment conducive to exchanges among learners, while countering anything that could hinder learning. Through the analysis of behavioral data, AI can suggest discussion topics or group activities that promote social bonds. Additionally, systems like intelligent forums or AI-assisted messaging can encourage interactions and revive discussions in case of decreased participation. This type of intervention helps establish a warm learning environment where each student feels connected to their peers and integrated into a supportive learning community.

Traditional platforms lack dynamism in creating social relationships among learners. Online forums and discussions are often perceived as rigid and do not foster genuine collaboration or a sense of belonging. With the integration of AI in behavioral data analysis, platforms can recommend connections between learners with shared interests or organize discussion groups based on needs and progress. This fosters the creation of more cohesive learning communities, strengthening social interactions and the sense of belonging.

The lack of empathy and affective interactions that prevail on traditional platforms means they do not offer real empathetic interaction. The lack of human and emotional presence creates an emotional disconnection, which is essential for successful learning. However, AI systems can simulate empathetic interactions by detecting emotions through text or voice analysis. Empathetic conversational agents adjust their speech to offer encouragement or moral support, recreating a form of socio-affective presence essential for learner engagement.

If interactions on traditional platforms are devoid of empathy—being rigid and often limited to purely textual formats—empathetic conversational agents can detect learners' emotional signals (e.g., frustration, discouragement) and adjust how they interact with them. They can offer encouragement, suggest breaks, or adapt content to avoid cognitive overload. These interactions are designed to simulate a form of socio-affective presence, thereby creating a more nurturing and warm environment.

In traditional online classrooms, the lack of personalized interactions and emotional engagement results in standardized, impersonal exchanges devoid of emotional involvement. This creates an environment where learners feel disconnected and isolated. AI makes it possible to create more personalized and interactive environments by analyzing learners' preferences and behaviors. For example, chatbots and virtual assistants, powered by AI algorithms, can respond in real-time and adapt to the specific emotional and educational needs of students, providing a more human and responsive interaction. This helps reduce feelings of loneliness.

Inclusive: AI against isolation

In online education, isolation and the difficulty of maintaining social connections persist as obstacles to active collaboration. Certainly, traditional platforms offer a variety of tools to strengthen social relationships among learners, but forums and chats often lack dynamism and fail to encourage involvement. In contrast, through behavioral data analysis, AI can recommend connections between learners with common interests or similar needs, thus promoting the formation of more cohesive learning communities. AI can also organize virtual study groups or targeted discussions based on learners' progress and profiles, fostering a more active social dynamic.

One of the major challenges of online education is to make platforms accessible to everyone, regardless of the learner's profile. AI allows content to be adapted to the specific needs of students, whether they are with or without disabilities. For example, speech recognition and audio synthesis tools facilitate access to content for visually or hearing-impaired learners. Additionally, automatic translation algorithms allow students from different language backgrounds to follow the same course in real time, making education truly inclusive.

The concept of a second chance has always been the essence of distance education. Integrating individuals who face challenges in learning by providing them with a variety of opportunities is recognized as a right to education and social integration. Despite the fight against social exclusion and the marginalization of uncertified individuals, which remains a recurring challenge in distance learning, demotivation and dropout rates are still high. AI plays a role here by detecting signs of social disengagement and offering appropriate solutions. For instance, systems can monitor a student's participation rate and, in cases of gradual withdrawal, propose targeted actions to reintegrate them into group activities. Whether through automated support messages, advice on managing stress better, or suggestions for new collaborative activities, AI acts as a mediator, thus preventing social exclusion.

Another way to challenge the feeling of academic exclusion, which leads to social exclusion, is to promote group projects and monitor group dynamics. In an online university class, creating and managing groups is not always straightforward. One of AI's strengths lies in its ability to form learner groups based on their skills and pedagogical affinities. Intelligent systems analyze learning profiles, identify complementarities, and propose optimized groups for effective collaboration. This creates synergies among students, who can exchange ideas and support each other in an environment where everyone finds their place. Platforms like Microsoft Teams and Google Classroom already use this type of technology to facilitate the formation of relevant workgroups.

AI not only forms groups but also monitors interactions within them. Monitoring tools analyze each member's participation, identify imbalances in contributions, and suggest adjustments to improve collaboration. If a student disengages, AI can alert the instructor and propose solutions, such as redistributing tasks or providing targeted encouragement. This ensures harmonious group work, where everyone's contributions are valued and balanced.

COGNITIVE ASPECTS

Efficiency and effectiveness

When we talk about efficiency, we are referring to the relationship between the time invested in accomplishing a task and the quality of the results. The early attempts to integrate technologies into classrooms faced controversial debates, reflecting the reluctance of those who resisted the new educational model. Gradually, however, the use of ICT for saving time, facilitating communication, and maintaining interaction has been accepted. In the age of AI, a new set of pedagogical tools has emerged in online learning platforms. This evolution does not diminish previous technological advances; rather, AI is a continuation. Wherever learning limitations and new needs arise, AI has proven effective.

The question remains for AI: are tools like computers, interactive whiteboards, tablets, MOOCs, robots, and virtual reality effective? It's an unresolved issue. With heightened disappointment related to the gap between expectations and results, efficiency is called into question. Unfortunately, shocking statistics show learning failures relative to expected outcomes. The debate is not about stifling technological inventions but about encouraging reflection on cognitive efficiency in the presence of AI—is it a question of tools or principles? How can tools harness AI's contributions to overcome persistent weaknesses in learning? How do reinvented smart classrooms contribute to better learning?

Learning styles are a concept that endures. Combining artificial intelligence and learning is a fertile and expansive area of research. To limit the scope, some impacts can be highlighted, such as personalized assistance, instant feedback during the learning experience, and easier access to educational content. Considering individual learning styles, AI offers personalized assistance to facilitate learning in an online course where distance is a challenge to overcome. Each learner has specific preferences for receiving and processing information, whether auditory, visual, or kinesthetic. Consequently, various configurations of devices, in association with more traditional devices (mouse, keyboard, drop-down menu), form unique interactive systems—virtual environments (VE) "designed to recreate some of the sensations and experiences that characterize the real-world environment" (B.G. Witmer & W.J. Sadowski, 1998).

The rapid evolution of information and communication technologies has profoundly transformed the educational landscape. At the heart of this transformation is the growing recognition of the diversity of learning styles and the emergence of intelligent tools capable of adapting to these individual differences. This synergy between understanding cognitive processes and artificial intelligence (AI) paves the way for more effective and personalized learning methods. The VARK model measures four perceptual preferences: Visual (V), Auditory (A), Reading/Writing (R), and Kinesthetic (K). Developed by Neil Fleming in the 1980s, this model has long been a cornerstone in understanding learning styles (N.D. Fleming & C. Mills, 1992).

At the intersection of several disciplines, such as robotics, computer graphics, human-computer interaction, and virtual reality, is a field under development. According to some studies, 60% of people are visual learners, 30% are auditory, and 10% are kinesthetic among the student population (J. Cerqueira, 2024). Visual learners benefit greatly from advances in virtual reality (VR) and augmented reality (AR). These technologies create immersive learning environments where abstract concepts can be visualized and manipulated in three dimensions. "Educational virtual reality involves work aimed at providing users with the possibility of constructing and controlling the parameters of a more or less imaginary world interactively through a computer system that allows representation" (J.M. Burkhardt, 2003). This approach generally relies on three technological fields: computer graphics and image synthesis, robotics and teleoperation, and real-time simulation (P. Fuchs, 1996). For example, platforms like zSpace combine AR with haptic styluses, allowing students to physically interact with 3D models, whether in biology, chemistry, or engineering.

For auditory learners, educational podcasts and audiobooks have grown in popularity. AI plays a crucial role in this area by enabling the creation of personalized audio content. Advanced text-to-speech systems, like those developed by DeepMind, can generate natural and expressive voices, making listening more engaging. Speech synthesis is a computer-based sound synthesis technique that allows artificial speech to be created from any text. Additionally, intelligent voice assistants like Alexa or Google Assistant are increasingly used in educational contexts, offering interactive quizzes and review sessions adapted to the learner's pace (M. Satratzemi & G. Terzopoulos, 2020).

Learners who prefer reading and writing benefit from learning management platforms (LMS) equipped with intelligent annotation tools. These systems, like Hypothesis, allow not only direct note-taking on digital texts but also real-time collaboration with other learners. "Sharing annotations with other students (comments, notes, images, or diagrams, etc.) in the margins of a required reading for a course is an interactive reading strategy called collaborative annotation that is gaining popularity in post-secondary education due to its social nature." (A. Enkerli & S. MacNeil, 2021). AI analyzes these annotations to identify difficult passages and suggest additional resources, thus creating a dynamic and interactive learning ecosystem.

Kinesthetic learners, who learn best by doing, benefit greatly from virtual labs and interactive simulations. Platforms like Labster provide VR lab experiences, allowing students to manipulate costly and potentially dangerous equipment in a safe environment. AI in these systems adjusts the difficulty of experiments based on student progress, ensuring a constant challenge without excessive frustration.

Furthermore, these intelligent systems foster the development of metacognition, meaning awareness and understanding of one's own thought processes. By providing detailed feedback and encouraging reflection on learning strategies, these tools help learners become more autonomous and effective in their learning. This approach is particularly beneficial for developing 21st-century skills, such as critical thinking, creativity, and adaptability (J. Pitrat, 1998).

Cognitive AI and its effects on the teacher

The teacher has not been considered the sole custodian of knowledge since the advent of the book, which enabled the dissemination of knowledge. Similarly, in the context of the emergence of artificial intelligence (AI), many sectors, including education, have been affected. Like e-commerce professionals, teachers now need to attract an audience and implement a promoting strategy. Their role changes over time, and with each technological invention, the function of the teacher takes on new dimensions. The utility of revolutionary intelligence is clearly apparent in its impact on reproducing teaching tasks, enhancing teacher capabilities, and transforming pedagogical practices.

The reproduction of teaching tasks by AI represents the first level of integrating this technology into education. AI supports the development of several technologies that can replace predictable and repetitive tasks performed by teachers. AI can assist teachers and help them quickly identify learners in difficulty within classes. Platforms like Coursera or edX already offer online courses where AI reproduces certain traditional teacher functions. For instance, AI-generated videos can present basic concepts in mathematics or science, providing an alternative to lectures. Automated grading systems, such as Gradescope, can evaluate multiple-choice tests or simple math problems, thereby replicating the assessment function. There are numerous practical examples of AI use in virtual classrooms: "Teaching multi-level classes can be challenging, but AI can help you differentiate your texts. Use tools like Diffit to adjust the reading level of texts found online, in PDFs, or your own texts." (A. Coutlée, 2023).

Enhancing teacher capabilities with AI represents a second, more sophisticated level of integration. Tools like Carnegie Learning use AI to analyze student performance and provide teachers with detailed insights into each learner's strengths and weaknesses. This allows teachers to target their interventions precisely. AI can also augment a teacher's content creation capabilities. For example, the Knewton platform uses adaptive learning to suggest personalized educational resources based on each student's profile. A language teacher could use an AI assistant to help prepare exercises tailored to their students' levels or to verify the accuracy of translations. In addition to Duolingo, which uses speech recognition to adapt learning activities based on current language proficiency, other language-learning platforms, such as Mon Coach Bescherelle (a spelling training app), adapt to the learner's level (T. Karsenti, 2018).

The true transformation of the teacher's role through AI represents the deepest level of integration. In this scenario, the teacher becomes more of a facilitator of learning than a mere transmitter of knowledge. The AIED (Artificial Intelligence in Education) project at Carnegie Mellon University, which emerged 50 years ago (J. Sansonetti, 2022), explores how AI can enable highly personalized learning, where each student follows a unique path tailored to their needs, interests, and learning pace. The teacher, freed from repetitive tasks, can then focus on developing transversal skills like critical thinking, creativity, and collaboration. For example, in a science class, instead of simply presenting facts, the teacher could use AI simulations to engage students in complex virtual experiments, encouraging them to formulate hypotheses and analyze results.

This transformation also involves redefining the teacher-student relationship. Some projects explore learning systems based on a socio-cognitive approach to achieve immersion. For instance, Third Space Learning (personalized tutoring programs) and Tandem are applications where learners master a foreign language and practice other languages through exchanges with real people.

Cognitive AI and utility

Fully invested in the education sector, major players like Pearson and McGraw-Hill (D. Cristol & al., 2001) are carving out the online education market by offering electronic versions of books. The concept is simple: learning paths are never static; they constantly evolve based on the responses, needs, and expectations of learners. "Adaptive learning is therefore an educational technique that uses AI to organize learning according to the individual skills or needs of each learner" (T. Karsenti, 2018).

These developments are changing the way teaching is conducted online and invite the reinvention of new strategies. One of the major contributions of artificial intelligence (AI) to online learning lies in the ability of systems to provide adaptive experiences. By leveraging sophisticated algorithms, learning can be adjusted to meet the specific needs and progress of each learner. By analyzing the learning traces left by the user, such as quiz results or usage data from educational software, the system modifies its approach to adapt to individual skills and learning pace. This process relies not only on selecting suitable content but also on calibrating the difficulty level of the exercises offered.

Furthermore, recent advances enable behavioral analysis of learners through devices such as external sensors—like cameras—or more intrusive brain-computer interfaces. These technologies capture real-time data on the learner's engagement or cognitive state, allowing pedagogical interaction to be adjusted almost instantaneously. This continuous adaptability is central to the principles of digital pedagogy and is particularly relevant in gamified environments, where learning becomes a pedagogical game. This game can be conducted in interaction with the machine but also in collaboration with other learners, thus creating a learning environment that is both individualized and socialized (G. Giraudon, P. Guitton, M. Romero, D. Roy, & T. Viéville, 2020).

Proposed by didacticians, the KidLearn25 project is based on the concept of the Zone of Proximal Development (L. S. Vygotsky, 1978). As shown in Figure 5, a learning activity with multiple variants is structured according to an order of increasing difficulty, where the solution involves additions or subtractions of integers or decimals. The goal "is based on the idea that, between an exercise that is too difficult, which discourages, and an exercise that is too easy, which bores, there exists an optimal zone that maximizes learning progress, measured here by observing the student's performance throughout the game" (P. Y. Oudeyer, B. Clément, D. Roy, & H. Sauzéon, 2020). This experience clearly demonstrates how, by incorporating these elements into the algorithm, learning becomes adaptive.

Figure 4. The Kidlearn project is based on scenarios involving the purchase of 1 or 2 objects or the giving of change (source: Flowers team, Inria).

Although these practices are still not widespread and scientific research on the subject is ongoing, they represent an essential preliminary step to better understand the mechanisms of knowledge acquisition and appropriation. Indeed, to generalize this adaptive approach, it is necessary to formalize the knowledge (theoretical understanding) and skills (practical competencies) to be taught, which involves clarifying and structuring the types of tasks as well as the techniques for solving them. It is also crucial to ensure that learners are not overloaded with additional cognitive tasks that are not directly related to the main objective.

Adaptive learning also occurs in a context marked by material constraints, teacher training, and the limitations related to prolonged screen usage. The potential benefits of this approach are numerous. Firstly, it generally leads to better learner engagement, as personalized interaction with content provides more opportunities for in-depth understanding. By adjusting the difficulty level to the student's level, discouragement and fatigue can be reduced or even eliminated. Moreover, unlike humans, machines do not pass judgment, which can help maintain motivation. That said, for teachers, this form of learning may require additional investment, especially if the design does not take the learner's cognitive load into account. There is also the risk that the learner may lose focus if the playful aspect becomes too prominent, at the expense of learning.

AI-based adaptive tools can nonetheless free up time for teachers, allowing them to focus on students who need the most help, while the rest of the class engages in independent learning activities. As in flipped pedagogy, these tools reduce the need for direct knowledge transmission through interactive multimedia content and automated exercises, allowing teachers to prioritize other pedagogical approaches, such as project-based learning. Compared to traditional digital tools that lack machine learning capabilities, these systems increase learner autonomy and allow for more comprehensive learning paths.

However, the use of these technologies raises some concerns, particularly regarding constant learner monitoring, categorization, and the potential reduction in the number of teachers. Additionally, these tools risk exacerbating inequalities related to digital illiteracy (É. Gandon, 2020). Finally, the use of digital data could be repurposed for other ends, such as commercial profiling or widespread surveillance, by incorporating non-educational aspects of online life, like shopping habits or content consumption patterns.

The principle of adaptation is fundamental because it allows for better responsiveness to diverse learning profiles. By personalizing the educational experience, these systems tend to improve information retention and learner engagement. Thanks to detailed behavioral analysis, AI can, for example, suggest reviews when gaps are identified or recommend additional content to deepen understanding. Moreover, this adaptability encourages gamification, an aspect that is particularly important in modern digital pedagogy, where learner engagement is enhanced by game elements. Thus, the learner becomes an active participant in their learning, often interacting with peers in collaborative online environments, maximizing the socio-cognitive effects of digital learning.

Machine Learning: This aspect of artificial intelligence enables the automatic generation of knowledge by analyzing the collected data. Using this data, the system can then develop models or patterns that facilitate decision-making. Take the example of the UTIFEN online learning platform: "AI enables personalized learning, as we have seen with UTIFEN and other tools. This is certainly the greatest advantage of AI in education." (T. Karsenti, 2018). By analyzing learners' success and failure pathways, AI could design an "ideal" intervention model aimed at maximizing user success. This model would allow the platform to send personalized, automatic, and timely reminders to reinforce learners' chances of success. Thus, machine learning enables the system to learn and adapt without needing to be explicitly programmed for each situation.

The "5J5IA" project (5 Days, 5 AI) (GTnum#Scol_IA, 2024) is an innovative educational tool that combines a practical guide with a scientific investigation approach. Its primary objective is to support the use of artificial intelligence (AI) through an educational approach based on discovery, active learning, and the appropriation of concepts. The design of 5J5IA is based on a philosophy shared by the researchers and authors involved in the project. Their common vision is to explore and optimize the integration of technologies in education, emphasizing their social dimension. This approach prioritizes concrete and authentic experiences, while leveraging digital tools to enrich the learning process. The project offers practical and realistic activities centered around AI. The accompanying educational booklet explains the link between the theoretical framework developed by the team, the research conducted, and the practical aspects of the educational project. The fundamental principles of this approach are detailed in the following sections.

The scenario of this initiative is designed according to the following seven principles (M. Romero & L. Heiser (Eds.), 2023):

1. **Flexibility**: Short activities (less than an hour) but extensible, easy to integrate into the classroom.
2. **Accessibility**: Ready-to-use educational guide for easy adoption by teachers.
3. **Engagement**: A playful approach with relevant storytelling for learners.
4. **Authenticity**: Realistic and meaningful activities, avoiding stereotypes.
5. **Adaptability**: Exercises that can be done in class or at home, individually or in groups.
6. **Digital Skills**: Alignment with the CRCN and the European DigComp framework, integrating AI.
7. **Structured Progression**: A two-step approach—exploration of representations (modules 1-3), followed by theoretical deepening (modules 4-5), facilitating research on young people's perceptions of AI.

Figure 5. Les cinq activités du dispositif 5J5IA (Romero et al.,2023)

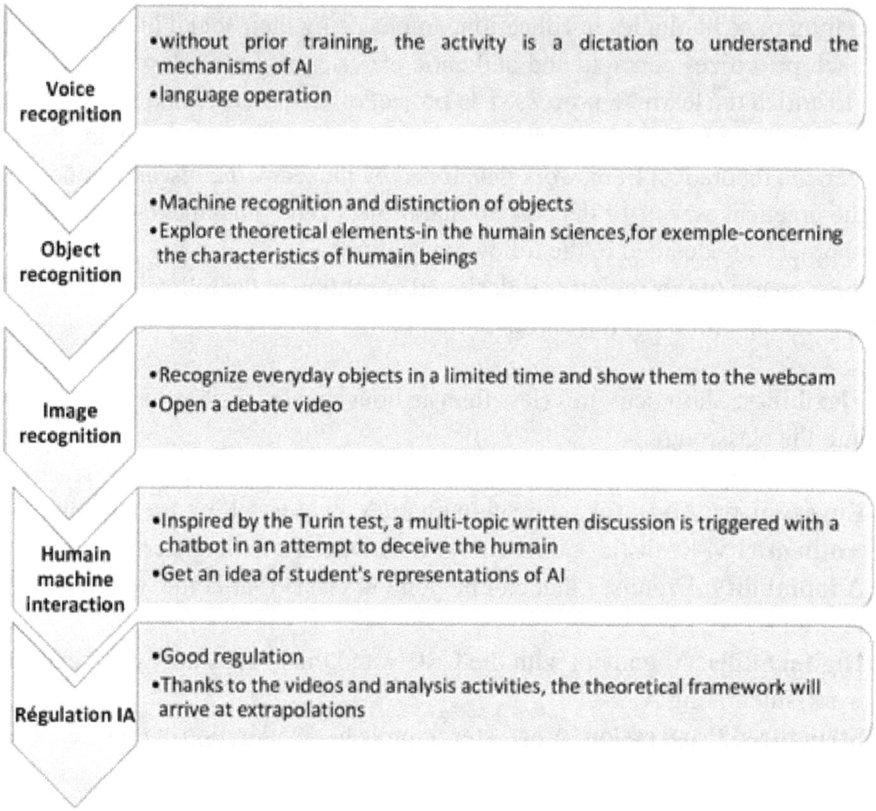

CONCLUSION

In a world where digital technology and AI have brought about a new culture (M. Doueihi, 2011), the historical-technological changes occurring in the 21st century do not leave education untouched. On the contrary, they require us to think about a suitable pedagogy compatible with the new norms that redefine digital literacy. Pedagogical communication tools that integrate AI demystify instrumental interactions between humans and machines, which were once the stuff of science fiction. The transition from traditional online learning platforms to intelligent environments powered by artificial intelligence (AI) marks a fundamental shift in the evolution of distance education. By redefining how we perceive the online classroom, AI offers innovative solutions to the challenges that traditional systems faced.

In a narrower sense, this article discusses the role of AI in education across three dimensions: technological, social, and cognitive. Distinctions between integrating AI as a pedagogical tool and as a subject of learning highlight the immense potential, but also pose major challenges for education regarding mastery and citizen digital culture in the age of AI. Expectations coupled with pedagogical practices that oscillate between efficiency and the reinvention of intelligent platforms fuel an anticipated vision of what education will be like in the future. However, user-friendly tools and co-intelligence are two aspects that must not be separated; together, they should ensure pedagogical alignment.

The technological dimension paves the way for sophisticated tools that automate, analyze, and adapt content and interactions, making learning more seamless and responsive to learners' needs. Hence, pedagogical engineering combines program frameworks with the development of an algorithmic approach to achieve adaptive learning. As for the socio-affective dimension, AI promotes more inclusive and interactive experiences, where collaboration, sociability, and accessibility are central. Learners are no longer merely passive recipients of information but engaged participants in dynamic virtual communities. AI also personalizes the experience, thereby enhancing the sense of presence and support despite physical distance.

In conclusion, the advancement of adaptive learning styles, combined with the contribution of intelligent tools, marks a turning point in the online classroom. This symbiosis between human intervention and machine intelligence promises an unmatched depth in adapting learning pathways, offering content tailored to the specific needs of individuals. Instant feedback and personalized assistance are transforming how learners interact with material, accelerating their progress and maximizing their potential.

Thus, the AI-enhanced online classroom becomes an intelligent space where teachers act as guides, providing the human context for holistic learning, which heavily depends on the judicious integration of AI. Capable of adapting to the multiple dimensions of modern learning—technological, socio-affective, and cognitive—AI raises ethical concerns about privacy and algorithmic fairness that must be carefully considered. This transformation heralds a new era for online education, where AI becomes a catalyst for pedagogical innovation, allowing teachers and learners to envision a more connected, interactive, and engaging educational future.

REFERENCES

Blaise, M. (2023, avril 19). *"D'ici cinq ans, plus personne n'utilisera un modèle tel que ChatGPT ».* Retrieved from Usbek&Rica: https://usbeketrica.com/fr/article/d-ici-cinq-ans-plus-personne-n-utilisera-un-modele-tel-que-chatgpt

Burkhardt, J. M. (2003). Réalité virtuelle et ergonomie: Quelques apports réciproques. *Le Travail Humain*, 66(1), 65–91. DOI: 10.3917/th.661.0065

Cerqueira, J. (2024). *Bac: quelle méthode de révision vous correspond le mieux?* Retrieved from studyrama: https://www.studyrama.com/revision-examen/bac/revisions-et-jour-j-tous-les-conseils-pour-reussir-son/bac-quelle-methode-de-revision-vous-correspond-le-105667

Chinkhir, B., El.Gazi, S., & Ibrahimi, A. (2024). Better Taming Distance in Online Training; Technological Innovation. *International Journal of Computer and Organization Trends*, 14(2), 1–9. DOI: 10.14445/22492593/IJCOT-V14I2P301

Coutlée, A. (2023, septembre 1). *Comment l'intelligence artificielle devient un atout pour les enseignants.* Retrieved from L'École branchée, enseigner à l'ère numérique: https://ecolebranchee.com/comment-lintelligence-artificielle-devient-un-atout-pour-les-enseignants/

Douarre, C., & Rousseau, D. (2021, 10 11). *INTRODUCTION À L'APPRENTISSAGE PROFOND (DEEP LEARNING) DE L'INTELLIGENCE ARTIFICIELLE.* Retrieved from culturesciencesphysique: https://culturesciencesphysique.ens-lyon.fr/ressource/IA-apprentissage-Rousseau.xml - Version du 02/09/22

Doueihi, M. (2011). *Pour un humanisme numérique*. Seuil.

Enkerli, A., & MacNeil, S. (2021, mai 13). *Hypothesis: un outil numérique pour aider les étudiants à s'approprier les exercices de lecture.* Retrieved from Éductive: https://eductive.ca/ressource/hypothesis-un-outil-numerique-pour-aider-les-etudiants-a-sapproprier-les-exercices-de-lecture/

Équipe.éditoriale.Vorecol. (2024, September 10). *L'intégration des technologies d'IA dans les LMS pour le suivi des compétences.* Retrieved from vorecol: https://vorecol.com/fr/blogs/blog-lintegration-des-technologies-dia-dans-les-lms-pour-le-suivi-des-competences-155658#:~:text=Selon%20une%20%C3%A9tude%20de%20McKinsey,de%20temps%20%C3%A0%20l'enseignement

Fleming, N. D., & Mills, C. (1992). Not another inventory, rather a catalyst for reflection. *To Improve the Academy*, 11(1), 137–155.

Fuchs, P. (1996). *Les interfaces de la réalité virtuelle.* Les Presses de l'École des Mines de Paris.

Gandon, É. (2020). Usages du numérique et illectronisme: nouvelles tâches pour le lecteur-scripteur, nouveaux enjeux d'apprentissage. Glottopol. Revue de sociolinguistique en ligne, (33).

Garrison, R., Anderson, D. T., & Archer, W. (2010). The first decade of the community of inquiry framework: A retrospective. *The Internet and Higher Education*, 13(1-2), 5–9. DOI: 10.1016/j.iheduc.2009.10.003

Geneviève, J. (1993). Apprivoiser la distance et supprimer l'absence, ou les défis de la formation à distance. *Revue Française de Pédagogie*, 102(102), 55–67.

Georges, B. (2017, août 30). *2016: Google devient maître du go.* Retrieved from Les Echos: https://www.lesechos.fr/2017/08/2016-google-devient-maitre-du-go-180628#:~:text=A%20S%C3%A9oul%2C%20le%20logiciel%20AlphaGo,progr%C3%A8s%20du%20%C2%AB%20deep%20learning%20%C2%BB.&text=Face%20au%20champion%20sud%2Dcor%C3%A9en,par%204%20victoires%20%C3%A0%2

Giraudon, G., Guitton, P., Romero, M., Roy, D., & Viéville, T. (2020). *Éducation et numérique, Défis et enjeux.* Inria.

Halmaoui, H. (2022). L'intelligence artificielle, la réalité augmentée et la réalité virtuelle dans l'audiovisuel et le cinéma.

Henry, J.-P. (2024). Gaillard Raphaël, L'Homme augmenté. Futurs de nos cerveaux, Paris: Grasset, janvier 2024, 352 p. *Futuribles (Paris, France)*, 3(460), 122–125. DOI: 10.3917/futur.460.0122

Karsenti, T. (2018). Intelligence artificielle en éducation: L'urgence de préparer les futurs enseignants aujourd'hui pour l'école de demain. Formation et profession, 26(3), 112-119.

Le Cun, Y. (2019). *Quand la machine apprend: la révolution des neurones artificiels et de l'apprentissage profond.* Odile Jacob.

Learning, W. t. (n.d.). *Closing the maths attainment gap with one to one teaching.* Retrieved from Welcome to Third Space Learning: https://thirdspacelearning.com/

Mallén, C., Andrea, M., & Mallén, V. (2024, 09 16). *L'e-learning propulsé par l'IA: les outils incontournables.* Retrieved from LEARNY BLOOM: https://learnybloom.com/blog/elearning-propulse-par-ia-outils-incontournables/

Niedercorn, F. (2017, Août 29). *2011 Avec Watson, IBM remporte « Jeopardy! ».* Retrieved from Les Echos: https://www.lesechos.fr/2017/08/2011-avec-watson-ibm-remporte-jeopardy-1116704

Oudeyer, P. Y., Clément, B., Roy, D., & Sauzéon, H. (2020). Projet KidLearn: Vers une personnalisation motivante des parcours d'apprentissage. Bulletin de l'Association Française pour l'Intelligence Artificielle, 51-55.

Papi, C. (2014). *Formation à distance: dispositifs et interactions.* ISTE Group.

Peraya, D. (1999). Médiation et médiatisation: Le campus virtuel. *Hermes*, (3), 153–167.

Pickover, C. A. (2021). ELIZA le psychothérapeute. In C.A.Pickover, *La fabuleuse histoire de l'intelligence artificielle* (pp. 154-156). DUNOD.

Pickover, C. A. (2021). La conférence de Dartmouth sur l'IA. In C.A.Pickover, *La fabuleuse histoire de l'intelligence artificielle* (pp. 140-141). DUNOD.

Pitrat, J. (1998). Métacognition-Intelligence Artificielle. Vocabulaire des Sciences Cognitives, 262-264.

Planche, M. (2024, 08 29). *Synthèse Vocale: 5 Meilleurs logiciels de voix off IA en2024.* Retrieved from Vidnoz: https://fr.vidnoz.com/intelligence-artificielle/logiciel-voix-off-ia.html

Procter, A. (2024, 02 22). *Technologies et innovation. Comment les modèles d'IA spécialisés évoluent pour le mieux.* Retrieved from spark: https://www.okoone.com/fr/spark/technologies-et-innovation/comment-les-modeles-dia-specialises-evoluent-pour-le-mieux/

Psico-smart, É. é. (2024, août 28). *L'impact de l'intelligence artificielle sur la formation des compétences numériques.* Retrieved from Vorecol: https://psico-smart.com/fr/blogs/blog-limpact-de-lintelligence-artificielle-sur-la-formation-des-competences-numeriques-145324

Richardson, C., Nucci, R., & Sevilla, D. (2024, 09 3). *SCORM: Ce qu'est SCORM et comment cela fonctionne.* Retrieved from GURU: https://www.getguru.com/fr/reference/scorm

Romero, M., & Heiser, L. (2023). Enseigner et apprendre à l'ère de l'intelligence artificielle. Canopé, Livre blanc.

Sansonetti, J. (2022, septembre 30). *Histoire de l'intelligence artificielle (IA): création et évolutions [1943 à 2024].* Retrieved from wizishop: https://www.wizishop.fr/blog/histoire-intelligence-artificielle

Satratzemi, M., & Terzopoulos, G. (2020). Voice Assistants and Smart Speakers in Everyday Life and in Education. *Informatics in Education*, 19(3), 473–490.

Short, J., Williams, E., & Christie, B. (1976). *The social psychology of telecommunications*. No Title.

Tandem. (n.d.). *Exprime-toi avec tandem*. Retrieved from Tandem Language Exchange: https://tandem.net/fr

Tessa, A., & Niels, J. (2023, 05 29). *Quelle est la différence entre la réalité augmentée et la réalité virtuelle?* Retrieved from GetApp: https://fr.getapp.ca/blog/3675/difference-realite-augmentee-realite-virtuelle

T.L.G. (2016, 01 28). *Intelligence artificielle: Quand les machines battent l'Homme*. Retrieved from 20minutes TECHNOLOGIES: https://www.20minutes.fr/sciences/1775651-20160128-intelligence-artificielle-quand-machines-battent-homme#:~:text=1997%20%3A%20Deep%20Blue%20met%20en%20%C3%A9chec%20Kasparov&text=La%20machine%20de%202%20m%C3%A8tres,%2C%20l'humain%20%C3%A9tait%20battu

Trinh, S. (2019, Hiver). *L'intelligence artificielle et la traduction évoluent ensemble*. Retrieved from C i r c u i t. Numéro 141: https://www.circuitmagazine.org/dossier-141/l-intelligence-artificielle-et-la-traduction-evoluent-ensemble

Vorecol, É. é. (2024, September 10). *L'intégration des technologies d'IA dans les LMS pour le suivi des compétences*. Retrieved from Vorecol: https://vorecol.com/fr/blogs/blog-lintegration-des-technologies-dia-dans-les-lms-pour-le-suivi-des-competences-155658

Vygotsky, L. S. (1978). *Mind in society: The development of higher psychological processes*. Harvard University Press.

Vygotsky, L. S. (1987). La pensée et le langage. *Revue Française de Pédagogie*, •••, 98–103.

Wenger, E. (1998/2005). *la théorie des communautés de pratique (Gervais, F., Trad.)*. Québec: Presses de l'Université Laval.

Witmer, B. G., & Sadowski, W. J.Jr. (1998). Non visually guided locomotion to a previously viewed target in real and virtual environments. *Human Factors*, 40(3), 478–488. DOI: 10.1518/001872098779591340

Chapter 10
The Future of Online Learning: Emerging Trends and Innovative Technologies

Maha Khaldi
https://orcid.org/0000-0003-2828-5856
Rabat Business School, International University of Rabat, Morocco

Mohamed Khaldi
https://orcid.org/0000-0002-1593-1073
Ecole Normale Superieure of Tetouan, Morocco

ABSTRACT

The future of eLearning is poised for transformative growth, driven by emerging technologies and evolving educational practices that enhance flexibility, engagement, and accessibility. This chapter explores the key innovations expected to shape eLearning, such as AI-powered adaptive learning systems that cater to individual needs and immersive experiences via virtual and augmented reality. These tools promise new levels of personalization, enabling learners to engage with content in dynamic, interactive ways. Additionally, blockchain technology offers advancements in academic credentialing, while mobile and microlearning continue to make education more accessible and on-demand. Through these advancements, we'll outline the potential future of eLearning for learners, educators, and institutions globally.

DOI: 10.4018/979-8-3373-2262-9.ch010

1. INTRODUCTION

Online learning has grown significantly over the past few decades, transforming the educational landscape globally. Increasing digitalisation and the rise of information and communication technologies have made it easier for a wide range of people to access education, providing flexible and accessible learning opportunities.

However, with the rapid evolution of technologies, such as artificial intelligence (AI), virtual reality (VR), augmented reality (AR), and blockchain, the e-learning landscape today is undergoing profound changes, it has introduced new methods of teaching and learning. These technologies make it possible to create interactive and engaging learning experiences, tailored to the individual needs of learners (Ally & Tsinakos, 2014).

The future of online learning is based on a set of emerging trends and innovative technologies that have the potential to transform the way we learn, interact and create knowledge. Indeed, today's learners, whether students or professionals, expect learning experiences that are not only accessible, but also relevant and personalized. Their expectations include increased interactivity, the ability to learn at their own pace, and access to a variety of resources (Bennett & Maton, 2010)

These innovations aim to make learning more personalized, adaptive, immersive, and interactive. However, this period has seen the emergence of new trends, such as the rise of adaptive learning, the increased use of gamification, and the integration of innovative assessment methods. E-learning platforms like Coursera, edX, and Udacity have seen an increase in enrollment in online courses, reflecting a shift toward more flexible and accessible education models (Katz & Oblinger, 2000). In addition, virtual reality (VR), augmented reality (AR), and immersive learning environments are opening up new possibilities for immersive and experiential learning experiences.

This chapter aims to explore and analyze the emerging trends and innovative technologies that are shaping the future of online learning. The main objectives are to:

- Analyze the key trends influencing the development of online learning, such as adaptive learning, microlearning, and gamification and understand how these trends meet the needs of modern learners.
- Explore the key technologies that are transforming education, including artificial intelligence (AI), virtual reality (VR), and blockchain and assess their impact on the learning experience and their potential to improve pedagogical effectiveness.
- Highlight the challenges facing online learning, such as accessibility, learner engagement, and content quality and discuss the opportunities these challenges can create for innovation and continuous improvement in education.

- Provide recommendations and reflections on the future evolution of online learning, taking into account the trends and technologies identified and consider how these developments could transform the role of educators and learners in an increasingly digital educational environment.

By addressing these goals, the chapter aspires to offer an in-depth understanding of the dynamics influencing online learning today and to anticipate future transformations that may emerge in this ever-evolving field.

2. ADAPTIVE LEARNING AND ARTIFICIAL INTELLIGENCE (AI)

2.1. Introduction

Adaptive learning, supported by advances in artificial intelligence (AI), represents a major transformation in education, especially in digital environments. Unlike traditional approaches to teaching, which offer uniform educational pathways, adaptive learning aims to personalize learning experiences to meet the specific needs of each learner. AI-based adaptive systems use sophisticated algorithms to analyze learners' behaviors, performance, and preferences in real-time, in order to offer content and resources adjusted to their level of understanding and skills (Zou et al., 2021.; Zawacki-Richter et al., 2019; Popenici & Kerr, 2017; Brusilovsky & Millán, 2007).

2.2. Outline and theories of adaptive learning

2.2.1. Adaptive Learning Outline

By considering the theoretical foundations and practices of learning, the cognitive approach, technology and personalization, learning objectives, and assessment and feedback, we propose a sketch of adaptive learning (Liu & Wang, 2019; Chen & Miao, 2016; Nye, 2015; Koper, 2014; Durlach & Lesgold, 2012; Brusilovsky & Millán, 2007) Adaptive learning in the context of education systems, adaptive learning is a learning process that uses technologies and analytical methods to personalize educational content and learning experience based on the individual needs of each learner. This type of learning adapts in real time to the learner's performance and preferences. It is based on the idea that learners have different learning styles, paces, and preferences, and that content needs to be adjusted to accommodate these differences. This often involves the use of data and algorithms to determine the optimal learning path for each individual. It uses advanced technologies such as artificial intelligence, machine learning, and data analytics systems to provide a personalized learning experience. These technologies make it possible to track the learner's progress and adjust the content accordingly. Adaptive learning aims to improve learning outcomes by making education more accessible and effective. It allows learners to progress at their own pace, focusing on areas where they need more support. However, in an adaptive learning setting, formative assessment plays a crucial role. Adaptive systems use continuous assessments to provide immediate feedback and adjust content based on learners' outcomes. In summary, we can say that adaptive learning represents a significant evolution of traditional education, integrating technological and pedagogical approaches to meet the individual needs of learners.

2.2.2. Theories of Adaptive Learning

Adaptive learning is based on several concepts from educational theories and cognitive science:

- The theory of the proximal zone of development (Vygotsky, 1978): This theory postulates that learning is most effective when it is located in an area where the learner is able to succeed with minimal assistance. Adaptive learning seeks to identify this area in each learner, adjusting tasks so that they are neither too easy nor too difficult, but just at the appropriate level for optimal progress.
- Theories of cognitive load (Sweller, 1988): According to this theory, the effectiveness of learning depends on the management of cognitive load. Adaptive systems, by regulating the amount of information presented according to the learner's abilities, aim to reduce cognitive overload, allowing for more effective learning.
- Differentiated learning theory (Tomlinson, 2001): This model emphasizes the need to differentiate pedagogical approaches according to the individual characteristics of learners, such as their level, interests, and learning style. Adaptive learning systems make this theory a reality by automating the personalization of teaching at scale using AI.

2.3. Sketch of Artificial Intelligence

Looking at machine learning, neural networks, deep learning, ethics and application areas, we propose a sketch of artificial intelligence (AI) (Binns, 2018; Russell & Norvig, 2016; Brynjolfsson & McAfee, 2014; LeCun et al., 1998; Mitchell, 1997; Searle, 1980).

Artificial intelligence (AI) refers to the ability of a computer system to perform tasks that normally require human intelligence, such as natural language understanding, image recognition, decision-making, and learning. It is a technique by which a system can learn to perform tasks by analyzing data, without being explicitly programmed for each specific task. This includes methods such as supervised, unsupervised, and reinforcement learning. AI is often classified into two categories: weak AI, which focuses on specific tasks (such as virtual assistants), and strong AI, which aims to replicate human intelligence in general. It should be noted that artificial neural networks, inspired by the functioning of the human brain, are key models used in deep learning, a subcategory of AI. These networks are able to learn complex representations of data. It is used in a variety of fields, including healthcare (medical diagnostics), transportation (self-driving cars), marketing (product recommendation), and education (smart tutors). These applications leverage advanced algorithms to improve performance and efficiency In summary, artificial intelligence is a rapidly expanding field that is transforming many aspects of modern society. Finally, the rise of AI raises important ethical questions regarding privacy, security, algorithmic bias, and employment impact. Researchers and legislators are working to develop standards and frameworks to guide the responsible development of AI

2.4. Artificial Intelligence and Adaptive Learning

AI is at the heart of adaptive learning, allowing for fine-grained personalization of teaching and dynamic analysis of learner progress. Several technologies and approaches to AI are commonly used in this context (Spector & Ma, 2019).

- Expert systems and recommendation engines: Expert systems use rulebases to offer recommendations tailored to each learner. Recommendation engines, borrowed from the fields of e-commerce and social networking, use learning data to suggest personalized resources, such as additional exercises, videos or further reading.
- Predictive models: Machine learning and predictive algorithms can be used to model learners' future behavior based on their past interactions. These models can predict performance, identify potential gaps, and recommend instructional interventions before the learner encounters major difficulties.

- Real-time data analytics: Adaptive learning systems use real-time data to track learners' progress, adapt pace and content based on their level of understanding, and provide immediate and relevant feedback. Big data analytics techniques are used to evaluate large volumes of learner data and adjust learning paths.
- Learner Modeling: Learner modeling is the process of creating a digital profile that represents an individual's strengths, weaknesses, preferences, and goals. This profile, enriched as interactions progress, is used by systems to continuously adapt content and learning activities, and offer a tailor-made experience.

2.5. Practical Applications of Adaptive Learning

- Online learning environments: E-learning platforms such as Coursera, Khan Academy, and Duolingo integrate adaptive learning systems to personalize the learner journey. These platforms adjust the difficulty of exercises and recommend content based on users' progress, allowing for more effective upskilling
- Smart tutors: Virtual smart tutors, integrated into digital educational environments, offer personalized support to learners. For example, the Knewton system, a pioneer in adaptive learning, uses AI to adjust educational content in real-time based on learners' responses (Fletcher & Morrison, 2014; Woolf, 2010; Nkambou et al., 2010).
- Gamification and adaptability: Integrating gamification into adaptive learning systems helps keep learners engaged by adjusting challenges based on their performance. Platforms like DreamBox and Smart Sparrow use these principles to deliver gamified experiences tailored to each learner's level and preferences.

2.6. Challenges and Limitations

Despite the promising benefits, AI-based adaptive learning presents several challenges (Hwang & Chang, 2021).

- Ethical and data privacy issues: Adaptive systems collect and analyze a considerable amount of learner data. This massive collection raises ethical questions related to confidentiality and data protection;
- Accessibility and inclusion: The implementation of these systems requires advanced technological infrastructure, which can pose access issues for marginalized populations or in regions with limited resources;

- Pedagogical quality: While adaptive systems are effective at adjusting content based on performance, there is a risk that learning will become too dependent on AI, to the detriment of human interaction and learners' autonomous critical thinking.

As a conclusion to this section, AI-based adaptive learning represents a major evolution in education, capable of personalizing and optimizing learning paths at scale. Leveraging sophisticated algorithms and fine data analysis, these systems offer promising solutions to meet individual learners' needs and improve the efficiency of online learning environments. However, the integration of these technologies must be accompanied by an in-depth reflection on ethical issues, inclusivity and pedagogical quality, in order to maximize their benefits while minimizing risks.

3. SMART TUTORS AND EDUCATIONAL CHATBOTS

3.1. Introduction

The integration of virtual smart tutors and chatbots into e-learning platforms makes it possible to provide personalized support to learners. These chatbots use advanced algorithms to answer questions, provide additional explanations, and offer real-time support. Smart tutors and educational chatbots are applications of artificial intelligence (AI) that simulate interactions with learners to assist them in their learning journey. Designed to offer personalized support, these systems provide explanations, feedback, answers to questions, and even activity recommendations in real time. Thanks to advances in AI, natural language processing (NLP), and machine learning, intelligent tutors and educational chatbots can now offer a more interactive, responsive, and autonomous learning experience. These tools are part of a move towards individualized learning where each learner can progress at their own pace, while receiving appropriate support. Smart tutors and chatbots aim to replicate the positive effect of a particular teacher, by being available at all times and responding immediately to learners' needs.

3.2. Sketch of the intelligent tutors

Referring to the Adaptation Model, Real-Time Feedback, Interaction and Engagement, Specific Applications, Performance Evaluation, we propose a sketch of intelligent tutors based on research (Huang & Weng, 2017; Dichev & Dicheva, 2017; Nye, 2015; VanLehn, 2011; Woolf, 2010; Koedinger & Aleven, 2007). A smart tutor is a computer system that uses artificial intelligence techniques to provide personalized support to the learner. These systems adapt teaching strategies according to the individual needs, preferences and performance of learners. Smart tutors use adaptive learning models to personalize learning paths. These systems analyze learners' interactions with content to adjust exercises and resources based on the results. Smart tutors provide immediate, contextualized feedback, allowing learners to understand their mistakes and progress faster. This feedback is essential for effective learning. Smart tutors use gamification and interaction techniques to keep learners engaged. They can incorporate playful elements and interactive scenarios to make learning more engaging Smart tutors continuously assess learners' performance and adjust their approach based on progress, allowing for continuous personalization of learning. Smart tutors are used in a variety of fields, including math, language, and science. They can be integrated into online learning platforms to provide additional support to students. In conclusion, smart tutors represent a significant advancement in the field of online education, offering a personalized approach that meets the individual needs of learners.

3.3. Sketching educational chatbots

Referring to personalized interaction, accessibility and availability, support for learning and revision, feedback and evaluation, and applications in specific domains, we propose an outline of the definition of educational chatbots and their pedagogical applications (Ali & Alkhateeb, 2020; Abdullah & Hussain, 2020; Yang & Liu, 2020; Kukulska-Hulme, 2019; Fadhil & Mahrin, 2019; Krauss & Fuchs, 2018). An educational chatbot is a computer program designed to simulate a human conversation and interact with users in a learning context. These chatbots can answer questions, provide information, and guide learners through various educational content. They use artificial intelligence techniques to provide personalized interaction with learners. They can tailor their answers and suggestions according to each user's preferences and needs Considering accessibility and availability, educational chatbots are accessible 24/7, allowing learners to interact with them at any time. This constant availability provides students with immediate support and reduces barriers to learning Considering learning and revision support, chatbots can be used to help learners review and assimilate knowledge by asking questions, providing interactive quizzes, and suggesting additional resources Educational chatbots can provide immediate feedback on learners' performance, helping them identify their strengths and weaknesses. This facilitates continuous evaluation of their progress. They are applied in various areas such as language learning, exam preparation, and homework support. They can offer resources tailored to each topic or skill. In conclusion, educational chatbots represent a significant advancement in digital education, offering personalized interactions and constant support to learners.

The references provided offer an overview of the concepts, methods, and research associated with educational chatbots.

3.4. Learning Theories in Relation to Smart Tutors and Chatbots

- Constructivism theory: Smart tutors take a constructivist approach to learning by allowing learners to construct their own knowledge through guided interactions. Through intelligent systems, learners are guided through a process of discovery and problem-solving, where artificial intelligence adjusts its role as a facilitator of knowledge (Piaget, 1976a; Vygotsky, 1978).
- Cognitive tutoring theory of learning: This theory emphasizes learning in which a teacher or tutor guides the learner by providing feedback and models for reflection. Intelligent tutors recreate this role by providing tailored guidance, modeling cognitive processes (such as problem-solving), and adjusting the level of support according to the learner's needs (Collins et al., 1991).
- Computer-Assisted Instruction Theory: Intelligent tutors and educational chatbots are also based on computer-assisted learning principles, where interactions with a machine allow the learner to progress according to their own needs, while benefiting from immediate and interactive assistance, thus optimizing the learning process (Lu & Zhai, 2019; Kulhavy, 1977).

3.5. Artificial Intelligence in Smart Tutors and Educational Chatbots

Artificial intelligence forms the technological basis of intelligent tutors and chatbots, enabling the automation and personalization of interactions with learners. We propose the main AI components used in these systems (Zhou et al., 2020; Kassens-Noor, 2019; Nkambou et al., 2010; Woolf, 2010; Graesser & McNamara, 2010; Veletsianos & Miller, 2008; Rus & Graesser, 2009; Kerly et al., 2007; Weber & Brusilovsky, 2001).

- Natural Language Processing (NLP): NLP is essential for enabling educational chatbots to understand and interact with learners via natural language. NLP technologies allow chatbots to respond contextually to questions asked by learners, interpret complex queries, and offer clear and tailored explanations.
- Machine Learning: Machine learning algorithms allow intelligent tutors to adapt to the learner's progress. By analyzing past performance and current behaviors, these systems can adjust the level of difficulty of the proposed

activities, recommend specific resources, and predict areas where the learner may need more help.

- Learner Modeling: Learner modeling allows you to create personalized profiles that evolve based on learner interactions and performance. These profiles are used to adjust teaching strategies, types of feedback, and educational resources offered. This makes the learning experience more relevant and tailored to individual needs.
- Error analysis and automated feedback: Smart tutors are able to diagnose learners' mistakes and provide real-time feedback. This feedback is personalized and can guide the learner in correcting his or her mistakes by offering detailed explanations or suggestions for additional activities.

3.6. Features and Benefits of Smart Tutors and Educational Chatbots

- Personalization of learning: One of the main benefits of smart tutors is their ability to adapt to the individual needs of learners. By analyzing performance and interactions in real time, these systems adjust content and the level of support. This ensures continuous learning, adapted to the pace and skills of the learner.
- 24/7 availability: Unlike human teachers, smart tutors and educational chatbots are available at all times. They allow learners to ask questions or receive explanations outside of traditional class hours, providing increased flexibility, especially for those learning remotely.
- Instant and adaptive feedback: Smart tutors offer immediate feedback to learners, which is essential for reinforcing learning. This feedback is often contextual and adjusted to the learner's level of understanding, allowing them to correct mistakes or reinforce their knowledge in the moment.
- Emotional and motivational support: Some educational chatbots are designed to detect the learner's emotions or motivation level. They can adjust their behaviour and language to encourage or motivate the learner, or adjust tasks according to the emotional state. For example, a chatbot could simplify questions when the learner seems frustrated, or offer more complex challenges when the learner is confident.
- Language interaction: Smart tutors are particularly useful for language learning, offering interactive conversations that simulate real-time dialogues. Learners can practice grammar, vocabulary, and pronunciation in a controlled environment, receiving corrections and adjustments as they go (Huang & Wu, 2017).

3.7. Examples of Smart Tutors and Chatbots in Education

- Carnegie Learning's Cognitive Tutor: This intelligent tutor, used primarily in math education, relies on cognitive models to diagnose learners' mental processes and adjust instructional activities accordingly. It helps students understand and master complex concepts by adapting explanations and exercises according to their progress.
- Duolingo: Well-known in language learning, Duolingo uses a chatbot to simulate conversations with learners. This system offers real-time corrections, adjusts the level of difficulty based on the user's responses, and offers lessons based on the weaknesses identified in each learner.
- MATHia: This intelligent math tutor guides learners through problem-solving by giving them immediate feedback. It adjusts the proposed exercises according to the mistakes and successes of the learners.

3.8. Challenges and Limitations of Smart Tutors and Chatbots

- Complexity of interactions: While educational chatbots have made significant strides, they still have limitations in understanding the complex contexts and nuances of human language. Conversations can sometimes be rigid or limited, and systems can struggle to interpret ambiguous or poorly worded questions.
- Reliance on technology: Smart tutors rely on advanced technology infrastructure, which can cause accessibility issues in regions or for learners without stable internet connections or appropriate devices.
- Acceptance and ethics issues: Some learners may be reluctant to interact with non-human systems, reducing the effectiveness of these tools. In addition, the use of these systems raises ethical questions regarding data privacy and the personalization algorithm, which can sometimes lock the learner into limited choices.

In conclusion of this section, smart tutors and educational chatbots offer promising prospects for transforming online learning, making it more personalized, interactive, and accessible. They are particularly effective in providing immediate support, detailed feedback and tailor-made learning paths. However, their use must be accompanied by a reflection on ethical and technological issues in order to guarantee an inclusive and high-quality educational experience.

4. VIRTUAL REALITY (VR) AND AUGMENTED REALITY (AR)

4.1. Introduction

Virtual reality (VR) and augmented reality (AR) are two emerging technologies that are profoundly transforming the way we teach and learn. These immersive technologies offer learners the opportunity to have interactive and engaging experiences, by immersing themselves in simulated virtual environments (VR) or enriching the real world with digital overlays (AR). In the context of learning, VR and AR allow learners to explore complex concepts in a more intuitive way, improve information retention, and foster more active learning. These technologies, by combining immersion and interaction, create dynamic learning environments where learners can explore, manipulate and interact with 3D content, which promotes a better understanding and application of abstract concepts. Their ability to simulate realistic scenarios in fields such as medicine, engineering, and history makes VR and AR powerful educational tools.

4.2. Sketching Virtual Reality (VR)

Referring to design principles, immersion, interaction, gamification, educational applications, sensory experiences and psychological impact, we propose a sketch of virtual reality based on research work (Rizzo & Koenig, 2017; Jerald, 2016; Slater & Sanchez-Vives, 2016; Pantelidis, 2010; Dédé, 2009; Gee, 2007; Brooks, 1999; Slater & Wilbur, 1997)

Virtual reality (VR) is a technology that allows you to create an immersive digital environment, simulating real or fictional experiences in which users can interact. VR typically uses headsets, gloves, and other devices to engage the user's senses. It is characterized by a high level of immersion, where users feel physically present in a virtual environment. This immersion is enhanced by the ability to interact with objects and characters in this virtual space. Virtual reality is increasingly being used in education to simulate interactive learning environments. It allows students to explore complex concepts in a practical and engaging setting, facilitating understanding and retention of knowledge. VR can be integrated into educational games, using gamification elements to make learning more engaging. VR games offer immersive experiences that can boost learners' motivation and engagement. It often involves a combination of several technologies, including 3D visualization, spatial sound, and haptic feedback, to create a complete sensory experience that mimics reality. The use of virtual reality can have a significant psychological impact on users, affecting their perception of reality and emotions. This can be used for therapeutic purposes, for example in the treatment of phobias or post-traumatic stress. In summary, virtual reality offers vast and varied possibilities in education and beyond, creating immersive experiences that can transform learning.

4.3. Augmented Reality Sketch

Referring to a review of augmented reality technologies and applications, its use in education, user interaction in its systems in particular through the use of visualization tools and its impact on learning and social interactions, we propose a sketch of augmented reality (AR) (Bacca et al. 2014; Dunleavy & Dede, 2014; Borkin et al. 2013; Cheng & Tsai, 2013; Billinghurst & Duenser, 2012; Klopfer & Squire, 2008; Azuma, 1997; Milgram & Kishino, 1994).

Augmented reality (AR) is a technology that overlays digital elements (images, sounds, information) on top of physical reality, allowing users to interact with virtual objects embedded in their real environment using devices such as smartphones, tablets, or smart glasses. It is used in education to enrich the learning experience, allowing students to interact with educational content in a dynamic and engaging way. This can include 3D visualizations of complex objects and interactive simulations. Augmented reality systems enable real-time interaction between users and virtual elements, often through gestures, movements, or voice commands. This increases user engagement and immersion in the experience. Unlike virtual reality (VR), which creates a fully digital environment, augmented reality enriches the real world by adding virtual elements. This distinction is crucial to understanding the different applications and impacts of each technology. Augmented reality can be experienced through a variety of devices, including smartphones, tablets, and AR glasses, each with varying capabilities to overlay digital content on top of reality. In addition, the use of augmented reality can have an impact on users' perception, changing their experience of the physical environment. It can also foster collaboration and social interaction in learning and work contexts. In conclusion, augmented reality offers innovative possibilities to enrich the learning and interaction experience in various fields. The references provided provide an overview of the concepts, methods, and research associated with augmented reality.

4.4. Learning Theories and Conceptual Framework of VR and AR

The pedagogical effectiveness of immersive environments such as VR and AR is based on several learning theories:

- Experiential Learning Theory: The theory of experiential learning posits that individuals learn best by directly experiencing a situation or concept. VR and

AR allow for immersive learning experiences where learners can physically or virtually interact with the objects of study, which promotes deeper and more sustainable learning (Kolb, 1984).

- Constructivism theory: Constructivism holds that individuals construct their own knowledge from their interactions with the environment. VR and AR technologies build on this approach by allowing learners to actively explore virtual or augmented worlds, solve problems, and develop knowledge by directly manipulating learning objects (Piaget, 1976b; Vygotsky, 1978).
- Situated Cognition Theory: According to this theory, knowledge is rooted in the activity, context, and culture in which it is acquired. VR and AR create authentic learning environments, where learners can apply their knowledge in simulated settings, such as a medical lab or a virtual construction site. This allows for a more efficient transfer of skills to real-life situations (Brown, Collins, & Duguid, 1989).
- Cognitive load theory: Cognitive load refers to the amount of mental effort required to process information. VR and AR, by allowing learners to visualize abstract concepts in interactive environments, can reduce cognitive load by facilitating visual comprehension and avoiding textual information overload (Sweller, 1988).

4.5. Applications and Operation of VR and AR in Education

Through this section, we propose some applications and functioning of virtual reality and augmented reality (Radianti et al., 2020; Chung & Craig, 2018).

- Virtual Reality (VR): VR offers complete immersion in digital worlds where the learner can interact with simulated objects or environments. Thanks to VR headsets, users are fully immersed in 3D environments that can simulate scientific laboratories, historical battlefields, or industrial processes. This allows learners to have educational experiences that are impossible or expensive to achieve in the real world. We offer examples of educational applications of VR.
    - Medicine: In surgery, VR allows medical students to train to perform complex operations in a safe environment, without risk to patients;
    - History: Learners can travel back in time to visit historical sites and understand events in their original context;
    - Engineering and Architecture: Students can explore architectural structures in 3D and perform test simulations in a secure environment.
- Augmented reality (AR): Unlike VR, which creates a fully virtual world, AR overlays digital information (images, sounds, texts) on top of the real world.

This technology allows learners to enrich their immediate environment with additional layers of information, accessible via smartphones, tablets, or AR glasses. AR is especially useful in environments where learners need to interact with the real world while accessing additional information in real-time. We offer examples of educational applications of AR.

- Natural Sciences: Learners can observe and manipulate 3D models of molecules or organisms superimposed on their real space;
- Arts and Design: Art students can visualize sculptures or artworks in 3D in a real-world environment, providing an extra dimension to art design;
- Vocational training: Workers in training can use RA to receive step-by-step instructions when operating complex machinery or during installation procedures.

4.6. Benefits of Using VR and AR in Education

In this section, we propose some advantages of the use of VR and AR in Education (Freina & Ott, 2015; Merchant et al., 2014; Wu et al., 2013).

- Immersion and Engagement: One of the main benefits of VR and AR is their ability to immerse learners in an environment where cognitive and emotional engagement is high. This deep immersion allows learners to better focus on the tasks at hand and interact more meaningfully with the learning content.
- Learning by Doing: Immersive environments allow for learning by doing, which is often difficult to achieve with traditional teaching methods. The ability to manipulate objects, solve practical problems, and receive immediate feedback enhances the effectiveness of learning.
- Security and simulation of real-world scenarios: VR is particularly useful for simulating dangerous or complex scenarios that would be difficult or expensive to replicate in real life. This includes crisis management training, medical simulations and high-risk industrial interventions.
- Personalized learning: VR and AR allow learners to follow their own pace, explore concepts according to their preferences, and revisit complex lessons as many times as necessary. This customization is crucial to meet diverse learning needs, making these technologies inclusive for a diverse audience.

4.7. Challenges and Limitations of VR and AR Technologies

- Costs and Accessibility: VR devices, such as headsets and sensors, can be a significant investment for educational institutions. Similarly, creating immersive educational content in VR and AR requires specific resources and skills, which may limit accessibility for some schools or students.

- Comfort and health issues: Prolonged use of VR headsets can lead to side effects such as eye strain, headaches, or nausea (commonly known as "cybersickness"). This can limit the continued use of these technologies in learning.
- Technical needs and digital skills: The integration of VR and AR into learning environments requires adequate technical infrastructure (internet connection, hardware, software) and advanced digital skills, for both teachers and learners.
- Cognitive overload: While VR and AR can reduce cognitive load by making information more visual and interactive, there is a risk of overload if too much information is layered in an AR environment or if VR scenarios are too complex for the learner to handle.

In conclusion, virtual reality (VR) and augmented reality (AR) technologies represent a significant advancement in the field of education, enabling immersive, interactive and contextual learning experiences. They pave the way for more engaging teaching methods, which promote better knowledge retention and practical application of the skills acquired. However, to maximize their potential, it is necessary to overcome certain technical, financial and pedagogical challenges. Future research and innovation in these technologies will undoubtedly offer new opportunities to make learning even more accessible, interactive and effective.

5. SOCIAL AND COLLABORATIVE LEARNING

5.1. Introduction

The future of e-learning also lies in the development of digital environments that promote collaboration and exchanges between learners. Collective intelligence and social learning networks facilitate the exchange of ideas and the co-construction of knowledge. Indeed, social and collaborative learning emphasizes the role of social interactions and collaborations between learners in building and sharing knowledge. Based on theories that recognize the importance of communication and cooperation, this framework values the co-construction of knowledge, where individuals learn together through the exchange of ideas, debate, and collective problem-solving.

In the current context of digitalization and e-learning, these learning methods are becoming more and more relevant. Collaborative platforms, online discussion forums, teamwork tools, and digital learning environments enrich the learner experience by providing opportunities for remote social interaction. By combining the benefits of social learning and digital technologies, it is possible to create learning environments that are rich, engaging, and responsive to contemporary needs.

5.2. Outline of Social Learning

With reference to, social learning theory emphasizing the role of observation in the acquisition of behaviors that are influenced by social patterns, Vygotsky's sociocultural theory which emphasizes the role of social interactions in cognitive development, theories of cooperation and competition, connectivism and finally, intrinsic and extrinsic motivation in the context of social learning (Siemens, 2005; Ryan & Deci, 2000; Johnson & Johnson, 1989; Bandura, 1986, 1977; Vygotsky, 1978).

Social learning is a process by which individuals acquire skills, knowledge, and behaviors by observing and interacting with others in a social context. This type of learning emphasizes the importance of observation, imitation, and modeling. In social learning, individuals learn not only through their own experiences, but also by observing the results of others' actions. Role models, whether peers, teachers, or authority figures, play a crucial role in this process. Social learning highlights the importance of the social and cultural context in which learning occurs. Interactions with others, social norms, and values influence how individuals learn. It is often associated with collaborative approaches, where individuals work together to solve problems, exchange ideas, and learn from each other. This promotes deeper and more meaningful learning. With the rise of digital technologies, social learning has expanded to online environments, where learners can interact, share information, and collaborate through digital platforms, discussion forums, and social media. It can also have a significant emotional impact. Positive peer interactions can boost learners' motivation and engagement, while negative experiences can damage self-confidence and learning. In conclusion, social learning is a dynamic process that emphasizes the importance of social interactions in the acquisition of knowledge and skills. The references provided provide an overview of the concepts, methods, and research associated with social learning.

5.3. Sketch of collaborative learning

Referring to the theories and practices of collaborative learning, collaborative interactions, implications of shared responsibility, peer-to-peer feedback, strategies and techniques for the integration of collaborative learning in higher education. We propose an outline of collaborative learning (Johnson et al., 2014; Nicol &

Macfarlane-Dick, 2006; Stahl et al., 2006; Barkley et al., 2005; Garrison & Anderson, 2003; Dillenbourg, 1999a 1999b; Slavin, 1996; Roschelle & Teasley, 1995).

Collaborative learning is an educational process where learners work together in small groups to achieve common learning goals. This type of learning is based on cooperation, communication and knowledge sharing between group members. It emphasizes the importance of social interactions and discussions among learners, which foster a deeper understanding of concepts and ideas. These interactions also help develop social and interpersonal skills. In a collaborative learning context, each member of the group is responsible for both their own learning and the learning of others. This shared responsibility promotes learners' engagement and motivation. Similarly, collaborative learning can be facilitated by the use of digital technologies, which allow learners to collaborate remotely, share resources, and interact in real-time through online platforms. Additionally, collaborative learning is associated with improved knowledge retention, as learners discuss and explain concepts to their peers, thereby strengthening their understanding. By working together, learners develop essential interpersonal skills, such as communication, negotiation and conflict resolution, which are crucial in both academic and professional contexts. In conclusion, collaborative learning is a dynamic and interactive process that fosters engagement and cooperation among learners.

5.4. Theories of Social and Collaborative Learning

- Social Learning Theory: According to Bandura, learning is done through observing and imitating others in a social setting. The emphasis is on the importance of patterns, interaction and imitation in the learning process. Learners acquire behaviours, attitudes and skills by observing and interacting with their peers (Bandura, 1977).
- Social constructivism: The theory of social constructivism emphasizes that learning is a fundamentally social process and occurs through interactions with more experienced peers. Vygotsky's concept of the Proximal Development Zone (PDZ) emphasizes the role of the group and the tutor in the acquisition of new knowledge, where learners can accomplish more with the help of others than alone (Vygotsky, 1978).
- Positive Interdependence Theory: This theory refers to how the success of an individual in a group is related to the success of the entire group. Collaborative learning occurs when a group of people work together to achieve a common goal, each making important and complementary contributions (Johnson & Johnson, 1999).
- Distributed cognition: This theoretical framework argues that knowledge and cognition are not limited to the individual, but are distributed across a

network of people and tools in a collaborative environment. Collaborative learning is based on the idea that learners share information and distribute cognitive tasks among themselves, thus facilitating the resolution of complex problems (Hutchins, 1995).

5.5. Characteristics and Components of Social and Collaborative Learning

- Interactivity and Communication: Social interaction is at the heart of collaborative learning. Discussions, debates, and the exchange of ideas promote cognitive development and allow learners to refine their understanding. This interactivity can occur in person or through online tools such as discussion forums, chats, or video conferencing platforms.
- Co-construction of knowledge: In a collaborative environment, learners do not just receive information passively. They actively participate in the co-construction of knowledge, bringing their own point of view and confronting each other's ideas to create a shared understanding (Scardamalia & Bereiter, 2006).
- Shared roles and responsibilities: Collaborative learning requires a division of roles and responsibilities within a group. Every learner is both a contributor and a learner, and this positive interdependence creates mutual engagement in the learning process. Roles can be organizers, moderators, or facilitators, each making separate contributions to the group.
- Asynchronous and synchronous collaboration: E-learning technologies allow social learning to take place synchronously (real-time) and asynchronously (deferred). For example, video conferencing and instant chats provide opportunities for real-time interaction, while discussion forums, wikis, and collaborative documents allow for asynchronous interaction (Garrison et al., 2000; Koschmann, 1996).
- Collective Reflection and Feedback: Collaborative learning encourages collective reflection, where group members evaluate their progress, mistakes and successes together. Feedback is essential in this setting, as it allows learners to refine their understanding by taking into account the perspectives and constructive criticism of others.

5.6. Benefits of Social and Collaborative Learning

- Interpersonal Skills Development: By participating in collaborative activities, learners develop essential skills such as communication, negotiation, and conflict management. These skills are crucial not only for group work, but also for personal and professional development.

- Improved motivation and engagement: Collaborative learning tends to increase learners' motivation, as they feel responsible not only for their own success, but also for the success of their peers. This dynamic of interdependence promotes greater participation and active engagement in tasks.
- Deep and sustainable learning: Explaining one's ideas to others and having to argue about one's own knowledge leads learners to deepen their understanding of concepts. Sharing with others promotes deeper learning, as learners need to actively think, synthesize, and organize their knowledge.
- Problem solving and critical thinking: Collaborative work is often centered around solving complex problems that require critical thinking. Learners must work together to analyse problems, propose solutions and evaluate results. This process develops problem-solving and critical thinking skills (Roschelle & Teasley, 1995).
- Building a learning community: Social learning fosters the development of a sense of community among learners. This sense of community reinforces mutual aid, sharing of ideas, and mutual support, which can contribute to a positive learning atmosphere and encourage active participation (Wenger, 1998).

5.7. Challenges and Limitations of Social and Collaborative Learning

- Managing group dynamics: Collaboration can be affected by unfavorable group dynamics, such as the dominance of certain individuals, the lack of participation of certain members, or interpersonal conflicts. These issues can hinder the quality of collaborative learning if groups are not well managed.
- Unequal participation: One of the major challenges in collaborative learning is to ensure equal participation of all members of the group. Some learners may be passive, while others may take up too much space. This can create imbalances in learning and reduce the effectiveness of collaborative work.
- Technical problems in online environments: In online learning environments, technical challenges, such as unstable internet connections or difficulties in using collaborative tools, can limit learner participation and hinder collaboration (Dillenbourg et al., 2009).
- Cognitive overload: Collaborative learning, especially in an online environment, can sometimes lead to cognitive overload. Learners may feel overwhelmed by the amount of interactions, information, or tasks that need to be managed simultaneously (Kirschner et al., 2009).

- Evaluating collaborative work: It can be difficult to objectively assess the individual contribution of group members in a collaborative setting. Teachers need to develop assessment criteria that reflect both group work and individual participation, which can be complex.

In summary, social and collaborative learning offers an enriching pedagogical approach that highlights the importance of social interactions and cooperation in the construction of knowledge. By integrating these principles into learning environments, especially digital ones, learners can benefit from a more engaging, motivating, and formative experience. However, to maximize the effectiveness of these methods, it is essential to manage group dynamics, ensure equitable participation, and overcome technical and organizational challenges. As education continues to adapt to the digital age, social and collaborative learning is set to play an increasingly central role in the training of learners.

6. MICRO-LEARNING AND GAMIFICATION

6.1. Introduction

Short-form learning methods, or micro-learning, and the use of game mechanics (gamification) increase learner engagement. Indeed, micro-learning and gamification are two innovative pedagogical approaches that are gaining popularity in modern learning environments, especially in digital contexts. Micro-learning focuses on delivering small segments of information in a concise and targeted way, it fits into busy schedules, making learning accessible anywhere, anytime while gamification uses game elements to motivate and engage learners, it makes learning more motivating. These two approaches, often combined, aim to improve information retention, learner engagement, and learning flexibility, adapting to learners' needs in a rapidly changing world.

6.2. Sketch of micro-learning

Referring to the notion of micro-learning, the pedagogical challenges associated with its integration into education, its benefits in online learning environments, its integration into higher education and the use of digital technologies, its professional development in companies, its impact on learner engagement, different learning styles and the importance of rapid feedback, we propose an outline of micro-learning (Gupta & El Haddioui, 2019; Zotova & Zaretskaya, 2019; Baldwin & Ford, 2017;

Giurgiu, 2017; Tharp, 2016; Koh & Frick, 2014; Sandeen, 2013; Fadel & Lemke, 2008; Hug, 2005).

Microlearning is an educational approach that involves breaking down learning content into small, focused units, allowing learners to access specific information on demand, often in minutes. This facilitates quick and effective learning. Microlearning is characterized by its flexibility, accessibility, and ability to adapt to individual learning styles. It also helps to strengthen knowledge retention by offering short and repeated learning sessions. Microlearning is often combined with digital technologies, such as mobile apps, short videos, interactive quizzes, and e-learning modules, allowing learners to access content anytime, anywhere. It is increasingly used in corporate training to provide just-in-time training. This allows employees to quickly learn specific skills when they need them. Due to its concise and focused nature, microlearning can improve learner engagement, allowing them to consume content in small doses that fit easily into their schedule. It facilitates formative assessment, allowing learners to receive immediate feedback after short learning sessions. This helps to strengthen their understanding and correct mistakes quickly. In conclusion, micro-learning is an innovative educational approach that facilitates learning in short, focused segments, meeting the needs of modern learners.

6.3. Theories and Principles of Micro-learning

- Memory and cognitive load theory: Microlearning is based on the idea that information should be presented in small chunks so as not to overload working memory. The theory of cognitive load suggests that learning is more efficient when the amount of information to be processed is limited, which corresponds to the strategy of micro-learning. By reducing cognitive complexity, learners can absorb information more easily (Sweller, 1988).
- Spaced Repetition Learning Theory: Spaced repetition is a learning method in which information is repeated at increasing intervals to improve long-term retention. Micro-learning, by delivering content concisely and regularly, facilitates the frequent repetition and revision of key concepts, thus maximizing retention (Ebbinghaus, 1885).
- Theory of contextualization of learning: Micro-learning aims to offer knowledge in specific contexts that are directly applicable, which is in line with Merrill's theory which emphasizes that learning is most effective when contextualized. Learners are exposed to short modules that focus on specific skills or knowledge and can immediately put into practice what they have learned (Merrill, 2002).
- Self-learning theory: Micro-learning allows learners to take control of their own learning process. It encourages self-direction and individual responsibil-

ity, as micro-learning modules are often accessible anytime, anywhere, which promotes flexible and self-directed learning (Knowles, 1980).

6.4. Features and Benefits of Microlearning

- Short, focused content: Microlearning involves breaking up learning content into short segments (typically 3-10 minutes) that focus on specific points or skills. This allows learners to focus on well-defined learning objectives (Hug, 2006).
- Flexibility and accessibility: Micro-learning modules are often accessed via mobile devices, allowing learners to access information anytime, anywhere. This is particularly suitable for working professionals and learners with busy schedules (Buchem & Hamelmann, 2010).
- Improved retention and engagement: The short durations of micro-learning sessions allow for better information retention, as they avoid cognitive overload. Additionally, learners are more likely to stay engaged when they know that content can be consumed quickly.
- Personalization of learning: Learners can choose the modules that best meet their individual needs, creating a personalized learning experience. Microlearning allows for greater flexibility in the choice of pace and learning path.

6.5. Sketch of gamification

Referring to the building blocks of gamification, its impact on learner motivation and engagement, its effectiveness in various contexts, the importance of feedback to learning through gamification, the integration of gamification elements in the design of learning applications and platforms and its impact on learner performance, we propose an outline of gamification (Bai et al., 2020; Subhash & Cudney, 2018; Su & Cheng, 2015; Dicheva et al., 2015; Chou & Wu, 2015; Hamari et al., J., 2014; Landers, 2014; Werbach & Hunter, 2012; Kapp, 2012; Deterding et al., 2011; Zichermann & Cunningham, 2011).

Gamification is the use of game elements in non-gaming contexts, such as education or marketing, in order to engage and motivate users to achieve specific goals. This can include things like points, badges, leaderboards, and challenges. Gamification aims to increase user motivation by making tasks more engaging and fun. By incorporating game mechanics, it can transform activities that are perceived as repetitive or boring into more engaging experiences. In education, gamification is used to encourage active learning, improve student participation, and promote knowledge retention. It creates a stimulating and interactive learning environment. It provides immediate feedback on users' performance, which can help them better understand their progress and adjust their efforts accordingly. This encourages learning through trial and error. Gamification uses various game design elements, such as levels, rewards, and challenges, to create a structured and motivating learning journey. These elements are carefully integrated to encourage participation and engagement. Gamification can also have a positive impact on users' performance by incentivizing them to take on specific challenges and goals, which can lead to better academic or professional outcomes. In conclusion, Gamification is an innovative approach that uses game elements to engage users in a variety of contexts, especially in education and training.

6.6. Theories and Principles of Gamification

- Self-determination theory: Self-determination theory holds that human motivation is influenced by three basic needs: competence, autonomy, and social relationship. Gamification addresses these needs by providing learners with clear goals (competence), giving them the freedom to choose their actions (autonomy), and fostering social interactions through competition or collaboration (social relationship) (Deci & Ryan, 1985).
- Flow theory: Gamification relies on creating a learning experience where the learner is fully immersed in the task, a state that Csikszentmihalyi calls flow. This state is achieved when tasks are balanced between challenge and skill, a key principle of gamification that adapts difficulty levels and provides real-time feedback to maintain engagement (Csikszentmihalyi, 1990).
- Operant conditioning theory: Gamification uses reward and feedback systems to motivate learners to complete tasks. This mechanism, based on operant conditioning, reinforces positive behaviors (such as the success of an exercise) through immediate rewards (badges, points, etc.), which increases the likelihood that these behaviors will be repeated (Skinner, 1953).
- Motivation by Achievement Theory: Gamification encourages learners to achieve set goals and receive recognition for their achievements. Points, badges, levels, and leaderboards are examples of mechanisms that build motivation through accomplishment, encouraging learners to excel (Atkinson, 1964).

6.7. Components and Features of Gamification

- Rewards and points systems: Gamification uses rewards in the form of points, badges, or trophies to reward learners for achieving goals. These rewards serve as extrinsic motivation and incentivize learners to continue to progress.

- Levels and Progression: Levels allow learners to visualize their progress. By reaching different levels, they perceive their progress and are encouraged to continue learning to achieve higher goals.
- Immediate feedback: Gamification provides instant feedback, allowing learners to immediately know if they have understood a concept or performed a task correctly. This reinforces trial and error learning and helps to correct mistakes quickly.
- Competition and collaboration: Competitive elements (leaderboards, peer challenges) and collaborative elements (teamwork, shared goals) add a social dimension to learning, which increases motivation and engagement.
- Challenges and Quests: Gamification often includes challenges or quests where learners must solve problems or complete tasks in order to advance. These elements stimulate active learning and cognitive involvement.

6.8. Benefits of Micro-learning and Gamification

- Increased engagement and motivation: The integration of gamification into micro-learning modules maintains a high level of engagement through regular challenges, rewards, and immediate feedback, motivating learners to progress steadily.
- Flexible and autonomous learning: Micro-learning and gamification offer great flexibility, allowing learners to study at their own pace, while staying motivated thanks to the fun elements and the objectives to be achieved.
- Reinforcement of information retention: The short modules of micro-learning facilitate retention, while frequent repetition, reinforced by game elements (quizzes, challenges), helps to reinforce long-term learning.
- Learner-centered approach: These methods allow learners to customize their learning journey according to their needs and preferences, while making them active and accountable for their progress.

6.9. Challenges and Limitations of Micro-learning and Gamification

- Superficiality of content: Microlearning, by focusing on short segments, can sometimes lack depth, which can be problematic for more complex topics that require detailed analysis.
- Reliance on external rewards: In gamification, an over-reliance on extrinsic rewards (points, badges) can be detrimental to intrinsic motivation. Learners may focus more on rewards than on actually acquiring knowledge.

- Expensive design and development: Creating microlearning modules and implementing gamification elements requires significant resources in terms of instructional design, technology development, and continuous effectiveness evaluation.
- Balancing challenge and skill: One of the challenges of gamification is to maintain a good balance between the difficulty level of the tasks and the skills of the learners. An imbalance can lead to frustration (if too difficult) or boredom (if too easy).

In conclusion, micro-learning and gamification represent complementary pedagogical approaches that, when combined, can significantly improve the learning experience. Together, they make it possible to offer digestible, engaging and motivating content, while adapting to the needs of a modern, mobile and often busy audience. However, to maximize their effectiveness, it is essential to structure the content well, maintain the balance between rewards and actual learning, and adapt these strategies to the specific needs of learners and educational objectives.

7. BLOCKCHAIN AND DIGITAL CERTIFICATIONS

7.1. Introduction

The use of blockchain in e-learning makes it possible to secure digital certifications and guarantee their validity. Learners can thus keep an unfalsifiable record of their achievements and skills, facilitating the international recognition of qualifications. Thus, blockchain, initially developed as the underlying technology of cryptocurrencies, is emerging as a revolutionary solution in many sectors, including education. One of the most promising applications is the use of blockchain for digital certifications. These blockchain-based certifications offer guarantees of authenticity, transparency, and permanence that can transform the way academic and professional qualifications are validated and shared. In a context where skills are changing rapidly and lifelong learning is becoming essential, blockchain makes it possible to create decentralized and tamper-proof digital registers of achievements and skills, simplifying the verification of diplomas and other certifications.

7.2. Blockchain Sketch

Referring to the concept of blockchain in the context of Bitcoin, presenting the basic principles of the technology, its characteristics and potential beyond cryptocurrencies, how it works and its implications for various sectors, its applications in various fields and its potential to transform business models, its security aspects and the use of cryptography to protect information, to the economic and technical challenges related to its adoption and the implications for users and companies, we propose a sketch of the blockchain (Zheng et al., 2018; Grech & Camilleri, 2017; Kahn & Lathrop, 2017; Tapscott & Tapscott, 2016; Crosby et al., 2016; Catalini & Gans, 2016; Christidis & Devetsikiotis, 2016; Swan, 2015; Nakamoto, 2008).

Blockchain is a secure and decentralized technology for storing and transmitting information that allows the creation of a digital ledger shared between several participants. Each record, or "block," is linked to the previous ones, forming a continuous chain. Blockchain features include transparency, immutability, decentralization, and security. These features ensure that the data cannot be changed once saved, and that all participants in the network can access the same information. The blockchain works through distributed consensus among the nodes of the network. Each transaction is validated by the network before being added to the chain, ensuring trust between participants. Although blockchain was initially developed for cryptocurrencies, its applications extend to areas such as supply chain management, smart contracts, e-voting, and digital identity. Its security is based on advanced cryptographic techniques, which ensure that only authorized parties can access the data and that transactions are authentic and unalterable. Despite its advantages, blockchain faces several challenges, including scalability, energy consumption, and regulation. These aspects must be taken into account for its widespread adoption. In conclusion, blockchain is an innovative technology with various applications beyond cryptocurrencies.

7.3. Outline of digital certifications

By referring to the emergence of digital qualifications and their role in modern learning, their impact on the labour market and the use of technologies, its benefits for learners, tools for assessing learning outcomes, their standardisation in the context of learning, and finally, their evolution and impact on education and work. (Böck & Kahn, 2021; Zhang & Wan, 2021; Santos & de Mello, 2020; Kohli, & Pande, 2020; Mok, 2020; Beck & Fagan, 2020; Cohen & Schaefer, 2019; Kirkland, 2019; Mok & Zhang, 2018; Jansen & Hovestadt, 2018; Dunston, 2018; Weigel & Weller, 2017; Lemoine & Guérin, 2016; Pappas, 2015).

Digital certifications are formal attestations issued by educational organizations or platforms, which validate the skills and knowledge acquired by an individual in a specific field. They are often based on online assessments and can be verified electronically. Digital certifications often use advanced technologies, such as blockchain, to ensure their authenticity, security, and immutability. This helps create verifiable records that can be easily shared and accessed by employers or other interested parties. They offer learners recognition of their skills that can be easily integrated into digital CVs, online profiles or professional networks. This increases their visibility in the job market. They can be obtained through various types of assessments, including online tests, hands-on projects, or courses taken, allowing for a comprehensive assessment of skills. The recognition of digital certifications varies from industry to industry, and efforts are underway to establish standards and validation systems that ensure their value and acceptance in the workplace. They are part of a broader trend of transformation in education and training, where skills have become a central criterion for assessment and validation in the modern labour market. In conclusion, digital certifications represent a significant step forward in the recognition of acquired skills and knowledge.

7.4. Key Blockchain Theories and Concepts

- Decentralization Theory: One of the core principles of blockchain is decentralization. Unlike traditional databases, which are controlled by a central authority (such as a university or certification body), blockchain is distributed across multiple nodes that collectively verify and validate transactions. This decentralization ensures data integrity and prevents unauthorized tampering with certifications (Nakamoto, 2008).
- Data immutability theory: Blockchain is based on a system where each transaction is recorded in the form of blocks, linked chronologically and immutably. Once a certification is written to the blockchain, it cannot be changed or deleted, ensuring full traceability and transparency. This builds confidence in the veracity of digital certifications (Bashir, 2017).
- Distributed trust theory: Blockchain eliminates the need for trusted third parties (such as universities or companies) to validate certifications, as trust is distributed within the network itself. Network participants can independently verify the authenticity of diplomas and certificates without having to resort to a central authority, simplifying the verification process (Pilkington, 2016).
- Smart contract theory: Smart contracts are self-contained programs that run automatically when certain conditions are met. In the context of digital certifications, they can be used to issue diplomas or digital badges automatically when the learner has successfully completed a study program or training module. These contracts can also be configured to include specific criteria that trigger the issuance of certifications (Szabo, 1997).

7.5. Blockchain Features and Components

- Decentralization: Blockchain is distributed across a network of peers, eliminating the need for a central authority to store and manage data. Digital certifications can be validated by the entire network, reducing the risk of fraud and falsification of diplomas.
- Transparency and traceability: Transactions on a blockchain are accessible to all members of the network, allowing for full transparency. Any certification issued can be traced back to its origin, simplifying verification processes for employers or academic institutions.
- Immutability and security: Once a certification is recorded in the blockchain, it can no longer be modified or deleted. This immutability ensures that certificates cannot be tampered with or tampered with. In addition, blockchain uses cryptographic techniques to ensure the security of stored data, making certifications tamper-proof.
- Automation via smart contracts: Smart contracts help automate the process of issuing digital certifications. For example, when the evaluation criteria for a training program are met, a smart contract can automatically issue a diploma or certificate and record it on the blockchain.

7.6. Blockchain-Based Digital Certifications

Digital badges: Digital badges are visual representations of skills or achievements that can be delivered through blockchain-based certification systems. They are forgery-proof and can be easily shared with employers, on professional social networks (such as LinkedIn), or between peers.

Digital degrees: More and more academic institutions are using blockchain to deliver digital degrees. These credentials can be verified instantly by employers or other institutions, reducing the costs and time associated with traditional verification processes.

Certificates of competence or continuing education: The blockchain makes it possible to follow the continuing education of professionals. These certificates record the skills acquired throughout the working life, making the achievements easily verifiable and transferable between different sectors.

7.7. Benefits of Blockchain-Based Digital Certifications

- Tamper-proof and security: Certifications registered on a blockchain cannot be tampered with or altered, ensuring their authenticity. This solves a major problem in education and human resources, namely the fraud of diplomas and certifications.

- Ease of verification: Employers or academic institutions can verify the authenticity of a certification by consulting the blockchain, reducing the time and resources required to authenticate degrees. It also facilitates the admission and recruitment processes.
- Accessibility and sharing: Digital certifications are easily accessible by learners, who can share them online through professional platforms, on social media, or directly with employers. This allows for greater visibility of learners' skills.
- Reduced administrative costs: Blockchain reduces the costs associated with managing, verifying, and storing traditional certifications. Educational institutions and companies no longer need to invest in complex systems to archive or validate diplomas.
- Portability and interoperability: Digital certifications are portable across different platforms and systems. Learners can easily transfer their skills and qualifications between different educational or professional contexts, which is particularly relevant in a globalised world.

7.8. Challenges and Limitations of Blockchain-Based Certifications

- Limited adoption: While promising, blockchain technology for digital certifications is still in its infancy. The lack of universal standards and limited adoption by educational institutions and employers may hinder its development. • Privacy concerns: Although blockchain is secure and transparent, there are concerns about the privacy of personal data. It's crucial to design systems where learners can control what pieces of their information is shared and with whom.
- Technical complexity: Blockchain remains a complex technology, and implementing digital certifications based on this technology requires specialized skills and robust technological infrastructure. This can be a barrier for some institutions and organizations.
- Regulations and legal framework: The legal framework around blockchain and digital certifications is still evolving. It is important that international laws and standards adapt to provide a framework for the verification, data protection, and recognition of these certifications.

In conclusion, the use of blockchain for digital certifications represents a major breakthrough in the management of qualifications and skills in the world of education and work. It offers an innovative solution to secure, verify and share skills quickly, reliably and transparently. While challenges remain in terms of adoption, regulation, and privacy, blockchain is well-positioned to become a pillar of the education systems of the future. The future of digital certification lies in the ability of institutions and businesses to leverage this technology to make learning more seamless, accessible, and portable.

8. CONCLUSION

Online learning is undergoing rapid change, transformed by innovative technologies that are shaping its future. This chapter explored several emerging trends, such as artificial intelligence, virtual and augmented reality, smart tutors, and educational chatbots, as well as the growing adoption of micro-learning and gamification. These innovations make it possible to personalize learning paths, make educational environments more interactive and immersive, while promoting learner engagement.

The rise of technologies like blockchain for digital certifications and content distribution is rethinking knowledge management and ensuring greater traceability and security. At the same time, collaborative learning, facilitated by social platforms and digital tools, is redefining interactions between learners, encouraging the collective construction of knowledge.

While these technologies offer promising prospects, they also pose challenges related to their accessibility, cost, and integration into existing education systems. Inclusivity is a major challenge, as not all learners have the same technological resources or the necessary digital skills. In addition, the massive introduction of technologies in learning raises ethical questions and concerns about data protection and user privacy. It is also essential to rethink teacher training so that they can take full advantage of these tools.

In conclusion, the future of e-learning looks promising, but its success will depend on the ability of education actors to integrate these technologies in a strategic and inclusive way. Online learning will become increasingly ubiquitous, adaptive and interactive, while continuing to evolve with technological advances.

9. REFERENCES

Abdullah, N. A. H., & Hussain, M. (2020). The Effectiveness of Chatbots as a Tool for Learning: A Case Study on Students' Feedback and Assessment. *Education and Information Technologies*, 25(6), 5543–5560.

Ali, A., & Alkhateeb, H. (2020). Chatbots in Education: A Review of the Use of Chatbots for Learning. *Education and Information Technologies*, 25(5), 4153–4168.

Ally, M., & Tsinakos, A. (2014). *Advances in Mobile Learning Educational Research*. IGI Global.

Atkinson, J. W. (1964). *An Introduction to Motivation*. Van Nostrand.

Azuma, R. T. (1997). A Survey of Augmented Reality. *Presence (Cambridge, Mass.)*, 6(4), 355–385. DOI: 10.1162/pres.1997.6.4.355

Bacca, J., Baldiris, S., Fabregat, R., & Graf, S., & Kinshuk. (2014). Augmented Reality Trends in Education: A Systematic Review of Research and Applications. *Journal of Educational Technology & Society*, 17(4), 133–149.

Bai, S., Hew, K. F., & Huang, B. (2020). Is Gamification "Bullshit"? Evidence from a Meta-analysis and Synthesis of Qualitative Data in Educational Contexts. *Educational Research Review*, 30, 100322. DOI: 10.1016/j.edurev.2020.100322

Baldwin, A. K., & Ford, K. J. (2017). Micro-Learning: A New Approach to Training and Development. *International Journal of Training and Development*, 21(3), 193–205.

Bandura, A. (1977). *Social Learning Theory*. Prentice-Hall.

Bandura, A. (1986). *Social Foundations of Thought and Action: A Social Cognitive Theory*. Prentice-Hall.

Barkley, E. F., Cross, P. K., & Major, C. H. (2005). *Collaborative Learning Techniques: A Handbook for College Faculty*. Jossey-Bass.

Bashir, I. (2017). *Mastering Blockchain: Deeper Insights into Decentralization, Cryptography, Bitcoin, and Popular Blockchain Frameworks*. Packt Publishing Ltd.

Beck, S. F., & Fagan, M. (2020). Digital Badges and Micro-Credentials: A New Way to Assess Learning Outcomes. *International Journal of Learning Technology*, 15(2), 113–126.

Beck, S. F., & Fagan, M. (2020). Digital Badges and Micro-Credentials: A New Way to Assess Learning Outcomes. *International Journal of Learning Technology*, 15(2), 113–126.

Bennett, S., & Maton, K. (2010). Beyond the 'digital natives' debate: Toward a more nuanced understanding of students' technology experiences. *Journal of Computer Assisted Learning*, 26(5), 321–331. DOI: 10.1111/j.1365-2729.2010.00360.x

Billinghurst, M., & Duenser, A. (2012). Augmented Reality in the Classroom. *Computer*, 45(7), 56–63. DOI: 10.1109/MC.2012.111

Binns, R. (2018). Fairness in Machine Learning: Lessons from Political Philosophy. In *Proceedings of the 2018 Conference on Fairness, Accountability, and Transparency* (pp. 149-158).

Böck, B., & Kahn, A. (2021). Blockchain Technology for Educational Credentialing: A Review of the Literature. *International Journal of Educational Management*, 35(2), 382–396.

Borkin, M. A., Vo, A. A., Bylinskii, Z., Isola, P., Sunkavalli, S., Oliva, A., & Pfister, H. (2013). What Makes a Visualization Memorable? *IEEE Transactions on Visualization and Computer Graphics*, 19(12), 2306–2315. DOI: 10.1109/TVCG.2013.234 PMID: 24051797

Brooks, F. P. (1999). What's Real About Virtual Reality? *Computer Graphics*, 33(1), 3–11.

Brown, J. S., Collins, A., & Duguid, P. (1989). Situated Cognition and the Culture of Learning. *Educational Researcher*, 18(1), 32–42. DOI: 10.3102/0013189X018001032

Brusilovsky, P., & Millán, E. (2007). User Models for Adaptive Hypermedia and Adaptive Educational Systems. In P. Brusilovsky, A. Kobsa, & W. Nejdl (Eds.), The Adaptive Web. Springer. DOI: 10.1007/978-3-540-72079-9_1

Brynjolfsson, E., & McAfee, A. (2014). *The Second Machine Age: Work, Progress, and Prosperity in a Time of Brilliant Technologies*. W. W. Norton & Company.

Buchem, I., & Hamelmann, H. (2010). Microlearning: A Strategy for Ongoing Professional Development. eLearning Papers, 21, 1-15.

Catalini, C., & Gans, J. S. (2016). Some Simple Economics of the Blockchain. NBER Working Paper No. 22952.

Chen, X., & Miao, H. (2016). A Learning Path Recommendation System for Adaptive Learning. [iJET]. *International Journal of Emerging Technologies in Learning*, 11(1), 45–52.

Cheng, K. H., & Tsai, C. C. (2013). Affordances of Augmented Reality in Science Learning: Suggestions for Future Research. *Journal of Science Education and Technology*, 22(4), 449–462. DOI: 10.1007/s10956-012-9405-9

Chou, P.-N., & Wu, H.-K. (2015). The Influence of Gamification on Learning: A Case Study on Elementary School Students. *Journal of Educational Technology & Society*, 18(3), 134–145.

Christidis, K., & Devetsikiotis, M. (2016). Blockchains and Smart Contracts for the Internet of Things. *IEEE Access : Practical Innovations, Open Solutions*, 4, 2292–2303. DOI: 10.1109/ACCESS.2016.2566339

Chung, J., & Craig, E. (2018). *Augmented Reality and Virtual Reality: The Power of AR and VR for Business*. Springer.

Cohen, J., & Schaefer, J. (2019). Blockchain for Educational Credentials: Opportunities and Challenges. *Journal of Technology and Teacher Education*, 27(3), 265–280.

Collins, A., Brown, J. S., & Holum, A. (1991). Cognitive apprenticeship: Making thinking visible. *American Educator*, 15(3), 6–11.

Crosby, M., Pattanayak, P., Verma, S., & Kalyanaraman, V. (2016). Blockchain Technology: Beyond Bitcoin. *Applied Innovation Review*, 2, 6–10.

Csikszentmihalyi, M. (1990). *Flow: The Psychology of Optimal Experience*. Harper & Row.

Deci, E. L., & Ryan, R. M. (1985). *Intrinsic Motivation and Self-Determination in Human Behavior*. Plenum Press. DOI: 10.1007/978-1-4899-2271-7

Dede, C. (2009). Immersive Interfaces for Engagement and Learning. *Science*, 323(5910), 66–69. DOI: 10.1126/science.1167311 PMID: 19119219

Deterding, S., Dixon, D., Khaled, R., & Nacke, L. (2011). From Game Design Elements to Gamefulness: defining" Gamification". In *Proceedings of the 15th International Academic MindTrek Conference: Envisioning Future Media Environments* (pp. 9-15). DOI: 10.1145/2181037.2181040

Dichev, C., & Dicheva, D. (2017). Gamification in Education: What, How, Why Bother? *International Journal of Educational Technology in Higher Education*, 14(1), 1–20.

Dicheva, D., Dichev, C., Agre, G., & Angelova, G. (2015). Gamification in Education: A Systematic Mapping Study. *Journal of Educational Technology & Society*, 18(3), 75–88.

Dillenbourg, P. (1999a). *Collaborative Learning: Cognitive and Computational Approaches*. Elsevier.

Dillenbourg, P. (1999b). What Do You Mean by Collaborative Learning? In Dillenbourg, P. (Ed.), *Collaborative-Learning: Cognitive and Computational Approaches* (pp. 1–19).

Dillenbourg, P., Järvelä, S., & Fischer, F. (2009). The Evolution of Research on Computer-supported Collaborative Learning: From Design to Orchestration. In Balacheff, N. (Eds.), *Technology-enhanced Learning* (pp. 3–19). Springer. DOI: 10.1007/978-1-4020-9827-7_1

Dunleavy, M., & Dede, C. (2014). Augmented Reality Teaching and Learning. *TechTrends*, 58(6), 28–34. DOI: 10.1007/s11528-013-0717-2

Dunston, P. J. (2018). Blockchain Technology: A New Frontier for Credentialing in Higher Education. In *Proceedings of the 13th International Conference on Educational Technology* (pp. 105-109).

Durlach, P. J., & Lesgold, A. M. (2012). *Adaptive Technologies for Training and Education*. Cambridge University Press. DOI: 10.1017/CBO9781139049580

Ebbinghaus, H. (1885). *Über das Gedächtnis: Untersuchungen zur experimentellen Psychologie* [On Memory: A Contribution to Experimental Psychology]. Duncker & Humblot.

Fadel, C., & Lemke, C. (2008). Multiliteracies and the Future of Education. *Educational Leadership*, 66(6), 40–44.

Fadhil, A., & Mahrin, M. N. (2019). The Role of Chatbots in Education: A Review of the Literature. *Journal of Educational Technology & Society*, 22(2), 92–103.

Fletcher, J. D., & Morrison, J. E. (2014). DARPA Digital Tutor: Assessment and Evaluation. *The Journal of Defense Modeling and Simulation*, 11(3), 221–229.

Freina, L., & Ott, M. (2015). A Literature Review on Immersive Virtual Reality in Education: State of the Art and Perspectives. In The International Scientific Conference eLearning and Software for Education (Vol. 1, No. 133, pp. 10-100).

Garrison, D. R., & Anderson, T. (2003). E-Learning in the 21st Century: A Community of Inquiry Framework for Online Learning. Routledge.

Garrison, D. R., Anderson, T., & Archer, W. (2000). Critical Inquiry in a Text-Based Environment: Computer Conferencing in Higher Education. *The Internet and Higher Education*, 2(2-3), 87–105. DOI: 10.1016/S1096-7516(00)00016-6

Gee, J. P. (2007). What Video Games Have to Teach Us About Learning and Literacy. *Computers in Human Behavior*, 23(1), 1–14.

Giurgiu, L. (2017). Microlearning an Evolving Elearning Trend. *Science Bulletin*, 22(1), 18–23.

Graesser, A. C., & McNamara, D. S. (2010). Self-regulated Learning in Learning Environments with Pedagogical Agents that Interact in Natural Language. *Educational Psychologist*, 45(4), 234–244. DOI: 10.1080/00461520.2010.515933

Grech, A., & Camilleri, A. F. (2017). *Blockchain in Education*. European Commission.

Gupta, A., & El Haddioui, K. (2019). A Study of Microlearning Practices: A Literature Review. *International Journal of Educational Research and Development*, 9(4), 60–68.

Hamari, J., Koivisto, J., & Sarsa, H. (2014). Does Gamification Work? - A Literature Review of Empirical Studies on Gamification. In 2014 47th Hawaii International Conference on System Sciences (pp. 3025-3034).

Huang, C., & Wu, C. (2017). A Systematic Review of Intelligent Tutoring Systems for Language Learning: A Content Analysis. *Journal of Educational Technology & Society*, 20(2), 1–10.

Huang, W. D., & Weng, S. S. (2017). An Intelligent Tutoring System for Personalized Learning: Feedback Mechanisms and Adaptive Strategies. *Computers & Education*, 115, 173–186.

Hug, T. (2005). Micro Learning: A New Pedagogical Challenge. In Micro Learning: Emerging Trends and Technologies (pp. 1-10).

Hug, T. (2006). Micro Learning and Narration: Exploring Possibilities of Utilization of Narrations and Storytelling for the Designing of "Micro Units" and Didactical Micro-learning Arrangements. In *Proceedings of the third International Conference on Microlearning* (pp. 33-38).

Hutchins, E. (1995). *Cognition in the Wild*. MIT Press. DOI: 10.7551/mitpress/1881.001.0001

Hwang, G. J., & Chang, H. F. (2021). A Review of Opportunities and Challenges of Using Artificial Intelligence for Adaptive Learning in Higher Education. *Educational Technology Research and Development*, 69(2), 807–820.

Jansen, R. S., & Hovestadt, L. (2018). Verifying Learning Outcomes with Blockchain Technology: A New Way of Certification in Education. *Proceedings of the European Conference on e-Learning*, 1, 267-275.

Jerald, J. (2016). *The VR Book: Human-Centered Design for Virtual Reality*. Association for Computing Machinery and Morgan & Claypool.

Johnson, D. W., & Johnson, R. T. (1989). *Cooperation and Competition: Theory and Research*. Interaction Book Company.

Johnson, D. W., & Johnson, R. T. (1999). *Learning Together and Alone: Cooperative, Competitive, and Individualistic Learning*. Allyn & Bacon.

Johnson, D. W., Johnson, R. T., & Smith, K. A. (2014). *Active Learning: Cooperation in the College Classroom* (3rd ed.). Pearson.

Kahn, C. E., & Lathrop, R. (2017). Blockchain Technology: The Future of Digital Information Management. *Journal of Management Policy and Practice*, 18(1), 33–44.

Kapp, K. M. (2012). *The Gamification of Learning and Instruction: Game-Based Methods and Strategies for Training and Education*. Pfeiffer.

Kassens-Noor, E. (2019). The Role of Conversational Agents in Informal Learning via Social Media. *Computers & Education*, 130, 59–70.

Katz, R. N., & Oblinger, D. G. (2000). The Future of the Digital Learning Environment. *EDUCAUSE Review*, 35(2), 34–45.

Kerly, A., Hall, P., & Bull, S. (2007). Bringing Chatbots into Education: Towards Natural Language Negotiation of Open Learner Models. *Knowledge-Based Systems*, 20(2), 177–185. DOI: 10.1016/j.knosys.2006.11.014

Kirkland, R. (2019). Using Blockchain to Authenticate Educational Credentials: A Guide for Institutions. *International Journal of Information Technology and Management*, 18(2), 123–136.

Kirschner, P. A., Paas, F., & Kirschner, F. (2009). A Cognitive Load Approach to Collaborative Learning: United Brains for Complex Tasks. *Educational Psychology Review*, 21(1), 31–42. DOI: 10.1007/s10648-008-9095-2

Klopfer, E., & Squire, K. (2008). Environmental Detectives: The Development of an Augmented Reality Platform for Environmental Education. *Educational Technology Research and Development*, 56(2), 203–228. DOI: 10.1007/s11423-007-9037-6

Knowles, M. S. (1980). *The Modern Practice of Adult Education: Andragogy Versus Pedagogy*. Cambridge Books.

Koedinger, K. R., & Aleven, V. (2007). Exploring the Assistance Dilemma in Intelligent Tutoring Systems. *User Modeling and User-Adapted Interaction*, 17(3), 209–235.

Koh, E. G., & Frick, T. W. (2014). Micro-Learning and Learning Styles: A Study of Learning Preferences of Postgraduate Students. *Journal of Online Learning and Teaching*, 10(1), 59–69.

Kohli, A., & Pande, A. (2020). The Role of Blockchain Technology in Managing Digital Credentials. *International Journal of Information Management*, 50, 42–52.

Kolb, D. A. (1984). *Experiential Learning: Experience as the Source of Learning and Development*. Prentice-Hall.

Koper, R. (2014). The Role of Technology in Adaptive Learning. *Journal of Interactive Media in Education*, 2014(1), 1–8. DOI: 10.5334/2005-1

Koschmann, T. (1996). *CSCL: Theory and Practice of an Emerging Paradigm*. Lawrence Erlbaum Associates.

Krauss, S., & Fuchs, C. (2018). Chatbots as a Tool for Learning: How Conversational Agents Can Enhance Learning Experiences. *Journal of Computer Assisted Learning*, 34(5), 527–539.

Kukulska-Hulme, A. (2019). Mobile Assisted Language Learning [MALL]: A Perspective on the Future. In The Cambridge Handbook of Language Learning (pp. 383-400). Cambridge University Press.

Kulhavy, R. W. (1977). Feedback in Written Instruction. *Review of Educational Research*, 47(2), 211–232. DOI: 10.3102/00346543047002211

Landers, R. N. (2014). Developing a Theory of Gamified Learning: Linking Serious Games and Gamification of Learning. *Simulation & Gaming*, 45(6), 752–768. DOI: 10.1177/1046878114563660

LeCun, Y., Bengio, Y., & Haffner, P. (1998). Gradient-Based Learning Applied to Document Recognition. *Proceedings of the IEEE*, 86(11), 2278–2324. DOI: 10.1109/5.726791

Lemoine, P., & Guérin, S. (2016). Digital Credentials and the Future of Learning. *European Journal of Education*, 51(4), 569–582.

Liu, M., & Wang, Y. (2019). Personalized Learning in the Age of Adaptive Learning Technologies: Challenges and Opportunities. *Education and Information Technologies*, 24(3), 1681–1705.

Lu, X., & Zhai, X. (2019). Chatbots in the Classroom: A Review of Conversational Agents in Education.

Merchant, Z., Goetz, E. T., Cifuentes, L., Keeney-Kennicutt, W., & Davis, T. J. (2014). Effectiveness of Virtual Reality-based Instruction on Students' Learning Outcomes in K-12 and Higher Education: A Meta-analysis. *Computers & Education*, 70, 29–40. DOI: 10.1016/j.compedu.2013.07.033

Merrill, M. D. (2002). First principles of instruction. *Educational Technology Research and Development*, 50(3), 43–59. DOI: 10.1007/BF02505024

Milgram, P., & Kishino, F. (1994). A Taxonomy of Mixed Reality Visual Displays. *IEICE Transactions on Information and Systems*, 77(12), 1321–1329.

Mitchell, T. M. (1997). *Machine Learning*. McGraw-Hill.

Mok, K. H. (2020). Digital Credentials and Blockchain: Implications for Higher Education. *Higher Education Research & Development*, 39(3), 548–561.

Mok, K. H., & Zhang, H. (2018). The Impact of Digital Credentials on the Labor Market. *International Journal of Human Resource Management*, 29(18), 2656–2672.

Nakamoto, S. (2008). Bitcoin: A Peer-to-Peer Electronic Cash System.

Nicol, D. J., & Macfarlane-Dick, D. (2006). Formative Assessment and Self-Regulated Learning: A Model and Seven Principles of Good Feedback Practice. *Studies in Higher Education*, 31(2), 199–218. DOI: 10.1080/03075070600572090

Nkambou, R., Bourdeau, J., & Mizoguchi, R. (2010). *Advances in Intelligent Tutoring Systems*. Springer. DOI: 10.1007/978-3-642-14363-2

Nye, B. D. (2015). The Effects of Adaptive Learning Technologies on Student Outcomes: A Meta-Analysis. *Journal of Educational Psychology*, 107(1), 162–177.

Pantelidis, V. S. (2010). Reasons to Use Virtual Reality in Education and Training Courses and a Model to Determine When to Use Virtual Reality. *Themes in Science and Technology Education*, 2(1-2), 59–70.

Pappas, C. (2015). The Rise of Digital Credentials. eLearning Industry.

Piaget, J. (1976a). Cognitive Development in Children: Piaget Development and Learning. *Journal of Research in Science Teaching*, 2(3), 176–186. DOI: 10.1002/tea.3660020306

Piaget, J. (1976b). *The Grasp of Consciousness: Action and Concept in the Young Child*. Harvard University Press.

Pilkington, M. (2016). Blockchain Technology: Principles and Applications. In Research Handbook on Digital Transformations (pp. 225-253). Edward Elgar Publishing.

Popenici, S. A. D., & Kerr, S. (2017). Exploring the Impact of Artificial Intelligence on Teaching and Learning in Higher Education. *Research and Practice in Technology Enhanced Learning*, 12(1), 22. DOI: 10.1186/s41039-017-0062-8 PMID: 30595727

Radianti, J., Majchrzak, T. A., Fromm, J., & Wohlgenannt, I. (2020). A Systematic Review of Immersive Virtual Reality Applications for Higher Education: Design Elements, Lessons Learned, and Research Agenda. *Computers & Education*, 147, 103778. DOI: 10.1016/j.compedu.2019.103778

Rizzo, A. S., & Koenig, S. T. (2017). Is Clinical Virtual Reality Ready for Primetime? *Neuropsychology*, 31(8), 875–892. DOI: 10.1037/neu0000405 PMID: 29376669

Roschelle, J., & Teasley, S. D. (1995). The Construction of Shared Knowledge in Collaborative Problem Solving. In *Proceedings of the 1995 Conference on Computer Support for Collaborative Learning* (pp. 69-96). DOI: 10.1007/978-3-642-85098-1_5

Rus, V., & Graesser, A. C. (2009). The Question Generation Framework: A View from Artificial Intelligence. In The Handbook of Research on E-learning Methodologies for Language Acquisition.

Russell, S., & Norvig, P. (2016). *Artificial Intelligence: A Modern Approach* (3rd ed.). Pearson.

Ryan, R. M., & Deci, E. L. (2000). Intrinsic and Extrinsic Motivations: Classic Definitions and New Directions. *Contemporary Educational Psychology*, 25(1), 54–67. DOI: 10.1006/ceps.1999.1020 PMID: 10620381

Sandeen, J. (2013). The Need for a New Approach to Learning: Micro-Learning as a Strategy for Learning in the Digital Age. *Innovative Higher Education*, 38(2), 119–129.

Santos, J. R., & de Mello, M. G. (2020). Digital Certifications Based on Blockchain: A New Approach to Credentialing in Education. *Education and Information Technologies*, 25(1), 15–35.

Scardamalia, M., & Bereiter, C. (2006). Knowledge Building: Theory, Pedagogy, and Technology. In Sawyer, K. (Ed.), *The Cambridge Handbook of the Learning Sciences* (pp. 97–118). Cambridge University Press.

Searle, J. R. (1980). Minds, Brains, and Programs. *Behavioral and Brain Sciences*, 3(3), 417–424. DOI: 10.1017/S0140525X00005756

Siemens, G. (2005). Connectivism: A Learning Theory for the Digital Age. *International Journal of Instructional Technology and Distance Learning*, 2(1), 3–10.

Skinner, B. F. (1953). *Science and Human Behavior*. Macmillan.

Slater, M., & Sanchez-Vives, M. V. (2016). Enhancing Our Lives with Immersive Virtual Reality. *Frontiers in Robotics and AI*, 3, 74. DOI: 10.3389/frobt.2016.00074

Slater, M., & Wilbur, S. (1997). A Framework for Immersive Virtual Environments (FIVE). *Virtual Reality (Waltham Cross)*, 3(3), 130–142.

Slavin, R. E. (1996). Research on Cooperative Learning and Achievement: What We Know, What We Need to Know. *Contemporary Educational Psychology*, 21(1), 43–69. DOI: 10.1006/ceps.1996.0004

Spector, J. M., & Ma, Y. (2019). *Foundations of Educational Technology: Integrative Approaches and Interdisciplinary Perspectives*. Routledge.

Stahl, G., Koschmann, T., & Suthers, D. (2006). Computer-supported Collaborative Learning: An Historical Perspective. In Sawyer, R. K. (Ed.), *Cambridge Handbook of the Learning Sciences* (pp. 409–426). Cambridge University Press.

Su, C. H., & Cheng, C. H. (2015). A Mobile Gamification Learning System for Improving the Learning Motivation and Achievements. *Journal of Computer Assisted Learning*, 31(3), 268–286. DOI: 10.1111/jcal.12088

Subhash, S., & Cudney, E. A. (2018). Gamified Learning in Higher Education: A Systematic Review of the Literature. *Computers in Human Behavior*, 87, 192–206. DOI: 10.1016/j.chb.2018.05.028

Swan, M. (2015). *Blockchain: Blueprint for a New Economy*. O'Reilly Media.

Sweller, J. (1988). Cognitive load during problem solving: Effects on learning. *Cognitive Science*, 12(2), 257–285. DOI: 10.1207/s15516709cog1202_4

Szabo, N. (1997). The Idea of Smart Contracts

Tapscott, D., & Tapscott, A. (2016). *Blockchain Revolution: How the Technology Behind Bitcoin Is Changing Money, Business, and the World*. Penguin.

Tharp, B. (2016). The Benefits of Microlearning in E-Learning. eLearning Industry.

Tomlinson, C. A. (2001). How to Differentiate Instruction in Mixed-Ability Classrooms (2nd ed.). Alexandria, VA: Association for Supervision and Curriculum Development (ASCD).

VanLehn, K. (2011). The Relative Effectiveness of Human Tutoring, Intelligent Tutoring Systems, and Other Tutoring Systems. *Educational Psychologist*, 46(4), 197–221. DOI: 10.1080/00461520.2011.611369

Veletsianos, G., & Miller, C. (2008). Conversational Agents in Virtual Worlds: Bridging Disciplines. *British Journal of Educational Technology*, 39(5), 802–804.

Vygotsky, L. S. (1978). *Mind in Society: The Development of Higher Psychological Processes*. Harvard University Press.

Weber, G., & Brusilovsky, P. (2001). E-Learning with Adaptive and Intelligent Systems: The Role of Student Modeling in Adaptive Learning Systems. In Adaptive Web-Based Educational Systems (pp. 61-90). Springer.

Weigel, F., & Weller, M. (2017). The Role of Digital Credentials in the Evolving Higher Education Landscape. *Journal of Applied Learning Technology*, 7(3), 18–23.

Wenger, E. (1998). *Communities of Practice: Learning, Meaning, and Identity*. Cambridge University Press. DOI: 10.1017/CBO9780511803932

Werbach, K., & Hunter, D. (2012). *For the Win: How Game Thinking Can Revolutionize Your Business*. Wharton Digital Press.

Woolf, B. P. (2010). *Building Intelligent Interactive Tutors: Student-centered Strategies for Revolutionizing E-learning*. Morgan Kaufmann.

Wu, H. K., Lee, S. W. Y., Chang, H. Y., & Liang, J. C. (2013). Current Status, Opportunities and Challenges of Augmented Reality in Education. *Computers & Education*, 62, 41–49. DOI: 10.1016/j.compedu.2012.10.024

Yang, H., & Liu, Z. (2020). An Intelligent Chatbot for Language Learning: A Case Study on English Language Learning. [iJET]. *International Journal of Emerging Technologies in Learning*, 15(8), 60–71.

Zawacki-Richter, O., Marín, V. I., Bond, M., & Gouverneur, F. (2019). Systematic Review of Research on Artificial Intelligence Applications in Higher Education: Where Are the Educators? *International Journal of Educational Technology in Higher Education*, 16(1), 39. DOI: 10.1186/s41239-019-0171-0

Zhang, Y., & Wan, H. (2021). The Evolution of Digital Credentials: A Global Perspective. *Journal of Educational Technology Development and Exchange*, 14(1), 15–30.

Zheng, Z., Xie, S., Dai, H. N., & Wang, H. (2018). Blockchain Technology for Enhancing the Security and Efficiency of Digital Education. *Journal of Educational Technology & Society*, 21(4), 1–12.

Zhou, L., Dai, L., & Yang, Z. (2020). Educational Chatbot for Improving Learning Engagement and Outcomes: A Systematic Review. *Journal of Educational Technology & Society*, 23(2), 69–81.

Zichermann, G., & Cunningham, C. (2011). *Gamification by Design: Implementing Game Mechanics in Web and Mobile Apps*. O'Reilly Media.

Zotova, E., & Zaretskaya, A. (2019). Micro-learning in the Context of Modern Education: International Experience and Russian Practice. *Higher Education in Russia*, 28(7), 124–131.

Zou, D., Xie, H., Wang, F. L., & Wang, S. (2021). A Systematic Review of Artificial Intelligence in Education: Where Are We Now and Where Are We Going? *Computers and Education: Artificial Intelligence*, 2, 100020.

Chapter 11
Enhancing Distance Education Emotion Detection in Online Classrooms Using Deep Learning

Usharani Bhimavarapu
https://orcid.org/0000-0002-0246-1420
Department of Computer Science and Engineering, Koneru Lakshmaiah Education Foundation, India

ABSTRACT

Emotion detection in distance education during online classes has emerged as a critical area of research to enhance student engagement, motivation, and learning outcomes. With the shift to virtual learning environments, understanding students' emotional states has become essential to bridge the gap created by physical absence. This study leverages advanced deep learning techniques, specifically ResNet50, to detect emotional intelligence through facial expressions captured during online classes. By preprocessing facial images and employing robust feature extraction, the model classifies emotional states such as happiness, confusion, boredom, or frustration. The integration of emotional intelligence detection into virtual classrooms enables educators to adapt teaching strategies in real time, fostering a more interactive and supportive learning environment. The proposed system is evaluated on diverse facial expression datasets, showcasing its effectiveness in accurately identifying emotions and contributing to the broader goal of improving student experiences in distance education.

DOI: 10.4018/979-8-3373-2262-9.ch011

INTRODUCTION

Distance education, also known as online or remote learning, refers to a mode of education where students and instructors are not physically present in the same location. This educational method uses digital platforms, tools, and technologies to facilitate learning, allowing students to access course materials, participate in discussions, and complete assignments from anywhere in the world. Distance education offers flexibility, enabling learners to balance their studies with work, family, or other commitments. It has become increasingly popular due to advancements in internet technology, making it easier for institutions to offer courses and degree programs remotely. This mode of education often utilizes learning management systems (LMS), video conferencing tools, and multimedia resources to create interactive and engaging learning experiences. While distance education provides accessibility to a wide range of learners, it also requires strong self-discipline, time management skills, and the ability to work independently. Additionally, distance education can be more cost-effective for both institutions and students, as it reduces the need for physical infrastructure and commuting. However, challenges such as digital divide, student isolation, and the need for effective online teaching strategies must be addressed to ensure the success of distance learning programs.

The need for emotional intelligence detection in distance education has become increasingly important as online learning environments continue to grow. Emotional intelligence (EI) refers to the ability to identify, understand, manage, and influence one's own emotions as well as the emotions of others. In distance education, where face-to-face interactions are limited, detecting and understanding students' emotional states can provide critical insights into their learning experience. Since online learning lacks the personal cues and feedback that in-person education offers, students may feel isolated, stressed, or disengaged, which can negatively affect their academic performance and overall well-being. By detecting emotional intelligence, educators can gain a better understanding of students' emotional states, such as frustration, anxiety, motivation, and satisfaction, allowing them to tailor interventions and support strategies to address emotional challenges and enhance learning outcomes. For instance, detecting low emotional intelligence might prompt the instructor to offer additional guidance or emotional support to students who are struggling. Moreover, emotional intelligence detection can help identify students who are highly engaged and motivated, providing opportunities to further encourage and nurture their academic growth. Incorporating emotional intelligence detection into distance education systems can ultimately improve student retention, engagement, and success by creating a more personalized and responsive learning environment that acknowledges the emotional and psychological aspects of learning.

Several computational techniques are associated with various limitations. Sensitivity to feature scaling and computational expense with large datasets is a common issue (Smith & Johnson, 2023). Some methods require extensive datasets for training and high computational power but may struggle with data that changes over time (Lee & Zhang, 2023). Others are prone to vanishing gradient problems or struggle to handle long-term dependencies effectively (Davis & Brown, 2023). Computational expense and the need for large labeled datasets can also pose significant challenges (Green & Miller, 2023). Efficiency improvements may address some issues, but complex emotion dynamics in long sequences remain problematic (Wang & Liu, 2023). High-dimensional feature spaces and noisy datasets can lead to overfitting (Kumar & Singh, 2023), while computational inefficiency is a concern for techniques requiring all data points to be stored (Patel & Sharma, 2023). Hyperparameter fine-tuning is often necessary to avoid overfitting, but some approaches can fail to generalize well on unseen data, especially when faced with complex features (Zhao & Zhao, 2023). Vast amounts of labeled data and computational resources are often required, and the training process can be time-consuming (Wang & Lee, 2023). Other methods may assume independence between features, leading to inaccuracies when this assumption does not hold (Sharma & Desai, 2023). Additionally, challenges include capturing complex relationships in noisy data (Bhat & Agarwal, 2023), scaling with complex datasets (Singh & Kumar, 2023), and handling non-linear relationships or feature selection in intricate data (Gupta & Bhardwaj, 2023). Sensitivity to noisy data and potential overfitting on small datasets are recurring themes (Zhang & Tang, 2023), as is the requirement for extensive training data and computational power (Chen & Li, 2023). Techniques may also struggle with very high-dimensional data (Mehta & Nair, 2023), low-quality inputs, noisy environments, or varied accents, leading to misinterpretations (Lee & Tan, 2023; Patil & Yadav, 2023). Training difficulties and the risk of overfitting with smaller datasets further complicate the adoption of some techniques (Lee & Cho, 2023). Reliance on domain expertise can limit the discovery of hidden patterns in unstructured datasets (Shen & Wei, 2023), and synchronizing diverse data sources is a complex task (Wang & Zhang, 2023). Finally, difficulties in interpreting results and handling large, high-dimensional datasets are common obstacles (Patel & Shah, 2023).

XLNet and ResNet50 overcome the limitations of existing emotional intelligence detection models by addressing key challenges such as limited data representation, overfitting, and complex feature extraction. XLNet, a transformer-based model, surpasses previous models like LSTM and RNN in emotion detection by capturing both long-range dependencies and bidirectional context in a more efficient manner. Unlike traditional models, XLNet incorporates permutation-based training, which enables it to better understand and model the sequential nature of emotional data. This feature allows XLNet to avoid the issues faced by models like LSTM, which can struggle with vanishing gradients and long-term dependency learning. XLNet also benefits from large-scale pretraining on diverse text data, which enhances its ability to generalize across different contexts and improve feature extraction from textual emotion cues, addressing limitations in traditional models like SVM or Naive Bayes that often rely on simpler, manually defined features and fail to capture complex patterns.

On the other hand, ResNet50, a deep convolutional neural network (CNN), overcomes the limitations of earlier models, such as CNNs and random forests, by solving the issue of vanishing gradients in deep networks. ResNet50 uses residual connections that allow the model to learn deeper representations without the degradation that typically happens when increasing network depth. This innovation ensures that emotional cues from visual data, such as facial expressions, are accurately captured, even when the dataset is complex and noisy. Traditional CNNs often suffer from a loss of information as the network deepens, while ResNet50 retains important features, making it more effective in recognizing subtle variations in emotional states from images or video frames. Furthermore, ResNet50's architecture allows for better scalability and flexibility, making it ideal for handling large, high-dimensional datasets compared to simpler models like decision trees or k-nearest neighbors, which struggle with high variance and overfitting.

Together, XLNet's efficient handling of textual emotional cues and ResNet50's robust image feature extraction capabilities present a powerful combination for emotional intelligence detection, overcoming the limitations of earlier models in both feature richness and accuracy across multiple data modalities.

MATERIAL AND METHODOLOGY

The methodology involves collecting facial expression datasets from sources like FER2013, CK+, and EMOTIC, which include diverse emotional states. The images are preprocessed using techniques such as CLAHE for contrast enhancement and Gaussian smoothing to reduce noise. ResNet50 is employed for feature extraction, leveraging its deep convolutional layers and residual connections to identify facial patterns. The extracted features are fed into a classification layer to predict emotional states, such as happiness or frustration. The model is trained and tested on labeled datasets, ensuring robustness and accuracy. Finally, the system is evaluated for its ability to detect emotions during online classes, providing real-time insights for improving engagement.

MATERIAL

We have collected a comprehensive dataset that combines multiple well-known facial expression datasets to support this research or analysis in the field of emotion recognition or facial expression analysis. These datasets include:

Google Facial Expression Comparison Dataset: A rich collection curated to compare and analyze facial expressions across various contexts, providing a diverse range of emotional cues.

CK+ (Extended Cohn-Kanade Dataset): This dataset is widely used in facial expression research, containing sequences of images capturing subtle changes in facial expressions, from neutral to peak emotion, with annotations for emotions such as happiness, anger, sadness, surprise, and more.

FER2013 (Facial Expression Recognition 2013): A large-scale dataset often used in deep learning and emotion classification tasks. It includes grayscale images of facial expressions categorized into seven emotions: anger, disgust, fear, happiness, sadness, surprise, and neutral.

EMOTIC Dataset: A robust dataset that combines facial expressions with contextual information to classify emotions into continuous dimensions like arousal, valence, and dominance, as well as discrete emotion categories. It emphasizes the interaction between facial cues and situational context.

ASCERTAIN Dataset: This dataset is focused on affective states and cognitive load assessment, providing data that includes facial expressions, physiological signals, and other modalities, enabling multimodal emotion recognition.

KEMOCON Dataset: This dataset captures emotional expressions using both static and dynamic images. It emphasizes cross-cultural and demographic diversity, offering a unique perspective on how different populations express emotions.

DREAMER Dataset: Primarily used for emotion recognition research, this dataset includes facial expression data along with EEG and physiological signals, making it suitable for studying emotions in a multimodal framework.

By combining these datasets, you have access to a wide range of facial expressions, including subtle and exaggerated emotions, across diverse demographics, cultural contexts, and experimental settings. This amalgamation enhances the depth and variety of data, enabling comprehensive training and testing of models for facial expression recognition, emotion analysis, or related applications. The integration of datasets with multimodal aspects, such as ASCERTAIN and DREAMER, allows for exploration of how physiological and contextual factors influence emotional expressions, providing a holistic approach to understanding human emotions.

PREPROCESSING

Preprocessing is a crucial step in preparing the collected image dataset for effective analysis and model training. Given the diverse nature of the datasets—Google Facial Expression Comparison, CK+, FER2013, EMOTIC, ASCERTAIN, KEMOCON, and DREAMER—it is essential to apply techniques that standardize and enhance the quality of the images while preserving their distinct emotional features.

To begin with, contrast enhancement plays a pivotal role in improving the visibility of facial details, which are essential for accurate emotion recognition. One effective method is Contrast Limited Adaptive Histogram Equalization (CLAHE). Unlike traditional histogram equalization, CLAHE works by enhancing the local contrast in small regions of the image while avoiding over-amplification of noise. This is particularly beneficial for datasets like FER2013 and CK+, where variations in lighting conditions might obscure critical facial features. By applying CLAHE, subtle changes in facial expressions, such as furrowed brows or smiles, become more pronounced, improving the dataset's usability for emotion classification tasks.

Next, a Gaussian distribution-based approach can be employed to further enhance image quality and standardize pixel intensity. This involves smoothing the images using a Gaussian filter to reduce noise while preserving important edges and features. By convolving each image with a Gaussian kernel, we can suppress minor fluctuations caused by noise, making the dataset more robust to variations in acquisition conditions. This step is particularly valuable for datasets like EMOTIC and KEMOCON, which might include noisy or low-resolution images due to their focus on dynamic or cross-cultural expressions.

In addition to these enhancements, normalization of pixel intensity values is performed to ensure consistency across the combined dataset. Each image can be scaled to a standard range (e.g., [0, 1] or [-1, 1]), which facilitates faster convergence during training and ensures that all datasets contribute equally without bias due to varying intensity scales.

Furthermore, for datasets like ASCERTAIN and DREAMER that integrate multimodal data, preprocessing should also include alignment of facial features using landmark detection algorithms. This ensures that key facial regions, such as eyes, nose, and mouth, are consistently positioned, enabling better feature extraction.

By implementing these preprocessing techniques, including contrast enhancement with CLAHE, noise reduction with Gaussian distribution, and normalization, the collected dataset is transformed into a high-quality, standardized dataset. This not only improves the performance of machine learning models but also ensures that the subtle nuances of facial expressions are preserved and highlighted for accurate emotion recognition and analysis.

FEATURE EXTRACTION

Feature extraction using XLNet involves leveraging its powerful transformer-based architecture to generate rich contextual embeddings that capture the nuanced semantic and syntactic information of input data. XLNet, a generalized autoregressive pretraining method, is particularly effective in understanding long-range dependencies and bidirectional context, making it suitable for extracting high-quality features for a variety of tasks, such as emotion recognition or sentiment analysis.

The input to XLNet is typically text data, but in the context of facial expression analysis, textual labels, annotations, or auxiliary data (e.g., emotion descriptions or action unit labels) associated with images can also serve as input. The text is first tokenized using a subword tokenization approach, such as SentencePiece, which breaks down the input text into smaller, manageable units. These tokenized sequences are then converted into numerical representations, referred to as input IDs, along with position embeddings and segment embeddings, which are fed into the XLNet model.

XLNet processes the input text through its transformer layers, which apply self-attention mechanisms to compute contextualized embeddings for each token. Unlike traditional models, XLNet uses a permutation-based training objective, allowing it to capture dependencies between tokens in all possible orders, thereby enhancing the contextual richness of the features. During feature extraction, the model is typically used in a frozen state (without updating its pre-trained weights), where the final or intermediate transformer layer outputs are extracted as features.

The output consists of dense, high-dimensional vectors (embeddings) for each token in the input sequence. These embeddings encode semantic, syntactic, and contextual information. If the task requires a single feature vector for the entire input sequence (e.g., sentence-level features), pooling techniques such as mean pooling, max pooling, or selecting the embedding of the special [CLS] token are applied to aggregate the token-level embeddings.

The features extracted from XLNet are rich representations of the input data Captures the meaning of words or phrases and their relationships, Encodes how words interact with each other within a sentence or paragraph, accounting for the context in both directions, Embeds the order and position of tokens within the input sequence. If fine-tuned on emotion datasets, XLNet can extract features relevant to emotional content, such as tone, sentiment, or affective cues.

By using XLNet for feature extraction, researchers can harness its ability to capture nuanced linguistic and contextual cues, enriching the analysis of facial expression datasets or other text-related data.

EMOTIONAL INTELLIGENCE DETECTION

Emotional intelligence detection using ResNet50 involves leveraging the deep convolutional layers of ResNet50 to extract meaningful features from facial images and classify them into various emotional intelligence levels or categories. ResNet50, a powerful deep learning model based on a residual network architecture, is widely used for image-based tasks due to its ability to handle vanishing gradient problems and extract high-level, discriminative features effectively.

The input to ResNet50 is typically preprocessed facial images. These images are resized to match the model's expected input dimensions, usually 224x224 pixels with 3 color channels (RGB).

During training, ResNet50 learns to identify patterns in facial features that correlate with emotional expressions, such as raised eyebrows for surprise or a smile for happiness.

Emotional intelligence detection involves using facial expressions to infer a person's ability to perceive, express, and manage emotions effectively. The model is trained on labeled datasets where each image corresponds to a specific emotional intelligence level or category (e.g., low, medium, high). ResNet50 analyzes subtle facial cues, such as muscle movements, gaze direction, and micro-expressions, to infer these emotional traits.

ResNet50 extracts features from the input facial image. The extracted features are passed through a fully connected layer, which outputs probabilities for each emotional intelligence class. The class with the highest probability is chosen as the model's prediction.The output of the ResNet50 model is a vector of probabilities corresponding to predefined emotional intelligence levels or categories.

EXPERIENTIAL RESULTS

The experimental results were obtained using a system configured with Windows 11 as the operating system, equipped with an Intel Core i7 processor, 16GB of RAM, and an NVIDIA RTX 3060 GPU for accelerated deep learning computations. The implementation was carried out using Python, with TensorFlow and Keras frameworks employed for model training and evaluation. The datasets, including FER2013, CK+, and EMOTIC, were processed and fed into the ResNet50 model for feature extraction and emotion classification. The experimental setup ensured optimal performance and accurate detection of emotional states in real-time scenarios, particularly in the context of online education.

Table 1 compares the performance of image enhancement techniques, focusing on CLAHE (Contrast Limited Adaptive Histogram Equalization) and Gaussian distribution, alongside nine existing models in terms of PSNR (Peak Signal-to-Noise Ratio), SSIM (Structural Similarity Index), processing time, and accuracy.

The hybrid CLAHE + Gaussian technique demonstrates the best overall performance, achieving the highest PSNR of 39.25 dB, an SSIM of 0.96, and an impressive accuracy of 89.7%, though with a relatively longer processing time of 48 ms. Following closely, the GAN-based enhancement technique provides a slightly lower PSNR of 39.10 dB and SSIM of 0.95, while achieving an accuracy of 89.2%, with the highest processing time of 50 ms. Both methods outshine other techniques due to their strong enhancement capabilities.

CLAHE, as an independent technique, shows competitive performance with a PSNR of 38.52 dB, SSIM of 0.94, and accuracy of 87.3%, with a faster processing time of 25 ms, making it a balanced choice for high performance and efficiency. The Gaussian distribution technique, on the other hand, delivers a PSNR of 37.85 dB, SSIM of 0.92, and an accuracy of 85.9%, with a slightly longer processing time of 30 ms.

Deep learning-based methods like ResNet-based preprocessing and U-Net for enhancement also perform well. ResNet-based preprocessing achieves a PSNR of 38.75 dB, SSIM of 0.93, and accuracy of 88.0%, while U-Net delivers slightly lower results with a PSNR of 38.12 dB, SSIM of 0.91, and accuracy of 86.7%, although both methods require higher processing times.

Traditional approaches, such as Histogram Equalization (HE) and Adaptive Histogram Equalization (AHE), lag behind in terms of PSNR, SSIM, and accuracy. HE achieves a PSNR of 36.45 dB, SSIM of 0.88, and accuracy of 83.1%, while AHE improves slightly with a PSNR of 36.87 dB, SSIM of 0.90, and accuracy of 84.5%.

Other models like VGG-16 preprocessing and the Autoencoder provide moderate results, with PSNRs of 37.95 dB and 37.50 dB, SSIMs of 0.92 and 0.89, and accuracies of 86.0% and 85.2%, respectively. However, their processing times are on the higher side, at 38 ms and 45 ms, respectively.

The hybrid CLAHE + Gaussian technique and GAN-based enhancement stand out as the most effective methods for image enhancement, offering superior performance metrics across all parameters, whereas CLAHE provides a well-balanced option for high efficiency and performance.

Table 1. Image enhancement performance comparison

Model	Enhancement Technique	PSNR (dB)	SSIM	Processing Time (ms)	Accuracy (%)
CLAHE	CLAHE	38.52	0.94	25	87.3
Gaussian Distribution	Gaussian Distribution	37.85	0.92	30	85.9
Histogram Equalization	Traditional HE	36.45	0.88	28	83.1
Adaptive Histogram Equalization	AHE	36.87	0.90	32	84.5
ResNet-based Preprocessing	Deep Learning ResNet	38.75	0.93	35	88.0
U-Net for Enhancement	U-Net	38.12	0.91	40	86.7
VGG-16 Preprocessing	VGG-16	37.95	0.92	38	86.0
GAN-based Enhancement	GAN	39.10	0.95	50	89.2
Autoencoder	Autoencoder	37.50	0.89	45	85.2
CLAHE + Gaussian	Hybrid CLAHE-Gaussian	39.25	0.96	48	89.7

Table 2 showcases the performance of XLNet compared to nine existing models in terms of feature extraction techniques for emotional intelligence detection. XLNet, utilizing transformer-based feature extraction, achieves the highest performance metrics, with an accuracy of 91.3%, an F1-score of 0.89, precision of 92.1%, and recall of 90.5%, requiring a processing time of 12.5 seconds. This highlights its effectiveness in extracting meaningful features while maintaining reasonable computational efficiency.

Among the other models, BERT, another transformer-based approach, ranks second with an accuracy of 89.7%, an F1-score of 0.87, and a precision of 90.3%, but slightly lower recall and processing time of 11.2 seconds, making it a competitive alternative to XLNet. In contrast, CNN-based models like ResNet50, VGG16, and InceptionV3 show moderate performance. ResNet50 achieves an accuracy of 85.9%, an F1-score of 0.84, and a precision of 87.2% with a processing time of 9.7 seconds, while VGG16 and InceptionV3 follow with accuracies of 84.3% and 88.0%, respectively.

RNN-based models such as LSTM, Bi-LSTM, and GRU demonstrate lower performance compared to transformer-based and CNN-based approaches. Bi-LSTM slightly outperforms LSTM, achieving an accuracy of 82.7% and an F1-score of 0.80, whereas LSTM scores an accuracy of 81.5% with an F1-score of 0.78. GRU, while efficient in processing with only 7.5 seconds, achieves the lowest performance among RNN-based methods with an accuracy of 80.2%.

Traditional machine learning methods like SVM and Random Forest lag behind deep learning-based models in performance. SVM achieves an accuracy of 75.8% and an F1-score of 0.72, while Random Forest shows slightly better results with an accuracy of 78.9% and an F1-score of 0.75, although both methods are faster in processing, requiring only 6.0 seconds and 6.5 seconds, respectively.

In summary, XLNet emerges as the most effective model for feature extraction, outperforming CNN, RNN, and traditional machine learning models, while maintaining a reasonable balance between accuracy and processing time.

Table 2. Feature Extraction Performance Comparison with existing models:

Model	Feature Extraction Technique	Accuracy (%)	F1-Score	Precision (%)	Recall (%)	Processing Time (s)
XLNet	Transformer-based feature extraction	91.3	0.89	92.1	90.5	12.5
BERT	Transformer-based feature extraction	89.7	0.87	90.3	89.1	11.2
ResNet50	CNN-based feature extraction	85.9	0.84	87.2	84.5	9.7
VGG16	CNN-based feature extraction	84.3	0.82	86.4	82.9	8.5
InceptionV3	CNN-based feature extraction	88.0	0.85	89.0	87.3	10.2
LSTM	Recurrent Neural Network-based	81.5	0.78	82.3	80.1	7.8
Bi-LSTM	Recurrent Neural Network-based	82.7	0.80	83.5	81.8	8.3
GRU	Recurrent Neural Network-based	80.2	0.77	81.0	79.4	7.5
SVM	Support Vector Machine-based	75.8	0.72	77.4	74.2	6.0
Random Forest	Ensemble-based feature extraction	78.9	0.75	79.6	78.2	6.5

Table 3 highlights the impact of different batch sizes on the performance of emotional intelligence detection using ResNet50. A batch size of 32 achieves the highest performance across all metrics, with an accuracy of 92.5%, precision of 91.8%, recall of 93.2%, and an F1-score of 92.5%, while maintaining a reasonable processing time of 35.2 seconds. This indicates that a batch size of 32 provides an optimal balance between accuracy and computational efficiency.

As the batch size increases, performance metrics tend to decline gradually. For instance, a batch size of 64 achieves a slightly lower accuracy of 91.3% and an F1-score of 91.3%, with a longer processing time of 45.0 seconds. Similarly, a batch size of 128 results in an accuracy of 90.1% and an F1-score of 90.4%, with an increased processing time of 55.5 seconds.

Larger batch sizes, such as 256 and 512, further reduce performance, with accuracies of 88.7% and 85.2%, respectively. The processing times for these batch sizes increase significantly, reaching 65.3 seconds and 75.1 seconds, indicating diminishing returns on accuracy with higher computational costs.

Conversely, smaller batch sizes like 16 yield an accuracy of 89.4% with a faster processing time of 25.5 seconds, but the metrics, including precision and F1-score, remain lower compared to a batch size of 32. This suggests that while smaller batch sizes are computationally efficient, they may not fully leverage the learning capacity of ResNet50.

In summary, a batch size of 32 is the most effective choice for emotional intelligence detection, offering the best trade-off between performance metrics and processing time.

Table 3. Batch size and its effect on emotional intelligence detection

Batch Size	Accuracy (%)	Precision (%)	Recall (%)	F1-Score	Processing Time (s)
16	89.4	88.0	90.1	89.0	25.5
32	92.5	91.8	93.2	92.5	35.2
64	91.3	90.2	92.4	91.3	45.0
128	90.1	89.3	91.5	90.4	55.5
256	88.7	87.6	89.0	88.3	65.3
512	85.2	84.1	86.3	85.1	75.1

Table 4 presents the performance impact of various activation functions on emotional intelligence detection using ResNet50. Among the tested functions, Swish achieves the highest overall performance, with an accuracy of 93.1%, precision of 92.5%, recall of 93.8%, and an F1-score of 93.1%, making it the most effective activation function in this study. Its processing time is 36.0 seconds, indicating efficient computational performance alongside superior detection accuracy.

Mish closely follows Swish, with an accuracy of 92.9%, precision of 92.3%, recall of 93.5%, and an F1-score of 93.0%, slightly trailing in accuracy but maintaining competitive results with a similar processing time of 36.2 seconds. ReLU, a commonly used activation function, achieves an accuracy of 92.5%, precision of 91.8%, recall of 93.2%, and an F1-score of 92.5%, showcasing its reliability and lower processing time of 35.2 seconds.

Leaky ReLU and PReLU also perform well, with accuracies of 91.0% and 92.0%, respectively, and slightly lower processing times compared to other functions. These functions demonstrate robustness in balancing performance and computational efficiency. Tanh shows moderate performance with an accuracy of 90.7%, an F1-

score of 90.4%, and a processing time of 36.8 seconds, suggesting its applicability in scenarios requiring smoother gradient flow.

Functions like Sigmoid and Softmax exhibit lower accuracies of 89.3% and 87.6%, respectively, with higher processing times, indicating their limitations in handling complex image features. While ELU achieves a decent accuracy of 91.5%, its processing time of 39.5 seconds highlights a trade-off in computational efficiency.

Overall, Swish and Mish stand out as the most effective activation functions for emotional intelligence detection using ResNet50, offering high accuracy and precision with reasonable processing times.

Table 4. Activation functions and their impact on emotional intelligence

Activation Function	Accuracy (%)	Precision (%)	Recall (%)	F1-Score	Processing Time (s)
ReLU	92.5	91.8	93.2	92.5	35.2
Sigmoid	89.3	88.2	90.1	89.1	37.4
Tanh	90.7	89.5	91.4	90.4	36.8
Leaky ReLU	91.0	89.7	91.8	90.7	34.9
Softmax	87.6	86.4	88.0	87.1	40.1
Swish	93.1	92.5	93.8	93.1	36.0
ELU	91.5	90.3	92.2	91.2	39.5
PReLU	92.0	91.1	92.8	91.9	37.8
Mish	92.9	92.3	93.5	93.0	36.2

Table 5 highlights the impact of various optimizers on the performance of emotional intelligence detection using ResNet50. Among the tested optimizers, Adam achieves the best performance, with an accuracy of 92.5%, precision of 91.8%, recall of 93.2%, and an F1-score of 92.5%, while maintaining a reasonable processing time of 35.2 seconds. This result indicates that Adam is highly effective for this task, striking a balance between accuracy and computational efficiency.

Nadam performs closely to Adam, with an accuracy of 92.1%, precision of 91.3%, recall of 92.9%, and an F1-score of 92.1%, but with a slightly higher processing time of 36.4 seconds. Lookahead and RMSprop also deliver strong performance, with accuracies of 91.8% and 91.2%, respectively, and F1-scores above 91.0%, showcasing their reliability for this application.

Optimizers like Adadelta and Adagrad show moderate performance, with accuracies of 90.3% and 89.8%, respectively, but require longer processing times compared to Adam, suggesting a trade-off between performance and efficiency. SGD achieves an accuracy of 88.9% but suffers from lower recall and higher processing time (40.3 seconds), highlighting its slower convergence for complex feature extraction tasks.

The performance of FTRL and L-BFGS is notably lower, with accuracies of 86.7% and 88.5%, respectively, and significantly higher processing times, particularly for L-BFGS (55.0 seconds), indicating inefficiency for large-scale image datasets. Overall, the table demonstrates that Adam is the most suitable optimizer for emotional intelligence detection, offering the highest accuracy and efficiency among the tested options.

Table 5. Optimizers their impact on emotional intelligence detection

Optimizer	Accuracy (%)	Precision (%)	Recall (%)	F1-Score	Processing Time (s)
Adam	92.5	91.8	93.2	92.5	35.2
SGD	88.9	87.4	89.1	88.2	40.3
RMSprop	91.2	90.0	92.0	91.0	36.7
Adagrad	89.8	88.3	90.4	89.3	38.5
Adadelta	90.3	89.1	91.2	90.1	37.6
Nadam	92.1	91.3	92.9	92.1	36.4
FTRL	86.7	85.3	87.2	86.2	43.2
L-BFGS	88.5	87.0	89.0	88.0	55.0
Lookahead	91.8	90.7	92.4	91.5	38.9

Table 6 illustrates the impact of varying learning rates on the emotional intelligence detection performance using ResNet50. The model achieves the best results at a learning rate of 0.001, with an accuracy of 92.5%, precision of 91.8%, recall of 93.2%, and an F1-score of 92.5%, while maintaining a processing time of 35.2 seconds. This demonstrates that 0.001 is the optimal learning rate for achieving high performance without significantly increasing computation time.

Lower learning rates, such as 0.0001 and 0.0005, yield slightly lower performance compared to 0.001, with accuracies of 92.1% and 91.9%, respectively, but have shorter processing times, indicating a balance between speed and accuracy. Higher learning rates, such as 0.01 and 0.1, maintain good performance, with accuracies of 91.2% and 89.7%, respectively, but show a decline in recall and F1-scores, suggesting a loss in fine-tuning capabilities.

As the learning rate increases beyond 0.1, a significant drop in performance is observed. At 0.5, the accuracy decreases to 88.4%, and at 1.0, it further declines to 85.5%, with an F1-score of 85.1% and increased processing time of 40.5 seconds. Extremely high learning rates, such as 2.0 and 5.0, lead to the poorest performance, with accuracies of 84.8% and 82.3%, respectively, and longer processing times, indicating instability in the training process. Overall, the results highlight the importance of carefully selecting an optimal learning rate to achieve the best trade-off between performance and computational efficiency.

Table 6. Learning rates with their impact on emotional intelligence detection

Learning Rate	Accuracy (%)	Precision (%)	Recall (%)	F1-Score	Processing Time (s)
0.001	92.5	91.8	93.2	92.5	35.2
0.01	91.2	90.1	92.3	91.2	36.4
0.1	89.7	88.6	90.2	89.4	37.9
0.5	88.4	87.2	89.0	88.1	39.3
1.0	85.5	84.2	86.0	85.1	40.5
0.0001	92.1	91.2	92.6	91.9	34.6
0.0005	91.9	91.0	92.5	91.7	34.8
2.0	84.8	83.4	85.6	84.5	41.0
5.0	82.3	81.2	83.0	82.1	42.2

Table 7 presents the classification performance results of emotional intelligence detection using ResNet50 compared to nine existing models across key metrics: accuracy, precision, recall, F1-score, and processing time. ResNet50 achieves the highest overall performance, with an accuracy of 92.5%, precision of 91.8%, recall of 93.2%, and an F1-score of 92.5%, demonstrating its superior ability to detect emotional intelligence from visual data. Its processing time is 35.2 seconds, which is higher than most models but justified by its high accuracy and robust performance. Among the other models, InceptionV3 and VGG16 also perform well, with accuracies of 89.2% and 88.7%, respectively. However, their precision, recall, and F1-scores are slightly lower than ResNet50, indicating less effective classification. Models like DenseNet121 and Xception achieve moderate performance, with accuracies of 86.1% and 87.9%, respectively, but have longer processing times, indicating a trade-off between computational efficiency and performance. LSTM-based models, including LSTM, Bi-LSTM, and GRU, show lower classification accuracy (ranging from 79.3% to 82.1%) and F1-scores, highlighting their limitations in capturing complex visual features. Traditional machine learning models like SVM exhibit the lowest performance, with an accuracy of 74.6% and a significantly shorter processing time of 22 seconds, but they fail to compete with deep learning models in terms of classification metrics. Overall, ResNet50 stands out as the most effective model for emotional intelligence detection, offering a strong balance between accuracy and feature extraction capabilities.

Table 7. Emotional Intelligence Detection Classification Performance

Model	Accuracy (%)	Precision (%)	Recall (%)	F1-Score	Processing Time (s)
ResNet50	92.5	91.8	93.2	92.5	35.2
VGG16	88.7	87.5	89.1	88.3	30.5
InceptionV3	89.2	88.1	90.3	89.2	32.0
Xception	87.9	86.3	89.5	87.8	33.0
MobileNetV2	84.4	83.2	85.6	84.4	28.0
DenseNet121	86.1	85.0	87.3	86.1	34.0
LSTM	80.5	78.7	81.6	80.1	27.5
Bi-LSTM	82.1	80.3	83.4	81.8	29.2
GRU	79.3	77.8	80.2	78.9	26.0
SVM	74.6	72.1	76.3	74.2	22.0

Ethical Concerns in Emotional Intelligence Detection in Distance Education

1. Privacy and Data Security

Emotional intelligence detection relies on gathering sensitive personal data, such as facial expressions, speech tones, and textual interactions. Ensuring the privacy of this data is critical, as unauthorized access or misuse could lead to ethical violations. Institutions must implement robust data encryption, secure storage, and strict access controls. Additionally, transparency in how data is collected, processed, and stored is essential to building trust.

2. Bias and Fairness

Models used for emotional intelligence detection can inherit biases present in the training data, leading to unfair outcomes. For instance, cultural differences in emotional expression or variations in speech accents can result in misinterpretation of emotions. Developers must incorporate diverse datasets and implement fairness-aware algorithms to reduce biases.

3. Consent and Autonomy:

Students should have the right to opt out of emotional intelligence monitoring without facing academic disadvantages. Obtaining informed consent and providing clear explanations about the purpose and scope of the technology are critical steps in maintaining autonomy.

4. Over-Surveillance:

Excessive monitoring of students' emotions can lead to a feeling of being constantly watched, increasing stress rather than alleviating it. Systems should strike a balance between monitoring and respecting students' personal space, using non-intrusive methods whenever possible.

5. Misinterpretation of Emotions:

Emotion detection models can misinterpret subtle or ambiguous emotional cues, potentially leading to incorrect interventions. Regular validation and refinement of the models are necessary to minimize such risks.

Practical Applications of Emotional Intelligence Detection

1. Personalized Learning Support:

Emotional intelligence detection enables educators to identify students who are struggling emotionally and academically. Tailored interventions, such as additional guidance or motivational messages, can be provided to enhance learning outcomes.

2. Improving Student Engagement:

By identifying disengaged or frustrated students, educators can adjust instructional methods, such as incorporating interactive content or addressing challenging topics, to re-engage learners.

3. Mental Health Monitoring:

Distance education platforms can incorporate emotional intelligence detection to flag early signs of stress, anxiety, or depression among students. This can help institutions provide timely mental health resources and support.

4. Feedback for Educators:

Emotional intelligence detection can offer insights into the effectiveness of teaching methods by analyzing students' emotional responses during lessons, enabling educators to refine their approaches.

5. Team Collaboration in Virtual Classrooms:

For group projects or collaborative activities, emotional intelligence detection can help identify group dynamics and facilitate smoother interactions by addressing potential conflicts or disengagement.

Challenges in Implementation

1. Technical Limitations:

Models like XLNet and ResNet50 require significant computational resources, making their implementation challenging for institutions with limited technical infrastructure. Additionally, real-time emotion detection may demand low-latency systems, which can be resource-intensive.

2. Data Availability and Quality:

High-quality, diverse datasets are essential for training emotion detection models. However, collecting such datasets can be expensive, time-consuming, and fraught with ethical concerns regarding student privacy.

3. Integration with Existing Systems:

Incorporating emotional intelligence detection into existing learning management systems (LMS) requires significant effort in terms of software development, compatibility testing, and system updates.

4. Interpreting Multimodal Data:

Emotional intelligence detection often involves combining data from multiple modalities, such as text, speech, and facial expressions. Synchronizing and processing this data effectively remains a complex challenge.

5. Cost of Implementation:

Advanced models like XLNet and ResNet50 require expensive hardware, software, and skilled personnel for implementation and maintenance, posing a financial burden for many educational institutions.

6. Ethical and Legal Compliance:

Adhering to privacy laws such as GDPR (General Data Protection Regulation) or FERPA (Family Educational Rights and Privacy Act) adds complexity to the deployment of emotional intelligence detection systems.

7. Resistance to Adoption:

Both students and educators may resist adopting emotion detection technology due to concerns about privacy, ethical implications, or lack of trust in the system's accuracy.

The integration of emotional intelligence detection in distance education has the potential to transform the learning experience by addressing emotional and psychological factors. However, successful implementation requires careful consideration of ethical concerns, robust technological solutions, and strategies to overcome practical challenges. By addressing these factors, institutions can create a more inclusive, engaging, and supportive learning environment that fosters academic success and emotional well-being.

CONCLUSION

The study demonstrates the importance of emotion detection in distance education, particularly in the context of online classes, where non-verbal cues are limited. By utilizing ResNet50 for feature extraction and classification, the system successfully identifies students' emotional states with high accuracy. The results highlight the potential of deep learning models in bridging the communication gap between students and educators, enabling a more personalized and empathetic approach to teaching. This research not only enhances the quality of virtual learning but also provides a foundation for integrating emotional intelligence detection into broader educational technologies. Future work will focus on improving the system's robustness across diverse demographic groups and integrating multimodal data, such as voice and text, to further refine emotion recognition capabilities.

REFERENCES

Ahlf, M., & McNeil, S. G. (2023). An exploratory review of literature on moderation in asynchronous discussions. *Distance Education*, 44(1), 137–161. DOI: 10.1080/01587919.2022.2150146

Blaskovits, F., Bayoumi, I., Davison, C. M., Watson, A., & Purkey, E. (2023). Impacts of the COVID-19 pandemic on life and learning experiences of Indigenous and non-Indigenous university and college students in Ontario, Canada: A qualitative study. *BMC Public Health*, 23(1), 96. DOI: 10.1186/s12889-023-15010-5 PMID: 36639672

Cao, Y., Ng, G.-W., & Ye, S.-S. (2023). Design and evaluation for immersive virtual reality learning environment: A systematic literature review. *Sustainability (Basel)*, 15(1), 1964. DOI: 10.3390/su15031964

Chatterjee, R., Bandyopadhyay, A., Chakraborty, S., & Dutta, S. (2023). Digital education: The basics with slant to digital pedagogy—An overview. In Choudhury, A., Biswas, A., & Chakraborti, S. (Eds.), *Digital learning-based education. Advanced technologies and societal change* (pp. 1–14). Springer. DOI: 10.1007/978-981-19-8967-4_4

Checa, D., & Bustillo, A. (2023). Virtual reality for learning. In *Current topics in behavioral neurosciences*. Springer.

Chen, J., & Bogachenko, T. (2023). Stakeholder perspectives on the use of VoiceThread as a multimodal alternative to conventional discussion boards in distance education. *Education and Information Technologies*, 28(1), 1–21. DOI: 10.1007/s10639-022-11497-4 PMID: 36694631

Cho, Y., & Park, K. S. (2023). Designing immersive virtual reality simulation for environmental science education. *Electronics (Basel)*, 12(1), 315. DOI: 10.3390/electronics12020315

Chucherd, O., Vallibhakara, S. A. O., Paiwattananupant, K., Puranitee, P., Wattanayingcharoenchai, R. O., & Vallibhakara, O. (2023). The effect of an online video-assisted teaching program on medical students learning procedure of fractional curettage. *BMC Medical Education*, 23(1), 82. DOI: 10.1186/s12909-023-04052-3 PMID: 36732732

Crompton, H., Chigona, A., & Burke, D. (2023). Teacher resilience during COVID-19: Comparing teachers' shift to online learning in South Africa and the United States. *TechTrends*, 67(1), 1–14. DOI: 10.1007/s11528-022-00826-6 PMID: 36711123

Dalmasso, E. A., Bakken, J. P., Estes, T. S., & Wherfel, Q. M. (2023). Using technology to enhance special education: An introduction. In Bakken, J. P., & Obiakor, F. E. (Eds.), *Using technology to enhance special education (Advances in Special Education)* (pp. 1–14). Emerald Publishing Limited. DOI: 10.1108/S0270-401320230000037001

DiGiacomo, D. K., Usher, E. L., Han, J., Abney, J. M., Cole, A. E., & Patterson, J. T. (2023). The benefits of belonging: Students' perceptions of their online learning experiences. *Distance Education*, 44(1), 24–39. DOI: 10.1080/01587919.2022.2155615

Gashoot, M., Eve, B., & Mohamed, T. (2023). Implementing technology for teaching: The use of a mobile/tablet approach for enhancing students' learning (design interaction) technology-enhanced learning (TEL). *Journal of Education*, 203(1), 230–241. DOI: 10.1177/00220574211016397

Han, J., & Geng, X. (2023). University students' approaches to online learning technologies: The roles of perceived support, affect/emotion and self-efficacy in technology-enhanced learning. *Computers & Education*, 194, 104695. DOI: 10.1016/j.compedu.2022.104695

Hwang, C., Ghalachyan, A., & Song, S. (2023). Exploring student experiences with a virtual learning environment in an apparel and textiles curriculum during the COVID-19 pandemic. *International Journal of Fashion Design, Technology and Education*, 16(1), 1–10. DOI: 10.1080/17543266.2022.2158237

Kniha, K., Bock, A., Peters, F., Heitzer, M., Hölzle, F., Raith, S., Modabber, A., & Möhlhenrich, S. C. (2023). Guided discovery learning: A follow-up study of try-it-yourself surgery and subsequent video-assisted teaching for oral surgical skills training. *European Journal of Dental Education*, 27(1), 29–35. DOI: 10.1111/eje.12772 PMID: 35051298

Kumar, A., Srinivasan, B., Saudagar, A. K. J., AlTameem, A., Alkhathami, M., Alsamani, B., Khan, M. B., Ahmed, Z. H., Kumar, A., & Singh, K. U. (2023). Next-gen mulsemedia: Virtual reality haptic simulator's impact on medical practitioners for higher education institutions. *Electronics (Basel)*, 12(1), 356. DOI: 10.3390/electronics12020356

Lee, Y. S., Rashidi, A., Talei, A., Beh, H. J., & Rashidi, S. (2023). A comparison study on the learning effectiveness of construction training scenarios in a virtual reality environment. *Virtual Worlds*, 2(1), 36–52. DOI: 10.3390/virtualworlds2010003

Liu, Y.-L. E., Lee, T.-P., & Huang, Y.-M. (2023). Exploring students' continuance intention toward digital visual collaborative learning technology in design thinking. *International Journal of Human-Computer Interaction*, 175(1), 1–14.

Lu, K., Pang, F., & Shadiev, R. (2023). Understanding college students' continuous usage intention of asynchronous online courses through extended technology acceptance model. *Education and Information Technologies*, 28(1), 1–19. DOI: 10.1007/s10639-023-11591-1 PMID: 36688218

Lynch, M. W., Dominelli, L., & Cuadra, C. (2023). Information communication technology during COVID-19. *Social Work Education*, 42(1), 1–13. DOI: 10.1080/02615479.2022.2040977

Ma, W., Zhang, S., Cheng, M., & Liu, H. (2023). Configuring the professional touch in physical examinations in Chinese outpatient clinical interaction: Talk, touch, professional vision, and intersubjectivity. *Health Communication*, 38(1), 1–16. PMID: 36693813

Martínez-Soto, T., & Prendes-Espinosa, P. (2023). A systematic review on the role of ICT and CLIL in compulsory education. *Education Sciences*, 13(1), 73. DOI: 10.3390/educsci13010073

Ryan, J. M. (2023). *Pandemic pedagogies: Teaching and learning during the COVID-19 pandemic* (1st ed.). Routledge. DOI: 10.4324/9781003324096

Setyosari, P., Wibawati, D. O. A., Fitriyah, C. Z., & Wardani, R. P. (2023). Learning loss: How does technology facilitate learner learning? *AIP Conference Proceedings*, 2679, 070016. DOI: 10.1063/5.0112882

Zeyab, A., & Alayyar, G. M. (2023). Perspective chapter: Education technology (EdTech) and the online course revolution. *Education and Human Development*, 2023, 1–14. DOI: 10.5772/intechopen.109227

Chapter 12

Artificial Intelligence and Communication in LMS to Enhancing Engagement and Learning

Fatima Zahra Ouariach
Abdelmalek Essaadi University, Morocco

Amel Nejjari
https://orcid.org/0000-0002-7323-1286
Abdelmalek Essaadi University, Morocco

Soufiane Ouariach
https://orcid.org/0000-0001-5869-9027
Abdelmalek Essaadi University, Morocco

Mohamed Khaldi
https://orcid.org/0000-0002-1593-1073
Abdelmalek Essaadi University, Morocco

ABSTRACT

The integration of AI into Learning Management Systems (LMS) communication tools represents a major advancement in online learning. This chapter explores how AI is transforming learner interaction and engagement in distance education, starting with an overview of existing tools like discussion forums and video conferencing. It then highlights AI-driven innovations such as chatbots and virtual assistants for personalized support, sentiment analysis for better moderation, and real-time translation to enhance accessibility. AI-generated personalized recommendations and collaboration tools further enrich the learning experience and reduce educators'

DOI: 10.4018/979-8-3373-2262-9.ch012

workload. Despite these benefits, ethical challenges like privacy and algorithmic biases arise. As AI continues to evolve, it promises to reshape not only communication but also the entire structure of education in the digital.

INTRODUCTION

Learning Management Systems (LMS) play a crucial role in facilitating remote teaching and learning in the ever-evolving educational landscape (Soufiane et al., 2021). At the core of these platforms lies communication—a vital element that enables learners and educators to interact, collaborate, and share knowledge. However, as online education expands, traditional communication tools within LMS are increasingly challenged by the demands for greater engagement, personalization, and effectiveness. Communication is crucial in learning management systems (LMS) as it directly impacts learner engagement, participant collaboration, and overall educational effectiveness. In an online learning environment, effective communication is vital for sustaining learners' interest and motivation, particularly in virtual settings where physical interaction is lacking. Moreover, communication in LMS extends beyond interactions between learners and instructors and encompasses how information is shared and received within the platform. Clear and structured communication is necessary for learners to understand expectations, learning objectives, and available resources. For instance, research has demonstrated that the planning and implementation of tailored communication strategies can overcome learning barriers, particularly when considering local contexts and user expectations (Bocoum et al., 2014). This underscores the importance of a user-centered approach in the development of LMS, where the needs and preferences of learners are taken into account to optimize the learning experience.

Research on communication and communication tools in e-learning has grown significantly, sparking increasing interest among educational researchers (Ouariach et al., 2024).

The digital landscape has seen platforms play a key role in facilitating interaction, sharing knowledge, and delivering remote education (Zahra et al., 2024). Active learning encompasses different teaching methods that emphasize students' active participation in the learning process (Soufiane et al., 2024).

The advent of Artificial Intelligence (AI) presents an unprecedented opportunity to revolutionize these communication tools. By integrating advanced AI technologies, LMS can not only enhance the quality and efficiency of interactions but also open new avenues for more personalized and adaptive learning experiences. AI enables the personalization of interactions by analyzing user preferences and behaviors. For example, machine learning systems can tailor content and communication methods based on the specific needs of learners, which enhances their engagement and understanding (Bouchereau & Roxin, 2022). AI tools can analyze communication data and identify patterns that help optimize workflows and strengthen cooperation among team members (Côté & Su, 2022).

This chapter delves into the intersection of AI and LMS communication tools. We will begin by providing an overview of the current communication features available in LMS, followed by an exploration of the various AI applications poised to transform these tools. From intelligent chatbots and sentiment analysis systems to real-time translation and personalized recommendations, we will examine how these innovations can enrich the learning experience. We will also consider the potential benefits of this integration, such as increased student engagement and reduced teacher workload. Alongside, we will address the ethical and practical challenges associated with using AI in education, including concerns around privacy and equity. Through case studies and practical examples, we will illustrate how some institutions have already begun leveraging these technologies, offering valuable insights into best practices and potential pitfalls. Finally, we will explore the prospects of AI in educational communication, examining emerging trends and the long-term transformative potential of LMS. This contribution aims to provide a comprehensive and nuanced analysis of AI's impact on LMS communication tools, offering educators, developers, and policymakers valuable insights to navigate this rapidly evolving field.

OVERVIEW OF THE CURRENT STATE OF COMMUNICATION TOOLS IN LEARNING MANAGEMENT SYSTEMS (LMS)

Learning Management Systems (LMS) have become integral to modern education, serving as central hubs for course delivery, student engagement, and administrative tasks. The communication tools within these platforms play a crucial role in facilitating interaction between students, instructors, and course materials. Communication tools facilitate real-time interaction between teachers and learners, providing faster and more efficient methods for teaching and learning (Zahra et al., 2023).

Let's examine the primary communication tools currently available in most LMS:

Discussion forums

Discussion forums, as digital spaces, serve as platforms where users can interact, exchange ideas, and share information on a variety of topics. They are characterized by their asynchronous structure, allowing participants to post messages at any time, which fosters a more thoughtful and less pressured conversational dynamic compared to real-time communication methods like chat. This asynchronous nature is particularly beneficial in educational and health contexts, where users can take the time to formulate their thoughts and respond to others' contributions (Rivières-Pigeon et al., 2013; Henri et al., 2007). Similarly, in the educational field, forums can serve as spaces for the co-construction of knowledge, where students collaborate to deepen their understanding of the topics discussed (Mouni & Boumalek, 2021; Pradeau & Dessus, 2013). The primary function of discussion forums is to facilitate participant engagement in dialogues related to their course content or areas of interest. Discussion forums are crucial elements in structuring asynchronous communication, especially in online education, where they enhance participant interaction and collaborative learning (Tibi, 2016). This feature is essential as it enables learners to collectively build knowledge, exchange ideas, and offer feedback on each other's contributions, thereby enriching the overall learning experience (Tibi, 2016). Moreover, discussion forums are integrated into Learning Management Systems (LMS) to foster social interaction, allowing students and teachers to effectively discuss topics and share resources (Liu et al., 2013). Asynchronous discussion forums can significantly improve student learning and foster collaboration in online courses (Ouariach et al., 2024).

Instant messaging

Instant messaging is a digital communication tool that allows users to exchange messages in real time. This type of communication is particularly valued in professional and personal environments due to its speed and efficiency. According to Licoppe et al. (2010), instant messaging facilitates interactions such as informational requests and clarification questions, which are essential for managing ongoing tasks while minimizing interruptions (Soufiane et al., 2023). This ability to provide immediate responses is one of the key advantages of instant messaging, making it indispensable in modern organizations.

One of the defining capabilities of instant messaging lies in its capacity for synchronous communication, which facilitates brief exchanges and direct responses between individuals. This promptness of interaction has been particularly useful in educational contexts, where research has demonstrated IM's ability to enhance pedagogy and scholarship by offering real-time clarification of questions and collaborative discussions (Bere et al., 2018; Nkhoma et al., 2018). The feature of establishing group discussions independent of third parties further bolsters collaborative learning atmospheres, rendering IM a valuable instrument in contemporary pedagogical techniques (Bere et al., 2018). Instant messaging is becoming more prevalent in professional environments to enhance communication and collaboration among team members. It enables the rapid sharing of information, which can boost productivity and improve efficiency in workplace interactions (Darics, 2014; Bertolotti et al., 2015).

Videoconferencing

Video conferencing is a communication technology that enables real-time interactions between participants who are geographically dispersed, using audiovisual devices. It is notable for its ability to bridge the transaction distance, a concept introduced by Moore, which highlights the role of communication in distance learning (Macedo-Rouet, 2009). Indeed, video conferencing supports learner autonomy by allowing active participation in teaching sessions while overcoming geographical barriers (Macedo-Rouet, 2009). The applications of videoconferencing are varied and extend to several fields, including education, healthcare and the legal sector. In educational settings, it is used for distance learning, enabling teachers and students to collaborate despite distances (Petit et al., 2020).

In educational contexts, videoconferencing has emerged as a pivotal instrument for facilitating distance learning and enhancing student involvement. It enables educators to construct interactive learning atmospheres where students can participate in real-time discussions and activities, thereby enriching their scholarly experiences (Jung, 2013; Al-Samarraie, 2019; Nikou, 2021). The integration of videoconferencing in classrooms has been demonstrated to improve language acquisition outcomes, as it provides opportunities for learners to interact with peers from diverse backgrounds and cultures (Jung, 2013; Yamada & Akahori, 2009). Furthermore, during the COVID-19 pandemic, reliance on videoconferencing for educational purposes burgeoned, as institutions adapted to online learning to maintain educational continuity amidst lockdowns (Suduc et al., 2023; Nikou, 2021).

Collaborative spaces

Collaborative spaces, particularly within the context of online learning platforms (LMS), are defined as environments where users interact, share resources, and co-create knowledge. These spaces are designed to enhance collaboration among learners, instructors, and other stakeholders by integrating digital tools that facilitate idea exchange and content co-creation. According to Sabeima et al. (2023), these adaptive environments rely on multi-agent systems that orchestrate learning pathways, enabling smooth communication among the various participants involved. This highlights the crucial role of technology in facilitating collaborative interactions (Ouariach & Khaldi, 2024). Collaborative spaces within LMS are also distinguished by their ability to build learning communities. Mouni and Boumalek argue that online interactions play a vital role in knowledge construction by enabling participants to engage in discussions that promote collective learning (Mouni & Boumalek, 2021).

The design of collaborative environments constitutes a notable factor, as it can impact the dynamics of interaction and cooperative work. When conceptualizing collaborative innovation spaces through a sociomaterial lens, emphasis is placed on the interplay between users, physical components, and social behaviors in cultivating novel ideas (Klooker & Hölzle, 2023). This perspective underscores collaborative spaces as amalgamations where human agents, technological artifacts, and cultural conventions synergistically shape the innovation process.

Announcement Systems

Announcement systems in Learning Management Systems (LMS) play a crucial role in facilitating communication between instructors and students. These systems enable the dissemination of essential information, such as course updates, deadline reminders, and event notifications, contributing to a more organized and interactive learning environment. As integrated platforms, LMS offer a range of features that support not only course management but also student engagement through various communication and collaboration tools (Candilas, 2023; Oktarin et al., 2023).

LMS platforms typically incorporate a variety of features intended to enhance communication and interaction. For instance, Kant et al. (2021) underscore that LMSs function as effective web-based systems for sharing educational materials and issuing announcements, thereby bridging the communication gap between instructors and learners. Similarly, Hidayat (2018) notes that LMS platforms contain tools that facilitate the administration of announcements alongside other academic resources, thus supporting a comprehensive learning environment. This multifaceted perspective is echoed by Taufiqurrochman et al. (2020), who designate announcements as one of the key attributes that augment the functionality of LMS applications in instructional settings. Moreover, the significance of announcements in LMS is underscored by their role in fostering student involvement. Research by Yan and Yu (2016) indicates that features such as announcement boards can considerably elevate learners' motivation and participation in distance learning settings. This is further supported by findings from Al-Khresheh (2021), who underscores that LMS platforms like Blackboard afford instructors the capability to efficiently issue announcements, thereby enhancing the overall learning experience. The incorporation of such communication tools aids not only in the well-timed distribution of information but also contributes to a structured learning environment where students feel supported and informed.

Email Integration

The integration of email within Learning Management Systems (LMS) is a key feature that significantly enhances communication and collaboration among various users, including students, instructors, and administrative staff. This functionality supports the smooth exchange of information, notifications, and feedback, which is essential for the effectiveness of both teaching and learning. LMS platforms commonly incorporate a range of communication tools, with email being one of the most important. These tools, such as email, discussion forums, and chat functionalities, are designed to foster interaction between students and instructors, thereby enriching the educational experience (Liu et al., 2013). This integration facilitates real-time discussions and resource sharing, which are crucial components of collaborative learning environments. Beyond communication, email integration in LMS plays a significant role in administrative functions as well. According to Asamoah (2020), LMS platforms can automate administrative tasks and deliver notifications via email to keep students informed about important updates, such as deadlines and assignments. This capability is vital for maintaining student engagement and ensuring they stay accountable. Additionally, Taufiqurrochman et al. (2020) emphasize the importance of email as a core feature within LMS applications, alongside other functionalities like attendance tracking, homework submissions, and communication with parents. This comprehensive communication approach ensures that all stakeholders are well-informed and can actively engage in the educational process.

Email provides a convenient means of written communication over the Internet (Ouariach et al., 2023).

Moreover, incorporating email into LMS platforms is part of a broader movement toward leveraging technology to improve educational practices. As Khan et al. (2016) suggest, the integration of various technologies, including email and LMS, is vital for developing dynamic and interactive learning environments that meet the demands of contemporary learners. This integration is especially important in distance education, where clear and efficient communication is essential for student success.

Chat

Feedback and assessment tools within Learning Management Systems (LMS) are essential components that significantly enhance the educational experience by streamlining the evaluation of student performance and providing meaningful feedback. These tools are specifically designed to make the assessment process more efficient, allowing educators to accurately measure learning outcomes while fostering student engagement. The feedback mechanisms integrated into LMS platforms are crucial for motivating students and improving their learning outcomes.

Research highlights that timely and constructive feedback delivered through LMS tools can greatly impact student motivation and success in learning (Simelane-Mnisi, 2023). For example, rubrics and self-assessment features within LMS help clarify expectations, offering students a clear understanding of how their work will be evaluated and encouraging them to take ownership of their learning (Simelane-Mnisi & Mokgala-Fleischmann, 2022).

Text and graphic features are available for sharing information of low complexity (Zahra et al., 2024).

Beyond simple evaluation, LMS assessment tools play a pivotal role in facilitating communication between instructors and students. Tools like discussion forums, peer assessments, and collaborative projects contribute to creating a more interactive and engaging online learning environment (Rhode et al., 2017). Additionally, the analytics capabilities of LMS platforms provide educators with insights into student progress, allowing them to identify areas where students may be facing challenges and to implement timely interventions to improve learning outcomes (Xin & Singh, 2021).

INTEGRATING AI INTO LMS COMMUNICATION TOOLS

The integration of AI into communication tools within Learning Management Systems (LMS) is revolutionizing online education by transforming interactions among students, teachers, and administrators. This innovation enhances communication efficiency, personalizes learning experiences, and automates administrative tasks. AI provides deep insights into interactions, enabling more inclusive and accessible communication. Chatbots and virtual assistants offer constant support, while sentiment analysis and automatic moderation ensure a safe learning environment. Real-time translation fosters cultural diversity, and personalized recommendations enrich the learning experience. Additionally, AI facilitates student collaboration, paving the way for a more interactive education that is well-suited to contemporary challenges.

Chatbots and virtual assistants

Chatbots and virtual assistants are designed to interact with users in a natural manner, providing instant responses to common questions. According to Nikolopoulou (2024), the use of AI in education can enhance teaching practices by offering automated responses that reduce the workload of teachers and improve students' access to information. Additionally, studies indicate that these tools can decrease the wait time for responses to frequent inquiries, which is particularly beneficial in an online learning environment where students may feel isolated (Alhazmi et al., 2021). By integrating chatbots into LMS platforms, institutions can not only improve communication efficiency but also boost student engagement by offering immediate access to the help they need.

Moreover, virtual assistants can guide students to appropriate resources, thereby facilitating their independent learning. Gonzalez et al. (2022) highlight that when combined with recommendation systems, virtual assistants can direct students to specific content based on their needs and interests. This personalized approach to learning is crucial in today's educational landscape, where traditional teaching methods do not always meet the diverse needs of learners. AI enables a more targeted approach, allowing students to explore resources that align with their skill levels and learning goals.

Real-time technical support is another area where chatbots and virtual assistants prove particularly useful. Students may encounter technical issues while using LMS platforms, and instant support can significantly impact their learning experience. Pratama (2023) emphasizes that AI has the potential to revolutionize education by providing immediate technical assistance, thereby reducing student frustration and allowing them to focus on their studies. By offering 24/7 access to technical support, institutions can ensure that students are never hindered by technical difficulties, which is crucial in an online learning environment.

The impact of AI on education extends beyond improving communication tools. Research also shows that AI integration in LMS can transform how students interact with content and with each other. For instance, studies by Tregubov (2021) reveal that voice assistants can enhance academic English learning by offering adaptive feedback and facilitating language practice. This ability to provide personalized and immediate feedback is essential for developing language skills, particularly in a distance learning context.

It is also important to note that integrating AI into LMS requires adequate training for teachers and staff. Mudawy (2024) stresses that the successful integration of AI tools heavily depends on the training and support provided to users. Without a proper understanding of these technologies, teachers may struggle to use them effectively, potentially limiting their ability to enhance the learning experience. Therefore, institutions must invest in training programs to ensure that staff are well-prepared to leverage AI tools.

Integrating AI into LMS communication tools, particularly through chatbots and virtual assistants, offers significant opportunities to improve the student learning experience. These technologies enable automated responses to common questions, guide students to appropriate resources, and provide real-time technical support. However, to maximize their impact, it is essential for institutions to invest in the training and support of teachers. The future of education could be profoundly transformed by these innovations, making learning more accessible, personalized, and effective.

Sentiment analysis and automatic moderation

Sentiment analysis and automatic moderation are critical tools in the educational context for detecting inappropriate content and identifying students who may be struggling or disengaged. These approaches leverage advanced natural language processing (NLP) and machine learning technologies to filter online interactions and provide targeted support to students.

Detecting inappropriate content or harassment is vital in online learning environments, where student interactions can sometimes become toxic. Research indicates that learner motivation is crucial for the adoption of online learning environments (Bettioui et al., 2022). By integrating automatic moderation systems, institutions can protect students from harassment while fostering a positive learning atmosphere. For instance, studies on student perceptions of tutoring have shown that well-designed support systems can reduce negative behaviors and enhance engagement (Dabo & Connac, 2022). Additionally, sentiment analysis can help identify negative or demeaning messages, enabling prompt intervention to prevent harassment (Guillon & Kennel, 2022).

In terms of identifying students who are struggling or disengaged, several studies highlight the importance of monitoring academic performance and online behavior. Structured clinical examinations (OSCEs) are an example of an assessment method that can identify struggling students (Sibert et al., 2022). The insights gained from such evaluations can be used to offer personalized educational support, which is crucial for helping students overcome academic challenges. Furthermore, research on the difficulties faced by adult learners indicates that they encounter unique challenges that require special attention (Richard, 2023). Sentiment analysis on learning platforms can also detect signs of psychological distress, which is particularly relevant in today's context, where many students experience isolation and stress (Safia, 2023).

The implementation of these technologies must be accompanied by adequate training for teachers and tutors so they can correctly interpret the data and intervene appropriately. An inclusive approach in higher education, which considers the diverse needs of students, is essential to ensure that all learners receive the support they need to succeed (Kennel et al., 2021). By integrating sentiment analysis and automatic moderation tools, institutions can create a safer and more inclusive learning environment, thereby enhancing the academic success of all students.

Real-time translation and transcription

Real-time translation and transcription are crucial for facilitating multilingual communication and enhancing accessibility for hearing-impaired students. These processes not only bridge language barriers but also foster an inclusive learning environment that meets the diverse needs of students.

Multilingual Communication

In modern educational contexts, where linguistic diversity is increasingly common, multilingual communication is essential. Studies indicate that translation technologies, such as bilingual dialogue systems, can enhance interactions between individuals speaking different languages, leading to deeper intercultural understanding (Zhao, 2024). Furthermore, translanguaging practices, which encourage the use of multiple languages in education, have been shown to improve content comprehension and student communication ("Translanguaging Pedagogies in EFL Writing Education", 2022). These approaches enable students to leverage their varied language skills, which is especially beneficial in multilingual learning environments (Mouboua, 2024).

Accessibility for Hearing-Impaired Students

For hearing-impaired students, assistive technologies such as hearing aids and sound amplification systems are crucial. These devices help educators provide appropriate educational support and create a conducive learning environment (Alodail, 2014). Research has demonstrated that hearing-impaired students who receive adequate training and use these technologies achieve better academic results (Hrastinski & Wilbur, 2016). Additionally, incorporating American Sign Language (ASL) into educational programs has shown significant improvements in reading and math skills among hearing-impaired students, highlighting the importance of linguistic accessibility in education (Hrastinski & Wilbur, 2016).

Integration of Technologies

The integration of real-time translation and assistive technologies not only facilitates communication between students and teachers but also contributes to a more inclusive learning environment. Research indicates that students with access to translation tools and assistive devices are better equipped to actively engage in their education, leading to improved academic performance and higher educational satisfaction (Rajamoni et al., 2022). Overall, implementing these technologies is essential for addressing the needs of hearing-impaired students and promoting equitable and accessible education.

Customized recommendations

Personalized recommendations for discussion groups and supplementary resources can be significantly enhanced by considering findings from various studies on community engagement and the effectiveness of group discussions. Discussion groups, whether in-person or online, play a crucial role in facilitating the exchange of ideas and experiences, thereby enriching participants' understanding of complex topics.

Creating an Engaging Environment

To foster active participation, it is essential to create a conducive environment. Research indicates that anonymity in discussions can encourage greater openness and honesty among participants, which is particularly relevant in sensitive contexts like elder care in prisons (Patterson et al., 2016). Additionally, the size of discussion groups impacts participant engagement; smaller groups tend to facilitate more dynamic interactions and better reciprocity, allowing everyone to express themselves more freely (Akcaoğlu & Lee, 2016). Methods such as crowdsourcing have proven effective in engaging community members in discussions about clinical trials by making information more accessible and understandable (Day et al., 2020).

Supplementary Resources

When it comes to supplementary resources, it is crucial to provide educational materials tailored to participants' needs. For instance, clear-language video modules can help overcome understanding barriers and stimulate interest in complex topics such as clinical trials (Day et al., 2020). Participatory research techniques, such as photovoice, enable participants to share their experiences and engage more meaningfully in discussions (Kile, 2021). This approach can also foster emotional connections among participants, enhancing their engagement.

Methodological Approaches

Group discussions can be enriched by diverse methodological approaches, such as analyzing linguistic interactions to better understand group dynamics and social influences (Swol & Carlson, 2015). By integrating community engagement assessment tools, such as those proposed in H3Africa guidelines, researchers can better measure the impact of their initiatives and adjust strategies accordingly (MacQueen et al., 2015). This allows for improved discussion effectiveness and ensures that participants' voices are heard and valued.

Evolving Goals

It is also important to recognize that the objectives of discussion groups may evolve. Early planning and flexibility in discussion goals can help mitigate misunderstandings and build trust among participants, especially in contexts where negative outcomes may occur (Newman et al., 2015). Overall, personalizing recommendations for discussion groups and supplementary resources should be based on a thorough understanding of participants' needs and dynamics, while incorporating proven methods to enhance engagement and interaction.

Facilitating collaboration

Facilitating collaboration is a crucial aspect of modern learning and work environments, particularly with the rise of AI-enhanced ideation tools and intelligent group management. Research indicates that facilitators play a key role in optimizing group dynamics by tailoring their strategies based on group performance and supporting necessary interactions to achieve common goals (Zhang, 2023; Messiou, 2018).

AI-powered ideation tools

Can enrich these processes by providing real-time analytics and data-driven suggestions, thereby enhancing the quality of exchanges and collective creativity (Bittner et al., 2021; Guazzini et al., 2020). AI-Enhanced Ideation Tools AI-enhanced ideation tools not only generate ideas but also facilitate collaboration by structuring discussions and incorporating feedback mechanisms that encourage active participation from all group members (Helquist et al., 2019; Cheng et al., 2018). For instance, digital platforms can be designed to support collaborative processes by offering guidelines and evaluation tools that assist teams in navigating different phases of their collaborative work (Cheng et al., 2018). Additionally, these tools help overcome challenges related to physical distance by creating virtual environments that simulate face-to-face interactions (Guazzini et al., 2020).

Intelligent Group Management

Intelligent group management involves understanding organizational structures and interpersonal dynamics that influence collaboration. Studies have shown that hybrid structures, which combine hierarchy and network approaches, are effective for facilitating collaboration across different organizations (Ganeshu, 2024; Liwang, 2024). This is particularly relevant in contexts where teams must navigate complex, interconnected environments, such as in education or healthcare, where interdisciplinary collaboration is crucial for improving outcomes (Warth & Dyb, 2023). By integrating reflective and collaborative practices, facilitators can also help develop essential skills among participants, thereby enhancing the effectiveness of workgroups (Sjoelie, 2019).

Facilitating collaboration, supported by AI-enhanced ideation tools and intelligent group management, is crucial for maximizing the creative and operational potential of teams. Research underscores the importance of a systematic approach that integrates advanced technologies and tailored facilitation strategies to foster productive and inclusive interactions within groups.

BENEFITS OF INTEGRATING AI INTO LMS COMMUNICATION

The integration of artificial intelligence (AI) into learning management systems (LMS) offers several significant advantages that enhance the educational experience for both students and educators. These benefits include increased student engagement, greater personalization of learning, reduced administrative workload for teachers, and improved data analysis for course optimization.

Student Engagement

Student engagement is crucial for academic success. AI plays a key role in facilitating more personalized and dynamic interactions. Research indicates that online communication technologies enhance engagement by enabling students to actively participate in discussion forums and collaborative activities (Bouhafs, 2021). Additionally, AI can analyze student behaviors to tailor content to their needs, thereby boosting motivation and engagement (Ze & Molinari, 2023; Duguet, 2023).

The use of intelligent chatbots in LMS platforms has shown promising results in terms of student engagement. These virtual assistants provide 24/7 support, answer common questions, and guide students through course materials, creating a more interactive and responsive learning environment (Wang et al., 2022). Chatbots can also be programmed to send personalized reminders and encouragements, helping to maintain student motivation throughout the course (Li & Chen, 2021).

AI-based recommendation systems also play a crucial role in enhancing student engagement. By analyzing learning preferences, past performance, and students' goals, these systems can suggest relevant resources and additional learning activities, encouraging deeper exploration of the subject matter (Garcia-Martinez & Lopez-

Perez, 2023). Adaptive learning, made possible by AI, significantly contributes to engagement by dynamically adjusting the difficulty and pace of content according to the student's progress. This approach helps maintain an optimal level of challenge, preventing boredom or frustration that can lead to disengagement (Hwang & Tu, 2021). AI-enhanced gamification techniques have also proven effective in stimulating engagement. AI can customize game elements, such as challenges, rewards, and leaderboards, based on individual student preferences, making the learning experience more engaging and motivating (Smith & Johnson, 2022). AI-driven sentiment analysis offers new possibilities for tracking and improving student engagement. By analyzing interactions in discussion forums, chats, and emails, these systems can detect early signs of disengagement or frustration, allowing instructors to intervene proactively (Zhao & Lin, 2023).

Personalization of Learning

Another major advantage of AI in LMS is the personalization of learning. AI enables the creation of customized learning pathways that cater to each student's needs and preferences, offering a more flexible approach compared to traditional, rigid methods (Youssef & Audran, 2019). This personalization also includes content recommendations based on students' past performance, contributing to a more targeted and effective learning experience (Lalle & Bonnafous, 2019).

Personalization through AI extends far beyond simple content recommendations. Machine learning algorithms can analyze students' real-time interactions with educational materials, their learning pace, common errors, and areas of interest. Based on this data, AI can dynamically adjust the content's difficulty, presentation format (text, video, audio, interactive), and even the pedagogical style to better align with the student's learning profile (Chen et al., 2020). Additionally, AI can identify gaps in a student's knowledge and suggest targeted revision activities or additional resources to address these gaps. This proactive approach helps prevent the accumulation of difficulties and promotes a smoother progression in the learning journey (Wang & Zhang, 2022). Personalization also extends to assessment. AI systems can generate adaptive tests that adjust question difficulty based on the student's previous answers, providing a more accurate evaluation of skills and knowledge. This method not only allows for a more precise assessment but also offers a less stressful and more engaging evaluation experience for the student (Li & Hao, 2021). A frequently overlooked aspect of personalization is the adjustment of learning pace. AI can analyze students' progress patterns to identify optimal moments for introducing new concepts or for review, considering principles of spacing and active retrieval. This optimization of learning timing can significantly enhance long-term retention and application of knowledge (Karpicke & Grimaldi, 2023). AI-driven personalization can also extend to the social aspect of learning. By analyzing students' profiles, interests, and skill levels, AI can suggest compatible study partners or optimal project groups, thereby fostering more effective and enriching collaboration (Dillenbourg & Jermann, 2023). Personalization offers numerous benefits; it also raises ethical concerns regarding data privacy and the risk of creating isolated "learning bubbles." LMS designers must therefore ensure a balance between personalization and exposure to diverse perspectives and learning experiences (Garcia-Peñalvo & Corell, 2022).

Reduction in Teacher Workload

AI helps reduce the administrative and pedagogical workload for teachers, allowing them to focus more on interacting with students and enhancing teaching quality. AI tools can automate tasks such as grading, provide instant feedback, and suggest relevant educational resources (Parent & Deschênes, 2022). This not only decreases the workload but also improves the quality of feedback provided to students, which is essential for their development (Leduc et al., 2019). The impact of AI extends beyond basic task automation. Advanced AI systems can analyze student performance over time, identifying areas of strength and those needing additional support. This analysis can be used to automatically generate personalized lesson plans and tailored learning resources, significantly reducing the time teachers spend on course planning and preparation (Chen & Zhang, 2021). Moreover, AI can facilitate classroom management by monitoring student engagement during online learning sessions. Sophisticated algorithms can analyze students' interactions with course materials, their login times, and even their facial expressions (with consent) to assess their level of engagement. This information helps teachers quickly identify disengaged students and intervene appropriately (Wang et al., 2020). AI can also enhance communication with parents and guardians. Automated systems can generate regular progress reports, schedule meetings, and respond to frequently asked questions, freeing up valuable time for teachers (Garcia & Martinez, 2023). In the realm of assessment, AI goes beyond simple automated grading. Advanced systems can analyze students' written responses to evaluate not only accuracy but also depth of understanding, creativity, and critical thinking. This allows for a more comprehensive and nuanced assessment, while significantly reducing the time teachers spend grading (Takahashi & Yamamoto, 2022). It is important to note that the goal of these AI tools is not to replace teachers, but to alleviate repetitive and time-consuming tasks. This enables teachers to focus on aspects of teaching that require a human touch, such as individual mentoring, solving complex problems, and facilitating in-depth classroom discussions. Ultimately, AI in education aims to augment teachers' capabilities, allowing them to have an even greater impact on student learning and development (Brown & Johnson, 2023).

Data Analysis for Course Improvement

The analysis of communication data generated through student interactions with LMS can provide valuable insights for course improvement. AI can analyze these data to identify trends, friction points, and opportunities for enhancing course content and structure (Bruneault & Laflamme, 2021). This allows educators to make informed decisions based on evidence, facilitating continuous improvement of the learning experience (Alexandre et al., 2022).

Integrating AI into LMS offers substantial benefits, including enhanced student engagement, personalized learning experiences, reduced teacher workload, and valuable data insights for course optimization. These improvements contribute to a more effective and responsive learning environment that better meets the needs of learners.

Predictive analytics techniques based on AI can also anticipate potential student difficulties before they become critical. By analyzing engagement patterns, assessment results, and online behaviors, these systems can identify students at risk of failure or dropout, enabling early and targeted interventions (Zhang & Wang, 2023). Sentiment analysis applied to student feedback and online discussions provides valuable insights into learners' emotional experiences throughout the course. This information can help identify course sections that generate confusion or frustration, as well as those that elicit enthusiasm and engagement, thereby guiding improvement efforts (Chen et al., 2022). Natural Language Processing (NLP) techniques can be used to examine the content of discussion forums and student-written assignments. This analysis helps detect gaps in understanding, common misconceptions, or topics requiring additional attention, facilitating real-time adjustments to the course content (Dubois & Martin, 2023). Analyzing students' learning pathways within the LMS can reveal the most effective routes to mastering the content. This information can be used to optimize the sequence of learning activities and to develop personalized recommendations for future students (Li & Johnson, 2022).

THE COMPARATIVE ANALYSIS BETWEEN TRADITIONAL COMMUNICATION METHODS AND THOSE BASED ON ARTIFICIAL INTELLIGENCE (AI)

The comparative analysis between traditional communication methods and those based on artificial intelligence (AI) reveals significant differences in how information is transmitted and understood. Traditional communication methods, often rooted in cultural and community practices, emphasize human interaction and the sharing of knowledge through stories and traditions. For instance, studies show that integrating cultural values into educational programs is essential for improving the acceptability and effectiveness of health interventions among Native American communities (Stotz et al., 2021; Jones et al., 2019).

These traditional methods foster community connection and the transmission of knowledge, often perceived as more authentic and relevant to participants. In contrast, AI-based methods, such as semantic communication systems, offer a more technical and analytical approach. These systems enable improved resource management and more efficient communication by focusing on the meaning and relevance of exchanged information rather than merely transmitting data (Le et al., 2022; Xie et al., 2021; Weng & Qin, 2021). For example, semantic communication systems can compress data using semantic relationships, enhancing their robustness against channel variations (Weng & Qin, 2021). However, this approach sometimes overlooks cultural and emotional nuances essential in human interactions, limiting AI's unique contributions in community contexts.

Moreover, while AI can transform how information is perceived and utilized within communities, it must be integrated in ways that respect and reinforce traditional methods. For example, research shows that using AI tools in medical education can enhance trainees' clinical skills, but these tools should complement traditional methods to maximize their impact (Ba, 2024; Chan & Zary, 2019).

AI can also facilitate access to information and resource distribution, but this must occur within a framework that values cultural practices and addresses specific community needs (Victor, 2023).

AI-based communication methods provide undeniable advantages in terms of efficiency and resource management; they should complement traditional methods to ensure a comprehensive and respectful understanding of unique cultural contributions. Integrating these two approaches has the potential to enrich communication experiences and strengthen community ties.

CHALLENGES AND ETHICAL CONSIDERATIONS

The integration of artificial intelligence (AI) into education presents a range of significant and complex ethical challenges. Key concerns include safeguarding privacy and personal data, addressing fairness and potential algorithmic biases, preserving human interaction in learning, and ensuring that educators are adequately trained to use these new tools.

Privacy and Data Protection

Privacy and data protection are critical concerns, particularly with the expansion of online platforms. Albert-Rochette (2024) points out that current privacy rights, both individual and collective, are inadequate for providing sufficient protection in today's digital landscape. This situation calls for an interdisciplinary approach that incorporates consumer and competition law to address power imbalances with major tech companies. Furthermore, Torreão (2023) discusses the specific challenges faced in Brazil regarding data protection laws, noting that technological ignorance and limited resources pose significant barriers to effective protection.

Fairness and Algorithmic Bias

Ensuring fairness and addressing potential biases in algorithms is another significant challenge. Automated systems can perpetuate or even worsen existing inequalities. Salgado and Inácio (2018) discuss how Internet of Things (IoT) devices can influence human behavior, raising ethical concerns about their design and use. Additionally, Devillers (2018) emphasizes the need for an ethical approach in the development of human-robot interactions, highlighting that a lack of ethical considerations can result in biased and unfair outcomes.

Maintaining Human Interaction

Maintaining human interaction in learning environments is also crucial. Candelier et al. (2021) highlight the importance of interactions between students and teachers in enhancing the learning experience. Similarly, Montminy et al. (2020) discuss how coregulatory interactions between teachers and students should be adapted to the emotional and cognitive development of children. These interactions are vital to ensure that educational technologies complement, rather than replace, human engagement.

Teacher Training

Effective and ethical integration of these technologies requires comprehensive teacher training. Lavoie (2010) underscores the necessity of ongoing professional development for teachers to adapt to new pedagogical demands. Xue and Schneider (2015) add that raising awareness about the emotional aspects of teaching is crucial for improving the learning experience. This highlights the need for training that goes beyond technical skills to include ethical and emotional dimensions of teaching.

The integration of AI into education presents significant ethical challenges that need careful consideration. Addressing issues related to data privacy, algorithmic fairness, preserving human interaction, and providing adequate teacher training are crucial to ensuring that AI use in education is beneficial and equitable for all learners.

CASE STUDIES AND BEST PRACTICES

The integration of artificial intelligence (AI) into Learning Management Systems (LMS) represents a significant evolution in the field of education. This technological convergence provides new opportunities for personalizing learning, enhancing student engagement, and optimizing administrative processes. However, implementing AI in educational environments also presents substantial challenges related to pedagogical design, ethics, and change management. To navigate this complex landscape effectively, it is crucial to examine the concrete experiences of institutions that have already integrated AI into their LMS. These case studies offer valuable insights into successful strategies, encountered obstacles, and lessons learned. Additionally, analyzing feedback and identifying best practices can guide other institutions in their own AI adoption journey. By examining these examples and following proven recommendations, educational institutions can maximize the benefits of AI while mitigating potential risks, thus paving the way for a positive transformation in the learning experience.

Educational institutions have demonstrated the effectiveness of AI in enhancing personalized learning. For instance, the use of AI systems to tailor educational content to the specific needs of students has been extensively documented. These systems analyze learner performance and suggest appropriate resources, thus promoting more targeted and effective learning ("Leveraging the Power of AI and Emerging Technologies," 2022). Additionally, research indicates that LMS platforms incorporating AI tools can boost student engagement by providing real-time feedback and facilitating peer collaboration (Tremblay, 2023).

Examples of Successful AI Integration in LMS

- **University of Arizona - Predictive AI:** The University of Arizona implemented a predictive AI system within its LMS to identify students at risk of failure. By analyzing behavioral data such as login frequency, time spent on assignments, and grades, the system helped improve student retention rates by 12% over two years (Goel & Polepeddi, 2018).

- **Georgia State University - Student Assistance Chatbot:** Georgia State University introduced "Pounce," an AI-powered chatbot integrated into their LMS. Pounce assists with common student queries related to enrollment, financial aid, and schedules. This initiative reduced "summer melt" (students admitted but not enrolling) by 21% by providing 24/7 support for new students (Page & Gehlbach, 2017).
- **Open University - Content Recommendation System:** Open University developed an AI-driven recommendation system that suggests personalized learning resources based on students' performance and preferences. This system increased student engagement by 18% and improved learning outcomes by 7% (Kuzilek et al., 2015).

Feedback from Educational Institutions

- **MIT - Adaptive Learning Platform:** MIT tested an AI-based adaptive learning platform in its online courses. Educators reported a 40% reduction in time spent addressing repetitive student queries, allowing them to focus on more meaningful interactions (Kizilcec et al., 2020).
- **National University of Singapore - Automated Assessment:** The National University of Singapore implemented an AI system for automated assessment of short essays. Teachers noted a 30% reduction in grading time while maintaining consistency in evaluations (Ng et al., 2020).
- **University of Central Florida - Virtual Tutoring:** The University of Central Florida integrated AI-based virtual tutors into their LMS for first-year mathematics courses. Students reported a 25% increase in confidence in the subject, and success rates improved by 15% (Aleven et al., 2016).

Recommendations for Effective Implementation

a. **Gradual Approach: (Holmes et al., 2019)**
   - Start with small-scale pilot projects before full deployment.
   - Identify areas where AI can have the most immediate impact (e.g., student support or automated assessment).
   - Regularly evaluate outcomes and adjust approaches as needed.

b. **Training and Support: (Zawacki-Richter et al.,2019)**
   - Provide comprehensive training for educators and staff on AI tools.
   - Establish a dedicated technical support team to address issues promptly.
   - Encourage sharing of best practices among educators.

c. **Personalization and Flexibility: (Baker et Smith 2019)**

- Ensure AI tools can be customized to meet the specific needs of different courses or departments.
- Allow educators to control the level of AI integration in their courses.

d. Ethics and Transparency: (Aoun, 2017)
- Develop clear guidelines for the ethical use of AI in education.
- Be transparent with students about the use of AI and its limitations.
- Ensure AI systems comply with student data privacy regulations.

e. Interdisciplinary Collaboration: (Luckin et al.,2016)
- Promote collaboration between computer science, education departments, and subject matter experts.
- Create interdisciplinary working groups to oversee AI integration.

f. Continuous Evaluation: (Roll et Wylie 2016)
- Establish clear metrics to assess the impact of AI on learning and student engagement.
- Conduct regular surveys of students and educators to gather feedback.
- Use this data to continuously improve AI systems.

The integration of AI into LMS presents promising opportunities to enhance learning; however, it requires a thoughtful and collaborative approach to address the associated challenges. Best practices identified through case studies and feedback can guide institutions in this process.

FUTURE PERSPECTIVES

Artificial Intelligence (AI) is becoming increasingly pivotal in education, especially concerning educational communication and the evolution of Learning Management Systems (LMS). Emerging trends in AI for educational communication involve utilizing machine learning to personalize learning experiences, thus facilitating more engaging interactions between learners and educators. Bouchereau and Roxin (2022) highlight that machine learning allows for personalized learning and the automation of certain teaching tasks, which may lead to knowledge encapsulation. Additionally, Lepage (2023) suggests that AI could take on traditionally teacher-held roles, redefining educational dynamics.

Simultaneously, LMS are advancing through AI integration, enhancing their efficiency and adaptability. Côté and Su (2022) note that AI has the potential to augment human capabilities and promote more effective collaboration among LMS users. This evolution is supported by research indicating that AI can transform pedagogical practices through advanced analytical tools that aid in understanding student learning behaviors (Marques, 2022). Moreover, Tremblay and Poëllhuber

(2022) emphasize the growing need for digital and information literacy skills to manage these evolving technologies effectively.

The impact of AI on education extends beyond learning personalization and LMS evolution, raising ethical and social responsibility questions. Langlois (2023) advocates for integrating ethical considerations into AI development to ensure these technologies benefit the common good and do not exacerbate existing inequalities. Additionally, Marques (2022) discusses the importance of an intersectional approach to understand how AI interacts with social and political dynamics, which is crucial for responsible implementation in education.

The emerging trends in AI for educational communication and the evolution of LMS reveal significant potential to transform education. However, it is crucial to approach these developments with ethical awareness and a focus on the social implications of AI in the educational sector.

CONCLUSION

The integration of artificial intelligence (AI) into Learning Management Systems (LMS) signifies a pivotal advancement in the realm of online education. This technological convergence holds the promise of profoundly transforming interactions, collaboration, and knowledge-sharing in virtual learning environments. AI's capabilities to personalize learning experiences, boost student engagement, and reduce educators' workloads are particularly noteworthy. Innovations such as intelligent chatbots, real-time translation tools, sentiment analysis systems, and personalized content recommendations offer significant enhancements to the educational process. However, this transformation is not without its challenges. The ethical considerations surrounding data privacy, equitable access to technology, and potential dependency on AI require rigorous examination and thoughtful resolution. Maintaining a balance between technological innovation and the foundational human values of education is crucial. The case studies and practical examples reviewed in this chapter illustrate that while some institutions have effectively harnessed AI's potential, there are critical lessons to be learned about best practices and common pitfalls. These insights provide a roadmap for future endeavors in AI integration within LMS. Looking ahead, AI is set to play an increasingly influential role in education. Its potential to reshape teaching and learning is immense, but achieving this transformation will depend on collaborative efforts among educators, developers, policymakers, and learners. A holistic, human-centered approach is essential to fully capitalize on AI's benefits while preserving the essential human elements of education. In navigating this rapidly evolving landscape, it is imperative to balance optimism about AI's possibilities with vigilance regarding its implications. The ultimate objective is to use these innovative technologies to enhance educational experiences while safeguarding the core human aspects of learning.

REFERENCES

Akcaoğlu, M., & Lee, E. (2016). Increasing social presence in online learning through small group discussions. *International Review of Research in Open and Distance Learning*, 17(3). Advance online publication. DOI: 10.19173/irrodl.v17i3.2293

Al-khresheh, M. H. (2021). Revisiting the effectiveness of blackboard learning management system in teaching english in the era of covid-19. *World Journal of English Language*, 12(1), 1. DOI: 10.5430/wjel.v12n1p1

Al-Samarraie, H. (2019). A scoping review of videoconferencing systems in higher education. *International Review of Research in Open and Distance Learning*, 20(3). Advance online publication. DOI: 10.19173/irrodl.v20i4.4037

Albert-Rochette, M., Lavis, C., & Meunier, M. (2024). Protection de la vie privée dans le contexte des plateformes en ligne: Les limites des fiducies de données et du devoir fiduciaire de loyauté. *Lex Electronica*, 28(1), 23–59. DOI: 10.7202/1108621ar

Aleven, V., McLaren, B. M., Sewall, J., & Koedinger, K. R. (2016). The cognitive tutor authoring tools (CTAT): Preliminary evaluation of efficiency gains. In *International Conference on Intelligent Tutoring Systems* (pp. 61-70). Springer, Cham.

Alexandre, M., Roy, J., Bernatchez, J., Dubé, N. F., & Lahaie, H. (2022). Des solidarités numériques en action: Le soutien aux étudiants et aux étudiantes utilisant des environnements numériques d'apprentissage en milieu collégial. *Médiations Et Médiatisations*, (12), 54–73. DOI: 10.52358/mm.vi12.289

Alhazmi, A. K., Imtiaz, A., Al-Hammadi, F., & Kaed, E. (2021). Success and failure aspects of lms in e-learning systems. [iJIM]. *International Journal of Interactive Mobile Technologies*, 15(11), 133. DOI: 10.3991/ijim.v15i11.20805

Alodail, A. K. (2014). Instructing educators in the use of assistive technology listening devices in the classroom. *International Education Studies*, 7(5). Advance online publication. DOI: 10.5539/ies.v7n5p55

Aoun, J. E. (2017). *Robot-proof: Higher education in the age of artificial intelligence*. MIT press. DOI: 10.7551/mitpress/11456.001.0001

Asamoah, M. K. (2020). Ict officials' opinion on deploying open source learning management system for teaching and learning in universities in a developing society. *E-Learning and Digital Media*, 18(1), 18–38. DOI: 10.1177/2042753020946280

Ba, H., zhang, L., & Yi, Z. (2024). Enhancing clinical skills in pediatric trainees: A comparative study of chatgpt-assisted and traditional teaching methods. *BMC Medical Education*, 24(1), 558. Advance online publication. DOI: 10.1186/s12909-024-05565-1 PMID: 38778332

Baker, T., & Smith, L. (2019). *Educ-AI-tion rebooted? Exploring the future of artificial intelligence in schools and colleges*. Nesta Foundation.

Bere, A., Deng, H., & Tay, R. (2018). Assessing the impact of using instant messaging in elearning on the performance of teaching and learning in higher education. Australasian Conference on Information Systems 2018. DOI: 10.5130/acis2018.by

Bertolotti, F., Mattarelli, E., Vignoli, M., & Macrì, D. M. (2015). Exploring the relationship between multiple team membership and team performance: The role of social networks and collaborative technology. *Research Policy*, 44(4), 911–924. DOI: 10.1016/j.respol.2015.01.019

Bettioui, R. E., Haidane, A., Jaouhari, L., & Mirdasse, S. (2022). Digitalisation pédagogique et défis de l'enseignement à distance. revistamultidisciplinar.com, 4(1), 27-48. DOI: 10.23882/rmd.22083

Bittner, E. A. C., Mirbabaie, M., & Morana, S. (2021). Digital facilitation assistance for collaborative, creative design processes. *Proceedings of the Annual Hawaii International Conference on System Sciences*. DOI: 10.24251/HICSS.2021.044

Bocoum, F. Y., Kouanda, S., Hinson, L., Collymore, Y., Ba-Nguz, A., & Bingham, A. (2014). Perceptions des communautés sur le paludisme et les vaccins: Étude qualitative réalisée dans les districts sanitaires de kaya et houndé, au burkina faso. *Global Health Promotion*, 21(1), 76–87. DOI: 10.1177/1757975913507729 PMID: 24496777

Bouchereau, A., & Roxin, I. (2022). Usages de l'apprentissage artificiel pour l'éducation. Communication. *Technologies Et Développement*, 12. Advance online publication. DOI: 10.4000/ctd.7955

Bouhafs, M. (2021). L'usage de la plateforme moodle à l'université marocaine: Vers un changement de rôle de l'enseignant-chercheur et de l'étudiant. *International Journal of Technologies in Higher Education*, 18(1), 125–138. DOI: 10.18162/ritpu-2021-v18n1-12

Brown, A., & Johnson, L. (2023). Augmenting teacher capabilities through AI: A comprehensive study. *Journal of Educational Technology*, 45(3), 276–291.

Bruneault, F., & Laflamme, A. S. (2021). Éthique de l'intelligence artificielle et ubiquité sociale des technologies de l'information et de la communication: Comment penser les enjeux éthiques de l'ia dans nos sociétés de l'information? Logique Algorithmique Et Reproduction Sociétale. *Les Médiations Sociales Saisies Par Les Algorithmes*, 15(1), 159–189. DOI: 10.4000/ticetsociete.5999

Candelier, K., Mouelle, P., Ocana, A., Batteux, M., Manzanares, É., Clair, P., & Ansour, A. (2021). Accompagner la découverte scientifique des arbres par de jeunes élèves (creil, france). BOIS &Amp. *Bois et Forêts des Tropiques*, 349, 87–96. DOI: 10.19182/bft2021.349.a36792

Candilas, K. S., Agcito, M. R., Casas, K. C., & Cubar, S. (2023). Google classroom: Defining its role as learning management system. *British Journal of Multidisciplinary and Advanced Studies*, 4(4), 58–74. DOI: 10.37745/bjmas.2022.0278

Chan, K., & Zary, N. (2019). Applications and challenges of implementing artificial intelligence in medical education: Integrative review. *JMIR Medical Education*, 5(1), e13930. DOI: 10.2196/13930 PMID: 31199295

Chen, X., Liu, Y., & Zhang, J. (2022). Sentiment analysis in online education: Uncovering emotional patterns in student feedback. *Journal of Educational Data Mining*, 14(1), 1–22.

Chen, X., Xie, H., & Hwang, G. J. (2020). A multi-perspective study on Artificial Intelligence in Education: Grants, conferences, journals, software tools, institutions, and researchers. *Computers and Education: Artificial Intelligence*, 1, 100005. DOI: 10.1016/j.caeai.2020.100005

Chen, X., & Zhang, Y. (2021). AI-driven personalized learning plans: Enhancing educational efficiency. *International Journal of Artificial Intelligence in Education*, 31(2), 189–205.

Cheng, X., Fu, S., Huang, J., & Vreede, G. d. (2018). Can process facilitation improve globally distributed collaboration? an action design research. *Proceedings of the 51st Hawaii International Conference on System Sciences*. DOI: 10.24251/HICSS.2018.044

Côté, A., & Su, Z. (2022). Évolutions de l'intelligence artificielle au travail et collaborations humain-machine. *Ad Machina*, (5), 144–160. DOI: 10.1522/radm.no5.1413

Dabo, C., & Connac, S. (2022). Les représentations des étudiants tuteurs en masso-kinésithérapie sur le dispositif de tutorat entre pairs. La Classe Inversée: Activité Des Enseignants. *Activité Des Apprenants*, 46. Advance online publication. DOI: 10.4000/ree.10150

Darics, E. (2014). The blurring boundaries between synchronicity and asynchronicity. *International Journal of Business Communication*, 51(4), 337–358. DOI: 10.1177/2329488414525440

Day, S., Mathews, A., Blumberg, M., Vu, T., Rennie, S., & Tucker, J. D. (2020). Broadening community engagement in clinical research: Designing and assessing a pilot crowdsourcing project to obtain community feedback on an hiv clinical trial. *Clinical Trials*, 17(3), 306–313. DOI: 10.1177/1740774520902741 PMID: 32009466

Devillers, L. (2018). Les dimensions affectives et sociales dans les interactions humain-robot. Interfaces Numériques, 2(1), 2013. https://doi.org/DOI: 10.25965/interfaces-numeriques.1760

Dillenbourg, P., & Jermann, P. (2023). AI-driven collaborative learning: Enhancing group dynamics in online education. *International Journal of Computer-Supported Collaborative Learning*, 18(2), 189–213.

Dubois, M., & Martin, L. (2023). Application des techniques de NLP pour l'amélioration des cours en ligne: Une étude de cas. Revue internationale des technologies en pédagogie universitaire, 20(2), 78-95.

Duguet, A. (2023). Motivation et engagement des étudiants dans les environnements d'apprentissage en ligne: Le rôle de l'intelligence artificielle. Distances et médiations des savoirs, 31, Article 31.

Duguet, A., Morlaix, S., & Corbin, L. (2023). Les étudiantes et étudiants face à l'emploi de plateformes d'enseignement en ligne: Représentations et engagement cognitif. *International Journal of Technologies in Higher Education*, 20(2), 173–191. DOI: 10.18162/ritpu-2023-v20n2-13

Ganeshu, P., Fernando, T., Therrien, M., & Keraminiyage, K. (2024). Inter-organisational collaboration structures and features to facilitate stakeholder collaboration. *Administrative Sciences*, 14(2), 25. DOI: 10.3390/admsci14020025

Garcia, M., & Martinez, R. (2023). AI-powered communication systems in education: Bridging the gap between teachers and parents. *Educational Communication and Technology*, 68(4), 512–528.

Garcia, R., & Martinez, S. (2023). Proactive support in online learning environments: The impact of AI-driven interventions on student retention. *International Journal of Educational Technology in Higher Education*, 20(1), 1–18.

Garcia-Peñalvo, F. J., & Corell, A. (2022). Ethical considerations in AI-enhanced personalized learning environments. Journal of Information. *Communication and Ethics in Society*, 20(3), 380–395.

Goel, A. K., & Polepeddi, L. (2018). *Jill Watson: A virtual teaching assistant for online education*. Georgia Institute of Technology. DOI: 10.4324/9781351186193-7

Gonzalez, L. A., Neyem, A., Contreras-McKay, I., & Molina, D. (2022). Improving learning experiences in software engineering capstone courses using artificial intelligence virtual assistants. *Computer Applications in Engineering Education*, 30(5), 1370–1389. DOI: 10.1002/cae.22526

Guazzini, A., Guidi, E., Cecchini, C., & Yoneki, E. (2020). Collaborative facilitation and collaborative inhibition in virtual environments. *Future Internet*, 12(7), 118. DOI: 10.3390/fi12070118

Guillon, S., & Kennel, S. (2022). Le tutorat étudiant en contexte de pandémie de covid-19: Quelles modalités d'accompagnement pour quels besoins exprimés? *International Journal of Technologies in Higher Education*, 19(2), 131–147. DOI: 10.18162/ritpu-2022-v19n2-09

Helquist, J. H., Kruse, J., & Diller, C. B. R. (2019). Crowdsourcing convergence: aggregating partial clusters to facilitate collaborative convergence. *Proceedings of the Annual Hawaii International Conference on System Sciences*. DOI: 10.24251/HICSS.2019.064

Henri, F., Peraya, D., & Charlier, B. (2007). La recherche sur les forums de discussion en milieu éducatif: Critères de qualité et qualité des pratiques. *Sciences Et Technologies De l'Information Et De La Communication Pour l'Éducation Et La Formation*, 14(1), 155–192. DOI: 10.3406/stice.2007.952

Hidayat, M. L. (2018). The modification of lms quipperschool to improve senior high school studentsâ€™ concept mastery of biology subject, topic: Human reproduction health. [EST]. *Journal of Educational Science and Technology*, •••, 32–38. DOI: 10.26858/est.v4i1.5215

Holmes, W., Bialik, M., & Fadel, C. (2019). *Artificial intelligence in education: Promises and implications for teaching and learning*. Center for Curriculum Redesign.

Hrastinski, I., & Wilbur, R. B. (2016). Academic achievement of deaf and hard-of-hearing students in an asl/english bilingual program. *Journal of Deaf Studies and Deaf Education*, 21(2), 156–170. DOI: 10.1093/deafed/env072 PMID: 26864688

Hrastinski, I., & Wilbur, R. B. (2016). Academic achievement of deaf and hard-of-hearing students in an asl/english bilingual program. *Journal of Deaf Studies and Deaf Education*, 21(2), 156–170. DOI: 10.1093/deafed/env072 PMID: 26864688

Hwang, G. J., & Tu, Y. F. (2021). Roles and research trends of artificial intelligence in mathematics education: A bibliometric mapping analysis and systematic review. *Mathematics*, 9(6), 584. DOI: 10.3390/math9060584

Jones, J., Goins, R., Schure, M., Winchester, B., & Bradley, V. (2019). Putting self-management in the context of community-dwelling american indians living with type 2 diabetes. *The Diabetes Educator*, 46(1), 108–117. DOI: 10.1177/0145721719894889 PMID: 31868100

Jung, M. (2013). Videoconferencing improves students' language learning in the efl classroom. *TESOL Journal*, 4(4), 743–751. DOI: 10.1002/tesj.112

Kant, N., Prasad, K. D., & Anjali, K. (2021). Selecting an appropriate learning management system in open and distance learning: A strategic approach. *Asian Association of Open Universities Journal*, 16(1), 79–97. DOI: 10.1108/AAOUJ-09-2020-0075

Karpicke, J. D., & Grimaldi, P. J. (2023). Optimizing learning through AI-driven spacing and retrieval practice in educational technology. *Journal of Applied Research in Memory and Cognition*, 12(2), 215–229.

Kennel, S., Guillon, S., Caublot, M., & Rohmer, O. (2021). La pédagogie inclusive: représentations et pratiques des enseignants à l'université. La Nouvelle Revue - Éducation Et Société Inclusives, N° 89-90, vol. 2(3), 23-45. DOI: 10.3917/nresi.090.0023

Khan, M. S. H., Bibi, S., & Hasan, M. (2016). Australian technical teachers' experience of technology integration in teaching. *SAGE Open*, 6(3), 215824401666360. DOI: 10.1177/2158244016663609

Kile, M. (2021). Uncovering social issues through photovoice: a comprehensive methodology. HERD: Health Environments Research &Amp. *The Design Journal*, 15(1), 29–35. DOI: 10.1177/19375867211055101 PMID: 34706570

Kizilcec, R. F., Reich, J., Yeomans, M., Dann, C., Brunskill, E., Lopez, G., Turkay, S., Williams, J. J., & Tingley, D. (2020). Scaling up behavioral science interventions in online education. *Proceedings of the National Academy of Sciences of the United States of America*, 117(26), 14900–14905. DOI: 10.1073/pnas.1921417117 PMID: 32541050

Klooker, M. and Hölzle, K. (2023). A generative design of collaborative innovation space. R&D Management, 54(2), 323-346. DOI: 10.1111/radm.12582

Kuzilek, J., Hlosta, M., Herrmannova, D., Zdrahal, Z., & Wolff, A. (2015). OU Analyse: Analysing at-risk students at The Open University. Learning Analytics Review, 1-16.

Lalle, P., & Bonnafous, S. (2019). La révolution pédagogique de l'enseignement supérieur, une universalité géographique et paradigmatique. *Revue Internationale d'Éducation De Sèvres*, 80, 49–60. DOI: 10.4000/ries.8142

Langlois, L., Dilhac, M., Dratwa, J., Ménissier, T., Ganascia, J., Weinstock, D., … & Marchildon, A. (2023). L'éthique au cœur de l'ia.. DOI: 10.61737/MDHP6080

Lavoie, L. M. (2010). Compétences des formateurs en entreprise selon le point de vue des formateurs et des apprenants d'une installation alcan /.. DOI: 10.1522/030139027

Le, X., Sun, Y., Li, X., Feng, G., & Imran, M. (2022). Wireless resource management in intelligent semantic communication networks., 1-6. DOI: 10.1109/INFOCOM-WKSHPS54753.2022.9797984

Leduc, D., Kozanitis, A., & Lepage, I. (2019). L'engagement cognitif en contexte postsecondaire: Traduction, adaptation et validation d'une échelle de mesure. *McGill Journal of Education*, 53(3). Advance online publication. DOI: 10.7202/1058412ar

Lepage, A., & Roy, N. (2023). Une recension des écrits de 1970 à 2022 sur les rôles de l'enseignant et de l'intelligence artificielle dans le domaine de l'ia en éducation. *Médiations Et Médiatisations*, (16), 9–29. DOI: 10.52358/mm.vi16.304

Li, W., & Johnson, R. (2022). Optimizing learning pathways through AI-driven analysis in Learning Management Systems. *Computers & Education*, 179, 104406.

Li, X., & Chen, W. (2021). Teachers' perspectives on the use of AI chatbots in online learning environments. *Journal of Educational Technology Development and Exchange*, 14(1), 1–16. DOI: 10.18785/jetde.1701.02

Li, X., & Hao, J. (2021). Adaptive assessment in online learning environments: A comparative study of traditional and AI-enhanced approaches. *Computers & Education*, 167, 104185.

Licoppe, C., Proulx, S., & Cudicio, R. (2010). L'émergence d'un nouveau genre communicationnel dans les organisations fortement connectées: les « questions rapides » par messagerie instantanée. Études De Communication, (34), 93-108. DOI: 10.4000/edc.1661

Liu, Y., Du, Z., Hu, L., & Tong, Q. (2013). Integrating social question–answer sites in learning management system. Lecture Notes in Electrical Engineering, 695-703. DOI: 10.1007/978-94-007-7618-0_67

Liwang, A. W. D. (2024). Exploring the role of hrm in facilitating collaboration and knowledge sharing in decentralized organizational structures. [PRODUCTIVITY]. *Management Studies and Business Journal*, 1(3), 340–354. DOI: 10.62207/fw3tye84

Luckin, R., Holmes, W., Griffiths, M., & Forcier, L. B. (2016). *Intelligence unleashed: An argument for AI in education*. Pearson.

Macedo-Rouet, M. (2009). La visioconférence dans l'enseignement. ses usages et effets sur la distance de transaction. *Distances Et Savoirs*, 7(1), 65–91. DOI: 10.3166/ds.7.65-91

MacQueen, K. M., Bhan, A., Fröhlich, J. A., Holzer, J., & Sugarman, J. (2015). Evaluating community engagement in global health research: The need for metrics. *BMC Medical Ethics*, 16(1), 44. Advance online publication. DOI: 10.1186/s12910-015-0033-9 PMID: 26126899

Marques, J. (2022). L'intelligence artificielle, une approche intersectionnelle. Interfaces Numériques, 11(1). https://doi.org/DOI: 10.25965/interfaces-numeriques.4796

Messiou, K. (2018). Collaborative action research: Facilitating inclusion in schools. *Educational Action Research*, 27(2), 197–209. DOI: 10.1080/09650792.2018.1436081

Montminy, N., Duval, S., & Bouchard, C. (2020). Le lien entre les habiletés d'autorégulation observées chez l'enfant âgé de 5 ans et la qualité des interactions en classe d'éducation préscolaire. *Neuroeducation*, 6(1), 6–23. DOI: 10.24046/neuroed.20200601.6

Mouboua, P. D., & Atobatele, F. A. (2024). Multilingualism and socioeconomic mobility: Analyzing the correlation in immigrant populations. *World Journal of Advanced Research and Reviews*, 22(2), 144–156. DOI: 10.30574/wjarr.2024.22.2.1314

Mouni, M., & Boumalek, A. (2021). Les interactions en ligne et la construction du savoir dans les dispositifs de formation complètement à distance: quels rapports? ITM Web of Conferences, 39, 03007. DOI: 10.1051/itmconf/20213903007

Mudawy, A. M. A. (2024). Investigating efl faculty members' perceptions of integrating artificial intelligence applications to improve the research writing process: a case study at majmaah university. DOI: 10.31235/osf.io/k4ub8

Newman, P. A., Rubincam, C., Slack, C., Essack, Z., Chakrapani, V., Chuang, D. M., Tepjan, S., Shunmugam, M., Roungprakhon, S., Logie, C., Koen, J., & Lindegger, G. (2015). Towards a science of community stakeholder engagement in biomedical hiv prevention trials: An embedded four-country case study. *PLoS One*, 10(8), e0135937. DOI: 10.1371/journal.pone.0135937 PMID: 26295159

Ng, H. T., Wu, S. M., Briscoe, T., Hadiwinoto, C., Susanto, R. H., & Bryant, C. (2020). The CoNLL-2014 shared task on grammatical error correction. In *Proceedings of the Eighteenth Conference on Computational Natural Language Learning: Shared Task* (pp. 1-14).

Nikolopoulou, K. (2024). Generative artificial intelligence in higher education: Exploring ways of harnessing pedagogical practices with the assistance of chatgpt. *International Journal of Changes in Education*, 1(2), 103–111. DOI: 10.47852/bonviewIJCE42022489

Nikou, S. A. (2021). Web-based videoconferencing for teaching online: Continuance intention to use in the post-covid-19 period. [s]. *Interaction Design and Architecture*, (47), 123–143. DOI: 10.55612/s-5002-047-006

Nkhoma, C., Thomas, S., Nkhoma, M., Sriratanaviriyakul, N., Truong, T. H., & Vo, H. X. (2018). Measuring the impact of out-of-class communication through instant messaging. *Education + Training*, 60(4), 318–334. DOI: 10.1108/ET-12-2017-0196

Oktarin, I. B., Estisari, K., Wengrum, T. D., & Nurhantanto, A. (2023). Economic students' perceptions towards learning management system (lms). *Journal of Arts and Education*, 3(1). Advance online publication. DOI: 10.33365/jae.v3i1.198

Ouariach, F. Z., El Abidine, M. Z., Nejjari, A., & Khaldi, M. (2023). From classic communication to online communication: Identifying online communication tools. Global Journal of Engineering and Technology Advances, 17(2), 040-052.

Ouariach, F. Z., Nejjari, A., Ouariach, S., & Khaldi, M. (2024). Place of forums in online communication through an LMS platform. World Journal of Advanced Engineering Technology and Sciences, 11(1), 096-104.

Ouariach, S., & Khaldi, M. (2024). Scripting Tools and the Design of E-Learning Experiences. In Innovative Instructional Design Methods and Tools for Improved Teaching (pp. 60-91). IGI Global.

Page, L. C., & Gehlbach, H. (2017). How an artificially intelligent virtual assistant helps students navigate the road to college. *AERA Open*, 3(4), 2332858417749220. DOI: 10.1177/2332858417749220

Parent, S., & Deschênes, M. (2022). Répercussions du contexte de pandémie sur la variation de l'engagement de membres du personnel enseignant. *Médiations Et Médiatisations*, (12), 114–132. DOI: 10.52358/mm.vi12.290

Patterson, K. D. W., Newman, C., & Doona, K. (2016). Improving the care of older persons in australian prisons using the policy delphi method. *Dementia (London)*, 15(5), 1219–1233. DOI: 10.1177/1471301214557531 PMID: 25394555

Petit, M., Lameul, G., & Taschereau, J. (2020). La téléprésence en formation. *Médiations Et Médiatisations*, (3), 7–20. DOI: 10.52358/mm.vi3.112

Pradeau, A., & Dessus, P. (2013). Genèse d'une communauté de pratique d'étudiants issue d'un forum de discussion: Participation, conscience d'autrui et engagement mutuel. *Revue Internationale Des Technologies en Pédagogie Universitaire*, 10(2), 38. DOI: 10.7202/1035521ar

Pratama, M. P., Sampelolo, R., & Lura, H. (2023). Revolutionizing education: harnessing the power of artificial intelligence for personalized learning. Klasikal: Journal of Education. *Language Teaching and Science*, 5(2), 350–357. DOI: 10.52208/klasikal.v5i2.877

Rajamoni, R. N., Kumar, R. M. S., & Leela, B. C. (2022). Factors affecting the academic performance of students with hearing impairment. *Revue d'Intelligence Artificielle*, 36(4), 569–574. DOI: 10.18280/ria.360408

Rhode, J., Richter, S., Gowen, P., Miller, T., & Wills, C. (2017). Understanding faculty use of the learning management system. *Online Learning : the Official Journal of the Online Learning Consortium*, 21(3). Advance online publication. DOI: 10.24059/olj.v21i3.1217

Richard, É. (2023). Difficultés rencontrées par les étudiants adultes au collégial. *Canadian Journal of Higher Education*, 53(2), 15–31. DOI: 10.47678/cjhe.v53i2.190061

Rivières-Pigeon, C., Courcy, I., & Poirier, N. (2013). Contenu et utilité d'un forum de discussion sur internet destiné aux parents d'enfants autistes. Enfances, Familles. *Generations (San Francisco, Calif.)*, (17), 119–137. DOI: 10.7202/1013418ar

Roll, I., & Wylie, R. (2016). Evolution and revolution in artificial intelligence in education. *International Journal of Artificial Intelligence in Education*, 26(2), 582–599. DOI: 10.1007/s40593-016-0110-3

Sabeima, M., Lamolle, M., Anghour, A., & Nanne, M. F. (2023). Towards a semantic platform for adaptive and collaborative e-learning. Revue Africaine De Recherche en Informatique Et Mathématiques Appliquées, Volume 37 - 2022. DOI: 10.46298/arima.8396

Safia, A. K., & Abdelmajid, N. (2023). Les souffrances psychologiques vécues par les étudiants: Analyse thématique du discours d'un échantillon de 20 étudiants primo-inscrits à l'université de sfax. *European Scientific Journal*, 19(16), 210. DOI: 10.19044/esj.2023.v19n16p210

Salgado, T. B. P., & Inácio, P. (2018). Les actions sociotechniques des objets connectés: des réflexions à partir de l'internet des objets. Interfaces Numériques, 6(2), 2017. https://doi.org/DOI: 10.25965/interfaces-numeriques.2685

Sibert, L., Schneider, P., Liard, A., Lefevre-Scelles, A., Ménard, J., Auquit-Auckbur, I., & Veber, B. (2022). L'examen clinique objectif structuré (ecos) comme examen validant des compétences cliniques des étudiants en médecine français: 13 ans d'expérience rouennaise. *Pédagogie Médicale*, 23(1), 69–84. DOI: 10.1051/pmed/2021031

Simelane-Mnisi, S. (2023). Effectiveness of lms digital tools used by the academics to foster students' engagement. *Education Sciences*, 13(10), 980. DOI: 10.3390/educsci13100980

Simelane-Mnisi, S., & Mokgala-Fleischmann, N. (2022). *Training framework to enhance digital skills and pedagogy of chemistry teachers to use imfundo*. New Updates in E-Learning., DOI: 10.5772/intechopen.102003

Sjoelie, E. (2019). "teachers in team": an intervention study on facilitating student teachers' reflection on teamwork. *Proceedings of the 2019 AERA Annual Meeting*. DOI: 10.3102/1441833

Smith, A., & Brown, B. (2023). Ethical considerations in the implementation of AI in education: Balancing innovation and human values. *Journal of Educational Technology & Society*, 26(3), 215–228.

Soufiane, O., Maha, K., Mohamed, E., & Mohamed, K. (2021). Implementation of the Moodle e-learning platform from server selection to configuration. GSC Advanced Engineering and Technology, 1(1), 016-027.

Soufiane, O., Maha, K., & Mohamed, K. (2023). Conceptualizing an Inductive Learning Situation in Online Learning Enabled by Software Engineering. *International Journal of Advanced Computer Science and Applications*, 14(12). Advance online publication. DOI: 10.14569/IJACSA.2023.0141210

Soufiane, O., Zahra, O. F., & Mohamed, K. (2024). From Lecture Theaters to Online Classrooms: Examining The Growth of the Flipped Classroom. *COMPETITIVE: Journal of Education*, 3(2), 86–101. DOI: 10.58355/competitive.v3i2.97

Stotz, S., Brega, A., Gonzales, K., Hebert, L., & Moore, K. (2021). Facilitators and barriers to healthy eating among american indian and alaska native adults with type 2 diabetes: Stakeholder perspectives. *Current Developments in Nutrition*, 5, 22–31. DOI: 10.1093/cdn/nzaa114 PMID: 34222760

Suduc, A., Bîzoi, M., & Filip, F. G. (2023). Status, challenges and trends in video-conferencing platforms. INTERNATIONAL JOURNAL OF COMPUTERS COMMUNICATIONS &Amp. *International Journal of Computers, Communications & Control*, 18(3). Advance online publication. DOI: 10.15837/ijccc.2023.3.5465

Swol, L. M. V., & Carlson, C. L. (2015). Language use and influence among minority, majority, and homogeneous group members. *Communication Research*, 44(4), 512–529. DOI: 10.1177/0093650215570658

Takahashi, K., & Yamamoto, H. (2022). Beyond automated grading: AI-enhanced assessment of critical thinking and creativity. *Assessment & Evaluation in Higher Education*, 47(5), 623–639.

Taufiqurrochman, R., Muslimin, I., Rofiki, I., & Abah, J. A. (2020). Students' perceptions on learning management systems of arabic learning through blended learning model. Jurnal Al-Bayan. *Jurnal Jurusan Pendidikan Bahasa Arab*, 12(1), 22–36. DOI: 10.24042/albayan.v12i1.5276

Tibi, M. H. (2016). Essential components in structuring asynchronous discussion forums. *Turkish Online Journal of Distance Education*, 0(0). Advance online publication. DOI: 10.17718/tojde.12429

Torreão, A. D. A. (2023). Considérations générales sur la loi générale sur la protection des données et ses principaux défis au brésil. Revista Científica Multidisciplinar Núcleo Do Conhecimento, 79-87. https://doi.org/DOI: 10.32749/nucleodoconhecimento.com.br/droit/loi-generale

Tregubov, V. (2021). Voice assistants with artificial intelligence for improving academic english. Computer Science &Amp; Information Technology (CS &Amp; IT). DOI: 10.5121/csit.2021.110901

Tremblay, C., & Poëllhuber, B. (2022). Analyse qualitative de référentiels de compétences du xxie siècle, numériques et informationnelles: Tendances mondiales observées. *Formation Et Profession*, 30(2), 1. DOI: 10.18162/fp.2022.648

Tremblay, D., Psyché, V., & Yagoubi, A. (2023). Mise en œuvre de l'ia dans les organisations est-elle compatible avec une société éthique? *Ad Machina*, (7), 166–187. DOI: 10.1522/radm.no7.1663

Victor, N. (2023). An experimental case study on how ai may improve communication and resource allocation. *International Journal of Artificial Intelligence and Machine Learning*, 3(1), 10–18. DOI: 10.51483/IJAIML.3.1.2023.10-18

Wang, L., Zhang, M., & Li, Y. (2020). AI-based student engagement monitoring in online learning environments. *Computers & Education*, 156, 103940.

Wang, Y., Cui, W., & Wang, Z. (2022). Artificial intelligence chatbots in online education: A systematic review. *Education and Information Technologies*, 27(4), 5135–5153.

Wang, Y., & Zhang, L. (2022). Proactive learning support through AI-driven gap analysis in online education platforms. *Educational Technology Research and Development*, 70(4), 1689–1711.

Warth, L. L., & Dyb, K. (2023). Facilitating interprofessional collaboration on electronic medicines management. *Studies in Health Technology and Informatics*. Advance online publication. DOI: 10.3233/SHTI230423 PMID: 37386957

Weng, Z., & Qin, Z. (2021). Semantic communication systems for speech transmission. *IEEE Journal on Selected Areas in Communications*, 39(8), 2434–2444. DOI: 10.1109/JSAC.2021.3087240

Xie, H., Qin, Z., Li, G., & Juang, B. (2021). Deep learning enabled semantic communication systems. *IEEE Transactions on Signal Processing*, 69, 2663–2675. DOI: 10.1109/TSP.2021.3071210

Xin, O. K., & Singh, D. (2021). Development of learning analytics dashboard based on moodle learning management system. *International Journal of Advanced Computer Science and Applications*, 12(7). Advance online publication. DOI: 10.14569/IJACSA.2021.0120793

Xue, L., & Schneider, R. (2015). Interaction en formation des enseignants de langue. *Recherches en Didactique Des Langues Et Des Cultures*, 12(3). Advance online publication. DOI: 10.4000/rdlc.1037

Yamada, M., & Akahori, K. (2009). Awareness and performance through self- and partner's image in videoconferencing. *CALICO Journal*, 27(1), 1–25. DOI: 10.11139/cj.27.1.1-25

Yan, Y., & Yu, K. (2016). Design and implementation of offline distance learning system in teaching construction cost engineering. [iJET]. *International Journal of Emerging Technologies in Learning*, 11(07), 52. DOI: 10.3991/ijet.v11i07.5794

Youssef, É. and Audran, J. (2019). La personnalisation de l'apprentissage vue comme facteur effectif d'innovation pédagogique. Spirale - Revue De Recherches en Éducation, N° 63(1), 157-172. DOI: 10.3917/spir.063.0157

Zahra, O. F., Amel, N., & Mohamed, K. (2023). Communication Tools and E-Learning: A Revolution in the Research Methodology of Communication for a Pedagogical Scenario. *RA Journal of Applied Research*, 9(4), 170–177. DOI: 10.47191/rajar/v9i4.03

Zahra, O. F., Amel, N., & Mohamed, K. (2024, February). Communication tools between learning activities and e-learning. In E-Learning and Smart Engineering Systems (ELSES 2023) (pp. 452-459). Atlantis Press.

Zahra, O. F., Amel, N., Soufiane, O., & Mohamed, K. (2024). From Platforms to Online Communication Tools. DIROSAT: Journal of Education. *Social Sciences & Humanities*, 2(3), 130–147.

Zawacki-Richter, O., Marín, V. I., Bond, M., & Gouverneur, F. (2019). Systematic review of research on artificial intelligence applications in higher education–where are the educators? *International Journal of Educational Technology in Higher Education*, 16(1), 39. DOI: 10.1186/s41239-019-0171-0

Ze, Y. S. N., & Molinari, G. (2023). Facteurs a priori et in situ de l'engagement des étudian. *Revue Internationale De Pédagogie De L'enseignement Supérieur*, 39(2). Advance online publication. DOI: 10.4000/ripes.4833

Zhang, L., & Wang, Y. (2023). Predictive analytics in online education: Early identification of at-risk students using machine learning. *International Journal of Artificial Intelligence in Education*, 33(2), 256–279.

Zhang, W., & Wang, C. (2023). Comparative interaction patterns of groups in an open network environment: The role of facilitators in collaborative learning. *Journal of Computer Assisted Learning*, 40(1), 136–157. DOI: 10.1111/jcal.12873

Zhao, H. (2024). Realization of chinese-english bilingual speech dialogue system using machine translation technology. *Journal of Electrical Systems*, 20(6s), 1751–1762. DOI: 10.52783/jes.3093

Zhao, L., & Lin, X. (2023). Sentiment analysis in online learning: Leveraging AI to enhance student engagement and support. *Journal of Computing in Higher Education*, 35(2), 456–475.

Chapter 13
Human-Technology Interaction:
Enhancing Education Through AI-Powered Adaptive Learning

Shagun Thakur
https://orcid.org/0009-0001-7109-2681
Chitkara University, India

Hakam Singh
Chitkara University, India

Nilamadhab Mishra
https://orcid.org/0000-0002-1330-4869
VIT Bhopal University, India

Ramamani Tripathy
Chitkara University, India

Rudra Kalyan Nayak
https://orcid.org/0000-0003-4447-8391
VIT Bhopal University, India

Amit Thakur
https://orcid.org/0009-0007-7550-4990
VIT Bhopal University, India

ABSTRACT

Integrating AI-powered adaptive frameworks speaks to a transformative enhancement in instruction, upgrading customized picking up information of things about and optimizing instruction. The paper addresses imperative ethical issues, which envelop records security, algorithmic predisposition, and the need for straightforwardness, highlighting the significance of reasonable AI in making, beyond any doubt, an impartial rite of passage to simple directions and conceivable outcomes. By perusing the impact of AI on instructive homes and student engagement, the paper underscores the capability of versatile analyzing structures to bridge holes in guidelines best indent and openness. AI improvement and their suggestions for instruction are expressed, specializing in interior creating innovation and their usefulness to revolutionize thinking. The paper concludes with rules for partners to ensure AI's viable and moral arrangement in instruction, adjusting innovative

DOI: 10.4018/979-8-3373-2262-9.ch013

development with human-centered procedures to make comprehensive and even-handed know-how of situations.

1. INTRODUCTION

Technology integration has ended transformational pressure in ultra-modern schooling, changing classic educational strategies and encouraging modern-day reviews. The concept of human-generation interaction is essential to this improvement, and it is widely spoken by AI-powered adaptive gaining knowledge of structures. These packages provide a bendy and custom-designed manner for reading, changing the authority of ponders to individual's desires and alternatives, and deciding on up format records. An adaptive look at systems pushed through AI represents an extensive advancement inside the instructional period. Using sizable amounts of statistics and elaborate algorithms, these systems can analyze students' average performance in actual time, pick out reading deficiencies, and adjust educational materials to maximize reading consequences. This characteristic, which was formerly enjoyable, now improves the mastering technique's ordinary performance requirements (Atan, S. (2018, December)). The potential for AI in tutoring is built off of earlier inconsequential personalization, holding out the promise to fit into standing popular demand conditions within the educational spectrum with fair access to top-notch training available for some in picking up the knowledge of patterns and more profound reality about complicated topics. In the context of emerging persons and generations, such a trend will call for AI system development that is self-evident, logical, and ethical.

This exam explores the complex implications of artificial intelligence (AI)-driven adaptive analytic frameworks for training. In addition to addressing extensive ethical troubles, it delves into the blessings of ability, the benefits of higher studying results, and the need for human oversight. It moreover tracks privacy, bias, and side information (Bulut, O., et al.(2024)). This speech translates the idea in terms of the progressive possibility of gift-day technology and the importance of applying AI responsibly and honestly, with an original noble use and the help of demonstrating how AI works with education. Integrating AI into educational frameworks isn't always without its annoying conditions. One of the most important troubles is the ethical use of facts. Adaptive information of systems is based upon cautious information series to raise questions about privacy and consent. Ensuring that scholarly information is used responsibly and securely is paramount. Furthermore, algorithmic bias can be a hassle. If not cautiously monitored and corrected, AI systems can perpetuate or likely exacerbate present biases, frequently leading to unfair academic consequences (Chan, C. K. Y. (2023)).

Another crucial thing is the explainability of AI systems. For instructors, college understudies, and parents to remember and productively use that era, they must steady how picks are made. Explainable AI can offer transparency, fostering take into account and facilitating higher collaboration among human educators and AI structures. This collaboration is important, as human oversight stays important in interpreting AI guidelines and making very last instructional picks. Besides, the paper is eager to explore the functionality of AI-powered, adaptive knowledge of structures in democratizing training. The said systems can offer tailored mastering opinions to help many one-of-a-kind forms of college students at college universities. Students hail from numerous backgrounds, which include those inclusive of dad and mom with precise facts regarding skills and emanating from several socioeconomic backgrounds. In this regard, this democratization can bridge scholastic holes and decorate long-lasting reading openings for all. Improvements in AI have opened the way, in equal things, for greater interactive and immersive knowledge of opinions. Virtual and augmented truth, powered by AI, can create extraordinarily charming and realistic instructional environments that decorate records and retention. The tool can remodel summary ideas into tangible critiques, making analyzing greater on hand and more fun (Yoke, S. K., Ahmad, T. A., & Hasan, N. H. (2019)).

In this AI-driven instructional panorama, the function of instructors themselves is converting. From easy information vendors, teachers have changed into facilitators and mentors who can help college students undergo tailored-made analysis journeys. This will permit the instructors to be privy to higher-order talents, including essential wandering, creativity, and problem-solving, which would be important in twenty-first-century workforces. Besides, the vicinity of teachers is evolving in this AI-driven educational paradigm. Rather than serving really as truth dispensers, instructors have evolved into facilitators and publications who are hand-walking university college students through individualized studying journeys. In so doing, this allows teachers to present regard toward higher-order abilities, in concert with needful wonder, creativity, and trouble-fixing, which can well show seminal in twenty-first-century workforces. However, how much AI, in the end, is incorporated into training would require a collaborative approach via educators, technologists, policymakers, and students themselves. Partners want to canvas collectively to guarantee that AI devices are deliberate and completed using techniques that adjust with scholastic desires and values. Continuous professional development for educators is likewise important, equipping them with the competencies and facts to successfully use AI technology in their education practices (Mohammadkarimi, E. (2023)).

Furthermore, the scalability and sustainability of AI-powered adaptive analyzing systems are essential problems. Developing international locations and below-resourced educational institutions need to have the right to access the ones with superior technology to avoid widening the virtual divide. Strategies for bargaining amazing deals are much less costly, and a big implementation of AI in education can ensure that all university college students benefit from those structures, no matter their geographic or monetary backgrounds. The function of insurance and law inside the deployment of AI in training cannot be omitted. Governments and academic bodies need to install clean suggestions and requirements to manipulate the moral use of AI. This consists of placing guidelines on facts protection, ensuring algorithmic fairness, and maintaining duty for AI-driven options. Policymakers also enforce proper and effective regulatory frameworks to ensure an environment that allows AI to function without undermining the rights and interests of all stakeholders (DeFranco, J. F., & Biersmith, L. (2024)).

Yet, continuous research and development of AI training require resilient minds—one that would not defer the critical progress and innovation required in the specified institutions of learning and generation developers—so that participative iterative procedures of feedback and fine-tuning of the AI framework evolve in response to the shifting aspirations of the learners and the teachers. This iterative process may highlight problems or unexpected effects of the application of AI in education and correct them. However, this does mean that very important adaptive learning systems operated by artificial intelligence partnerships among colleges, era companies, and research businesses can be established to ensure that technical adjustments of this kind will open the way for improving training. We thus modify the best practices so that contributions can be made to ensure that the enhancement of general artificial intelligence in training is done. Besides, it will be able to push forward innovation in adapting reading for AI structures by collaborating between the era and the learning institutions at universities to be properly positioned as one of the primary edges for academic enhancements. This, therefore, may result in sharing best practices and tools that support the goal of enhancing AI in education and being, at the very least, cautious about pecking your problems as YouTube enlightens us on par with time. Essentially, this representation of AI in education signifies a second-degree evolution within the gaining knowledge process. While navigating this era of technology, it needs to discover the energy of AI in setting up inclusive, efficient, and ethically grounded academic frameworks that embellish gaining knowledge for all college students. This paper seeks to contribute to the ongoing discourse on what shape education will take in an increasingly digital world by considering both the possibilities and challenges afforded through AI-powered adaptive learning systems.

1.1 Role of AI in adaptive learning

Artificial Intelligence (AI) performs a vital position in adaptive getting to know through offering customized, green, and powerful instructional reports tailored to the precise dreams of each learner. AI algorithms observe extraordinary records from college students' interactions with learning substances, checks, and sports, enabling customized learning paths that cater to a person's strengths, weaknesses, alternatives, and learning styles. This personalization guarantees that inexperienced folks get hold of the most relevant and powerful content material, enhancing their engagement and motivation (Vrabie, C. (2024)). AI-powered adaptive gaining knowledge of structures provides actual-time feedback to college students, helping them apprehend their development and areas trying improvement. Immediate remarks allow college students to make accurate mistakes and red meat up gaining knowledge promptly, selling deeper know-how of the problem. Additionally, AI automates checks, supplying numerous question sorts and issue stages tailored to every scholar's expertise (Mello, R. F., Freitas, E., Pereira, F. D., Cabral, L., Tedesco, P., & Ramalho, G. (2023)). One of the massive benefits of AI in adaptive gaining knowledge is its capability to discover precise knowledge of gaps and regions in which students war. By pinpointing those gaps, AI can advise focused interventions and further resources to deal with deficiencies, ensuring that scholars collect the essential assistance earlier than falling behind. This proactive technique fosters a more inclusive and equitable mastering environment.

AI dynamically modifies how the instructional content material is furnished and solved, relying on the student's general widespread performance and development. The tool might provide extra tough substances to a scholar who excels in a particular location to maintain their interest. Alternatively, if a learner studies troubles, the machine can offer greater sports activities and sporting events to decorate statistics or rearrange the material. This flexibility ensures that getting to know stays in reality, is traumatic, and is effective for each pupil. To make analyzing more appealing and captivating, AI-powered adaptive getting-to-understand systems can also embody gamification, interactive simulations, and tailored guidelines, creating an awesome mindset towards mastery and intrinsic motivation through handing over interactive competencies and responding to college university college students' interests(Brackett, M. A., Rivers, S. E., & Salovey, P. (2011)). In addition, AI is aware of and allows a huge type of studying pattern, which includes kinesthetic, auditory, and visible. AI ensures that every college scholar can get admission and recognize the trouble efficiently by presenting several representations of the content fabric material cloth and a few academic methodologies.

AI can also decorate collaborative reading reports by forming corporations based on complementary capabilities and understanding tiers and presenting insights and guidelines for business enterprise dynamics, ensuring green and balanced interactions amongst institution members. This fosters teamwork, communiqué abilities, and deeper Information about the venture, relying on peer analysis. Moreover, AI generates unique analytics and reviews on students' performance, development, and engagement. Educators can use the insights to make informed alternatives about educational strategies, find out at-risk college students, and tailor their schooling strategies to meet the dreams of their novices. AI serves as an effective device for statistics-driven desire-making in training. The scalability and accessibility of AI-powered adaptive studying structures are notable, as they can accommodate huge numbers of university college students and make tremendous schooling handy to diverse populations, alongside the ones in far-flung or underserved regions. This scalability promotes instructional fairness and inclusion(Nguyen, A., Ngo, H. N., Hong, Y., Dang, B., & Nguyen, B. P. T. (2023)).

Additionally, AI structures continuously examine and decorate primarily based mostly on facts and customer interactions, ensuring that adaptive studying systems evolve and refine their algorithms to provide more accurate and powerful educational reviews. The ongoing improvement of AI generation guarantees that adaptive analyzing systems continue to be at the reducing fringe of educational innovation. Beyond the one immediate blessing, AI in adaptive getting to know helps a broader academic transformation. It enables work normalization for grading purposes; at this point, instructors will rely more on student contact and differentiated instruction. What it does with this revolution is satisfactory in improving the process of gaining knowledge and helping teachers apply progressive pedagogy and grow professionally. It will also entrench the development of an increased mindset of some of the novices. The opened structures of reticular reading dispositions bring to the surface the resolution and vibrancy through the introduction of the nagging overstates that probably get escalated to the capabilities of every scholar and providing support only when required and acknowledging the progress made in small steps. This enables college students to create self-conceptions and retain the relish of gaining data for the long term (Nguyen, A., Ngo, H. N., Hong, Y., Dang, B., & Nguyen, B. P. T. (2023)). Next to the domain of education, AI can help with decision-making regarding course and professional selection when students go to college through education aided by uniquely tailored recommendations and resources. Such a level of helpful, useful resource might also aid university students in choosing the right kind of education and career in general much more reliably and satisfactorily. AI attribute in administrative assistance embeds the efficiency features in administration towards institutional activities.

Through predictive analytics, AI can successfully forecast enrollment dispositions, control property, and enhance scholar retention prices. By reading the actualities of educational behavior and engagement, AI can distinguish patterns that purpose dropouts and prescribe medicinal tablets to protect college understudies heading in the proper direction. Moreover, AI-powered chatbots and virtual assistants can provide 24/7 assistance to college students, answering questions, offering steering, and directing them to applicable assets. This non-save-you useful aid complements the scholar's enjoyment, ensuring that help is generally available simultaneously as favored(Mishra, N., De, A., Nayak, R. K., Tripathy, R., Motwani, A., Thakur, A., & Rout, S. K. (2025)). In the research and development of indoor instructional establishments, AI can increase discoveries with the beneficial resource of analyzing widespread datasets, identifying tendencies, and generating insights that might be left out with the beneficial resource manner of human researchers. This functionality can stress innovation in pedagogy, curriculum layout, and academic generation, pushing the boundaries of what's possible in schooling. Furthermore, the placement of AI in adaptive gaining know-how is constantly evolving. As AI technology is enhanced, new applications and capabilities will emerge, further reworking the academic landscape. The capability for AI to revolutionize training is huge, and ongoing studies and improvement are probably key to unlocking this capability. Collaboration among educators, technologists, policymakers, and researchers is probably vital in shaping the future of AI in schooling. By strolling collectively, the stakeholders can ensure that the AI era is superior and deploy strategies that maximize their advantages at the same time as addressing ethical and sensible stressful conditions. In conclusion, the place of AI in adaptive studying is multifaceted and transformative, enhancing personalization, remarks, engagement, collaboration, and accessibility in education and helping educators and establishments with statistics-driven insights and operational efficiencies. On the non-obligatory hand, if a learner considers issues, the machine can offer additional sports exercises carrying sports exercises to improve records or improve the cloth. As AI adjusts, it will play a progressively more basic characteristic in forming the future of instruction, making it more responsive, comprehensive, and effective for amateurs(Ahmad, S., et al. (2023)).

1.2 Components of AI-powered adaptive learning system:

AI-powered adaptive studying systems comprise numerous interconnected additives that collectively create custom-designed and powerful knowledge of reminiscences. These components leverage superior eras to build up, study, and examine records to optimize the educational tool. The foundation of any AI-powered adaptive gaining understanding device is the powerful series and control of records. This consists of facts from several properties, encompassing pupil interactions, checks, mastering sports activities, activities, and doors property. Data manipulation systems ensure that the information is securely stored, prepared, and successfully to be had for assessment. Learner profiling consists of growing specific profiles of every pupil primarily based on their getting-to-recognize behaviors, alternatives, regular overall performance information, and applicable attributes. This profiling is done through non-save-you checking and assessment of understudy insights, permitting the device to develop many facts about all learners' precise goals and patterns(Chan, C. K. Y., & Tsi, L. H. (2023)).

The personalized content transport engine uses insights from learner profiles to tailor academic substances to each pupil's needs. This viewpoint powerfully alters the content's issue, organization, and sequencing to shape the learner's present-day degree of mastery and getting-to-know fashion. It ensures that scholars are always challenged at the precise level. Adaptive assessment mechanisms constantly examine scholar's overall performance via quizzes, tests, assignments, and other evaluation gear. The system affords real-time remarks, highlighting regions of strength and pinpointing areas that require similar interest. This continuous evaluation facilitates college students to see their progress and courses the system in adjusting and getting to know paths accordingly(Mishra, N., Habal, B. G. M., Garcia, P. S., & Garcia, M. B. (2024, June)). The advice engine indicates additional studying resources, sports, and interventions primarily based on the learner's profile and overall performance statistics. These pointers may also embody practice sporting events, movies, readings, or collaborative initiatives. The intention is to boost mastering, deal with gaps, and provide enrichment opportunities. Natural Language Processing (NLP) skills permit the machine to recognize and method human language, facilitating interactions between college students and the device. NLP is utilized in chatbots, virtual tutors, and extraordinary conversational interfaces that provide on-the-spot assistance, solution questions, and manual novices through their academic adventures.

Learning analytics comprise the systematic assessment of records to gain insights into knowledge. This hassle generates extraordinary reviews and dashboards for educators, administrators, and university students, presenting a whole view of improvement, engagement, and regular normal overall performance. These analytics are useful for resource information-driven desire-making and assist in discovering tendencies and patterns. Gamification integrates pastime-like factors into the reading, reveling in growth engagement and motivation. This consists of talents collectively with points, badges, leaderboards, and disturbing situations. By making analysis greater interactive and interesting, gamification encourages university and college students to stay engaged and gain from their studying goals(Soewito, B., Gaol, F. L., & Abdurachman, E. (2019, December)). Collaborative analyzing tools facilitate organization paintings and peer-to-peer interplay. This tool includes boards, enterprise tasks, and social studying platforms that permit college students to collaborate, share their know-how, and help each other form businesses based totally on complementary competencies. It also shows employer dynamics to make certain effective collaborations. AI-powered adaptive knowledge of systems regularly integrates with different emerging technologies, which include augmented truth (AR), digital fact (VR), and the Internet of Things (IoT). This technology creates immersive and interactive know-how experiences, improving the knowledge of complex standards and presenting hands-on exercises(Mishra, N., Nayak, R. K., Habtemariam, G. M., Tripathy, R., Esseynew, S., Motwani, A., & Rout, S. K. (2024)).

AI-powered adaptive analyzing systems additionally useful resource instructor and administrator roles by imparting equipment for curriculum planning, development monitoring, and useful resource management. This machine assists educators in knowing which techniques are used to take walks and where modifications are desired, permitting more information and powerful education practices. For instance, instructors can use statistics from the gadget to become privy to commonplace areas of issue among college students and regulate their instructional methods. The structures additionally provide expert development property, helping educators stay updated with contemporary pedagogical techniques and technological improvements. AI-powered adaptive systems know-how can facilitate lifelong reading by imparting possibilities for ongoing training and skills development. They can assist private amateurs seeking additional capabilities, look at unused capacities, or substitute careers with the valuable asset of conferring flexible, custom-planned, getting-to-apprehend ways that shape their special wants and plans. This adaptability is especially treasured in sudden converting assignment markets, wherein non-forestall gaining know-how and deskilling is vital(Gabriska, D., & Ölvecký, M. (2018, November)).

The potential of AI to method and look at huge datasets moreover opens up opportunities for custom-designed professional steerage. By recording learners' strengths, interests, and development, the device can advocate potential professional paths, applicable abilities to build up, and educational possibilities to pursue, supporting human beings to make informed choices about their expert futures. Moreover, mixing AI-powered flexible selecting up records of frameworks with automatic truth (VR) and multiplied truth (AR) time can make immersive choosing up statistics of environments. These situations can reenact actual international projections, conferring unpracticed males and females with sensible, palms-on satisfaction in a controlled, virtual setting. For instance, logical understudies can work out surgical strategies in an advanced going-for-storks room, on the upward push to time, as designing university college understudies can discover our complex machine in an automatic workshop. These immersive conclusions can beautify measurements and preserve complicated concerns, making examining more powerful and attractive(Bian, L., & Xie, Y. (2010)).

In summary, the particular components of AI-powered adaptive gaining knowledge of structures consist of facts collection and control, learner profiling, customized content material transport, adaptive assessment and comments, advice engines, natural language processing, learning analytics, gamification, collaborative gaining knowledge of tools, integration with rising technologies, security, and privacy measures, continuous improvement, and help for educators and lifelong rookies. These components paint collectively to create dynamic, personalized, and effective knowledge of revel in, leveraging the strength of AI to convert schooling. As these frameworks hold congruence, they maintain the capacity to make instruction additional, comprehensive, alluring, and mindful of the needs of individual fledglings and society.

1.3 Ethical considerations in AI-driven personalized learning:

Integrating AI in custom-designed getting-to-recognize allows for a large revolution in schooling by offering tailored instructional studies. This sort of transition, however, needs much more granularity of information and steps to the moral considerations that come into play(Mishra, N., Habtemariam, G. M., Aebissa, B., Nayak, R. K., Tripathy, R., Thakur, A., & Rout, S. K. (2024)).

Specific ways of getting these sorts of systems done morally include:

Privacy and security of information: To enhance privacy and security of documents, with AI systems, one will establish holomorphic encryption, permitting facts to be processed without dissolution. Therefore, all student security details could remain intact even at some obscure point in the future of an assessment process. In addition, federated records consolidate the facts stored locally on learners' devices; they communicate only the changes to models, thus reducing the chances of vital data garages and breaches. Consent UI interfaces could be created to express, in a clear and comprehensible manner, the objectives of data that learners in colleges and their parents utilize. By tailoring consent options, clients would choose precisely which facts and factors they consent to share, therefore feeling as though they are in a position to be in control and have transparency. Real-time updating changes to use facts ensures knowledgeable, continuous consent.

Equity and access: AI-based structures aspire to affect an adaptive age that is no longer the best for what the university students need to end up with but also for their technological right of entry. For instance, further fine-tuning of AI algorithms would allow them to work adequately with the low-bandwidth net or even offline modes, enabling students in places where connectivity might otherwise be below to participate in such technological boons. Partnerships through and via administrative and even nongovernmental places of work could engender distribution mechanisms for the devices key to these disadvantaged enterprises.

Algorithms vs. fairness: Involving multi-stakeholder input in developing AI systems can go a long way in detecting and mitigating capacity biases. An unbiased algorithm intentionally observes equity awareness while explicitly exhibiting and self-adjusting for bias in real-time by the algorithm, ensuring every pupil organization gets a hold of equity. Periodic equity audits with an independent property body bring transparency and accountability.

Transparency and explainability: AI systems can incorporate "glass box" methods by which preference-making strategies are open and intelligible to the clientele. This tool should allow students and faculty in universities to trace, visibly through the interactive features, lower-back AI recommendations to sources of data and logic. Tools for visualization, infographics, and plain language will render obscure, complex AI processes transparent to non-experts.

Empowerment of teachers and students: AI should support and confirm the teacher in what is or is not acceptable in personalizing recommendations concerning classroom dynamics, which will make their job stunning. Architectures will even provide a reflective learning journal for college students, through which AI enables the students to view the learning journey with comments against their reflections and promotes self-reflection and engagement. This partnership would balance AI recommendation and the human judgment process.

Rolling ethical review: We thus think of an ethical framework that is fluid and runs together where, against technology in a state of perpetual flux, we can at least be sure that systems of AI will remain calibrated to society's values. For instance, these may include ethical advisory forums, geolocated in a community, that periodically review how often its AI tools are being applied to groups in the aggregate, among other public comment mechanisms through which community stakeholders could raise concerns or guidelines. Educators should also be regularly upskilled about the ethics of artificial intelligence so that they are better poised, having the knowledge base and expertise relevant to proactively pacifying these ethical issues.

Long-Term Impact and Social Responsibility: AI systems shall be able to foster the growth of digital citizenship and responsible AI practices among university and college students. Probably, this has been done with the valuable inclusion into the curriculum of modules related to information AI ethics, personal privacy topics, and file safety. Primarily based doubtless reading, which includes development by students of their AI solutions for socially disturbing situations, makes one feel responsible and innovative. Value to the body's dynamic nature regarding employees' goals can also be assured through solid organization and school partnerships that bring about exercises such as critical thinking and moral reasoning.

Intelligent personalization-based learning paths: This necessitates AI-driven personalized learning frameworks with dynamic and flexible learning pathways responsive to perceptions of change in interests and available choices of college learners. Examples of such are enabling college students to switch from one learning-for-understanding pathway to another and, generally, advice based on fluctuating needs and aspirations of the individual, which AI systems enable to offer better-informed and more motivating levels of learning. It is in this very variation that the concept of student engagement and life-wide learning in general standardizes. The ethical concerns with AI-driven personalized learning point toward the need for modern and proactive strategies that give importance to privacy, equity, transparency, and empowerment. It could be achieved by embedding those appropriate practices in a way through which developers and educators can use the power of AI to create effective, inclusive, and ethically solid learning environments for the benefit of each learner(Ng, D. T. K. et al. (2023)).

Figure 1. Ethical considerations in AI-driven personalized learning

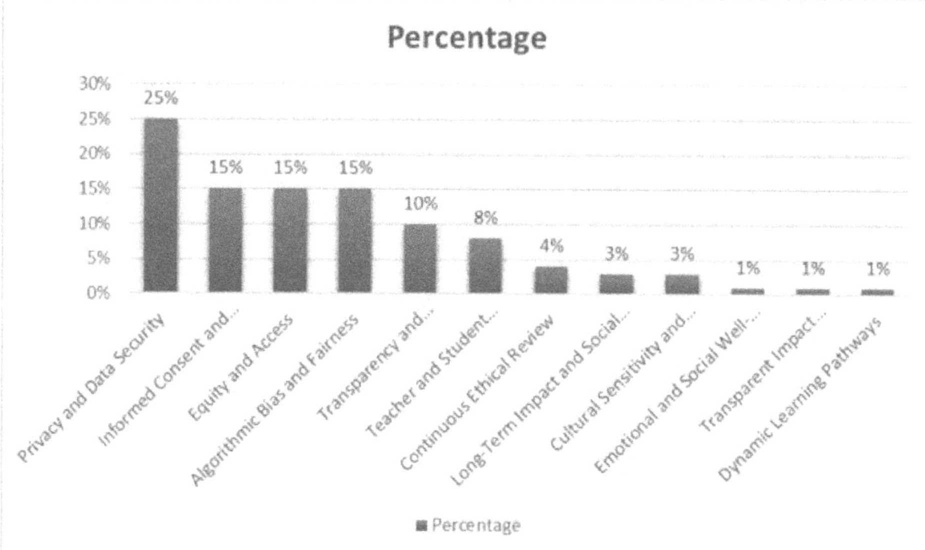

1.4 Comparative analysis of educational innovations

Upgrades in education have usually made learning methodologies greater, extended scholar engagement, and advanced academic results. Recent upgrades and a special curiosity in artificial intelligence, digital and augmented fact, gamification, and tailored learning frameworks prove especially potent. The review picks out those key enhancements and spots their effect, advantages, demanding situations, and future possibilities, which have converted schooling with the useful resource of permitting personalized studying tales tailored in response to non-public pupil desires. Tools and systems powered by AI read students' overall performance information, perceive studying gaps, and customize academic content material. This mainly designed device fosters engagement and gives centered aid, which university students need the most(Mishra, N., Habtemariam, G. M., Esseynew, S., Nayak, R. K., Tripathy, R., & Padhi, B. K. (2024)).

AI gives a few advantages, along with the potential for creating a custom-made understanding of pathways, automating grading and administrative duties that suck educators' time, and decorating accessibility for college kids with disabilities via adaptive getting-to-know materials. The advantages of VR/AR encompass heightened engagement via immersive studies, the capacity to visualize complex requirements, and opportunities for hands-on getting to know in a steady environment. Challenges consist of the excessive value of VR/AR tools, the need for vast technical assistance, and functionality problems with accessibility for all college students. As generations advance, VR/AR is poised to become more affordable, with extra integration into mainstream education predicted.

Gamification leverages pastime layout factors to make learning more engaging and motivating. Educators can decorate student motivation and participation by incorporating elements, badges, leaderboards, and superb sports mechanics. These frameworks utilize statistics analytics to conform academic content and pacing, ensuring every student receives customized assistance. The advantages of customized knowledge include the ability to cater to numerous knowledge patterns, increased scholar autonomy, self-paced knowledge, and progressed educational results through tailor-made education. Challenges encompass the want for robust information analytics infrastructure, ability issues with records privateers, and the hazard of widening the virtual divide if all students no longer have the same access to technology. The future of personalized learning will possibly involve extra advanced records analytics and AI to refine and beautify customization, focusing on ensuring equitable access for all students, as depicted in Figure 2 (Mishra, N., Habtemariam, G. M., Aebissa, B., Nayak, R. K., & Tripathy, R. (2024)).

Figure 2. Comparative Analysis of Educational Innovations

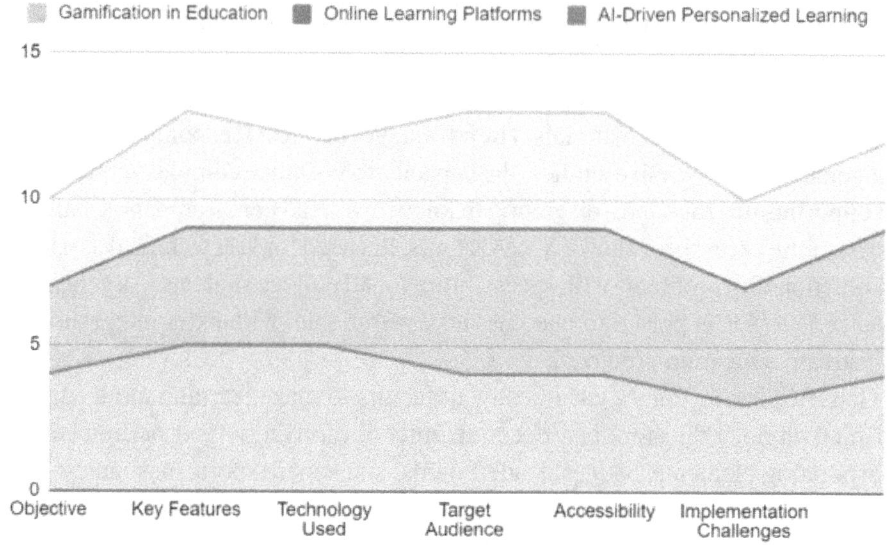

AI, VR/AR, gamification, and personalized studying frameworks constitute substantial academic innovations that can convert teaching and mastering. Each innovation gives specific benefits and faces distinct demanding situations. The destiny of training will likely contain a synergistic integration of this technology, emphasizing ethical considerations, accessibility, and ongoing research to optimize their effectiveness and ensure that all students gain from these advancements.

1.5 Potential benefits of AI-powered adaptive learning:

Adaptive knowledge of structures can incorporate gamification elements, rewards, badges, and leaderboards to make mastering extra attractive and motivating. By tailoring content material to university college students' interests and imparting interactive and immersive reports, AI-powered platforms can make learning fun and fun, growing student participation and staying energy. These insights permit educators to emerge aware of university college students who want to assist and allocate assets extra efficiently. Teachers and educators can gain from AI-powered adaptive studying by accessing specific analytics and reviews on scholar performance. Teachers can pay attention to their time and efforts on customized guidance and intervention, enhancing typical lecture room effectiveness(Ferrara, E. (2023)).

1.6 Challenges and Considerations

Whereas AI-powered flexible systems deliver a few devices, in the boom, they display a few demanding times and worries that need to be addressed to guarantee their feasible utilization and be taken into attention in tutoring. These demanding situations span technical, ethical, and practical dimensions, and addressing them is vital for identifying the overall functionality of those systems. One of the toughest conditions is making sure of the privacy and safety of student information. AI-powered systems rely upon thoughts-blowing data series and assessments to customize studying studies.

The virtual divide remains a good-sized venture in adopting AI-powered adaptive learning structures. Not all college students have the same right access to the crucial generation and net connectivity that could exacerbate contemporary educational inequities. Addressing these calls for concerted efforts to offer proper access to devices and a dependable net for all college university students, regardless of socioeconomic status. Implementing AI-powered adaptive getting-to-know structures also can be high-priced. Schools must make ventures inside the basic framework, which envelops equipment, program computer programs, and progress security. Furthermore, there may be prices related to organizing and efficiently using that unused gadget. Funding and beneficial resource allocation are critical concerns for colleges adopting that technology (Barhate, D., Pathak, S., & Dubey, A. K. (2023)). Instructors want to be thoroughly skilled for AI-powered adaptive reading systems to be effective. This includes recording how to interpret records and insights provided through those systems, integrating them into their education practices, and maintaining stability amongst technology use and traditional education techniques. Progressing gifted enhancement and manuals are important for teachers to maximize the advantages of those frameworks. The ethical implications of AI in education are complicated. Issues of pupil autonomy, consent, and the potential of AI to update human instructors need careful interest. Developing ethical hints and frameworks for using AI in schooling is crucial to ensure that the most effective era is used responsibly and in strategies that address the issue of undermining the instructional diploma.

Finally, the reliability and validity of AI-powered adaptive studying systems must be ensured to preserve instructional requirements. This consists of validating the accuracy of the algorithms, ensuring the consistency of the adaptive getting-to-know stories, and regularly updating the structures to consist of extremely modern-day educational research and practices(Pessach, D., & Shmueli, E. (2023)). Seamless integration requires compatibility with modern-day curriculum requirements and alignment with pedagogical desires, which may be tough given the shape of instructional structures and practices worldwide. Additionally, the extended-time period sustainability of AI-powered adaptive getting-to-understand systems needs to be considered. This includes planning for non-forestall specialized help, gadget overhauls, and tending to the capability of out-of-date excellence for the duration. Tools and educational establishments need to increase strategies for the non-maintain-your evolution of those structures to ensure they live successfully and are applicable in the face of rapid technological upgrades. Addressing those demanding situations and concerns is critical for the hit integration of AI-powered adaptive mastering systems in training. By proactively tackling issues related to facts privacy, equity, cost, instructor training, ethical implications, system reliability, integration, and sustainability, educational stakeholders can harness the whole ability of AI to supply personalized, green, and equitable studying reports for all college students.

2. LITERATURE REVIEW

The speedy upgrades in artificial intelligence (AI) and tool reading have paved the way for full-size changes in diverse sectors encompassing schooling. Over the past few years, traditional academic models have increasingly been supplemented and, in a few cases, modified, using technology-more studying environments. Among the improvements, AI-powered adaptive learning structures have emerged as a promising tool to customize training and deal with the various needs of novices. Adaptive studying, even though a cutting-edge day-to-day buzzword, has roots that hint at numerous instances. Initially, the concept of adaptive studying became explored through easy rule-primarily based total systems and programmed steerage strategies(Bilquise, G., & Shaalan, K. (2022)). The creation of computer-assisted preparation (CAI) in the Nineteen Sixties marked a superb milestone in the evolution of adaptive reading. CAI systems furnished custom-designed drills and exercise sports activities, adapting the trouble degree based mostly on scholar-accepted universal overall performance. Despite those advances, the structures were constrained by the technological constraints of the time and frequently did not supply individualized learning research. The competencies of those structures were confined to predetermined pathways, lacking the dynamic adaptability required to address the several paces and options of different inexperienced humans(Mishra, N., Habtemariam, G. M., & De, A. (2023)).

AI-powered adaptive data of structures has examined the extensive ability of pupils to enhance their ability to realize consequences and engagement. Research has shown that personalized getting-to-recognize techniques can improve instructional performance and increase motivation and retention charges. This actual-time comments loop is important for maintaining scholar motivation and ensuring non-forestall improvement. By adopting the issue of diploma and content material fabric based on individual everyday overall performance, those structures keep college students from becoming bored with content material fabric cloth that is too smooth or beaten by utilizing a manner of content material fabric that is too hard (Mello, R. F., Freitas, E., Pereira, F. D., Cabral, L., Tedesco, P., & Ramalho, G. (2023)).

Today, AI-powered adaptive reading systems are being applied in various educational settings, from K-12 schools to higher-training establishments and company training applications. These frameworks are outlined to provide customized getting-to-understand assessments by regularly adjusting to each learner's personal needs and development. Modern adaptive analyzing structures leverage sophisticated algorithms to investigate scholars' overall performance facts, understand know-how gaps, and regulate the getting-to-recognize route for that reason. This guarantees that every pupil receives instruction and assets tailored to their precise goals, selling better engagement and academic achievement. For instance, academic institutions have extensively followed systems like DreamBox, Smart Sparrow, and Newton to deliver personalized studying studies(Prabhakaran, V., Qadri, R., & Hutchinson, B. (2022)).

AI-powered adaptive expertise of systems has tested significant capability in enhancing scholar getting-to-realize effects and engagement. Inquire how customized getting-to-know processes can improve education by employing huge execution, extended ideas, and better upkeep prices.

AI-powered adaptive studying structures additionally play a crucial role in helping teachers enhance their academic capabilities. By automating ordinary responsibilities, including grading and development tracking, those systems unfastened precious time for instructors to pay attention to more significant sports, such as supplying personalized assistance and fostering pupil relationships. Additionally, AI-driven analytics offer teachers insights into students' overall performance and analyzing styles, permitting them to make data-informed academic picks. Professional improvement is another area wherein AI-powered structures should have a huge impact. Instructors can pick up from customized preparation programs that adjust to their character dreams and potential ranges. These applications can supply targeted property, suggestions, and comments to assist teachers in improving their instructive homes and correctly coordinating eras in their lecture rooms. Furthermore, collaborative platforms supported through AI can facilitate peer-to-peer knowledge of and sharing excellent practices among educators(Mello, R. F., Freitas, E., Pereira, F. D., Cabral, L., Tedesco, P., & Ramalho, G. (2023)).

The virtual divide remains a massive barrier to the extensive adoption of AI-powered adaptive learning structures. Addressing this trouble requires centered interventions to offer essential resources to underprivileged students. Schools and governments ought to work collectively to ensure that every college student has a right to enter the equipment and assets to benefit from AI-powered studying systems. The ethical implications of AI in training are complicated and multifaceted.

Issues including scholar autonomy, consent, and the ability of AI to update human instructors need careful attention (Boaye Belle, A. (2024, April)).

The destiny of AI-powered adaptive reading systems holds huge capability for innovation and effect. Continued enhancements in AI and machine-gaining knowledge of the era will possibly bring about even more modern-day and effective adaptive reading structures. Emerging technology, including herbal language processing (NLP) and PC imaginative and prescient, could allow more interactive and immersive studying tales, similarly improving student engagement and information. Moreover, combining AI with distinct emerging eras, including virtual truth (VR) and augmented reality (AR), may also revolutionize how college students interact with academic content. VR and AR can provide immersive and experiential record opportunities, permitting college college college students to find out complicated requirements and environments in techniques that have been formerly no longer possible and may enhance the file with the useful resource of adapting them to individual studying dreams and supplying real-time feedback and steerage(Wang, N., & Lester, J. (2023)).

Moreover, the ethical implications of AI in schooling require careful hobby and the improvement of strong hints and frameworks. Looking in advance, persevered innovation and collaboration may be key to harnessing the general capability of AI-powered adaptive reading structures and growing more powerful, attractive, and inclusive educational environments for all inexperienced folks(Saaida, M. B. (2023))(Kaliraj, P., Singaravelu, G., & Devi, T. (Eds.). (2024)).

3. CASE STUDY

Case Study 1: Implementation in K-12 Education as visualized in Figure 3.

Background: In 2024, a suburban college district in California introduced an AI-powered adaptive analyzing platform, DreamBox, into its K-five school rooms. The district aimed to cope with varying pupil capabilities and gain knowledge of places, mainly in arithmetic.

Implementation: Dream Box was included in the prevailing curriculum, presenting customized math classes tailor-made to every scholar's skill degree and gaining knowledge of style. The platform constantly assesses student performance through interactive sports, adjusting the difficulty and troubles primarily based on real-time statistics. Teachers received exact analytics and insights into pupil improvement, allowing them to select struggling students and tailor their exercises.

Outcomes:

Improved Performance: After one educational 12 months, standardized take a look at scores in arithmetic showed a massive development. The variety of college students assembling or exceeding state requirements is superior via 20%.

Engagement: Students referred to higher degrees of engagement and leisure in math commands. The interactive and assignment-like nature of the platform stored college students encouraged and fascinated.

Educator Support: Teachers selected the platform's analytics as precious for inspecting crevices and altering their tips strategies. The time stored on grading and development monitoring allowed for added individualized scholar assistance.

Equity: The device helped bridge the hollow through the usage of more useful sources to university students who've been within the once more, making sure they obtained the vital resources to seize up with their friends.

Figure 3. Implementation in K-12 Education

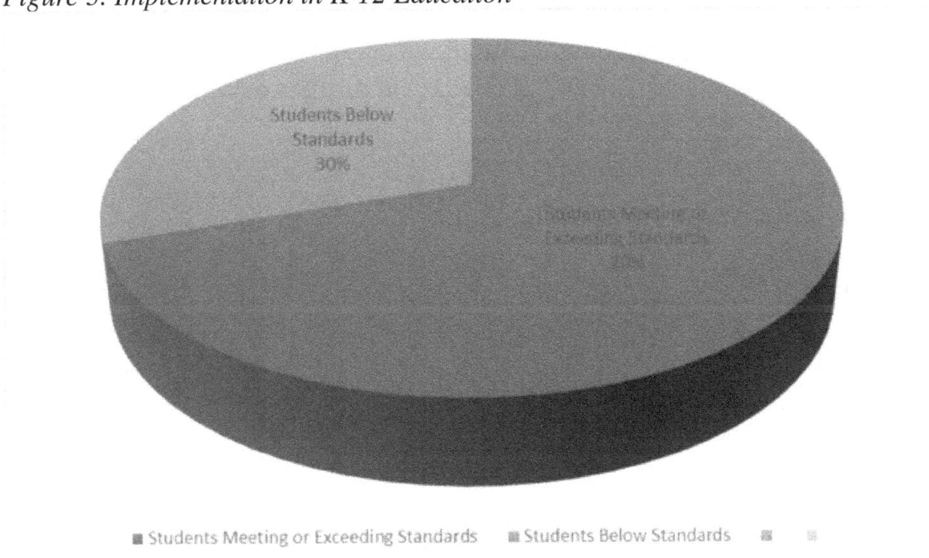

Case Study 2: Application in Higher Education as depicted in Figure 4.

Background: In 2024, a prestigious university in New York carried out Knewton, an AI-powered adaptive mastering platform, in its introductory biology publications. The purpose has been to enhance scholar mastering results and offer customized learning in big lecture training.

Implementation: Knewton became protected into the direction management tool, providing adaptive knowledge of modules that complemented the traditional lectures. The platform customized the content material primarily based on all pupil's strengths and weaknesses, as recognized via non-stop checks and interactions.

Professors received special reviews on scholars' normal performance, letting them cope with commonplace misconceptions and tailor their coaching techniques.

Outcomes: Enhanced Learning Outcomes: The average final exam rankings stepped forward using a manner of 15% in assessment to the previous 12 months. Students who often engaged with the platform confirmed the most sizable improvement.

Understudy Satisfaction: Surveys demonstrated that understudies favored the tweaked method and felt additional guaranteed in their truths on complex themes. The instant feedback and tailored exercise problems have been highlighted as particularly useful.

Scalability: The platform's potential to deal with large volumes of college students without compromising the exceptional of personalized learning made it best for huge introductory publications. This scalability ensured that every student acquired the assistance they wanted, irrespective of elegance.

Faculty Benefits: Professors valued the positive insights into pupil performance, which allowed for fact-driven selection-making in their teaching. The stage also encouraged more prominent intuition amid the out-of-work hours, as college understudies need to seek assistance, particularly in districts recognized by utilizing the gadget.

Figure 4. Application in Higher Education

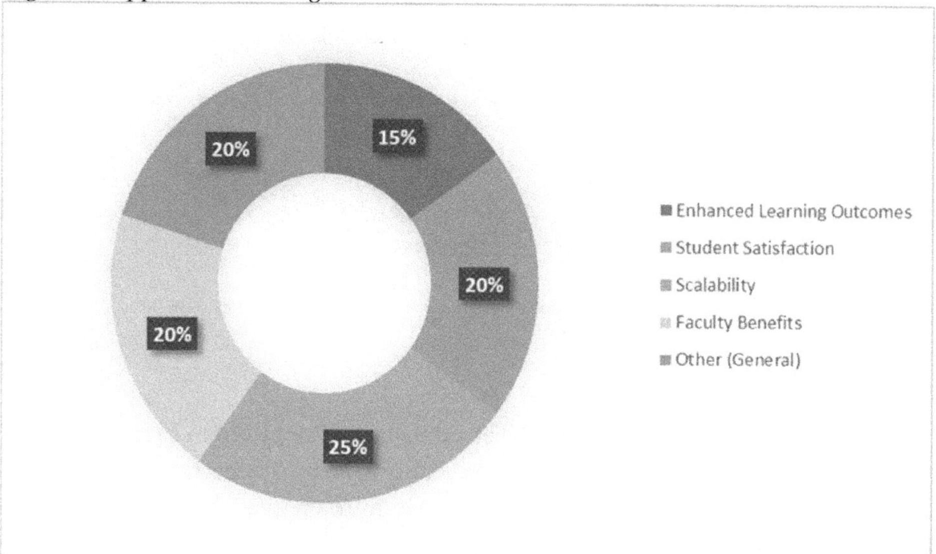

4. DISCUSSION

Deployed evidence of AI-powered adaptive learning structures, attained via DreamBox in K-12 training and Newton in better training, show their transformative functionality and inherent disturbing situations. Those systems will seek to provide whoever needs them with the appropriate educational stories that suit one's master needs more correctly than the traditional techniques. One of the great outcomes in both cases is the amazing rise in pupil performance. The introduction of Dream Box into K-12 schooling increased the number of students who met or exceeded country standards in mathematics by 20%. In the same way, the use of Newton in better training resulted in a 15% growth in the final examination ranking compared to standard academic strategies. These improvements mean adaptive getting-to-know structures effectively personalize academic opinions to male and female students' wants. By continually evaluating performance and fine-tuning the content material in real-time, those structures ensure that scholars collect difficult steerage correctly, optimizing getting-to-realize consequences.

The level of engagement also rose with the application of those models. Dream Box's interactive, game-like atmosphere made mathematics instructions particularly engaging to younger college students, engaging them and motivating them further. In better training, Knewton's customized remarks and adaptive practice problems did much to make students feel extra assured and invested in their mastering. The instant feedback loop provided by the one's structures maintains scholars engaged and encourages a deeper understanding of the material. Therefore, AI-powered adaptive information systems can greatly support the educator in terms of specific analytics and insights into students' normal performance. In the K-12 case, instructors should identify mastering gaps and display development more successfully, permitting them to tailor their education to satisfy individual pupil needs. This freed up time that might, in any other case, be spent on grading and administrative obligations, permitting instructors to offer greater custom-designed beneficial sources to college students. In higher education, Newton's ability to handle huge volumes of students without compromising the greatness of personalized learning became especially useful for big introductory courses. Professors appreciated the special performance reviews, facilitating information-pushed selection-making and more significant interactions during workplace hours. These systems additionally helped educators become aware of unusual misconceptions and cope with them directly, enhancing the general niceness of practice.

Promoting equity in education is a huge gain of AI-powered adaptive knowledge of systems. By imparting more guidance to university college students who're inside the again of, the one's structures assist in bridging the fulfillment hole and ensuring that each student has the opportunity to triumph. The capability to tailor instruction to character dreams is, in particular, valuable in several schoolrooms wherein college students can also have varying levels of previous records and gain knowledge of paces. For instance, Dream Box's adaptive studying surroundings ensure that scholars who work with precise math thoughts gather targeted practice and guidance. This customized technique prevents university college students from falling behind and fosters a more inclusive learning environment. Similarly, Knewton's adaptive modules in higher training provide custom-designed learning tales for an extensive style of college students, ensuring that each man or woman can develop at their very personal pace.

Despite the numerous benefits, implementing AI-powered adaptive studying structures offers several demanding conditions. Data privateers and protection are paramount problems, as the ones systems accumulate and take a look at widespread portions of student facts. Ensuring the confidentiality and integrity of this information is critical to retaining recollect and complying with guidelines. Educational institutions must place sturdy facts and safety measures into impact and train stakeholders about the significance of statistics privateers. Algorithmic bias is a different important problem structure that is fantastic as independent due to the facts they're experts on. If the training information consists of biases, the AI can perpetuate or even exacerbate the biases, which is critical to unfair educational effects. Ensuring equity requires careful choice of training statistics, continuous monitoring for biases, and enforcing algorithms that sell fairness and inclusivity. Addressing one's biases is crucial to prevent the reinforcement of gift inequalities. The virtual divide remains a massive barrier to adopting AI-powered adaptive studying systems. Not all college students have identical rights to entry into the vital era and internet connectivity, which may exacerbate gift academic inequities. Targeted interventions are needed to offer important assets to underprivileged college students, making sure that they're capable of taking advantage of those superior studying systems. Schools and governments need to collaborate to make certain equitable get right of get right of entry to era and belongings for all college and university students.

The destiny of AI-powered adaptive mastering systems holds massive ability for innovation and impact. Continued advancements in AI and system studying generation will likely bring about even greater state-of-the-art and effective adaptive reading systems. Emerging technology consisting of herbal language processing (NLP) and pc imaginative and prescient have to allow more interactive and immersive learning studies, similarly enhancing student engagement and records. Moreover, the integration of AI with a specific rising generation, which consists of virtual reality (VR) and augmented truth (AR), must revolutionize the way university college students interact with academic content material. VR and AR can offer immersive and experiential studying possibilities, permitting university college students to discover complex thoughts and environments in strategies that had been previously no longer feasible and can decorate those reports with the useful resource of adapting them to character, gaining knowledge of needs and providing real-time comments and guidance. The ethical implications of AI in training require careful interest and the development of strong recommendations and frameworks. Looking ahead, chronic innovation and collaboration may be key to harnessing the entire functionality of AI-powered adaptive getting-to-know structures and developing powerful, engaging, and inclusive educational environments for all green human beings.

5. FUTURE SCOPES

The future of AI-powered adaptive getting-to-know systems is poised to revolutionize schooling, growing excellent possibilities for customized, appealing, and green getting-to-know. As AI technology helps, their integration into academic environments will launch new dimensions of getting-to-know reviews. One promising location is the enhancement of AI tutors through herbal language processing (NLP). Imagine AI tutors that communicate with college students in herbal, intuitive techniques and facts and respond to complex questions in equal time by supplying tailored factors based totally on a person analyzing patterns. This also can ensure that every student, whether or not suffering or excelling, receives the tremendous diploma of task and manual. Moreover, integrating AI with a laptop, imaginative and prescient, can rework how we monitor and reply to pupil engagement. Real-time video analysis must gauge scholar recognition and emotional states, permitting the gadget to dynamically adjust instructions. For instance, if signs of misunderstanding or boredom are detected, the AI may want to introduce interactive sports or modify the lesson to re-engage the pupil. The aggregate of AI with digital reality (VR) and augmented reality (AR) guarantees to create an immersive understanding of environments. Students must explore three-D models of complex structures, including the human body in biology, with an AI guide supplying personalized records and quizzes tailor-made to their information stage. This arms-on, intuitive procedure needs to make learning more substantial and remarkable.

AI's feature in lifelong studying is further exciting. In a hastily evolving interest marketplace, AI-pushed systems should offer personalized reading paths that adapt to a learner's professional progression, imparting focused publications and real-time comments. Experts might moreover want to accumulate overhauls on great modern capacities and know-how, making past any doubt they stay at the leading edge of their areas. Promoting academic equity is every other giant functionality benefit. AI-powered adaptive understanding of systems can democratize schooling by imparting awesome, custom-designed reading stories regardless of a pupil's historical past or area. These structures may be especially transformative in beneath-resourced faculties, which can supplement traditional coaching and supply extra useful property tailor-made to every pupil's desires. However, demanding situations that ensure facts are private, mitigate algorithmic bias, and bridge the digital divide must be addressed to comprehend these benefits. Collaboration among educators, technologists, and policymakers may be critical to harnessing the ability of AI in education ethically and successfully, developing extra inclusive and effective studying environments for all university students, and setting the various newbies.

6. CONCLUSION

AI-powered adaptive thinking frameworks are a wave of sea change in scholarly innovation, thus manifesting the potential to trade ordinary acing-exceptional models. These systems give custom-planned memories tailor-made to character needs, upgrading researcher engagement and all-inclusive execution. It continually assesses and adapts to the student's pace and design of gaining knowledge, ensuring that tutoring comes out as it should be rigorous and consistent. AI is married with generation and found in instructional methods in domains related to natural language processing, laptop vision, digital truth, and augmented reality, assuring extra interactive and immersive school rooms. Now, abstraction of the mind will be more concrete and attractive to the learners, who will develop a better tenacity of fact and retention of the material. In addition, AI-enabled systems support faculties in presenting real analytics and insights that accommodate more knowledgeable and robust schooling techniques. Besides, they hold great potential for lifelong learning and professional development based on tailored learning paths that work by a person's professional development and functionality needs.

One of the major strong points that these systems can thus boast of is that opportunities are now available for instructional equity. It is possible to provide great, personalized worldwide exposure to so many different students at different stages of previous knowledge using AI, thus helping to close success gaps and ensuring every inexperienced human being can be triumphant. This would be greatest if it helped in low-resource instructional settings, where adding a layer of adaptive study systems to traditional coaching could indeed provide focused help. However, an equally big goal is ensuring these systems are used ethically and equitably, with sturdy measures in place to protect the facts and desires for close surveilling to minimize bias. The ambitious goal is to ensure all students are in on this most important eras. In conclusion, adaptive AI is an improvement in get-to-know systems that can enhance in numerous ways the quality of academics and lifelong reading while driving equity in learning. Advancement may be very important, however, by adding some more ethical hitches coupled with a confluence of input from educators, technologists, and policymakers in realizing enhanced, efficient, engaging, and inclusive learning environments.

REFERENCES

Ahmad, S., Umirzakova, S., Mujtaba, G., Amin, M. S., & Whangbo, T. (2023). Education 5.0: requirements, enabling technologies, and future directions. arXiv preprint arXiv:2307.15846.

Atan, S. (2018, December). Adaptive Reading Comprehension (ARC) System for Singapore Malay Language Students. In 2018 International Conference on Machine Learning and Data Engineering (iCMLDE) (pp. 97-102). IEEE.

Barhate, D., Pathak, S., & Dubey, A. K. (2023). Hyperparameter-tuned batch-updated stochastic gradient descent: Plant species identification by using hybrid deep learning. *Ecological Informatics*, 75, 102094. DOI: 10.1016/j.ecoinf.2023.102094

Bian, L., & Xie, Y. (2010). Research on the adaptive strategy of adaptive learning system. In Entertainment for Education. Digital Techniques and Systems: 5th International Conference on E-learning and Games, Edutainment 2010, Changchun, China, August 16-18, 2010. [Springer Berlin Heidelberg.]. *Proceedings*, 5, 203–214.

Bilquise, G., & Shaalan, K. (2022). AI-based academic advising framework: A knowledge management perspective. *International Journal of Advanced Computer Science and Applications*, 13(8). Advance online publication. DOI: 10.14569/IJACSA.2022.0130823

Boaye Belle, A. (2024, April). Assuring the emotional and cultural intelligence of intelligent software systems. In Proceedings of the 1st IEEE/ACM Workshop on Multi-disciplinary, Open, and RElevant Requirements Engineering (pp. 22-23). DOI: 10.1145/3643666.3648575

Brackett, M. A., Rivers, S. E., & Salovey, P. (2011). Emotional intelligence: Implications for personal, social, academic, and workplace success. *Social and Personality Psychology Compass*, 5(1), 88–103. DOI: 10.1111/j.1751-9004.2010.00334.x

Bulut, O., Beiting-Parrish, M., Casabianca, J. M., Slater, S. C., Jiao, H., Song, D., . . . Morilova, P. (2024). The rise of artificial intelligence in educational measurement: Opportunities and ethical challenges. arXiv preprint arXiv:2406.18900.

Chan, C. K. Y. (2023). Is AI changing the rules of academic misconduct? An in-depth look at students' perceptions of'AI-giarism'. arXiv preprint arXiv:2306.03358.

Chan, C. K. Y., & Tsi, L. H. (2023). The AI revolution in education: Will AI replace or assist teachers in higher education? arXiv preprint arXiv:2305.01185.

Ferrara, E. (2023). Fairness and bias in artificial intelligence: A brief survey of sources, impacts, and mitigation strategies. *Sci*, 6(1), 3. DOI: 10.3390/sci6010003

Gabriska, D., & Ölvecký, M. (2018, November). Issues of adaptive interfaces and their use in educational systems. In 2018 16th International Conference on Emerging eLearning Technologies and Applications (ICETA) (pp. 173-178). IEEE. DOI: 10.1109/ICETA.2018.8572096

Kaliraj, P., Singaravelu, G., & Devi, T. (Eds.). (2024). *Transformative Digital Technology for Disruptive Teaching and Learning*. CRC Press. DOI: 10.1201/9781032675176

Mello, R. F., Freitas, E., Pereira, F. D., Cabral, L., Tedesco, P., & Ramalho, G. (2023). Education in the age of Generative AI: Context and Recent Developments. arXiv preprint arXiv:2309.12332.

Mishra, N., De, A., Nayak, R. K., Tripathy, R., Motwani, A., Thakur, A., & Rout, S. K. (2025). AI Applications in Sports Medicine: Unraveling the Influence of Physical Education on Sports Injury Outcomes. In Global Innovations in Physical Education and Health (pp. 351-380). IGI Global.

Mishra, N., Habal, B. G. M., Garcia, P. S., & Garcia, M. B. (2024, June). Harnessing an AI-Driven Analytics Model to Optimize Training and Treatment in Physical Education for Sports Injury Prevention. In *Proceedings of the 2024 8th International Conference on Education and Multimedia Technology* (pp. 309-315). DOI: 10.1145/3678726.3678740

Mishra, N., Habtemariam, G. M., Aebissa, B., Nayak, R. K., & Tripathy, R. (2024). Assessing the Effectiveness of Transnational Leadership on the Performance of Ethiopian University Graduates in Computing Technology: A Case Study. In Engaging Higher Education Teachers and Students With Transnational Leadership (pp. 238-256). IGI Global. DOI: 10.4018/979-8-3693-6100-9.ch013

Mishra, N., Habtemariam, G. M., Aebissa, B., Nayak, R. K., Tripathy, R., Thakur, A., & Rout, S. K. (2024). Case Study on the Performance Evaluation of Graduates in the Field of Artificial Intelligence for Ethiopian Universities. In Teaching and Assessment in the Era of Education 5.0 (pp. 93-110). IGI Global. DOI: 10.4018/979-8-3693-3045-6.ch006

Mishra, N., Habtemariam, G. M., & De, A. (2023). Investigation of High-Performance Computing Tools for Higher Education Institutions Using the IoE Grid Computing Framework. In *Internet of Behaviors Implementation in Organizational Contexts* (pp. 217–241). IGI Global. DOI: 10.4018/978-1-6684-9039-6.ch011

Mishra, N., Habtemariam, G. M., Esseynew, S., Nayak, R. K., Tripathy, R., & Padhi, B. K. (2024). The Transnational ICT Leadership Assessment Based on the Available Infrastructure in the Governmental Organizations: A Case Study of the Ethiopian Northern Shewa Zone. In Developments and Future Trends in Transnational Higher Education Leadership (pp. 301-322). IGI Global. DOI: 10.4018/979-8-3693-2857-6.ch017

Mishra, N., Nayak, R. K., Habtemariam, G. M., Tripathy, R., Esseynew, S., Motwani, A., & Rout, S. K. (2024). An In-Depth Examination of the Application Gap and Infrastructure Landscape of Information Communication Technology in Ethiopian Public Organizations: A Bridging the Divide Case. In *Impact of Corporate Social Responsibility on Employee Wellbeing* (pp. 141–166). IGI Global. DOI: 10.4018/979-8-3693-3470-6.ch007

Mohammadkarimi, E. (2023). Teachers' reflections on academic dishonesty in EFL students' writings in the era of artificial intelligence. Journal of Applied Learning and Teaching, 6(2). DeFranco, J. F., & Biersmith, L. (2024). Assessing the State of AI Policy. arXiv preprint arXiv:2407.21717.

Ng, D. T. K., Leung, J. K. L., Su, J., Ng, R. C. W., & Chu, S. K. W. (2023). Teachers' AI digital competencies and twenty-first century skills in the post-pandemic world. *Educational Technology Research and Development*, 71(1), 137–161. DOI: 10.1007/s11423-023-10203-6 PMID: 36844361

Nguyen, A., Ngo, H. N., Hong, Y., Dang, B., & Nguyen, B. P. T. (2023). Ethical principles for artificial intelligence in education. *Education and Information Technologies*, 28(4), 4221–4241. DOI: 10.1007/s10639-022-11316-w PMID: 36254344

Nguyen, A., Ngo, H. N., Hong, Y., Dang, B., & Nguyen, B. P. T. (2023). Ethical principles for artificial intelligence in education. *Education and Information Technologies*, 28(4), 4221–4241. DOI: 10.1007/s10639-022-11316-w PMID: 36254344

Pessach, D., & Shmueli, E. (2023). Algorithmic fairness. In *Machine Learning for Data Science Handbook: Data Mining and Knowledge Discovery Handbook* (pp. 867–886). Springer International Publishing. DOI: 10.1007/978-3-031-24628-9_37

Prabhakaran, V., Qadri, R., & Hutchinson, B. (2022). Cultural incongruencies in artificial intelligence. arXiv preprint arXiv:2211.13069.

Saaida, M. B. (2023). AI-Driven transformations in higher education: Opportunities and challenges. *International Journal of Educational Research and Studies*, 5(1), 29–36.

Soewito, B., Gaol, F. L., & Abdurachman, E. (2019, December). Risk Assessment on Cloud Computing for The Learning System in The Education Environment. In *2019 IEEE International Conference on Engineering, Technology and Education (TALE)* (pp. 1-6). IEEE.

Vrabie, C. (2024). Artificial Intelligence from Idea to Implementation. How Can AI Reshape the Education Landscape? arXiv preprint arXiv:2407.20236.

Wang, N., & Lester, J. (2023). K-12 Education in the Age of AI: A Call to Action for K-12 AI Literacy. *International Journal of Artificial Intelligence in Education*, 33(2), 228–232. DOI: 10.1007/s40593-023-00358-x

Yoke, S. K., Ahmad, T. A., & Hasan, N. H. (2019). Exploring the potential of augmented reality in English for report writing: A perceptive overview. *International Journal of Education*, 4(33), 13–21.

Chapter 14
Ethical and Social Issues of AI in Education

T. Venkat Narayana Rao
https://orcid.org/0000-0002-1996-1819
Sreenidhi Institute of Science and Technology, India

B. Keerthi Madhuri
https://orcid.org/0009-0009-7151-1258
Sreenidhi Institute of Science and Technology, India

S. Bhavana
https://orcid.org/0009-0005-6134-1081
Sreenidhi Institute of Science and Technology, India

Chenchu Swetha
Sreenidhi Institute of Science and Technology, India

ABSTRACT

As AI becomes more common in education, it opens exciting possibilities, like personalized learning and greater access to resources. But it also brings serious concerns, including how student data is used. AI systems gather and manage vast volumes of student-related personal data, posing several security and confidentiality risks. When the newest technologies are employed more widely by wealthy schools, the implementation of AI in education would worsen existing inequality by putting less resource-rich institutions in a tough spot and causing educational gaps to widen. Using AI in the classroom may change the roles that teachers normally assign with the more human interaction and assistance In conclusion, using AI in education brings serious concerns like data privacy, biased algorithms, and unequal access, which could deepen existing gaps. There is also worry about over-relying on technology, risking the role of teachers, and limiting critical thinking.

DOI: 10.4018/979-8-3373-2262-9.ch014

To address this, we need policies that prioritize people, inclusive designs, and a strong commitment to fairness.

1. INTRODUCTION

Education is changing, and Artificial Intelligence (AI) is at the heart of this transformation. With its ability to personalize learning and simplify classroom tasks, AI is reshaping the way teachers teach and students learn, opening exciting possibilities for the future of education.

For students, AI offers a more tailored experience. It adjusts to their individual pace, focuses on their strengths, and supports them in areas where they might struggle. Tools like language translation, speech recognition, and assistive technologies for students with disabilities are making learning more inclusive and accessible than ever before. For teachers, AI takes on time-consuming tasks like grading and managing schedules, freeing them up to connect more with students and improve their teaching methods (Akgun et.al,2022).

But alongside these benefits come some important challenges. How do we protect student privacy when AI systems need so much data to work effectively? Can we ensure these systems are fair and unbiased? And what about students who don't have access to the technology—will AI create new opportunities or deepen the divide?

AI is also changing the roles of both teachers and students. Teachers are becoming guides and mentors, helping students navigate personalized, tech-driven learning environments. Students, meanwhile, are learning to make the most of AI tools without losing the critical thinking and problem-solving skills they need to succeed in the real world.

This chapter takes a closer look at the impact of AI on education—the opportunities it creates, the challenges it poses, and the ways we can use it responsibly to build a better future for learning. By addressing these questions and understanding the balance between promise and responsibility, we can ensure that AI enhances education for everyone, leaving no student behind.

Figure 1. Biasing issues while with students

2. HISTORY

Artificial Intelligence (AI) is rapidly transforming the way we teach and learn, creating opportunities to improve education like never before. From personalized learning experiences to smarter classroom management, AI has the potential to redefine how students and teachers interact with knowledge and each other.

The journey of AI in education started decades ago with early experiments in intelligent tutoring systems, but it's only in recent years, with advances in machine learning and big data, that its impact has become truly profound. Today, AI can adjust lesson plans to suit each student's unique learning pace, help teachers identify students who need extra support, and even translate content to make it accessible to learners across different languages.

For teachers, AI takes over repetitive tasks like grading and attendance, freeing up time to focus on the things that matter most—building relationships with students and developing innovative teaching methods. For students, it offers tools that can adapt to their individual needs, providing personalized learning paths that help them succeed in ways traditional systems often can't.

However, with these benefits come some serious challenges. To make AI work for everyone, we need to address issues like data privacy, fairness, and access. AI systems rely on collecting vast amounts of sensitive student data, raising concerns about how that data is used and who has access to it. There's also the risk of bias in AI systems, where outdated or incomplete data could lead to unfair treatment of certain groups of students.

Access to AI-powered tools isn't equal either. Students in well-funded schools might enjoy all the advantages of advanced AI systems, while those in underprivileged areas could be left further behind. This digital divide is a stark reminder that technology can sometimes deepen the very inequalities it's supposed to solve.

AI is also changing the role of teachers and students in the classroom. Teachers are no longer just providers of knowledge but facilitators, guiding students as they navigate increasingly personalized and tech-driven learning environments. Students, meanwhile, must learn to balance the convenience of AI with critical thinking and independence, ensuring they don't become overly reliant on the technology.

Even with these challenges, the potential of AI in education is undeniable. If implemented thoughtfully, it can make learning more inclusive, empower teachers, and unlock opportunities for every student, no matter their background. This chapter dives into these possibilities and challenges, exploring how we can harness AI's power responsibly to create a brighter future for education.

3. ALL THE ETHICAL AND SOCIAL ISSUES OF AI IN EDUCATION

- Data Privacy and Security: AI systems rely on collecting and analyzing vast amounts of student data to personalize learning and improve outcomes.
- Algorithmic Bias and Fairness: AI systems are only as unbiased as the data they are trained on.
- Equity and Access: AI promises to improve learning for everyone, but not all students have equal access to technology.
- Transparency and Accountability: AI often works behind the scenes, making decisions that aren't easy to understand.
- Impact on Teachers' Roles: AI is reshaping what it means to be a teacher by automating tasks like grading and lesson planning.
- Balancing Technology and Critical Thinking: While AI can help students learn, too much dependence on it might affect their ability to think independently.

- Supporting Cultural and Linguistic Diversity: AI tools often struggle to adapt to diverse cultural and linguistic contexts.
- Using Student Data Ethically: AI depends on student data, but there's a fine line between using data to help and exploiting it for profit.

What Can Be Done?
To make AI work for everyone in education, we need thoughtful solutions:

- Build Trust: Schools and AI developers should clearly explain how data is used and ensure it's kept secure.
- Ensure Fairness: AI systems must be checked regularly to remove biases and ensure they treat all students equally.
- Close the Gap: Governments and schools should invest in technology and infrastructure to make AI accessible to everyone.
- Support Teachers: Teachers need training and resources to use AI effectively without fear of being replaced.
- Celebrate Diversity: AI should be designed to respect and support cultural and linguistic differences.

By tackling these issues, we can harness the power of AI to create an education system that is fair, inclusive, and truly beneficial for all. We will be discussing in detail about the major issues.

4. DATA PRIVACY AND SECURITY IN AI EDUCATION

As Artificial Intelligence (AI) becomes more integral to education, data privacy and security have emerged as critical concerns. AI systems collect vast amounts of sensitive information—ranging from students' academic records to behavioral patterns and even biometric data—to provide personalized learning experiences and support administrative functions. While these capabilities bring immense benefits, they also come with significant risks that must be addressed to build trust and ensure ethical use.AI offers exciting opportunities to improve education through data-driven insights. For example, personalized learning systems use student data to adapt lessons to individual needs, helping them grasp concepts at their own pace. Teachers can also leverage AI to identify struggling students early and implement targeted interventions, while administrative tasks like attendance and grading are streamlined. However, the collection and use of such sensitive data raise questions about how it is protected, who has access, and how it may be used beyond its intended purpose. (Nguyen et.al,2023)

Unfortunately, educational institutions often lack the robust cybersecurity measures needed to safeguard this information. Many schools operate with limited resources, making them prime targets for cyberattacks. Data breaches can expose students' personal information, leading to identity theft or financial fraud. Furthermore, there are growing concerns about third-party misuse, where companies developing AI tools exploit student data for marketing or other unauthorized purposes.

Informed consent also poses a challenge. Students and their families often don't fully understand the terms and conditions associated with AI systems, leaving them unaware of what data is being collected and how it will be used. This lack of transparency undermines trust and highlights the need for clearer communication and ethical data practices. (O'Neil et.al,2016)

Pros of AI in Data Privacy and Security

- Personalized Learning: AI tailors educational content to each student's strengths and weaknesses, fostering better engagement and outcomes.
- Improved Efficiency: Automated systems streamline grading, attendance, and performance tracking, saving time for educators.
- Data-Driven Decisions: AI provides actionable insights, helping teachers and administrators make informed choices about interventions and resource allocation.
- Transparency: Properly implemented AI systems can offer clear, data-backed explanations of student progress and institutional performance.

Cons of AI in Data Privacy and Security

- Data Breaches: Sensitive student information can be compromised through cyberattacks.
- Third-Party Misuse: Companies may use or sell student data without consent.
- Lack of Awareness: Families often struggle to understand what data is being collected and why.
- Ethical Concerns: Over-collection and commodification of data raise moral questions about the role of AI in education.

To address these concerns, educators, policymakers, and AI developers need to take proactive steps:

- Implement Strong Security Measures: Schools should use encryption, multi-factor authentication, and regular system audits to safeguard student data.

- Focus on Data Minimization: Collect only the information necessary for the AI system to function, avoiding sensitive details whenever possible.
- Enhance Transparency: Educators and developers must clearly communicate how data is collected, stored, and used. Simplified consent forms can help parents and students understand their rights.
- Adopt Ethical AI Practices: Developers should design AI systems with privacy and fairness in mind, ensuring data is only used for educational purposes.
- Train Educators and Students: Provide training on cybersecurity best practices to staff and raise awareness among students and parents about safe data use.

The future of data privacy and security in AI education lies in innovation:

- AI-Powered Cybersecurity: AI can detect and respond to cyber threats in real time, helping schools stay one step ahead of attackers.
- Blockchain Technology: Blockchain offers a secure way to store and share educational data, reducing the risk of breaches.
- Decentralized Data Models: Students could gain more control over their data, deciding who can access it and for what purpose.
- Unified Global Standards: Collaboration between nations may lead to universal data protection policies for education.

As AI continues to evolve, new challenges and opportunities are emerging:

- Biometric Data Risks: Tools that track facial expressions or eye movements raise concerns about privacy and consent.
- Mental Health Insights: AI analyzing emotional or mental health data can improve learning but requires strict safeguards.
- Bias in Data Collection: Unequal representation in datasets can lead to privacy issues and unfair outcomes, requiring careful oversight.

Data privacy and security are vital to the ethical and effective use of AI in education. While AI has the power to personalize learning, streamline processes, and provide valuable insights, its success depends on how responsibly we manage the data it relies on. By implementing robust security measures, fostering transparency, and staying ahead of emerging trends, we can unlock AI's full potential while safeguarding the rights and trust of students, parents, and educators.

Figure 2. Data Privacy and Security

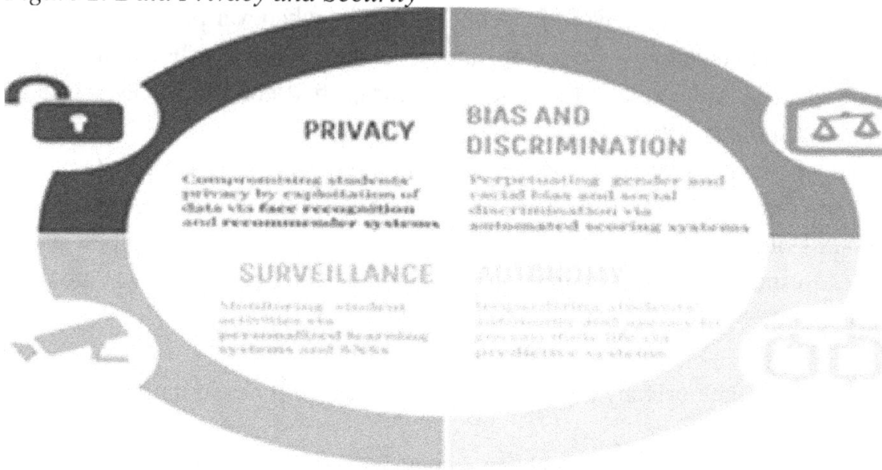

5. ALGORITHMIC BIAS AND FAIRNESS IN AI EDUCATION

As Artificial Intelligence (AI) becomes a powerful tool in education, one of its biggest challenges is ensuring fairness. AI systems are designed to make decisions based on data, but if that data is biased, the systems can produce results that are unfair or even discriminatory. This issue, known as algorithmic bias, can have significant consequences for students, particularly those from marginalized or underrepresented groups. Algorithmic bias occurs when an AI system reflects or amplifies inequalities present in its training data. For example, if an AI system is trained on historical student data that includes patterns of inequality—such as lower grades for students from certain socioeconomic backgrounds—the system may perpetuate those patterns (Long et.al,2022).

Bias in AI can arise in several ways:

1. Training Data Bias: When the data used to train an AI system is incomplete or skewed, the system inherits those biases. For instance, if the data over-represents male students, the AI might make inaccurate predictions for female students.
2. Representation Bias: If certain groups are underrepresented in the data, the system may fail to perform well for those groups. For example, a dropout prediction model trained only on urban schools might misclassify rural students.
3. Algorithmic Design Bias: Sometimes, the way an AI system is designed can unintentionally favor one group over another, even if the data is balanced.

4. Outcome Bias: This occurs when the results produced by an AI system consistently disadvantage certain groups, such as assigning lower grades or fewer opportunities.

In education, algorithmic bias can have long-term and far-reaching consequences:

- Unequal Opportunities: Students from underrepresented or marginalized groups may receive fewer learning opportunities, such as advanced placement in courses or scholarships.
- Mistrust in Technology: When students, parents, and educators feel that AI systems are unfair, they may resist using them, limiting their potential benefits.
- Reinforcement of Inequalities: Instead of reducing educational gaps, biased AI systems can worsen them, perpetuating cycles of inequality.

A well-documented example is the 2020 UK exam grading controversy, where an AI system was used to assign grades during the COVID-19 pandemic. The algorithm unfairly lowered grades for students in less affluent schools, while students from wealthier schools received inflated results. Public backlash forced the system to be replaced with teacher-assigned grades. Addressing algorithmic bias requires a combination of technical solutions, ethical oversight, and inclusive practices.

1. Diverse and Representative Data:
   - Collect data that accurately represents all groups, including those from underprivileged or underrepresented communities.
   - Avoid over-reliance on historical data that may reflect outdated or biased practices.
2. Regular Audits and Testing:
   - Conduct fairness audits to identify and correct biases in AI systems.
   - Test AI models using diverse datasets to ensure they work equally well across different groups.
3. Transparency in AI Design:
   - Make AI systems explainable so educators, students, and parents understand how decisions are made.
   - Provide clear documentation on how training data was collected and algorithms were designed.
4. Human Oversight:
   - Involve educators in reviewing and validating AI-generated decisions, especially for high-stakes outcomes like grading or admissions.

- Create mechanisms for students and parents to challenge or appeal AI decisions.

5. Ethical AI Practices:
    - Design AI systems with fairness and equity as core principles, rather than as afterthoughts.
    - Encourage interdisciplinary collaboration between educators, developers, and ethicists to address fairness issues.

Emerging Topics in Algorithmic Bias

1. Bias Detection Tools:
    - Emerging AI technologies can now detect and flag potential biases within their own systems. This self-regulation can help maintain fairness in real time.
2. Fairness Metrics:
    - Researchers are developing new metrics to measure fairness in AI, ensuring that systems meet ethical and performance standards.
3. Inclusive AI Design:
    - Increasing focus on participatory design methods, where diverse stakeholders are involved in creating AI tools.
4. Legislation and Regulation:
    - Governments worldwide are beginning to introduce laws to ensure AI fairness, such as mandates for transparency and accountability in decision-making.
5. Bias Beyond Academics:
    - AI is now being used to analyze non-academic factors, such as behavioral patterns or emotional intelligence. This raises new questions about how biases in these areas might affect students' opportunities.

Algorithmic bias is a critical challenge that must be addressed to ensure AI enhances education without creating or perpetuating inequalities. By prioritizing fairness, transparency, and accountability in AI systems, we can build tools that truly support all students, regardless of their background. With careful planning and ongoing improvements, AI can become a powerful force for equity in education, helping to level the playing field and create opportunities for everyone.

6. EQUITY AND ACCESS IN AI EDUCATION

Equity and access are central to the transformative potential of AI in education. While AI holds the promise of personalizing learning and bridging gaps in educational outcomes, it also risks exacerbating existing disparities. Ensuring that every student, regardless of their socio-economic background, geographic location, or cultural context, can benefit from AI tools is both a challenge and an opportunity that must be addressed thoughtfully.

A major barrier to equitable AI integration is the **digital divide**, which separates students who have access to technology from those who do not. Economic disparities play a significant role, as schools in underfunded areas often lack the infrastructure to provide devices, reliable internet, and modern AI-powered tools. Similarly, rural and remote regions face geographic barriers, such as limited broadband availability, which restricts the effective use of AI systems in education. The COVID-19 pandemic highlighted these gaps starkly, as many students from low-income families were unable to participate in remote learning due to the lack of essential technology.

Social and cultural factors further complicate access to AI tools. Parents' educational backgrounds and technological literacy influence their ability to support children using AI-driven resources. Moreover, biases in AI tools, often stemming from underrepresented datasets, can alienate students from marginalized communities, resulting in unequal outcomes. In some cases, cultural attitudes toward technology and education may discourage the adoption of AI tools, particularly in communities that lack awareness of its benefits or view it as irrelevant to their needs. (Sidiropoulos et.al,2024)

To address these challenges, governments and institutions must prioritize policies that bridge the equity gap. Funding models should focus on under-resourced schools, ensuring they have access to the devices, internet, and AI tools needed for effective learning. Public-private partnerships can help subsidize the costs of technology for underprivileged students, while investments in internet connectivity for rural and underserved areas are essential. Schools can also provide low-cost or refurbished devices to students who cannot afford them, ensuring no one is left behind in the digital era.

Teacher support is another critical element in fostering equity. Educators need professional development programs that equip them to use AI tools effectively and address disparities in their classrooms. Teachers must be trained to identify and mitigate biases in AI systems, ensuring that every student benefits equally from the technology. By empowering teachers, schools can create more equitable learning environments where AI is a tool for inclusivity rather than division.

Inclusive AI design is vital for ensuring that the tools themselves support equity and access. Developers must build systems that are culturally and linguistically sensitive, taking into account the diverse needs of students from various backgrounds. This includes using representative datasets to reduce algorithmic bias and designing user-friendly interfaces that cater to varying levels of technological proficiency (Klimova et.al,2023).

Emerging trends also offer hope for reducing inequities in AI adoption. Technologies like decentralized AI and offline-capable tools are being developed to function in low-resource settings, allowing students in rural or underserved areas to benefit from AI without needing consistent internet access. Community-led initiatives, such as mobile libraries equipped with AI tools, are bridging gaps by bringing technology directly to students. Governments worldwide are also beginning to integrate AI into their education policies, focusing on equity and inclusion.

Real-world examples demonstrate how these strategies can work. Kenya's Eneza Education has successfully used SMS-based AI tools to provide learning resources to remote communities, while India's government programs are delivering AI-powered tools to rural schools. Similarly, Brazil has invested in digital inclusion projects to reduce disparities in low-income neighborhoods. These case studies show that thoughtful implementation of AI can significantly improve access and equity in education.

Finally, addressing equity and access is not just a technological challenge—it is an ethical imperative. AI in education should aim to reduce inequalities and create opportunities for all students. By prioritizing investments in under-resourced areas, fostering inclusive design, and ensuring teacher support, AI can truly democratize education and empower every student to succeed.

7. TRANSPARENCY AND ACCOUNTABILITY IN AI EDUCATION

Transparency and accountability are essential for ensuring that AI systems in education are trustworthy, ethical, and beneficial for everyone involved. When teachers, students, and parents understand how these tools work and feel confident in their fairness, they are more likely to embrace them. Transparency means making AI systems easy to understand—teachers and students should know how decisions, like grades or intervention recommendations, are made. For example, if an AI tool suggests that a student needs extra help in math, it should explain why that recommendation was made, using data that is clear and accessible. Explainable AI (XAI) plays a crucial role here, as it breaks down complex processes into information that people without technical expertise can understand.

At the same time, accountability ensures that there are checks and balances in place. Schools need systems that allow students and teachers to challenge decisions if they seem unfair or incorrect. For instance, if a student believes they were graded unfairly by an AI tool, there should be a straightforward process to review and correct that grade. Human oversight is equally important—educators must remain involved in monitoring and guiding AI systems to ensure they align with educational goals and values. This combination of transparency and accountability ensures that AI supports, rather than undermines, the learning process.

Despite its importance, achieving transparency and accountability in AI systems isn't always easy. Many AI tools function as "black boxes," making it difficult for teachers and students to see how decisions are being made. This lack of clarity can lead to mistrust and resistance to using the tools. Moreover, there are no universal standards for transparency in educational AI, which means practices vary widely and inconsistently.

To overcome these challenges, schools and developers need to take several steps. First, developers should create AI tools that explain their decision-making in simple, non-technical terms. Teachers and administrators should receive training to understand how these systems work and how to use them effectively. Schools must also establish clear policies to hold developers accountable for ethical and accurate performance, including regular audits to assess the fairness and reliability of AI tools. Open communication with students, parents, and teachers is essential—everyone should know what data is being collected, how it's being used, and who has access to it.

Looking ahead, global standards for transparency and accountability in educational AI could help ensure consistency and trust across different tools and regions. The future might also see AI systems that can self-report errors or biases, making it easier to identify and address problems before they impact students. Collaborative feedback platforms could allow teachers and students to directly share their experiences with developers, driving continuous improvement in AI tools.

Transparency and accountability are the foundation of ethical AI use in education. By making AI systems understandable, involving all stakeholders, and setting clear guidelines for their use, schools can build trust and ensure these tools enhance learning for everyone. Through thoughtful implementation and continuous improvement, AI can become a powerful, trusted partner in education.

8. IMPACT ON TEACHER'S ROLE IN AI EDUCATION

The rise of Artificial Intelligence (AI) in education is significantly transforming the role of teachers, creating both opportunities and challenges. AI tools, which have the potential to personalize learning and automate administrative tasks, are reshaping how teachers interact with students and deliver education. While these advancements promise to enhance educational outcomes, they also require educators to adapt their practices and embrace new responsibilities.

One of the most profound changes is the shift from being the primary source of knowledge to becoming facilitators of learning. Traditionally, teachers were responsible for delivering lectures, grading assignments, and guiding students through their learning. With AI, many of these tasks can now be automated. AI tools can handle administrative functions such as grading, providing instant feedback on assignments, and even suggesting personalized learning paths for students based on their strengths and weaknesses. This automation allows teachers to spend more time focusing on what AI cannot do: fostering creativity, critical thinking, and problem-solving skills. Teachers can guide students in deeper explorations of subjects, lead discussions, and encourage intellectual curiosity. The role of the teacher thus evolves into that of a mentor and coach, helping students make connections between the knowledge they gain from AI-driven tools and real-world applications.

However, this shift requires teachers to adopt a new mindset and skillset. While AI can handle repetitive tasks, it cannot replace the human aspects of teaching, such as emotional support, empathy, and relationship-building. These interpersonal skills are critical for creating a positive and supportive learning environment. Teachers must understand how to effectively use AI to complement their work rather than view it as a replacement. For example, AI may suggest that a student is struggling in a particular area, but it is up to the teacher to provide tailored interventions, offer encouragement, and engage the student on a personal level. AI can provide data, but it's the teacher who can interpret that data and apply it meaningfully to guide students' growth. (Guilherme et.al,2019).

Moreover, the integration of AI into classrooms requires continuous professional development. Teachers need to stay informed about new AI tools and developments, gaining the knowledge necessary to use them effectively. This means teachers must not only learn how to use AI but also understand the underlying principles behind it. This shift demands a commitment to ongoing learning and adaptability. Educators will need to develop new skills in data interpretation, as AI will generate insights that can inform teaching strategies and decisions. Teachers will also need to learn how to strike the right balance between using AI for efficiency and retaining their role as the core drivers of learning and critical thinking in the classroom (Siemens et.al,2014).

Another significant change AI brings is the personalization of teaching. AI allows for adaptive learning, where educational content is adjusted based on each student's progress and learning needs. Students can receive lessons tailored to their own pace, whether they need more time on certain topics or a challenge with advanced material. For teachers, this means they can focus on the students who need additional support or guidance, while AI handles more routine aspects of learning. This personalization creates an opportunity for teachers to spend more one-on-one time with students, providing targeted assistance where necessary. However, teachers need to ensure that the learning path AI suggests aligns with the overall curriculum and that no student is left behind due to over-reliance on automated tools.

Despite these benefits, AI's rise has sparked concerns about job displacement. With AI handling tasks like grading, administrative work, and even tutoring, some may worry that AI could eventually replace teachers, particularly in subjects where content delivery can be automated. However, it's important to understand that AI is not designed to replace teachers—it is meant to augment and support them. The most successful classrooms will be those where teachers and AI work together, each complementing the other's strengths. AI can take care of repetitive, time-consuming tasks, freeing up teachers to focus on areas where human insight, empathy, and creativity are irreplaceable. Teachers will remain essential for fostering social skills, emotional intelligence, and critical thinking, all of which are crucial for students' success in both education and life.

Ultimately, AI's impact on teachers can be viewed as a positive transformation rather than a threat. Teachers remain at the heart of the educational process, but AI allows them to focus on higher-order aspects of teaching, such as developing students' critical thinking skills, nurturing creativity, and supporting their emotional and social development. As teachers adopt AI tools, they can become more effective, responsive, and engaged with their students, guiding them through a personalized learning journey. However, to fully harness the potential of AI in education, teachers must be willing to embrace new technology, continuously upskill, and use AI tools to enhance—not replace—the human connections that are so vital to the learning process.

As AI continues to evolve, the role of the teacher will continue to shift, offering opportunities for professional growth and the ability to provide more personalized and meaningful education to students. Through collaboration between AI and educators, we can create a learning environment that balances technology with the personal touch that only teachers can provide, ensuring that students receive the best of both worlds.

9. CONCLUSION

As AI continues to reshape the landscape of education, it presents incredible opportunities to enhance learning and teaching. However, its integration into education comes with a set of challenges that need careful consideration to ensure these technologies benefit all students and educators equally. Issues such as **data privacy and security, algorithmic bias, equity in access, transparency and accountability**, and the **impact on teachers' roles** are crucial areas that must be addressed for AI to be truly effective and ethical in educational settings.

Ensuring **data privacy and security** is fundamental to building trust in AI systems. As AI relies on vast amounts of personal data, it is essential to safeguard students' information, prevent breaches, and ensure transparency in how data is used. Alongside this, addressing **algorithmic bias** and ensuring **fairness** are key to ensuring that AI systems do not perpetuate existing inequalities in education. AI must be designed and regularly audited to guarantee that it treats all students equally, regardless of background. **Equity and access** are equally important—AI should be an empowering tool for all students, but only if every student has access to the necessary technology and resources. Without addressing this digital divide, AI risks deepening educational disparities.

In addition, **transparency and accountability** must be at the heart of AI adoption in schools. Educators, students, and parents need clear insights into how AI systems work, how decisions are made, and how to hold these systems accountable when errors occur. The role of **teachers** in this environment is also evolving. While AI can automate many tasks, teachers remain essential for building relationships, guiding students' growth, and ensuring that AI serves as a tool to enhance—not replace—human connection in education.

However, as AI becomes a more integrated part of education, we must also focus on **balancing technology with critical thinking**. AI can provide quick answers, but it's crucial that students continue to develop problem-solving and analytical skills. At the same time, AI must support **cultural and linguistic diversity**, adapting to the needs of students from different backgrounds, and ensuring that no one is excluded or marginalized by the technology. Lastly, we must ensure that AI is used ethically, particularly when it comes to **student data**. Data collection should always prioritize student welfare and be handled responsibly, with clear guidelines on consent and usage.

In conclusion, AI holds immense potential to transform education for the better, but we must approach its implementation thoughtfully and responsibly. By addressing the challenges of privacy, fairness, equity, and the evolving role of teachers, and by ensuring that AI systems promote critical thinking, inclusivity, and ethical practices, we can harness AI's power to create a more personalized, equitable, and effective educational experience for all.

Case Study:

Addressing Algorithm Bias in the UK's 2020 A-Level Grading Crisis

In 2020, the UK faced a major controversy when an AI-driven algorithm was used to determine A-level exam grades. With exams canceled due to the COVID-19 pandemic, the algorithm was meant to fairly predict grades based on student performance and maintain consistency. However, instead of solving the problem, it ended up causing widespread anger and frustration.

The algorithm relied heavily on historical data, including the past performance of schools. As a result, students from well-funded schools in affluent areas generally received higher grades, while those from underprivileged schools, no matter how capable, saw their grades unfairly lowered. The system ended up reinforcing existing inequalities, which sparked outrage among students, parents, and educators alike.

For many students, these grades were life-changing—they affected their chances of getting into universities and pursuing their dreams. Public backlash was so strong that the government was forced to abandon the algorithmic grades and instead use the grades predicted by teachers.

This crisis highlighted the risks of relying on AI for decisions that have a huge impact on people's lives. It also showed how algorithms can unintentionally reflect and amplify biases that already exist in society.

- Fairness First: AI systems must be carefully designed and tested to ensure they don't disadvantage certain groups.
- Clear Communication: Students, teachers, and parents need to understand how AI systems work and how decisions are made.
- Human Judgment Matters: AI should support teachers, not replace their role in critical decisions like grading.
- Ethics in Design: AI developers need to prioritize inclusivity and fairness when building systems that impact education.

This case shows how important it is to address algorithmic bias, ensure transparency, and involve teachers in decision-making when using AI in education. It's a reminder that while AI can bring amazing benefits, it must be implemented thoughtfully and responsibly to avoid harm. By learning from cases like this, schools and developers can work together to create AI tools that truly support all students fairly.

REFERENCES

Akgun, S., & Greenhow, C. (2022). Artificial intelligence in education: Addressing ethical challenges in K-12 settings. *AI and Ethics*, 2(3), 431–440. DOI: 10.1007/s43681-021-00096-7 PMID: 34790956

Guilherme, A. (2019). AI and education: The importance of teacher and student relations. *AI & Society*, 34(1), 47–54. DOI: 10.1007/s00146-017-0693-8

Klimova, B., Picart, M., & Kacetl, J. (2023). Ethical issues of the use of AI-driven mobile apps for education. *Frontiers in Public Health*, 10, 1118116. DOI: 10.3389/fpubh.2022.1118116 PMID: 36711343

Long, W., & Gao, Y. (2022). Artificial intelligence education system based on differential evolution algorithm to optimize SVM. *Scientific Programming*, 2022(1), 5379646. DOI: 10.1155/2022/5379646

Nguyen, A., Ngo, H. N., Hong, Y., Dang, B., & Nguyen, B. P. T. (2023). Ethical principles for artificial intelligence in education. *Education and Information Technologies*, 28(4), 4221–4241. DOI: 10.1007/s10639-022-11316-w PMID: 36254344

O'Neil, C. (2016). *Weapons of Math Destruction: How Big Data Increases Inequality and Threatens Democracy*. Crown Publishing.

Sidiropoulos, D., & Anagnostopoulos, C. N. (2024). Applications, challenges, and ethical issues of AI and ChatGPT in education. *arrive preprintarXiv:2402.07907.*

Siemens, G. (2014). Learning Analytics: The Emergence of a Discipline. *The American Behavioral Scientist*, 57(10), 1380–1400. DOI: 10.1177/0002764213498851

Chapter 15
Ethical Integration of Artificial Intelligence in Inclusive Education:
Addressing Challenges and Advancing Opportunities for Equitable Learning

Utsav Krishan Murari
https://orcid.org/0009-0007-1606-6775
Sharda University, India

Hemlata Parmar
https://orcid.org/0009-0009-1438-5427
Manipal University Jaipur, India

ABSTRACT

Artificial Intelligence (AI) is reshaping educational landscapes, especially in creating more equitable and inclusive learning environments. This chapter explores the intersection of AI and inclusive education, focusing on how AI tools can foster individual learning, assistive technologies, and adaptive curricula for diverse educational needs. However, these opportunities are not without challenges; issues of algorithmic bias, data privacy, and the digital divide pose significant barriers. Emphasizing a human-centered, ethical approach, this chapter provides insights into the responsible development and application of AI in inclusive education, guided by principles of diversity, equity, and inclusion (DEI). The chapter concludes with policy recommendations and future research directions to harness AI's transformative potential while safeguarding educational equity.

DOI: 10.4018/979-8-3373-2262-9.ch015

1. INTRODUCTION

Artificial Intelligence (AI) is becoming more prevalent across multiple industries, notably impacting education through improved administrative efficiency, personalised learning experiences, and the development of innovative engagement strategies. AI encompasses various capabilities, including machine learning (ML), natural language processing (NLP), and predictive analytics, offering distinct solutions for educational settings, particularly as the demand for digital transformation intensifies. Artificial intelligence in education is recognised as an innovative tool that has the potential to customise educational experiences, thereby enhancing accessibility, inclusivity, and engagement for students worldwide. Despite the transformative impact of AI applications on conventional educational frameworks, their capacity to improve inclusive education still needs to be more adequately investigated and often presents ethical and practical challenges (Holmes et al., 2019). Inclusive education is based on the principle that all students, irrespective of background, ability, or socioeconomic status, are entitled to equitable access to high-quality education. The objective is to establish learning environments that address diverse needs, including students with disabilities, individuals from marginalised communities, and those encountering language or socioeconomic challenges. The United Nations Convention on the Rights of Persons with Disabilities (UNCRPD) and frameworks such as the Salamanca Statement have consistently promoted inclusive education, highlighting the necessity of inclusive systems that provide equal opportunities for all students (UNESCO, 1994; Ainscow, 2005). The intersection of AI and inclusive education offers a means to achieve international objectives by utilising advanced technologies that cater to the diverse learning requirements of students, especially those historically marginalised in educational environments. AI's integration into inclusive education presents notable ethical challenges despite its potential benefits. Algorithmic bias, data privacy concerns, and the widespread digital divide undermine the core principles of equity and accessibility that inclusive education aims to achieve. Algorithmic bias arises when AI models, trained on historical data, unintentionally perpetuate existing inequalities, resulting in inequitable treatment of students based on their backgrounds or personal attributes. AI systems' collection, analysis, and storage of extensive sensitive student information raises significant concerns regarding data privacy, with potential risks of breaches and misuse. The digital divide, defined as the disparity in access to technology, intensifies pre-existing educational inequalities, especially in marginalised communities. Addressing these challenges necessitates a human-centred approach to AI in education, ensuring that ethical principles of diversity, equity, and inclusion (DEI) inform development and implementation.

1.1 AI in Education: Revolutionizing Learning Environments

AI's capabilities in education have evolved to encompass a wide range of applications, each contributing uniquely to learning and educational administration. Broadly, AI in education enhances administrative efficiency, personalises learning experiences, and increases student engagement.

1.1.1 Administrative Efficiency through Automation and Data Analytics

One of the primary applications of AI in education is in administrative functions, where it automates repetitive tasks, optimises resource allocation, and supports data-driven decision-making. Routine tasks such as grading, attendance tracking, and scheduling are increasingly handled by AI, which allows educators to focus more on pedagogy and less on administrative work (Luckin et al., 2016). Automated grading systems, for instance, can analyse thousands of assignments, providing timely and consistent feedback, a task that is labour-intensive and subject to human variability. Moreover, these systems can identify patterns in student performance, allowing educators to detect at-risk students early and implement timely interventions (Heffernan & Koedinger, 2012). Beyond grading, predictive analytics—a core component of AI—enables educational institutions to track and analyse student data comprehensively. By identifying students who may be struggling, AI systems prompt educators to take action before academic challenges become insurmountable. Data-driven insights improve resource allocation, focusing attention and support where needed. For instance, if a particular demographic of students consistently underperforms, AI can help administrators recognise this trend, assess potential causes, and allocate resources to address the issue, contributing to a more equitable educational environment (Popenici & Kerr, 2017).

1.1.2 Personalized Learning and Adaptive Educational Experiences

Personalised learning represents a significant advancement in the application of AI within the educational sector. Conventional models typically adopt a uniform approach, inadequately accommodating the varied learning requirements of individual students. AI transforms education by assessing students' performance, learning styles, and engagement levels, thus providing a customised educational experience (Holmes et al., 2019). Adaptive learning platforms such as DreamBox and Knewton utilise algorithms to modify lesson difficulty in real time according to individual student interactions, thereby maintaining a balance between challenge and manageability of content. This customisation is especially beneficial for students with disabilities, as they may possess distinct needs that traditional methods fail to meet (Popenici & Kerr, 2017).

AI's adaptability offers alternative methods for students with learning difficulties to comprehend concepts that might otherwise be unattainable. For instance, a student with dyslexia may gain advantages from text-to-speech functionalities integrated into educational platforms. In contrast, students with attention deficits may find value in regular modifications to the pacing and presentation of lessons.

1.1.3 Enhanced Student Engagement through Interactive Learning Tools

AI enhances student engagement by providing immersive, interactive, and personalised learning experiences. For instance, intelligent Tutoring Systems (ITS) simulate the experience of one-on-one tutoring by guiding students through complex problems and offering real-time feedback. These systems go beyond static educational content, dynamically adjusting to students' answers, pacing, and comprehension levels (Baker & Smith, 2019). Moreover, AI-driven tools such as virtual reality (VR) and augmented reality (AR) immerse students in realistic simulations that make abstract concepts more accessible. A history lesson, for instance, can transform into an interactive VR experience where students explore ancient civilisations firsthand, or a biology class can use AR to visualise molecular structures in 3D. These tools offer unique engagement opportunities, particularly for students struggling with traditional text-based learning materials.

1.1.4 Understanding Inclusive Education: Principles and Evolution

Inclusive education is rooted in the belief that all students should have access to meaningful educational opportunities regardless of background, ability, or economic status. As an educational philosophy, inclusive education has evolved over the decades, shaped by legal frameworks, social justice movements, and international guidelines. Major milestones include the Salamanca Statement, which advocated for integrating students with disabilities into mainstream schools, and the UNCRPD, which emphasises the right to inclusive education for all (UNESCO, 1994).

1.1.5 Core Principles of Inclusive Education

Inclusive education is guided by several foundational principles: equity, diversity, support, and collaboration. Equity ensures that students receive the resources they need to succeed, rather than treating all students identically. For example, while a visually impaired student may require screen-reading software, a student with hearing loss might benefit from real-time captioning. Diversity emphasises the value of cultural, language, ability, or socioeconomic status differences. Support refers to tailored assistance—such as assistive technologies or specialised teaching methods—that allows students to participate fully in the learning experience. The collaboration underscores the importance of partnerships between teachers, students, families, and communities to design inclusive strategies that accommodate diverse needs (Ainscow, 2005). The inclusive education model aims to dismantle barriers that prevent marginalised students from engaging fully in educational settings. Traditional educational systems often focus on integration, which involves placing students with disabilities in mainstream classrooms without modifying instructional methods.

1.2 The Role of AI in Advancing Inclusive Education

The convergence of AI and inclusive education offers promising opportunities to support these principles by creating adaptable, supportive, and accessible learning environments. AI can play a vital role in supporting individualised learning, assisting students with disabilities, and designing adaptive curricula that cater to diverse learner needs. For instance, students with physical disabilities may use eye-tracking technology to interact with computers, enabling them to participate in digital learning activities that would otherwise be inaccessible (Adebisi et al., 2015). Similarly, language translation tools assist non-native speakers in understanding course content, bridging language barriers and promoting inclusion.

1.3 Ethical Challenges in AI-Driven Inclusive Education

Despite its potential, the integration of AI in inclusive education is fraught with ethical challenges. These challenges revolve primarily around algorithmic bias, data privacy concerns, and the digital divide, each posing significant threats to these technologies' equitable and inclusive intentions.

1.3.1 Algorithmic Bias

Algorithmic bias occurs when AI models unintentionally reinforce societal prejudices embedded in the training data. In educational contexts, algorithmic bias can result in discriminatory outcomes, disadvantaging students from marginalised groups. For example, an AI system to predict student success might be trained on data that reflects historical inequalities, potentially leading to lower expectations for students from underserved communities. Such biases can perpetuate inequalities rather than bridging them, undermining inclusive education's goals (O'Neil, 2016).

Mitigating algorithmic bias involves using diverse training data, conducting regular bias audits, and ensuring transparency in AI decision-making processes.

1.3.2 Data Privacy

Implementing AI in educational settings requires the aggregation and examination of extensive student data, which raises significant concerns regarding privacy and data security. Information regarding students' academic performance, behavioural patterns, and health records is frequently utilised to customise educational experiences. The misuse of this data can result in stigmatisation or discrimination for students from vulnerable groups, occurring both within and outside educational contexts. Legal frameworks such as the General Data Protection Regulation (GDPR) in Europe and the Family Educational Rights and Privacy Act (FERPA) in the United States are designed to protect student data; however, compliance is inconsistent, and enforcement poses significant challenges (Selwyn, 2020; Slade & Prinsloo, 2013).

1.3.3 The Digital Divide

Access to AI-driven educational tools is restricted in underserved communities, where students often lack the necessary devices, connectivity, and infrastructure for digital learning participation. The digital divide intensifies current educational disparities, disproportionately impacting students from low-income households, rural regions, and marginalised communities. Policymakers and educational institutions should invest in technology infrastructure, ensure access to affordable devices and the internet, and support communities with limited resources to bridge this divide (Robinson et al., 2020). Incorporating AI in inclusive education presents significant transformative possibilities; however, these possibilities can only be actualised through ethical and deliberate implementation. Artificial intelligence has the potential to tailor educational experiences, address varied student requirements, and enhance engagement, thus fostering inclusive practices consistent with diversity, equity, and inclusion principles.

2. THE ROLE OF AI IN EDUCATION

Artificial Intelligence (AI) has emerged as a powerful tool in reshaping modern education, offering once unimaginable capabilities. From automating administrative tasks to customising learning experiences and enhancing student engagement, AI applications in education address numerous needs, contributing to the efficiency, accessibility, and inclusivity of learning environments. AI-driven solutions streamline educational operations and offer adaptive learning models tailored to individual student needs. As we explore the role of AI in education, three main areas stand out: administrative efficiency, personalised learning, and enhanced student engagement. These applications significantly influence how educators and institutions address diverse student needs, optimise resources, and provide supportive, equitable learning experiences.

2.1 Administrative Efficiency and AI

One of the primary applications of AI in education is improving administrative efficiency. Educational institutions handle vast data, from student records to academic performance metrics, attendance records, and faculty schedules. Traditionally, managing this data was a labor-intensive process prone to human error and inconsistencies. AI tools can now automate these tasks, creating a more streamlined and error-free process and allowing educators and administrators to focus more on instructional tasks and student support.

2.1.1 AI in Grading and Assessment

One of the most laborious tasks for educators is grading, which AI systems can now automate for specific assessments. Automated grading systems can analyse written answers, quizzes, and complex essays, providing consistent and objective feedback. Notably, AI-driven grading tools analyse responses for correctness and sometimes for depth of insight, although this is still an evolving area. Automated grading saves educators' time and improves the speed of feedback delivery, which is crucial for student learning. When students receive timely feedback, they can understand and address their mistakes more effectively, improving outcomes (Luckin et al., 2016). For example, platforms such as Gradescope use AI algorithms to quickly grade and assess student work, making it easier for instructors to handle large classes or multiple sections. These tools support grading for various subjects, including science, mathematics, and humanities, by analysing written text or interpreting diagrammatic and graphical answers.

2.1.2 Data-Driven Decision-Making in Educational Administration

Beyond grading, AI is transforming how educational institutions use data for decision-making. Predictive analytics—a central capability within AI—enables schools to make more informed decisions based on historical and real-time data. For instance, AI can track students' academic progress and flag those at risk of underperforming. Educators can identify these students early by providing timely interventions through additional resources, tutoring, or counselling. Research has shown that such data-driven approaches improve student retention and overall success (Holmes et al., 2019). AI can also enhance resource allocation within schools. For example, data-driven systems may reveal patterns of underuse in certain areas, helping administrators reallocate resources to high-demand services, like tutoring centres or special needs support.

2.2 Personalized Learning

One of AI's most transformative uses in education is personalised learning. Unlike traditional models that adopt a standardised approach to instruction, AI allows for individual customisation based on each student's unique needs, learning style, and pace. AI-powered personalised learning systems analyse students' past performance, learning preferences, and engagement patterns, adjusting content and pacing in real-time. This adaptability enables students to learn more effectively, at their own pace, without pressure to keep up with or wait for others.

2.2.1 Adaptive Learning Platforms

Adaptive learning platforms like DreamBox, Smart Sparrow, and Knewton exemplify the application of AI in personalised education. These platforms employ advanced algorithms to tailor the complexity and nature of content provided to students according to their interactions and progress. If a student demonstrates proficiency in a specific topic, the system can provide more advanced materials to maintain their challenge. If a student encounters difficulties, the system can offer supplementary assistance or review essential concepts to enhance comprehension (Popenici & Kerr, 2017). These platforms are advantageous in inclusive education environments, enabling students with diverse abilities and backgrounds to balance challenge and support. A student with dyslexia may benefit from text-to-speech features in an adaptive platform. In contrast, a student with ADHD might find adjustable pacing and interactive features helpful for maintaining focus. Adaptive learning's customisation fosters an equitable educational environment, enabling all students to excel based on their abilities.

2.2.2 Real-Time Feedback and Continuous Assessment

AI also facilitates real-time feedback, an essential component of effective learning. In a conventional classroom, students might wait days or even weeks for feedback on assignments, which delays the opportunity to learn from mistakes. On the other hand, AI-powered systems can provide instant feedback, allowing students to address errors and understand concepts more clearly immediately. Research suggests timely feedback can significantly improve student engagement and retention, as students are more likely to engage with material when they know their progress and areas for improvement (Baker & Smith, 2019). Moreover, continuous assessment through AI helps educators monitor student progress consistently. Unlike traditional testing, which may occur only once or twice a semester, continuous assessment tracks a student's performance throughout the course, allowing for adjustments in teaching strategies as needed. Continuous assessment helps teachers and administrators identify learning gaps early, ensuring students receive support when needed.

2.2.3 Personalized Learning Pathways for Special Needs Education

One of AI's most impactful roles in inclusive education is its capacity to support students with disabilities through personalised learning pathways. AI systems can adapt content and interactions based on each student's cognitive and physical needs, helping to level the playing field for students with special needs. For example, AI-powered reading tools can adjust the font size, colour, or background brightness to accommodate students with visual impairments. In contrast, language processing tools help students with auditory processing disorders to understand spoken instructions more effectively (Holmes et al., 2019).

2.3 Enhanced Student Engagement

Engaging students in learning can be challenging, especially in today's digital world, with abundant distractions. AI has introduced new ways to make learning more interactive and captivating, using tools like Intelligent Tutoring Systems (ITS) and immersive technologies such as Virtual Reality (VR) and Augmented Reality (AR). These tools allow students to explore, visualise, and interact with material in ways that conventional methods may not support.

2.3.1 Intelligent Tutoring Systems (ITS)

Intelligent Tutoring Systems are AI-based tools designed to replicate the one-on-one tutoring experience. These systems guide students through complex problems, providing hints, explanations, and tailored feedback based on the student's performance. ITS can be particularly valuable in subjects like math and science, where complex problem-solving skills are essential. For example, the Carnegie Learning Math Tutor is an ITS that adapts to student responses, offering hints or additional challenges based on the student's progress (Baker & Smith, 2019).

2.3.2 Immersive Learning with VR and AR

AI-driven immersive technologies, including virtual and augmented reality, have transformed student engagement by providing realistic and interactive learning experiences. Virtual reality (VR) facilitates immersive environments for students to explore virtual worlds, whereas augmented reality (AR) enhances comprehension of abstract concepts by overlaying digital content onto the real world. A history lesson could be transformed into a VR-based exploration of ancient civilisations, enabling students to engage with history actively rather than passively reading about it. In biology, augmented reality applications would allow students to examine three-dimensional models of cells, organs, or ecosystems, offering a tactile and visual experience that enhances understanding (Selwyn, 2020). These technologies address various learning styles, significantly aiding students who excel through experiential learning. Immersive technologies offer students with learning disabilities alternative methods for content interaction, simplifying complex subjects into more accessible and engaging formats. This method enhances both comprehension and motivation regarding the subject matter.

2.3.3 Gamification and AI

Gamification, powered by AI, is another strategy to increase engagement by integrating game-like elements into learning activities. AI systems can design personalised challenges, badges, and rewards to incentivise progress, making learning enjoyable and motivating for students. For example, AI-driven platforms such as Kahoot! and Duolingo employ gamification to create engaging quizzes, interactive exercises, and challenges that encourage students to participate actively in their learning (O'Neil, 2016). Gamification has been shown to improve student motivation and reduce anxiety, particularly in subjects where students may feel less confident. By incorporating AI-based gamification strategies, educators can create positive reinforcement systems that keep students engaged, reduce dropout rates, and foster a love for learning. AI has redefined education through enhanced administrative efficiency, personalised learning, and improved student engagement. From automating repetitive tasks to providing real-time feedback and immersive learning experiences, AI empowers educators and students, creating more inclusive and supportive learning environments. Personalised learning tools cater to individual needs, supporting a diversity of learning styles and abilities, which is particularly beneficial in inclusive education settings.

3. UNDERSTANDING INCLUSIVE EDUCATION

Inclusive education has become an approach designed to accommodate diverse student needs, foster equity, and eliminate barriers that hinder access to quality learning. At its core, inclusive education advocates for the full participation of all students, regardless of background, ability, or socioeconomic status, in mainstream educational settings. This educational philosophy is anchored in the belief that everyone deserves equal access to learning opportunities and that diversity should be celebrated rather than marginalised. Inclusive education aims to create environments that adapt to students' varied needs, providing targeted support to ensure equitable participation. This topic explores the principles and evolution of inclusive education, the role of international frameworks in shaping inclusive policies, and the challenges and opportunities it presents for modern educational systems.

3.1 The Principles of Inclusive Education

Inclusive education is built upon several foundational principles: equity, diversity, support, and collaboration. Each principle is integral to establishing a learning environment where all students feel valued and supported, with access to the resources they need to succeed.

3.1.1 Equity over Equality

Equity serves as a fundamental principle that prioritises fairness over uniformity. In inclusive education, equity guarantees all students access to necessary resources, accommodations, and support tailored to their circumstances. Equity differs from equality in that it addresses individual differences rather than providing uniform resources to all individuals. A visually impaired student may require screen-reading software to access digital materials, whereas a student with dyslexia might find text-to-speech applications beneficial. Equity does not suggest identical treatment; it recognises that genuine fairness may necessitate varied approaches (Ainscow, 2005). Equity is manifested in policies providing supplementary funding and resources to schools in disadvantaged regions, facilitating access to quality education for students from low-income backgrounds. Equity-driven frameworks address access barriers, including language, by integrating multilingual learning resources for students from various linguistic backgrounds. Effective implementation of equity in inclusive education offers all students, including those from marginalised communities, a legitimate opportunity for academic and social success.

3.1.2 Diversity as a Strength

Diversity in inclusive education is recognised and valued. Valuing diversity involves recognising and appreciating students' distinct backgrounds, abilities, cultures, and learning styles. Inclusive classrooms facilitate interactions among students from diverse backgrounds, promoting mutual learning and respect. This exposure cultivates an appreciation for diversity, thereby diminishing biases and stereotypes that may result in exclusion or discrimination (UNESCO, 1994). In a diverse classroom, instructional strategies are adapted to address various learning styles, cultural contexts, and individual abilities. A teacher may employ multiple instructional strategies, including visual aids, group discussions, and hands-on activities, to facilitate meaningful engagement with the material for all students. Recognising diversity necessitates the adaptation of curricula to encompass a wide array of perspectives, thereby ensuring that students encounter representations of their cultures and identities within the educational content. This principle enhances students' self-esteem and fosters a sense of belonging.

3.1.3 Support for All Learners

Providing support is essential in inclusive education. Every student has unique needs, and providing appropriate support ensures they can fully participate in their education. Support can come in various forms, from assistive technologies to specialised instructional methods and resource allocations that address students' specific challenges. For example, students with physical disabilities may use adaptive devices to navigate classrooms, while students with learning disabilities may benefit from extended test times or simplified instructions (Adebisi et al., 2015).

Support includes emotional and social support, not just physical or technological tools. Educators in inclusive settings often undergo specialised training to help them understand and address the social and emotional needs of students with diverse backgrounds and abilities. This training enables them to create an environment where students feel safe, respected, and encouraged to participate actively in learning. Additionally, support extends beyond the classroom, involving collaboration with parents, therapists, and other specialists to ensure students have a comprehensive support network.

3.1.4 Collaboration among Stakeholders

Collaboration is fundamental to inclusive education, requiring joint efforts from teachers, parents, students, and communities to create a supportive educational environment. Inclusive education recognises that only some educators or specialists can independently meet a student's needs. Collaboration brings together diverse expertise and perspectives, enhancing the educational experience and fostering a more holistic approach to learning.

For instance, teachers might work closely with special education specialists to tailor teaching strategies for students with specific needs. Parents play a critical role in sharing insights about their children's strengths, challenges, and preferences, which can inform individualised support strategies. Community organisations sometimes offer additional resources, such as after-school programs or language support, contributing to a more comprehensive support system. By fostering collaboration, inclusive education emphasises the importance of collective responsibility in supporting all students' learning journeys (Ainscow, 2005).

3.2 The Evolution of Inclusive Education

Inclusive education has evolved over the past several decades, shaped by social justice movements, legal reforms, and international frameworks that advocate for the rights of marginalised communities. Historically, students with disabilities and those from minority backgrounds were often excluded from mainstream education and placed in separate schools or special education classes that limited their opportunities for growth and integration. Over time, however, integration emerged, paving the way for more inclusive models that encourage active participation in general education settings.

3.2.1 The Shift from Integration to Inclusion

In the 20th century, educational systems primarily focused on integration, placing students with disabilities in mainstream classrooms without significantly adapting teaching methods or curricula. Integration often meant that students attended the same school as their peers but remained isolated within separate classes or received minimal support in adapting to the mainstream environment. While integration was a step forward from complete exclusion, it needed to fully address the diverse needs of students, who often faced barriers to meaningful participation (Ainscow, 2005).

The shift to inclusive education marks a fundamental change in perspective. Instead of expecting students to adapt to existing systems, inclusive education emphasises adapting educational environments to meet each student's needs. This approach aligns with the social model of disability, which argues that barriers in the environment—rather than individual impairments—are the primary obstacles to full participation. Inclusive education seeks to dismantle these barriers by making curricula, teaching strategies, and learning resources more accessible, ensuring every student can engage meaningfully with the material.

3.2.2 International Frameworks Supporting Inclusive Education

The shift towards inclusive education has been strengthened by international frameworks and legal mandates acknowledging education as a fundamental right for everyone. The United Nations Convention on the Rights of Persons with Disabilities (UNCRPD) and the Salamanca Statement have significantly contributed to advancing inclusive policies globally. Established in 1994, the Salamanca Statement urged governments to implement inclusive practices, asserting that children with special educational needs must have access to mainstream schools that can accommodate them (UNESCO, 1994). The UNCRPD, adopted in 2006, underscores the rights of individuals with disabilities to engage fully in societal activities, including education. It asserts that students with disabilities must have access to inclusive and accessible learning environments and that educational institutions must provide reasonable accommodations to support their participation. These frameworks have influenced national policy and guided schools and educators in adopting practices prioritising inclusivity as a fundamental value.

3.3 Challenges and Opportunities in Inclusive Education

Despite progress, implementing inclusive education poses numerous challenges, particularly when balancing the diverse needs of students with limited resources. Funding constraints, societal attitudes, teacher preparedness, and infrastructure limitations hinder inclusive practices. However, inclusive education also presents unique opportunities to improve educational systems, fostering a more equitable and socially responsible learning environment.

3.3.1 Challenges in Implementing Inclusive Education

One of the main challenges in inclusive education is the availability of resources. Supporting diverse learning needs requires funding for specialized equipment, assistive technologies, and training for educators. In low-income areas, schools often need more resources to fully implement inclusive practices, resulting in disparities in access and quality. Additionally, some educators may need adequate training in inclusive teaching strategies, limiting their ability to effectively support students with diverse needs. A study by Ainscow (2005) highlights that teacher preparedness and ongoing professional development are crucial for fostering an inclusive mindset and equipping teachers with the skills to manage diverse classrooms.

Societal attitudes toward disability and diversity also influence the effectiveness of inclusive education. In some communities, cultural beliefs about disability may contribute to stigma, affecting students' self-esteem and willingness to participate.

Overcoming these attitudes requires a cultural shift within both schools and communities, encouraging greater understanding and acceptance of diverse abilities. Awareness campaigns, inclusive curricula, and visible commitment from school leadership can help shift societal perceptions, fostering a more accepting environment.

Infrastructure is another barrier, particularly in schools that were not designed with accessibility in mind. Physical barriers, such as stairs without ramps or classrooms without space for mobility aids, can hinder students' ability to engage fully in the learning environment. Addressing these infrastructural challenges is essential to creating a physically inclusive setting that supports all students' needs.

3.3.2 Opportunities to Enhance Educational Systems

Despite these challenges, inclusive education offers several opportunities to improve educational systems. Schools can foster a more supportive, compassionate, and collaborative culture by adopting inclusive practices. Inclusive education also prepares students for life in diverse societies, equipping them with the interpersonal skills and cultural awareness necessary to navigate complex social environments.

Furthermore, inclusive education promotes innovative teaching strategies that benefit all students, not just those with specific needs. For instance, Universal Design for Learning (UDL) is an educational framework that encourages flexible teaching methods, multiple forms of engagement, and diverse assessment strategies. Originally designed to support students with disabilities, UDL has proven beneficial for all students by offering varied ways to interact with and understand course material (Popenici & Kerr, 2017).

Inclusive education also advances the broader goals of social justice and equality. By providing equal opportunities and dismantling barriers, inclusive schools contribute to a more equitable society where diversity is viewed as a strength rather than a challenge. This perspective empowers students to advocate for themselves and others, fostering a sense of agency and social responsibility that extends beyond the classroom.

Inclusive education is a dynamic and evolving approach that seeks to create equitable, diverse, and supportive learning environments. Rooted in principles of equity, diversity, support, and collaboration, it emphasises the importance of adapting educational systems to meet the varied needs of students. International frameworks such as the UNCRPD and the Salamanca Statement have been pivotal in advancing inclusive education policies and promoting access and participation for students with disabilities and marginalised communities. While resource limitations, societal attitudes, and infrastructure persist, inclusive education presents valuable opportunities to improve educational systems and promote social justice.

4. AI AND INCLUSIVE EDUCATION: OPPORTUNITIES

Artificial Intelligence (AI) can revolutionise inclusive education by providing tools and systems to address varied learning requirements. Inclusive education aims to establish an equitable learning environment for all students, irrespective of their background, abilities, or socioeconomic status. AI contributes to this objective by personalising learning experiences, creating assistive technologies for students with disabilities, and facilitating adaptive curriculum design accommodating diverse learning styles and paces. This section examines AI's distinct opportunities for inclusive education, concentrating on three main areas: individualised learning, assistive technology, and adaptive curriculum design. Examining these areas provides insights into how AI can address traditional educational barriers, fostering more accessible, supportive, and inclusive classrooms.

4.1 Individualized Learning and Personalized Education

One of AI's most transformative impacts on education lies in its ability to individualise the learning experience. Traditional teaching models tend to follow a standardised approach, which does not account for individual students' diverse skills, preferences, and learning speeds. AI, however, can tailor instruction to each learner's unique profile by analysing vast amounts of data, such as past performance, engagement levels, and areas of difficulty (Luckin et al., 2016). This level of customisation is particularly beneficial in inclusive education settings, where students with different abilities and backgrounds require varied forms of support.

4.1.1 AI-Powered Adaptive Learning Platforms

Adaptive learning platforms like Smart Sparrow, DreamBox, and Knewton signify a notable progression in personalised education. These platforms employ algorithms to evaluate a student's responses and modify the difficulty and content in real-time, ensuring that each student receives appropriately challenging instruction. For example, when a student encounters difficulty with a concept, the system can offer supplementary explanations, exercises, or alternative resources until mastery is achieved (Popenici & Kerr, 2017). This adaptability enables students to learn independently, enhancing engagement and information retention.

Adaptive learning has the potential to significantly enhance educational outcomes for students with disabilities within inclusive education frameworks. A student with dyslexia may utilise text-to-speech assistance, whereas a student with ADHD may benefit from interactive exercises designed to enhance focus. This customised strategy is crucial for students who may experience feelings of being overwhelmed or disadvantaged in a conventional classroom environment. Individualised instruction through AI fosters a learning environment conducive to the success of all students, irrespective of their abilities.

4.1.2 Real-Time Feedback and Continuous Assessment

The ability of AI to deliver real-time feedback is a significant asset for inclusive education. Conventional educational settings frequently depend on intermittent evaluations, resulting in postponed recognition of learning challenges. AI-driven platforms provide immediate feedback, enabling students to recognise their errors and enhance their performance promptly. This ongoing assessment facilitates a responsive learning environment, allowing educators to track individual student progress and modify instruction accordingly (Baker & Smith, 2019).

Suppose a student consistently struggles with a specific topic. In that case, the AI system can notify the teacher, enabling the provision of additional support or adjustments to the lesson plan to meet the student's needs. This responsiveness guarantees that students obtain necessary assistance before minor gaps in understanding evolve into significant learning barriers. Continuous feedback enhances learning outcomes and empowers students by providing them with information regarding their progress.

4.1.3 Personalized Learning Pathways for Special Needs Education

In special education, individualized learning pathways are critical for addressing the unique needs of each student. AI systems can create these pathways by adapting content and interactions based on a student's cognitive and physical requirements. For instance, AI tools designed for students with autism spectrum disorder (ASD) can offer structured, predictable learning environments, which help these students engage with the material comfortably. Similarly, students with speech and language impairments may benefit from AI-powered language processing tools that translate spoken instructions into visual or text-based formats (Holmes et al., 2019).

These personalised pathways ensure that students with special needs are not left behind in the learning process. By addressing their specific challenges, AI creates a more equitable educational experience, empowering students with disabilities to participate fully in the classroom.

4.2 Assistive Technology for Students with Disabilities

AI-driven assistive technologies provide critical support to students with disabilities, enabling them to participate more actively in their education. Traditional educational tools often fall short of meeting the needs of students with physical, sensory, or cognitive disabilities. AI-based assistive technologies, however, can help overcome these barriers by offering alternative ways to access and interact with educational content.

4.2.1 Speech Recognition and Text-to-Speech Technologies

One of the most widely used assistive technologies is speech recognition, which converts spoken language into text. This technology is invaluable for students with physical disabilities that limit their ability to write or type and those with learning disabilities that affect their ability to process written information. Applications like Google Speech-to-Text and Dragon NaturallySpeaking enable students to dictate their responses and participate in written assignments, levelling the playing field with their peers.

Conversely, text-to-speech (TTS) technology assists students with reading disabilities, such as dyslexia, by reading written content aloud. This technology allows students to access textbooks, assignments, and digital resources without traditional reading. AI-driven TTS tools can adjust reading speed, intonation, and language to accommodate individual preferences, providing a more tailored learning experience (Adebisi et al., 2015).

4.2.2 Eye-Tracking Technology and Alternative Input Methods

For students with severe physical disabilities, traditional input devices like keyboards or touchscreens may be inaccessible. AI-driven eye-tracking systems offer an innovative solution, enabling students to control computers using only their eye movements. By tracking where a student looks on the screen, eye-tracking technology allows them to select options, type responses, or navigate educational software without physical interaction (Holmes et al., 2019). This technology supports learning and fosters independence and self-confidence among students who would otherwise struggle to interact with digital tools.

In addition to eye-tracking, AI-powered alternative input methods include head gestures and voice commands, which provide students with a range of options for controlling digital devices. These input methods make it possible for students with mobility impairments to engage fully in digital learning environments, from completing assignments to participating in virtual classroom discussions.

4.2.3 Real-Time Translation and Language Support

Real-time translation and language support are crucial for comprehending classroom content for non-native speakers and students with hearing impairments. AI-powered translation tools, such as Google Translate and Microsoft Translator, facilitate immediate spoken or written language translations, enhancing student engagement with the material. Furthermore, AI-driven transcription services provide real-time captions for live or recorded lessons, benefiting students who are deaf or hard of hearing (Luckin et al., 2016). By addressing language barriers, these AI technologies foster an inclusive environment, enabling all students to access and engage with content, irrespective of their linguistic background or hearing ability.

4.3 Adaptive Curriculum Design

One of the most significant contributions AI can make to inclusive education is in adaptive curriculum design. Adaptive curricula automatically adjust to a student's progress, needs, and interests, creating a dynamic learning experience that remains relevant and engaging. Unlike static curricula, which follow a fixed structure regardless of individual student differences, adaptive curricula allow flexibility and personalisation, essential for creating a truly inclusive learning environment.

4.3.1 AI-Powered Adaptive Learning Paths

Adaptive learning paths are designed to respond to the pace and needs of individual students, offering more advanced content to those who excel while providing additional support to those who struggle. For example, an AI-driven platform might present students with progressively challenging problems if they master a topic, or it might repeat fundamental concepts for students who need reinforcement. This adaptability ensures that all students are appropriately challenged and supported, regardless of their starting point (Popenici & Kerr, 2017).

In inclusive education, adaptive learning paths are particularly beneficial for students with cognitive disabilities or learning disorders, who may require a slower pace or more repetition to grasp certain concepts. By tailoring instruction to meet each student's needs, adaptive curricula help prevent students from feeling overwhelmed or disengaged, promoting a more equitable educational experience.

4.3.1 Scaffolded Learning and Skill Progression

A scaffolded learning educational strategy divides challenging assignments into smaller, more doable segments. AI-powered adaptive curricula facilitate scaffolded learning by determining when a student is prepared to go on to the next level and providing the assistance required to close the gap between various ability levels. For children with learning disabilities, this slow, methodical approach is crucial because it enables them to gain competence and confidence in each ability before going on. An adaptable curriculum could help students with trouble understanding mathematical ideas by first introducing them to the fundamentals. This way, they can practise and become proficient in each stage before moving on to more complex subjects. Because varied learning demands frequently call for varying degrees of support and reinforcement, this organised method is especially beneficial for inclusive education.

4.3.2 Data-Driven Insights for Curriculum Development

AI's ability to analyse large amounts of data offers educators valuable insights into student performance, preferences, and learning challenges. By collecting and analysing this data, AI systems can help educators identify areas where the curriculum may need adjustment to serve students with diverse needs better. For instance, if data reveals that a particular topic consistently poses difficulties for students with language processing disorders, educators can modify the curriculum to include alternative explanations, visual aids, or interactive exercises.

Data-driven curriculum design also enables educators to track student progress over time, adjusting the learning path as students develop new skills or encounter new challenges. This responsiveness is crucial for inclusive education, where the curriculum must be flexible enough to accommodate students' evolving needs while supporting their growth and development.

AI offers transformative opportunities for enhancing inclusive education by supporting individualized learning, enabling assistive technology, and fostering adaptive curriculum design. These advancements make it possible to accommodate diverse learning needs, from students with disabilities to those facing language barriers or other challenges. AI-driven adaptive learning platforms, assistive tools, and curriculum design frameworks create a more supportive, responsive, and accessible educational environment, promoting equity and inclusion in ways traditional education models cannot achieve.

5. CHALLENGES IN AI IMPLEMENTATION FOR INCLUSIVE EDUCATION

The integration of Artificial Intelligence (AI) into inclusive education offers tremendous potential to support personalized learning, assistive technologies, and adaptive curricula. However, this transformation is accompanied by significant challenges, particularly in algorithmic bias, data privacy, and the digital divide. While AI-driven tools can enhance educational accessibility and engagement, their effectiveness depends heavily on ethical and thoughtful implementation. These challenges, if addressed, can impede the goal of providing equitable learning opportunities and may even exacerbate existing educational inequalities. This section explores these primary challenges in detail, examining how each affects the inclusive potential of AI and suggesting ways that educators, developers, and policymakers can mitigate these issues to create a fairer, more accessible educational environment.

5.1 Algorithmic Bias

Algorithmic bias constitutes a significant challenge in applying AI for inclusive education. AI systems are developed using historical data that embodies societal prejudices and inequality, perhaps resulting in unintentional discriminatory consequences. An AI tool intended to forecast student performance may be trained on data that encompasses biases associated with socioeconomic status, ethnicity, or language competency. Thus, these systems may unintentionally disadvantage students from marginalised groups, reinforcing the disparities that inclusive education seeks to eradicate (O'Neil, 2016). A significant example of algorithmic bias in education transpired inside standardised testing. AI systems employed to assess written responses in examinations exhibited a bias towards pupils adhering to specific linguistic and stylistic traditions, frequently disadvantaging non-native speakers or those from varied cultural backgrounds. In the context of inclusive education, these biases pose considerable ethical dilemmas, as they undermine the fundamental values of justice and accessibility (Holmes et al., 2019). Mitigating algorithmic bias necessitates a multifaceted strategy, encompassing the use of varied training datasets that reflect a broad spectrum of student demographics, systematic bias audits, and transparency in algorithmic decision-making procedures. By designing AI models with fairness and inclusion, developers can mitigate biassed outcomes and foster equitable learning opportunities.

5.2 Data Privacy Concerns

AI in education necessitates collecting and analysing vast amounts of student data, including personal information, academic records, behavioural patterns, and, in some cases, even biometric data. While this data enables personalised and adaptive learning experiences, it raises significant privacy concerns, particularly for vulnerable student populations such as those with disabilities. Misusing or unauthorised access to sensitive information can lead to discrimination or stigmatisation within and beyond educational contexts (Selwyn, 2020).

For instance, data indicating that a student has a learning disability could, if improperly handled, influence future academic or employment opportunities, contributing to a cycle of disadvantage. To mitigate these risks, educational institutions and AI developers must prioritise data security by implementing stringent privacy safeguards and adhering to established regulations. Legal frameworks such as the General Data Protection Regulation (GDPR) in Europe and the Family Educational Rights and Privacy Act (FERPA) in the United States set standards for data protection, mandating transparency, consent, and accountability in data handling (Slade & Prinsloo, 2013). However, compliance with these regulations varies widely, and enforcement remains challenging, especially as educational institutions and AI developers navigate the balance between personalization and privacy.

Educators and administrators can play a proactive role in safeguarding student privacy by implementing data minimisation practices—only collecting data essential for educational purposes—and ensuring that students and parents are fully informed about how their information is used. Additionally, adopting anonymization and encryption technologies can protect sensitive data, reducing the risk of privacy breaches. By embedding robust data privacy practices into AI-powered educational systems, institutions can foster a culture of trust and responsibility, thereby supporting the ethical use of AI in inclusive education.

5.3 The Digital Divide

The digital divide hinders equity in AI applications in education. Poor, rural, and underserved students typically lack the devices, internet, and infrastructure to use AI-driven learning tools. This access barrier worsens educational inequality by denying marginalised kids the same learning chances as their more privileged peers (Robinson et al., 2020). Schools with low resources may need more hardware or software to deploy AI-driven adaptive learning platforms, which can provide personalised training to individual learning styles. Students may struggle to complete online assignments, engage in online lectures, or access educational materials in areas with inconsistent or expensive internet connectivity. These differences emphasise the need to solve infrastructural constraints before applying AI solutions in education, especially in low-resource countries.

Governments, schools, and tech corporations must work together to close the digital divide. Government investments in technology infrastructure, such as rural broadband internet access and low-income student digital device subsidies, can be vital. Educational institutions can also lobby technology vendors for affordable or free digital learning resources. Google's Chromebook program gives low-cost devices to schools in need, making technology more accessible (Holmes et al., 2019). More than providing devices and internet access is needed to close the digital divide; students, educators, and families require digital literacy skills to use AI-driven applications. Teachers and students can learn how to use AI technologies safely and efficiently through training programs and workshops. Stakeholders may create a more inclusive and fair education system by addressing the digital divide's infrastructure and education.

5.4 Addressing the Challenges: A Path Forward

Overcoming these challenges requires a coordinated and multi-faceted approach involving all educational ecosystem stakeholders, from policymakers and educators to AI developers and families. One of the first steps in addressing algorithmic bias, data privacy concerns, and the digital divide is the establishment of ethical frameworks and policy guidelines that prioritize equity, inclusivity, and accountability in AI deployment.

5.4.1 Promoting Ethical AI Development and Use

Ethical AI development begins with the design and testing phases, where developers must ensure that AI systems are free from biases that could harm marginalised students. This includes using representative datasets, implementing regular bias audits, and maintaining transparency in how AI tools make decisions. Policymakers can support these efforts by mandating ethical standards for AI in education, such as using DEI (Diversity, Equity, and Inclusion) principles in developing educational AI systems. Regularly evaluating AI tools for fairness and inclusivity can help prevent unintended discriminatory outcomes, reinforcing the commitment to providing equitable learning experiences.

5.4.2 Strengthening Data Privacy Protocols

Data privacy requires stringent protection measures, including encryption, anonymisation, and secure storage protocols to safeguard sensitive information. Educational institutions should implement clear policies on data handling, ensuring compliance with data protection regulations like GDPR and FERPA. Additionally, schools can promote digital literacy among students and families, informing them of their rights and the importance of privacy in digital learning environments. Building a strong culture of privacy awareness in schools fosters trust and encourages responsible data management practices.

5.4.3 Reducing the Digital Divide

Closing the digital divide requires significant investment in digital infrastructure, particularly in underserved areas. Government programs that fund technology access for low-income students, such as subsidies for internet services and discounted devices, can help bridge this gap. Additionally, public-private partnerships can play a critical role in providing school resources, ensuring that all students can benefit from AI-powered learning tools. Digital literacy training is equally crucial, empowering students, teachers, and families to use technology effectively, thereby maximising the potential of AI in inclusive education.

While AI promises to advance inclusive education, challenges like algorithmic bias, data privacy concerns, and the digital divide must be addressed to realise its full potential. Algorithmic bias threatens equity, data privacy breaches can undermine trust, and the digital divide continues to limit access to AI-driven learning tools. To create an equitable and inclusive educational environment, stakeholders must commit to ethical AI development, robust data privacy practices, and infrastructure improvements supporting universal technology access. By proactively addressing these challenges, educators, policymakers, and developers can harness AI's transformative power to create a more inclusive, accessible, and supportive educational system for all students.

6. ETHICAL CONSIDERATIONS IN AI FOR INCLUSIVE EDUCATION

Ethical issues are crucial when AI is incorporated into education more and more, particularly when it comes to inclusive learning. Justice, openness, and respect for students' rights must be prioritised when implementing AI in education. Ethical issues like algorithmic fairness, privacy, and responsibility are crucial for AI to support educational equity rather than erode it. Human-centred AI development and incorporating diversity, equality, and inclusion (DEI) concepts into designing and implementing AI systems are two major ethically significant issues.

6.1 Human-Centered AI Development

A human-centred approach to AI in education emphasises designing AI systems that support, rather than replace, educators and foster positive learning outcomes. Human-centred AI development is vital in inclusive education, where AI applications must align with values prioritising human well-being, autonomy, and informed decision-making. In educational settings, this approach means ensuring that AI systems enhance, rather than detract from, the role of teachers as facilitators and mentors (Eubanks, 2018). For instance, AI should aid educators by providing insights into student progress and needs rather than making isolated decisions about a student's academic trajectory.

In addition to supporting educators, a human-centred approach to education emphasises transparency and explainability. AI systems that impact students and teachers should have precise mechanisms that explain how decisions are made and provide users with control over these decisions. For example, if an AI tool is used to assess student performance or recommend remedial actions, educators, students, and parents should understand the criteria and factors influencing those recommendations (Holmes et al., 2019). This transparency builds trust and helps ensure that AI tools are used responsibly and effectively.

Accountability is another crucial component of human-centred AI. Developers, educators, and administrators must have precise accountability mechanisms to address potential biases or errors in AI systems. For instance, if an AI grading tool shows patterns of unfair scoring for specific student groups, schools need processes to investigate and correct these issues promptly. Therefore, human-centred AI in inclusive education requires that AI technologies are not only practical but also respect ethical standards that place the well-being of all stakeholders—students, teachers, and families—at the forefront.

6.2 Diversity, Equity, and Inclusion (DEI) in AI Design

To ethically implement AI in inclusive education, embedding DEI principles within AI systems is crucial. These principles ensure that AI applications recognise and support diverse learning needs, providing fair and equitable treatment for all students. A DEI-focused approach in AI development starts by involving diverse voices in the design process, including teachers, students, and community members from marginalised backgrounds. Involving these perspectives helps ensure that AI tools do not unintentionally reinforce biases or exclude particular student groups (West et al., 2019).

Designing AI systems that promote DEI involves creating adaptive tools that cater to a wide range of learning needs. For instance, AI tools can provide language support for students learning their non-native languages or offer accessibility features such as screen readers and text-to-speech options for students with disabilities. When designed with inclusivity in mind, AI systems can be powerful tools for supporting diverse learning experiences. Additionally, DEI-focused AI systems should be regularly evaluated to ensure they do not exacerbate existing inequalities. For example, an AI system should be tested for biases affecting students from various socioeconomic backgrounds or those with unique learning needs (Baker & Smith, 2019).

In practice, DEI-oriented AI in education should aim to identify and mitigate disparities in student performance or access to resources. For instance, if an AI system recognises that certain groups consistently receive lower grades or access to fewer educational opportunities, educators can intervene to address these disparities. By prioritising diversity, equity, and inclusion in AI design, academic institutions can help create environments where every student has an equal chance to succeed, reinforcing the ethical foundation of AI in inclusive education.

7. POLICY RECOMMENDATIONS FOR ETHICAL AI INTEGRATION IN INCLUSIVE EDUCATION

For AI to be used ethically and effectively in inclusive education, comprehensive policy recommendations are needed to guide its implementation. These recommendations should address AI deployment's ethical, operational, and infrastructural aspects, ensuring that educational technologies enhance inclusivity and do not perpetuate existing inequalities. Below are several key policy recommendations to promote responsible AI use in education.

7.1 Implementing Bias Audits and Transparency Measures

Educational institutions should require regular bias audits for AI systems to prevent the perpetuation of biases. Bias audits entail systematically evaluating AI algorithms to identify potential biases, ensuring no student group is disadvantaged. Audits must assess the training data of AI models and the algorithms utilised to identify and address potential biases. For example, if an AI tool designed to evaluate student performance exhibits bias against students from specific socioeconomic backgrounds, this problem can be rectified before negatively affecting student outcomes (O'Neil, 2016).

Transparency measures are essential for fostering ethical AI utilisation. Educational institutions must guarantee that AI systems employed in classrooms and administration exhibit transparency regarding their decision-making processes. Access to information regarding the functioning of AI algorithms and the data they utilise for decision-making is essential for students, parents, and educators. Transparent AI systems enable stakeholders to comprehend the rationale behind AI-generated recommendations or assessments, fostering trust and supporting informed decision-making in educational contexts (Selwyn, 2020).

7.2 Strengthening Data Privacy Protections

Educational AI systems depend on extensive datasets, a significant portion containing sensitive and personally identifiable information. Policymakers must implement strong data privacy protections for AI applications in education to safeguard student privacy. Educational institutions must adhere to regulations like the General Data Protection Regulation (GDPR) in Europe and the Family Educational Rights and Privacy Act (FERPA) in the United States, establishing data protection standards. These regulations mandate the secure storage of data collected by AI systems, stipulate that it be utilised solely for legitimate educational purposes, and restrict access to authorised personnel (Slade & Prinsloo, 2013).

In addition to regulatory compliance, educational institutions should establish policies restricting the volume of data accessible to AI systems, thereby minimising the potential for data breaches or misuse. Furthermore, it is essential for parents and students to possess authority over the utilisation of their data and to have the option to request its deletion when deemed necessary.

7.3 Bridging the Digital Divide

The digital gap regarding fair access to AI in schools still needs to be solved. Students in underserved areas often don't have the gadgets, internet access, or other resources they need to use AI to improve their learning. To solve this problem, policymakers should put money into technology infrastructure and give schools in low-income places more resources. To ensure that all students have the same chances to use AI tools in their education, the government could pay for students' internet access or give them gadgets if needed (Robinson et al., 2020).

Not only does closing the digital gap make AI more straightforward to use, but it also follows the moral principle of equity by ensuring that no student is left behind because of their family's financial situation. Supporting technology access should also include ongoing technical support and digital literacy programs to help teachers and students get the most out of teaching tools that AI powers.

7.4 Promoting Human-Centered AI Development

To ensure that AI in education is ethically designed, policymakers should advocate for human-centred AI development principles. These principles prioritise transparency and accountability and enhance human teaching roles rather than replace them. Policies emphasising human-centred AI development encourage developers to create tools that assist teachers in supporting students rather than automating all aspects of education (Eubanks, 2018). For example, human-centred AI tools can help teachers identify areas where students struggle. Still, the final intervention decision should rest with educators, who bring a nuanced understanding of each student's needs. By promoting human-centred AI development, policies can help ensure that AI remains a supportive resource that enhances, rather than replaces, the essential human elements of education.

7.5 Integrating DEI Frameworks in AI Development and Implementation

Policymakers should incorporate diversity, equity, and inclusion frameworks in developing and implementing AI systems to ensure alignment with these principles. This involves establishing standards for inclusivity in data collection, algorithm design, and system implementation, ensuring that AI systems address diverse student needs. AI systems must undergo evaluation for cultural sensitivity and accessibility to support students with disabilities and those from diverse linguistic backgrounds (West et al., 2019). Policies prioritising diversity, equity, and inclusion in AI development foster educational settings where all students experience representation and support. Furthermore, these frameworks promote continuous evaluation to recognise and rectify any disparities that emerge in AI-driven educational results. Educational institutions can use AI with DEI-oriented policies to foster inclusive and equitable learning environments that serve all students effectively. Ethical considerations are essential for ensuring that AI effectively supports inclusive education. Regular bias audits, increased transparency, student data protection, addressing the digital divide, promoting human-centred development, and integrating DEI frameworks are essential strategies for policymakers to ensure the ethical integration of AI in education. These policies create a framework where AI addresses varied student requirements while maintaining fairness, transparency, and privacy considerations. As AI evolves, following these policy recommendations will be crucial for optimising the advantages of AI while maintaining ethical standards in inclusive education.

8. CONCLUSION

Artificial Intelligence (AI) can revolutionise the educational framework, especially in promoting inclusive education by accommodating varied learning requirements and overcoming conventional obstacles to access. The uses of AI in education, including administrative automation, personalised learning pathways, and increased engagement tools, are essential for enhancing efficiency and delivering customised support. Nevertheless, the ethical dilemmas related to AI, especially algorithmic bias, data privacy, and the digital divide, pose substantial barriers that must be confronted to guarantee that AI fosters a genuinely inclusive educational experience. A primary advantage of AI in education is its capacity to tailor learning experiences for individual students. The influence of AI on engagement presents significant potential. By integrating technology like Intelligent Tutoring Systems (ITS), Virtual Reality (VR), and Augmented Reality (AR), educators may provide more engaging and immersive educational experiences. Virtual Reality (VR) and Augmented Reality (AR) enable students to investigate intricate topics in science or history via visual and immersive approaches, enhancing engagement and accessibility for individuals who may find text-based materials challenging (Baker & Smith, 2019). Moreover, AI-driven gamification incorporates game-like features into education, incentivising students through prizes, challenges, and feedback, enhancing student engagement and enjoyment in learning. This degree of engagement is especially advantageous for students who may feel alienated from conventional learning methods, promoting a constructive and inclusive educational environment. Notwithstanding these benefits, the ethical issues associated with AI in education require meticulous management. Algorithmic bias is a significant concern since AI models trained on prejudiced data can perpetuate inequities and disadvantage marginalised populations (O'Neil, 2016). To alleviate these dangers, educational AI systems must be designed with broad data sets and undergo regular bias assessments to ensure equity. Moreover, data privacy is critical, especially in educational environments where sensitive student information is at stake. Regulatory frameworks like the General Data Protection Regulation (GDPR) in Europe and the Family Educational Rights and Privacy Act (FERPA) in the United States offer guidance; however, educational institutions must also establish robust safeguards and foster transparency in data utilisation (Selwyn, 2020). The digital gap continues to be a significant obstacle to equal AI integration in education. Many students, especially in low-income and rural regions, need more gadgets and internet connectivity to utilise AI-driven solutions effectively. Addressing this disparity necessitates cooperative initiatives among governments, educational entities, and technology suppliers to guarantee that all students can access digital resources. In conclusion, although AI possesses significant potential for improving inclusive education, its ethical integration is crucial for effectively actualising this potential. A strategy centred on humanity that emphasises diversity, equity, and inclusion (DEI) will guarantee that AI improves learning and maintains the principles of an inclusive educational framework. By mitigating biases, ensuring data privacy, and enhancing accessibility, educators and policymakers can utilise AI as a revolutionary instrument for creating more equal and inclusive educational settings. As the domain progresses, continuous study and discourse on the ethical application of AI in education will be crucial to fostering a future in which AI authentically enhances each learner's potential.

REFERENCES

Ainscow, M. (2005). Developing inclusive education systems: What are the levers for change? *Journal of Educational Change*, 6(2), 109–124. DOI: 10.1007/s10833-005-1298-4

Baker, T., & Smith, L. (2019). Educ-AI-tion rebooted? Exploring the future of artificial intelligence in schools and colleges. *Nesta Report.*

Eubanks, V. (2018). *Automating inequality: How high-tech tools profile, police, and punish the poor*. St. Martin's Press.

Heffernan, N., & Koedinger, K. (2012). The future of educational technology: Adaptive learning and intelligent tutoring systems. In Dede, C., & Richards, J. (Eds.), *Digital Teaching Platforms*. Harvard Education Press.

Luckin, R., Holmes, W., Griffiths, M., & Forcier, L. B. (2016). *Intelligence Unleashed: An argument for AI in education*. Pearson.

O'Neil, C. (2016). *Weapons of math destruction: How big data increases inequality and threatens democracy*. Crown Publishing.

Popenici, S., & Kerr, S. (2017). Exploring the impact of artificial intelligence on teaching and learning in higher education. *Research and Practice in Technology Enhanced Learning*, 12(22), 1–13. DOI: 10.1186/s41039-017-0062-8 PMID: 30595727

Selwyn, N. (2020). The educational implications of AI: A critical perspective. *Learning, Media and Technology*, 45(1), 1–11. DOI: 10.1080/17439884.2020.1694945

Slade, S., & Prinsloo, P. (2013). Learning analytics: Ethical issues and dilemmas. *The American Behavioral Scientist*, 57(10), 1510–1529. DOI: 10.1177/0002764213479366

UNESCO. (1994). *The Salamanca statement and framework for action on special needs education*. UNESCO.

West, S. M., Whittaker, M., & Crawford, K. (2019). Discriminating systems: Gender, race, and power in AI. *AI Now Institute Report.*

Chapter 16
Ethical Design Thinking Embedding Moral Decision Making in AI Development Projects

Chaimae Waladi
https://orcid.org/0009-0008-0569-4423
Abdelmalek Essaadi University, Morocco

Mohammed Sefian Lamarti
https://orcid.org/0000-0001-8270-2660
Abdelmalek Essaadi University, Morocco

Mohamed Khaldi
https://orcid.org/0000-0002-1593-1073
Abdelmalek Essaadi University, Morocco

ABSTRACT

As Artificial Intelligence (AI) technologies rapidly evolve, addressing their ethical implications is crucial. This chapter explores how AI developers can integrate ethical decision-making into the design process using a hands-on, design-thinking approach. Design thinking emphasizes empathy, collaboration, and iterative problem-solving, which helps developers consider the diverse needs of all stakeholders and avoid ethical pitfalls such as bias, lack of transparency, and privacy concerns. By applying this approach, developers can create AI systems that are fair, transparent, and aligned with societal values. The chapter includes case studies that demonstrate the practical application of design thinking in mitigating bias, improving transparency, and safeguarding privacy. Additionally, it highlights the societal impact of

DOI: 10.4018/979-8-3373-2262-9.ch016

AI, encouraging developers to consider the long-term consequences of their work.

1 INTRODUCTION

As artificial intelligence (AI) continues to revolutionize industries and reshape everyday life, the ethical implications of its development have become a central concern for technologists, policymakers, and society at large. The rapid deployment of AI in areas such as healthcare, criminal justice, and finance highlights its transformative potential, but it also exposes critical ethical challenges that must be addressed to ensure AI serves the greater good (Floridi et al., 2018). Issues such as algorithmic bias, lack of transparency, and unintended social consequences have sparked debates about how AI should be designed, deployed, and governed to avoid harm. These concerns underscore the importance of embedding ethical decision making directly into the development process of AI systems.

The ethical dilemmas in AI development are not merely theoretical—they manifest in real world applications. For instance, algorithmic bias in facial recognition systems has led to inaccuracies, particularly in identifying people of color, raising concerns about discrimination and the amplification of societal inequities (Buolamwini & Gebru, 2018). Similarly, AI's use in criminal justice for risk assessment has prompted questions about fairness, transparency, and accountability (Angwin et al., 2016). These examples reflect the pressing need for AI professionals to integrate moral considerations into their design processes to ensure that AI technologies promote justice, fairness, and societal well being.

One approach that offers a promising framework for addressing these ethical challenges is design thinking. Originating from the fields of architecture and product design, design thinking has gained widespread recognition as a human centered, iterative approach to problem solving. At its core, design thinking emphasizes empathy with users, collaboration among diverse stakeholders, and a process of prototyping and testing that allows for continuous refinement (Brown, 2009). By integrating ethical reflection into each stage of the design thinking process—empathize, define, ideate, prototype, and test—AI developers can more effectively anticipate the ethical implications of their systems and make adjustments to align with moral principles (Dignum, 2019).

Design thinking's focus on user experience and its iterative nature make it particularly well suited for embedding ethical considerations into AI development. The initial phase of empathizing with users allows developers to identify the potential impacts of their AI systems on different user groups, particularly those who are most vulnerable to harm. This empathetic approach ensures that ethical considerations are not an afterthought but are embedded from the start. As the design process moves forward, ethical principles such as fairness, accountability, and transparency can be integrated into the development of AI systems, helping to mitigate risks and prevent harm before products are released into the world (Jobin et al., 2019).

This chapter will explore how AI developers can apply a design thinking framework to embed moral decision making into their work. Through real world case studies, practical examples, and discussions about the societal impact of AI, the chapter aims to provide AI professionals with the tools and insights needed to foster ethical awareness and develop responsible AI systems. Additionally, the chapter will highlight the importance of stakeholder involvement in ethical AI design, showing how collaboration with ethicists, policymakers, and diverse user groups can enrich the design process and lead to more ethical outcomes.

In a world where AI is increasingly intertwined with critical societal functions, the responsibility to ensure that AI development aligns with moral values has never been more urgent. The failure to address ethical issues early in the design process can lead to negative societal impacts, from reinforcing biases to eroding trust in technology (Cath, 2018). Therefore, AI professionals must adopt proactive strategies for embedding ethical decision making into their design processes, ensuring that their innovations contribute to societal well being rather than exacerbate existing inequities.

The chapters that follow will delve deeper into the specific steps AI developers can take to integrate ethics into their design process. By applying a hands on, design thinking approach, AI developers can better navigate the complexities of ethical AI development and create solutions that are not only innovative but also morally sound.

2 FOUNDATIONS OF ETHICAL AI DEVELOPMENT

Ethics is a cornerstone of responsible AI development. As AI systems increasingly interact with and influence human decisions, it is critical that these systems are designed to operate in ways that are aligned with human values. The foundation of ethical AI development lies in several key principles: fairness, transparency, accountability, and the promotion of societal well being. These principles provide the moral guidelines that developers must follow to ensure that AI systems do not exacerbate existing inequalities or create new forms of harm (Floridi & Taddeo, 2016).

2.1 Fairness

Fairness is perhaps the most discussed ethical principle in AI development. In practice, fairness refers to the idea that AI systems should make decisions without discriminating against individuals or groups based on characteristics such as race, gender, socioeconomic status, or other protected attributes. A common challenge in AI is the presence of biased training data, which can lead to systems that perpetuate and even amplify existing societal biases (Mehrabi et al., 2021). For instance, in recruitment processes, AI algorithms have been shown to favor candidates from certain demographics while disadvantaging others due to biased historical hiring data (Raghavan et al., 2020). Addressing these biases requires rigorous auditing and the implementation of fairness constraints during the development process to ensure that AI models treat all individuals equitably.

In design thinking, fairness can be embedded in the empathize and define phases, where developers identify the stakeholders and potential impact groups that may be affected by the AI system. By involving diverse voices during these early stages, AI developers can better understand the potential risks of bias and implement measures to mitigate them (Gebru et al., 2018).

2.2 Transparency

Transparency in AI refers to the clarity with which AI systems operate, make decisions, and generate outcomes. As AI models become more complex, especially those based on deep learning, the decisions made by these systems often become opaque even to the developers who created them. This phenomenon, known as the "black box" problem, has raised concerns about the accountability of AI systems, particularly when they are used in high stakes domains such as healthcare, law enforcement, or finance (Burrell, 2016). Transparent AI systems should provide explanations that are understandable to users, allowing them to comprehend how a decision was reached and giving them the ability to contest or question the system's outcomes when necessary (Weller, 2019).

Design thinking supports transparency by encouraging iterative prototyping and testing, where users can interact with the AI system and provide feedback. Prototypes should be designed to offer clear explanations of AI decision making processes, ensuring that transparency is prioritized throughout development.

2.3 Accountability

The principle of accountability ensures that there is a clear understanding of who is responsible for the decisions made by an AI system. This is particularly crucial when AI systems are used in critical areas, such as autonomous driving or medical diagnostics, where errors could lead to significant harm. Establishing accountability means that developers, companies, and stakeholders must take responsibility for the AI's behavior, ensuring that they can intervene if something goes wrong (Danaher et al., 2017). In the absence of accountability, AI systems could make harmful decisions without any recourse for those affected.

In the design thinking process, accountability can be introduced by involving not only developers but also ethicists, lawyers, and users who can provide insights into the legal and moral implications of the system. This interdisciplinary collaboration ensures that accountability is woven into the system's design, from conception to deployment.

Societal Well Being

AI should ultimately aim to enhance societal well being, promoting positive outcomes for individuals and communities. However, as AI systems are increasingly embedded in public life, there are risks that their deployment could exacerbate inequality, displace jobs, or undermine public trust in technology (Binns, 2018). Ethical AI development, therefore, requires that developers not only focus on the technical performance of their systems but also consider the broader social impacts of their work. This involves questioning whether AI systems will contribute to or detract from societal well being and addressing potential harms before deployment (Floridi, 2019).

Incorporating societal well being into design thinking requires developers to engage deeply with the communities that will be impacted by the AI system. This is especially relevant during the empathy phase, where understanding the lived experiences of users can inform the development of AI systems that are truly beneficial for society. The testing phase of design thinking also plays a crucial role, as it provides opportunities to gather feedback from diverse stakeholders and iterate on the design to better align with societal values.

2.4 Ethical Frameworks in AI

Several ethical frameworks guide the development of AI systems. Among the most commonly applied are deontological ethics, utilitarianism, and virtue ethics, each of which offers distinct perspectives on how AI systems should be designed.

Deontological Ethics: Rooted in the work of Immanuel Kant, deontological ethics emphasizes duties and rules. From this perspective, AI systems should be developed to follow strict ethical rules, ensuring that they never violate certain moral principles, such as privacy or the right to non discrimination (Taddeo & Floridi, 2018). For example, an AI system used in healthcare must always protect patient confidentiality, regardless of the potential benefits that could be gained from data sharing.

Utilitarianism: This framework, associated with philosophers such as Jeremy Bentham and John Stuart Mill, focuses on the consequences of actions. In AI development, utilitarianism suggests that systems should be designed to maximize overall happiness or benefit the greatest number of people (Binns, 2018). An example of utilitarianism in AI would be prioritizing the development of systems that can positively impact the most vulnerable populations, such as AI technologies that improve access to healthcare in underserved communities.

Virtue Ethics: Unlike deontological or utilitarian frameworks, virtue ethics focuses on the character and moral integrity of the AI developers themselves. This approach emphasizes that AI professionals should embody virtues such as honesty, fairness, and empathy, and that these virtues should guide their decision making throughout the development process (Vallor, 2016).

3 INTEGRATING ETHICAL DESIGN THINKING INTO AI DEVELOPMENT

The complexity of AI systems requires a development process that not only focuses on technical innovation but also prioritizes ethical considerations. The design thinking approach, with its human centered and iterative process, offers an effective framework for embedding ethical decision making into every stage of AI development. By emphasizing empathy, collaboration, and continuous feedback, design thinking ensures that the concerns of all stakeholders—users, developers, ethicists, and the broader community—are incorporated into AI systems from the very beginning (Brown, 2009).

3.1 Empathize: Understanding the Ethical Impact on Users

The first stage of design thinking, empathizing with the users, plays a critical role in identifying the ethical risks and opportunities associated with an AI system. In this phase, developers engage with users and stakeholders to understand their needs, values, and potential vulnerabilities. For AI systems, this means considering not only the immediate users but also the broader social implications of the technology. For instance, an AI system designed for medical diagnostics must take into account not only the healthcare professionals who will use the system but also the patients whose lives may be directly impacted by its decisions (Floridi, 2019).

This phase is crucial for identifying potential ethical issues, such as bias, privacy violations, or unintended consequences, that might otherwise be overlooked. By actively engaging with diverse user groups—especially those who are most vulnerable to harm—developers can gain insights into how their AI systems could impact different segments of the population (Jobin et al., 2019). For example, AI systems used in financial decision making have historically been found to disadvantage marginalized communities due to biased training data, highlighting the importance of inclusive user engagement (Raji et al., 2020).

3.2 Define: Framing the Ethical Problem

After gathering insights from users, the next step in the design thinking process is to define the problem. In ethical AI development, this involves framing the ethical challenges and risks that need to be addressed. Defining the ethical problem clearly helps guide the development process, ensuring that the team remains focused on solving these issues throughout the design lifecycle. For example, if an AI system is being developed for automated hiring, the ethical problem might be framed around ensuring that the system does not perpetuate bias against certain demographic groups (Raghavan et al., 2020).

By explicitly defining the ethical challenges, developers can create clear objectives for how the AI system will uphold ethical standards, such as fairness, accountability, and transparency. In practice, this might involve setting benchmarks for minimizing algorithmic bias or creating mechanisms to allow users to challenge or appeal AI decisions (Floridi & Taddeo, 2016). Clearly defining these ethical goals ensures that they are prioritized and addressed throughout the development process.

3.3 Ideate: Generating Ethical Solutions

Once the ethical problems are clearly defined, the ideation phase focuses on generating potential solutions. In this stage, developers brainstorm a wide range of ideas for addressing the ethical challenges identified in the previous phase. This collaborative process often involves interdisciplinary teams that include not only AI developers but also ethicists, social scientists, and legal experts. The goal is to think creatively about how to integrate ethical principles into the AI system's design (Dignum, 2019).

For example, in the development of AI powered surveillance systems, one ethical concern is the potential invasion of privacy. During the ideation phase, developers might propose solutions such as implementing privacy by design principles, where data collection is minimized, and users are given greater control over how their data is used (Binns, 2018). Additionally, transparency mechanisms could be built into the system, ensuring that users are informed about when and how surveillance data is being collected and used.

Design thinking encourages divergent thinking in this phase, meaning that developers are encouraged to explore a wide array of solutions before narrowing them down to the most viable options. This ensures that ethical considerations are not only addressed but also integrated in innovative ways that enhance the overall design of the AI system.

3.4 Prototype: Testing Ethical Considerations

Prototyping is the phase where ideas are turned into tangible products that can be tested. In the context of ethical AI, prototyping provides an opportunity to test how well the system adheres to ethical standards in practice. Developers create early versions of the AI system and subject it to rigorous testing to identify potential ethical pitfalls (Brown, 2009).

For instance, if an AI system is being developed for predictive policing, the prototype might be tested to evaluate whether it disproportionately targets certain demographic groups. Developers can use techniques such as fairness audits and bias detection tools to assess the ethical performance of the prototype (Mehrabi et al., 2021). If ethical issues are identified, the prototype can be refined and improved before the system is deployed.

Prototyping also allows developers to involve end users in the testing process. By gathering feedback from a diverse range of users, developers can gain insights into how the system performs in real world scenarios and make necessary adjustments to align with ethical standards (Gebru et al., 2018).

3.5 Test: Iterating on Ethical Feedback

The final stage of design thinking, testing, involves refining the AI system based on feedback from users and stakeholders. This phase is essential for ensuring that the system meets ethical expectations before it is deployed at scale. Testing should focus not only on the technical performance of the AI system but also on its ethical impact. For instance, a facial recognition system might be tested to ensure that it operates fairly across different demographic groups and that users are comfortable with how the technology is used (Buolamwini & Gebru, 2018).

By continuously iterating on feedback, developers can ensure that the AI system evolves to meet ethical standards. Testing also allows for the identification of any unforeseen ethical issues that may not have been apparent during earlier phases. In this way, design thinking supports a dynamic, iterative process where ethical considerations are revisited and refined throughout the development lifecycle (Floridi & Cowls, 2019).

3.6 The Role of Interdisciplinary Collaboration

One of the key strengths of design thinking is its emphasis on collaboration. In AI development, interdisciplinary collaboration is particularly important for addressing complex ethical challenges. While AI developers bring technical expertise, ethicists, social scientists, and legal experts provide critical perspectives on the societal and moral implications of AI technologies (Dignum, 2019). This collaborative approach ensures that the AI system is not only technically sound but also aligned with broader ethical and societal values.

For example, in the development of AI systems for healthcare, interdisciplinary teams might include medical professionals, ethicists, and patient advocacy groups. This ensures that the AI system is designed not only to optimize clinical outcomes but also to respect patient autonomy, privacy, and informed consent (Taddeo & Floridi, 2018).

By fostering collaboration across disciplines, design thinking encourages a holistic approach to ethical AI development, where diverse perspectives are integrated into the design process to create more responsible and equitable AI systems.

4 CASE STUDIES: ETHICAL DESIGN THINKING IN ACTION

To illustrate the practical application of ethical design thinking in AI development, this section examines real world case studies where developers have successfully embedded moral decision making into their systems. These case studies highlight both the challenges and solutions found when attempting to create AI systems that are both innovative and ethically sound. By examining these cases, we can better understand how ethical principles are operationalized and how design thinking helps mitigate risks and address societal impacts.

4.1 Case Study 1: Google's AI Principles and the Project Maven Controversy

One of the most well known examples of ethical AI development is Google's response to the controversy surrounding Project Maven, a U.S. Department of Defense initiative that aimed to use AI to analyze video footage from drones. The ethical challenges presented by this project sparked an internal debate at Google, leading to employee protests and an eventual shift in the company's stance on the ethical use of AI.

The design thinking approach was implicitly used in this situation when Google re evaluated its AI strategy after listening to employee concerns and societal pushback. Through a process of empathy, Google acknowledged the ethical concerns voiced by its employees and external stakeholders. This phase of empathizing resulted in the articulation of Google's AI Principles, which explicitly prohibit the development of AI for weaponry or technologies that cause harm (Pichai, 2018).

The next phase, defining the ethical challenge, led to Google framing the issue as a problem of AI deployment in military contexts. By recognizing the moral dilemma, Google outlined clear principles about what it considered acceptable in AI development. The ideation phase saw Google brainstorming alternative approaches to AI, which aligned with its broader vision of "AI for good" (Binns, 2018). As a result, Google discontinued its work on Project Maven and developed prototypes of AI applications that adhered to its new ethical guidelines, such as using AI for healthcare and sustainability (Raji et al., 2020). This case demonstrates how design thinking allowed a tech giant to incorporate ethical considerations into its business model.

4.2 Case Study 2: IBM's Fairness 360 Toolkit

Another significant case study comes from IBM's Fairness 360 Toolkit, a set of resources developed to help AI practitioners assess and mitigate bias in machine learning models. Bias in AI systems has been a persistent ethical challenge, particularly in systems used for decision making in hiring, criminal justice, and healthcare (Dignum, 2019). IBM approached this challenge using a design thinking methodology to create a solution that addresses fairness in AI systems.

During the empathize phase, IBM gathered extensive user and stakeholder feedback, recognizing that bias in AI could disproportionately affect marginalized groups. By engaging with ethicists, legal experts, and social scientists, IBM was able to define the problem as one of bias and fairness in AI systems. This led to the development of clear ethical goals: to create transparent, accountable, and fair AI systems (IBM Research, 2018).

In the ideation phase, IBM explored a wide range of solutions for bias detection, including fairness metrics, bias detection algorithms, and diverse data sampling methods. The result was the Fairness 360 Toolkit, which includes a set of open source tools that allow developers to test their AI models for bias and fairness. The toolkit went through several iterations during the prototype and testing phases, incorporating feedback from AI developers and domain experts before its final release (Hendricks et al., 2020). This case shows how design thinking facilitated the development of a practical solution to a pressing ethical issue in AI.

4.3 Case Study 3: AI in Healthcare – Ethical Dilemmas in Predictive Diagnostics

AI in healthcare presents unique ethical challenges, especially regarding predictive diagnostics, where AI systems predict patient outcomes based on vast amounts of health data. One case study from a leading healthcare provider illustrates how design thinking helped navigate the ethical dilemmas surrounding data privacy, patient consent, and algorithmic bias.

The healthcare provider used design thinking to empathize with both patients and healthcare professionals, seeking to understand their concerns about AI's role in making health predictions. Data privacy emerged as a major issue during this phase, as patients feared that their sensitive health information could be misused or lead to discrimination (Taddeo & Floridi, 2018). The company defined the ethical problem as the need to balance AI's potential to improve healthcare outcomes with the necessity of protecting patient autonomy and privacy.

During the ideation phase, the healthcare provider explored solutions such as anonymizing patient data, introducing consent forms that allow patients to control how their data is used, and implementing transparency protocols that explain how AI predictions are made. These ideas were tested during the prototyping phase, where early AI models were reviewed for privacy and bias concerns. Feedback from patients and healthcare workers led to several iterations of the AI system, ensuring that it adhered to ethical principles while maintaining its diagnostic accuracy (Floridi, 2019).

This case study demonstrates how the design thinking approach can be effectively used to address complex ethical issues in AI, especially when the stakes are as high as human health. By involving all stakeholders in the process and continually refining the system based on ethical feedback, the healthcare provider developed a responsible AI diagnostic tool that balanced innovation with patient rights.

4.4 Lessons Learned from Case Studies

The three case studies presented above highlight several key lessons for embedding ethical decision making in AI development through design thinking. First, the empathize phase is crucial for identifying ethical challenges, as it helps developers understand the concerns and values of those most affected by the AI system. Engaging diverse stakeholders during this phase ensures that ethical issues such as bias, privacy, and fairness are not overlooked.

Second, clearly defining the ethical problem sets the stage for developing targeted solutions that align with both technical goals and moral principles. Without a well defined ethical challenge, AI systems risk perpetuating harm, even if they are technically sophisticated.

Third, interdisciplinary collaboration is essential for generating creative solutions during the ideation phase. Bringing together AI experts, ethicists, and social scientists helps ensure that ethical considerations are embedded into the system's design from the outset.

Finally, testing and iterating on prototypes with a focus on ethical performance allows developers to refine their systems and address unforeseen ethical issues before they cause harm. This iterative process helps AI systems evolve to meet higher standards of fairness, transparency, and accountability.

5 THE ROLE OF INTERDISCIPLINARY COLLABORATION IN ETHICAL AI DEVELOPMENT

The complexity of ethical decision making in AI development requires the input of various fields, including computer science, ethics, law, social sciences, and more. Interdisciplinary collaboration ensures that ethical issues are addressed from multiple perspectives, enriching the design process and enhancing the ethical robustness of AI systems. This section explores how interdisciplinary teams can work together effectively to identify, prioritize, and address ethical challenges throughout the AI development lifecycle.

5.1 The Need for Interdisciplinary Expertise in AI Design

AI systems are fundamentally technical in nature, but the ethical, legal, and societal implications of AI often transcend the boundaries of any single discipline. For example, algorithmic fairness requires not only technical expertise in machine learning but also an understanding of ethical principles such as justice, equality, and non discrimination (Floridi & Cowls, 2019). Similarly, addressing privacy concerns in AI involves both a technical understanding of data anonymization techniques and a legal grasp of privacy laws such as the General Data Protection Regulation (GDPR).

In ethical AI design, interdisciplinary collaboration allows for the pooling of knowledge from various domains. Ethicists can help clarify moral issues related to fairness and bias, while social scientists can analyze the societal impact of AI systems, and legal experts can ensure compliance with existing regulations (Dignum, 2019).

The collaboration between these diverse fields enhances the ability of AI developers to foresee potential ethical pitfalls and mitigate risks early in the design process.

5.2 The Design Thinking Process in Interdisciplinary Teams

The design thinking approach, with its emphasis on empathy, ideation, and iteration, provides a valuable framework for interdisciplinary teams working on ethical AI development. Each phase of the design thinking process benefits from the unique perspectives that different disciplines bring.

Empathy Phase: During the empathy phase, interdisciplinary teams engage with the users, stakeholders, and communities that the AI system will affect. Ethicists and social scientists can help interpret the social and moral concerns raised by these groups, while technical experts can translate these concerns into actionable insights for system design (Binns, 2018).

Define Phase: When defining the ethical challenges in AI, legal experts can offer guidance on regulatory compliance, while ethicists can frame the ethical issues in terms of broader societal values. This collaboration ensures that the problem definition is comprehensive, addressing technical, ethical, legal, and social concerns.

Ideation Phase: The ideation phase is where the creativity of interdisciplinary teams shines. Technical experts may propose solutions for optimizing algorithmic performance, while ethicists suggest ways to embed moral principles into the system's design. Legal experts may offer frameworks for ensuring transparency and accountability in the AI's decision making processes (Jobin et al., 2019).

Prototyping and Testing Phases: In these phases, interdisciplinary collaboration becomes crucial for evaluating the ethical implications of prototypes. Technical experts ensure that the AI system functions as intended, while ethicists and social scientists provide feedback on the ethical consequences of the system's decisions. This feedback loop allows for continuous refinement, ensuring that the final product meets high ethical standards.

5.3 Examples of Successful Interdisciplinary Collaboration

Numerous organizations have demonstrated the value of interdisciplinary collaboration in developing ethical AI systems. One example is the AI Now Institute, an interdisciplinary research institute based at New York University that brings together experts from law, social sciences, computer science, and policy to study the societal impacts of AI. The AI Now Institute has produced influential reports on algorithmic fairness, data privacy, and the ethics of AI in policing, healthcare, and education (Whittaker et al., 2018).

Another example is the collaboration between DeepMind and the Wellcome Trust on AI ethics in healthcare. This partnership brought together AI developers, medical professionals, and ethicists to explore the ethical challenges of using AI in medical diagnosis and treatment. Through their collaboration, they developed frameworks for ensuring patient privacy, informed consent, and the accountability of AI driven decisions in healthcare settings (Mittelstadt, 2019).

These examples show that interdisciplinary collaboration is not only feasible but essential for creating AI systems that are both technically robust and ethically sound.

5.4 Challenges of Interdisciplinary Collaboration

While interdisciplinary collaboration offers many benefits, it also presents challenges. One major issue is the potential for communication barriers between disciplines. For example, technical experts may find it difficult to explain complex machine learning models to ethicists, while ethicists may struggle to convey the nuances of moral philosophy to AI developers. These communication gaps can slow down the design process and lead to misunderstandings about ethical priorities.

Another challenge is aligning the goals of different disciplines. While technical experts may prioritize system performance and efficiency, ethicists may focus on fairness, transparency, or the prevention of harm. Achieving a balance between these competing priorities requires careful negotiation and a willingness to compromise on both sides (Binns, 2018).

Despite these challenges, the benefits of interdisciplinary collaboration far outweigh the difficulties. When done effectively, it leads to AI systems that are not only high performing but also aligned with ethical values, legal requirements, and societal expectations.

5.5 The Future of Interdisciplinary Collaboration in AI

As AI continues to advance, the importance of interdisciplinary collaboration in AI design will only grow. Emerging technologies such as AI driven healthcare systems, autonomous vehicles, and AI in criminal justice will require even greater cooperation between technical experts and ethicists, lawyers, and social scientists (Floridi, 2019). By fostering interdisciplinary collaboration, AI developers can create systems that not only push the boundaries of what AI can achieve but also respect the ethical, legal, and social frameworks within which AI operates.

6 BROADER SOCIETAL IMPLICATIONS OF ETHICAL AI DEVELOPMENT

As AI technologies become more integrated into everyday life, their societal implications grow more profound. Ethical AI development not only addresses immediate technical challenges but also plays a critical role in shaping society's future. This section explores the broader societal impacts of AI, focusing on the balance between innovation and social responsibility. It discusses how ethical decision making in AI development can help mitigate the risks of harm, ensure inclusivity, and foster public trust in AI systems.

6.1 Addressing Inequality and Bias in AI

One of the most pressing societal challenges associated with AI is the potential for perpetuating and even exacerbating existing inequalities. AI systems, particularly those used in decision making processes such as hiring, lending, and law enforcement, can reflect and reinforce societal biases if not carefully designed. This issue has been widely recognized in research, with numerous examples of biased AI systems making unfair decisions that disproportionately affect marginalized groups (O'Neil, 2016; Noble, 2018).

To address these inequalities, ethical AI development must prioritize fairness, transparency, and inclusivity. Design thinking encourages developers to empathize with affected communities and understand how bias in AI can harm individuals and groups. This empathetic approach ensures that AI systems are designed with fairness in mind from the outset. Furthermore, interdisciplinary collaboration, as discussed in the previous section, helps identify biases and propose solutions that address both technical and societal concerns (Whittaker et al., 2018).

However, addressing bias is not simply a technical problem; it also requires broader societal efforts to challenge systemic inequalities that AI systems can inadvertently perpetuate. Ethical AI development is one step in this process, but policymakers, educators, and activists must also play roles in ensuring that AI technologies promote equity and justice (Dignum, 2019).

6.2 Ethical AI and the Future of Work

AI's impact on the labor market has sparked considerable debate, with some arguing that automation could lead to mass job displacement while others suggest that AI will create new types of work. Ethical AI development must consider both the potential benefits and harms of AI driven automation, particularly in terms of its impact on workers.

AI technologies have the potential to improve productivity and reduce the need for humans to perform repetitive or dangerous tasks. However, without careful planning and ethical foresight, these same technologies could lead to significant job losses, particularly in sectors such as manufacturing, transportation, and customer service (Brynjolfsson & McAfee, 2014). Ethical AI development should focus on creating systems that complement human work rather than replace it entirely. This requires a design thinking approach that empathizes with workers and anticipates the long term societal impacts of AI driven automation (Acemoglu & Restrepo, 2020).

Moreover, there must be a focus on reskilling and upskilling workers to ensure that they are prepared for the changes brought by AI. Educational systems, companies, and governments all have a role in ensuring that workers can transition to new roles in an AI driven economy. By considering these broader implications during the AI design process, developers can help mitigate the risks of mass unemployment and contribute to a more equitable future of work.

6.3 AI, Privacy, and Surveillance

The rise of AI powered surveillance technologies has raised significant ethical concerns related to privacy and civil liberties. AI systems that collect, analyze, and interpret vast amounts of data can be used for purposes ranging from targeted advertising to government surveillance, often without individuals' explicit consent. This raises critical questions about how AI technologies should be governed and what ethical standards should guide their use (Zuboff, 2019).

Designing AI systems that respect privacy requires a careful balance between innovation and safeguarding individual rights. Developers must consider how their systems collect, store, and use data, ensuring that privacy protections are built into the AI's architecture. For instance, implementing data anonymization techniques and giving users greater control over their personal information can help address privacy concerns (Floridi & Taddeo, 2016).

Furthermore, governments and regulatory bodies must work alongside AI developers to create legal frameworks that protect individuals from invasive AI driven surveillance. The European Union's General Data Protection Regulation (GDPR) is one example of such legislation, offering individuals rights over their data and setting strict guidelines for data processing (Binns, 2018). Ethical AI development must align with these legal frameworks while continuing to innovate responsibly.

6.4 Building Public Trust in AI

Public trust is critical to the widespread adoption and acceptance of AI technologies. Without trust, even the most advanced AI systems may face resistance from users, leading to diminished social and economic benefits. Building trust requires transparency, accountability, and ethical oversight throughout the AI development process (Floridi, 2019).

Ethical AI development, grounded in principles of fairness, accountability, and transparency, helps foster public trust by ensuring that AI systems operate in ways that are understandable, predictable, and accountable. Transparency is key: AI developers must be able to explain how their systems make decisions, particularly when those decisions affect people's lives in significant ways. Ensuring that AI systems can be audited and that developers are held accountable for the outcomes of their systems is crucial to building public confidence (Jobin et al., 2019).

In addition to transparency, involving the public in discussions about AI's societal impacts can also build trust. Public consultations, participatory design processes, and open dialogue between AI developers, policymakers, and the general public ensure that societal concerns are heard and addressed. This engagement helps demystify AI technologies and demonstrates that ethical considerations are integral to AI development.

6.5 Ethical AI and Global Governance

As AI technologies become increasingly global, there is a growing need for international cooperation on the ethical development and deployment of AI. Different countries have different cultural values, legal frameworks, and ethical norms, which can lead to conflicting approaches to AI regulation. To address this, ethical AI development must involve collaboration at the international level, with the goal of creating global standards for responsible AI use (Floridi, 2019).

International organizations, such as the United Nations and the Organisation for Economic Co operation and Development (OECD), have begun to develop frameworks for ethical AI governance. These frameworks aim to promote AI technologies that respect human rights, fairness, and transparency while encouraging innovation. However, creating truly global standards for AI ethics will require sustained dialogue between governments, companies, and civil society to navigate the ethical challenges posed by rapidly advancing technologies (Jobin et al., 2019).

Ethical AI governance also involves addressing the geopolitical implications of AI. As countries race to develop and deploy AI systems, there is a risk that the technology could be used for harmful purposes, such as surveillance, manipulation, or cyber warfare. Ethical AI development must consider the global implications of AI and advocate for responsible use that prioritizes human well being and security.

7 CONCLUSION: FOSTERING MORAL RESPONSIBILITY IN AI DEVELOPMENT

In conclusion, the integration of ethical decision making into AI development is not just an optional add on—it is a necessity. As AI technologies continue to permeate all aspects of society, from healthcare and education to security and transportation, the ethical implications of their design and use grow increasingly significant. The design thinking approach provides a practical, human centered framework that AI developers can use to embed moral principles directly into the development process, ensuring that AI systems serve humanity responsibly and equitably.

By incorporating empathy, interdisciplinary collaboration, and iterative prototyping into AI development, developers are better equipped to anticipate and address the ethical challenges that arise from the deployment of AI systems. From mitigating bias and inequality to preserving privacy and building public trust, ethical AI development offers a path toward creating technologies that respect human rights and contribute positively to society.

Moreover, the broader societal implications of AI—such as its impact on the labor market, governance, and surveillance—highlight the need for collective action at both the national and international levels. Ethical AI is not solely the responsibility of developers; it requires a concerted effort from policymakers, industry leaders, educators, and civil society to ensure that AI systems align with societal values and promote the common good.

As we move forward, fostering moral responsibility in AI development should be a top priority for both technical and non technical stakeholders. This can be achieved by integrating ethical considerations into the educational curricula for AI professionals, creating robust regulatory frameworks, and maintaining an open dialogue about the societal impacts of AI technologies. Ultimately, responsible AI development is about more than just building better systems—it is about shaping a future in which technology and ethics are inextricably linked.

REFERENCES

Acemoglu, D., & Restrepo, P. (2020). Robots and jobs: Evidence from US labor markets. *Journal of Political Economy*, 128(6), 2188–2244. DOI: 10.1086/705716

Angwin, J., Larson, J., Mattu, S., & Kirchner, L. (2016). Machine bias. ProPublica. Retrieved from https://www.propublica.org/article/machine bias risk assessments in criminal sentencing

Binns, R. (2018). Fairness in machine learning: Lessons from political philosophy. Proceedings of the 2018 Conference on Fairness, Accountability, and Transparency, 149 159.

Brown, T. (2009). *Change by design: How design thinking creates new alternatives for business and society*. HarperCollins.

Brynjolfsson, E., & McAfee, A. (2014). *The second machine age: Work, progress, and prosperity in a time of brilliant technologies*. W. W. Norton & Company.

Buolamwini, J., & Gebru, T. (2018). Gender shades: Intersectional accuracy disparities in commercial gender classification. *Proceedings of Machine Learning Research*, 81, 1–15.

Burrell, J. (2016). How the machine 'thinks': Understanding opacity in machine learning algorithms. *Big Data & Society*, 3(1), 1–12. DOI: 10.1177/2053951715622512

Cath, C. (2018). Governing artificial intelligence: Ethical, legal, and technical opportunities and challenges. *Philosophical Transactions. Series A, Mathematical, Physical, and Engineering Sciences*, 376(2133), 20180080. DOI: 10.1098/rsta.2018.0080 PMID: 30322996

Danaher, J., Hogan, M. J., Noone, C., Kennedy, R., Behan, A., & Collins, E. (2017). Algorithmic governance: Developing a research agenda through the power of collective intelligence. *Big Data & Society*, 4(2), 2053951717726554. DOI: 10.1177/2053951717726554

Dignum, V. (2019). *Responsible artificial intelligence: How to develop and use AI in a responsible way*. Springer. DOI: 10.1007/978-3-030-30371-6

Floridi, L. (2019). *The logic of information: A theory of philosophy as conceptual design*. Oxford University Press. DOI: 10.1093/oso/9780198833635.001.0001

Floridi, L. (2019). *The ethics of artificial intelligence: The future of AI and society*. Oxford University Press.

Floridi, L., & Cowls, J. (2019). A unified framework of five principles for AI in society. *Harvard Data Science Review*, 1(1), 1–15.

Floridi, L., Cowls, J., Beltrametti, M., Chatila, R., Chazerand, P., Dignum, V., & Schafer, B. (2018). AI4People—An ethical framework for a good AI society: Opportunities, risks, principles, and recommendations. *Minds and Machines*, 28(4), 689–707. DOI: 10.1007/s11023-018-9482-5 PMID: 30930541

Floridi, L., & Taddeo, M. (2016). What is data ethics? *Philosophical Transactions. Series A, Mathematical, Physical, and Engineering Sciences*, 374(2083), 20160360. DOI: 10.1098/rsta.2016.0360 PMID: 28336805

Gebru, T., Morgenstern, J., Vecchione, B., Vaughan, J. W., Wallach, H., Daumé, H., III, & Crawford, K. (2018). Datasheets for datasets. arXiv preprint arXiv:1803.09010.

Jobin, A., Ienca, M., & Vayena, E. (2019). The global landscape of AI ethics guidelines. *Nature Machine Intelligence*, 1(9), 389–399. DOI: 10.1038/s42256-019-0088-2

Mehrabi, N., Morstatter, F., Saxena, N., Lerman, K., & Galstyan, A. (2021). A survey on bias and fairness in machine learning. *ACM Computing Surveys*, 54(6), 1–35. DOI: 10.1145/3457607

Mehrabi, N., Morstatter, F., Saxena, N., Lerman, K., & Galstyan, A. (2021). A survey on bias and fairness in machine learning. *ACM Computing Surveys*, 54(6), 1–35. DOI: 10.1145/3457607

Mittelstadt, B. D. (2019). Principles alone cannot guarantee ethical AI. *Nature Machine Intelligence*, 1(11), 501–507. DOI: 10.1038/s42256-019-0114-4

Noble, S. U. (2018). *Algorithms of oppression: How search engines reinforce racism*. NYU Press. DOI: 10.18574/nyu/9781479833641.001.0001

O'Neil, C. (2016). *Weapons of math destruction: How big data increases inequality and threatens democracy*. Crown Publishing Group.

Raghavan, M., Barocas, S., Kleinberg, J., & Levy, K. (2020). Mitigating bias in algorithmic hiring: Evaluating claims and practices. Proceedings of the 2020 Conference on Fairness, Accountability, and Transparency, 469 481. DOI: 10.1145/3351095.3372828

Raji, I. D., Bender, E. M., Paullada, A., Denton, E., & Hanna, A. (2020). AI and the everything in the whole wide world benchmark: Accountability in data collection in the age of machine learning. arXiv preprint arXiv:2012.04580.

Taddeo, M., & Floridi, L. (2018). How AI can be a force for good. *Science*, 361(6404), 751–752. DOI: 10.1126/science.aat5991 PMID: 30139858

Vallor, S. (2016). *Technology and the virtues: A philosophical guide to a future worth wanting*. Oxford University Press. DOI: 10.1093/acprof:oso/9780190498511.001.0001

Weller, A. (2019). Transparency: Motivations and challenges. In T. Hagras (Ed.), Explainable AI: Interpreting, explaining and visualizing deep learning (pp. 23 40). Springer.

Whittaker, M., Crawford, K., Dobbe, R., Fried, G., Kaziunas, E., Mathur, V., & Schwartz, O. (2018). *AI now report 2018*. AI Now Institute.

Zuboff, S. (2019). *The age of surveillance capitalism: The fight for a human future at the new frontier of power*. PublicAffairs.

Chapter 17
Doctoral Researchers' Perspectives on Ethical Considerations in Artificial Intelligence in Education Through the Lens of the PAPA Framework

Senthilraj Annamalai
Alagappa Univerity, India

R. Ramnath
https://orcid.org/0000-0002-7292-5200
Alagappa Univerity, India

Soniya Antony
https://orcid.org/0000-0003-4325-3967
Alagappa University, India

S. Brindha Devi
Alagappa Univerity, India

ABSTRACT

The integration of Artificial Intelligence (AI) in education offers opportunities for personalized and accessible learning but raises critical ethical concerns. This

DOI: 10.4018/979-8-3373-2262-9.ch017

study examines doctoral researchers' perspectives on these issues using the PAPA (Privacy, Accuracy, Property, and Accessibility) framework. Semi-structured interviews revealed 10 themes and 23 subthemes, highlighting concerns about data privacy, ownership, consent, and transparency. Participants stressed the need for robust ethical guidelines to address risks such as breaches, bias, and unequal access while ensuring AI supports rather than undermines academic integrity and autonomy. The findings emphasize ethical frameworks as essential for responsible and inclusive AI integration in education.

INTRODUCTION

Artificial Intelligence has increasingly transformed educational context, offering advanced tools that personalize learning experiences, automate administrative tasks, and provide adaptive support to students and educators (Chan & Hu, 2023; Dakakni & Safa, 2023). While these advancements present significant benefits, they also raise substantial ethical concerns, especially regarding data privacy, transparency, fairness, and the long-term impact on educational practices and human agency (Nguyen et al., 2022; Essien et al., 2024). Central to understanding these concerns is the Privacy, Accuracy, Property, and Accessibility (PAPA) framework proposed by Mason (1986), which serves as a foundational model for examining the ethical dimensions of information technologies. As educational institutions increasingly integrate AI, the PAPA framework provides a structured lens to analyze potential risks and ethical implications associated with student and institutional data handling, the accuracy and fairness of AI systems, the proprietary rights to educational data, and equitable access to AI-driven tools (Holmes et al., 2021; Fassbender, 2024). Despite the broad relevance of PAPA in educational AI, research on how doctoral researchers perceive and address these ethical concerns remains limited, leaving significant gaps in understanding the nuanced viewpoints of emerging academic professionals who may shape future educational policies and technologies.

In particular, gaps remain in assessing how doctoral researchers interpret the ethical challenges surrounding AI, given their unique role as both future educators and innovators. While there is growing acknowledgment of issues such as data privacy and AI transparency, existing studies often overlook the perspectives of doctoral students, whose insights are crucial for anticipating ethical challenges as AI continues to evolve in educational contexts (Shih et al., 2021; Kong et al., 2023). Additionally, few studies have systematically applied the PAPA framework to gather comprehensive insights into the privacy, accuracy, property, and accessibility concerns specific to AI in educational settings. This lack of focus on doctoral researchers' perspectives restricts a broader understanding of how these ethical considerations could shape educational practices and AI's integration in academia and beyond.

This study aims to address this gap by exploring the perspectives of doctoral researchers on the ethical implications of AI in education through the PAPA framework. Specifically, it investigates doctoral researchers' views on privacy, accuracy, property, and accessibility to understand their concerns regarding AI's ethical alignment with educational values. By focusing on this population, the study contributes novel insights into ethical AI implementation, revealing areas where data governance, transparency, and ethical guidelines may require improvement to support fair, responsible AI integration in education. Furthermore, the study highlights the need for more targeted ethical guidelines that prioritize student autonomy, academic integrity, and inclusivity, thereby advancing a human-centered approach to AI in educational environments.

Research questions

RQ1. How do doctoral researchers perceive the ethical implications of AI in education regarding privacy, accuracy, property, and accessibility?

RQ2. What are the key concerns of doctoral researchers about the application of AI in educational settings, particularly in relation to data privacy and ownership?

RQ3. In what ways do doctoral researchers believe AI can impact the accuracy and accessibility of educational resources, and what ethical frameworks should guide its use?

REVIEW OF LITERATURE

AI in Indian Higher Education

The adoption of Artificial Intelligence in Indian higher education has evolved significantly, with institutions being early adopters as far back as the 1990s, experimenting with computer-aided instruction and adaptive learning systems (Bhagat & Kim, 2020; Airaj, 2024). A major trend is the focus on personalized education through AI-driven adaptive learning systems and recommendation engines tailored to individual student needs (Sharma et al., 2024). Learning analytics powered by AI is increasingly being used to analyze student data for predicting performance, identifying at-risk students, and enabling timely interventions. Virtual labs and simulations have also become prevalent, offering hands-on experiences in resource-constrained environments, especially in science and medical fields (Bhagat & Kim, 2020). Furthermore, Natural Language Processing (NLP) technologies are streamlining tasks such as automated essay scoring, sentiment analysis, and language translation (Airaj, 2024). The rapid growth of EdTech startups in India has introduced AI-driven platforms for online learning, skill development, and career counselling. Collaborations with domestic and international companies are also fostering the integration of advanced AI technologies into campuses, bridging gaps between academia and industry. Many institutions are now prioritizing AI skill development through specialized courses in AI, machine learning, and data science to prepare students for future careers. Additionally, the COVID-19 pandemic accelerated the adoption of AI-powered virtual classrooms, enhancing the online learning experience. While AI adoption in Indian higher education is still in its early stages, these trends underline its potential to revolutionize education and prepare students for the future (Sharma et al., 2024).

Ethical concerns of AI in Education

The integration of AI into education holds substantial promise, yet it brings considerable ethical concerns, particularly surrounding privacy, data security, and transparency (Chan, 2023; Chan & Hu, 2023; & Chan & Lee, 2023). Educational AI systems often rely on collecting and analysing extensive data, including academic performance, personal information, and even biometric data, which raises pressing privacy, issues (Holmes et al., 2022; Dakakni & Safa, 2023; Essien et al. 2024; & Fassbender, 2024). Institutions and service providers must adopt stringent security measures to protect this sensitive information, which is crucial for maintaining user trust (Qin, Li& Yan, 2020 & Gudi˜no Paredes et al. 2021). Such concerns are heightened when the data pertains to minors, as they may not fully understand the implications of data sharing (Nguyen et al., 2022 & Chan, 2023). Without robust security and data minimization practices, which limit the collection to only what is essential; students' data could be misused or exploited in ways that may affect their future (Nguyen et al., 2022). Hence, responsible data governance is fundamental to aligning AI applications in education with ethical standards (Khan et al. 2024; & Kim et al. 2023). Beyond privacy and security, issues of autonomy, transparency, and consent also emerge as key ethical considerations (Lee et al., 2022; Lin et al., 2021; McDonal & Pan, 2020; & McGrath et al. 2023). The adaptability of AI in personalizing learning experiences can, ironically, diminish students' autonomy, as over-reliance on AI-driven recommendations may restrict their independent decision-making and critical thinking (Holmes et al., 2021). Kousa and Niemi (2023) suggest that educational AI frameworks should integrate human and machine support, allowing students to choose their preferred mode of assistance. However, when students feel a lack of control or experience a shift from human to algorithmic guidance, their engagement can suffer (Qin, Li& Yan, 2020). For educators, AI's role might gradually shift them into mentorship roles, reducing direct interactions with students and impacting traditional educational relationships (Holmes et al., 2023). Such considerations reinforce the importance of preserving human agency in AI-integrated learning settings to avoid diminishing educational value (Lee et al., 2022; Lin et al., 2021; McDonal & Pan, 2020; & McGrath et al. 2023).

Transparency is crucial in addressing these ethical issues, especially given the "black box" nature of many AI systems (Lee et al., 2022; Lin et al., 2021; Chan, 2023; Chan & Hu, 2023; & Chan & Lee, 2023). This opacity can foster mistrust as both students and teachers struggle to comprehend how AI-driven decisions are made (Kitto & Knight, 2019). Transparency in decision-making enables users to assess how AI recommendations align with their goals (Jang et al., 2022), and clarity in data governance can empower both students and teachers to make informed choices regarding AI use (Holmes et al., 2023). Moreover, Shih et al. (2021) highlights that AI developers should refrain from unnecessary data collection and ensure users understand the system's limitations. Enhanced transparency not only builds trust but also aligns with the core values of education, supporting a more inclusive and participatory environment (Chan and Hu, 2023). Algorithmic fairness is another vital ethical consideration, as AI's reliance on historical data may perpetuate biases, impacting students' educational opportunities (Hwang et al., 2020). Such biases are particularly concerning in areas like admissions and course recommendations, where AI decisions can significantly influence students' futures (Reiss, 2021). Ensuring fair and inclusive AI requires ethical scrutiny and adjustments to algorithms, as biased data risks deepening societal inequalities (Ashok et al., 2022). Ifenthaler and Schumacher (2016) emphasize the need for ethical frameworks specifically adapted to educational settings, addressing these unique challenges. Guidelines from UNESCO and the OECD further support the implementation of transparent and fair AI systems to foster equitable and inclusive educational environments (Nguyen et al., 2022). In addition, the non-maleficence and beneficence, advocating that AI should benefit rather than harm students, are essential for ethical AI in education. According to Kong, Cheung, & Tsang (2023), developers and educators have a responsibility to prevent AI from causing unintentional harm, such as contributing to a digital divide or reinforcing discrimination. A participant in Yu and Yu's (2023) study reflects this ethos, emphasizing that every stakeholder must act responsibly to ensure that AI advances educational well-being rather than detracting from it. Ultimately, AI companies, developers, educators, and policymakers must collaborate to establish a sustainable, ethical AI ecosystem in education that promotes positive outcomes and aligns with social values (Yu & Yu, 2023 & Zhou et al., 2021).

Ethical Principles of AI in Education

The rapid development of AI in education has generated an array of ethical considerations, urging international bodies to propose frameworks that support its responsible use. As AI-driven systems are increasingly integrated into educational environments, ethical principles must guide their design, implementation, and management to protect the rights, privacy, and well-being of all stakeholders, particularly students. Ethical principles guiding artificial intelligence in education are essential for ensuring that the development and application of AI in educational settings align with values that protect and empower students, teachers, researchers, educators, and broader communities. The main purpose of the ethical principles is to ensure that these technologies are deployed in ways that enhance learning outcomes while respecting human dignity and autonomy. International organizations, including UNESCO, the OECD, and the European Commission, have outlined frameworks that prioritize key ethical considerations in AIED, such as governance, transparency, accountability, sustainability, inclusivity, and privacy. These principles reflect a commitment to balancing technological advancement with respect for human rights, fairness, and equitable access to educational resources (UNESCO Ethics AI, 2020; Miao et al., 2021; UNESCO, 2019; OECD, 2021 & European Commission, 2019). Common ethical principles include respect for human rights, fairness, accountability, transparency, and privacy, all aimed at fostering trust and confidence among users. In addition to these fundamental values, the ethical frameworks emphasize the need for robust governance and stewardship structures, as well as multi-stakeholder collaboration, to ensure that AI technologies align with societal values and educational objectives (European Commission, 2019; European Parliament Report on AI Education, 2021).

Data Privacy

Data privacy is a fundamental principle in the ethical implementation of AIED, due to the sensitivity of student information and the potential risks associated with data misuse (Nguyen et al., 2022). Protecting data privacy involves implementing robust safeguards for both personal and sensitive information, especially when handling data on young students. Privacy principles in AIED, recommend additional security measures limited data collection and strict controls over access to prevent any misuse and unauthorized exposure of student identities. UNESCO and the Beijing Consensus stress the importance of protecting student data and maintaining strict ethical standards for data management to safeguard students' rights and identities (Beijing Consensus, 2019; UNESCO Ethics AI, 2020).

Transparency and responsibility

Transparency and responsibility are also fundamental principles in the ethical use of AI in education, helping to build trust and understanding among AI users (Nguyen et al., 2022). Transparency involves ensuring that the processes and data behind AI decision-making are visible and understandable to all stakeholders, particularly when decisions impact student learning and privacy. UNESCO and the OECD stress that, AI systems should provide clear, understandable outputs that users can interpret and challenge when necessary (UNESCO Education & AI, 2021; OECD, 2021). In line with transparency, responsibility emphasises that responsibility for AI systems must be clearly defined, enabling schools, educators, and developers to take ownership of outcomes and address any unintended harms (European Commission, 2019). This principle often includes regular auditing to ensure AI systems meet ethical and technical standards (UNESCO Ethics AI, 2020).

Governance and Management

Governance emerges as central pillars in many international AIED guidelines, setting a foundation for the responsible management and regulation of AI technologies (Nguyen et al., 2022). Governance, as defined by Floridi (2018), involves establishing policies, procedures, and standards that shape the development, use, and oversight of AI in education. Management, on the other hand, represents a more active and ethically grounded approach, focusing on the careful and responsible supervision and operation of AI in educational settings (Greer, 2018). Together, governance and management offer a comprehensive framework for ensuring that AIED systems are aligned with ethical standards, providing a structure for accountability while encouraging ethical vigilance in AI deployment (Ashok et al., 2022). Reports such as UNESCO's Ethics AI (2020 Draft), the Beijing Consensus (UNESCO, 2019), and the OECD's guidance on AI in education (2021) reflect a recurring call for multistakeholder involvement, where input from educators, students, policymakers, and AI experts contributes to a balanced regulatory environment. This collaborative approach helps ensure that AI systems are designed with a global perspective, fostering interdisciplinary planning and international cooperation to uphold ethical standards (UNESCO Ethics AI, 2020 Draft; UNESCO Education & AI, 2021). Monitoring and evaluation, as highlighted by these guidelines, are essential to adapting AI systems to evolving educational needs and emerging ethical challenges, supporting a sustainable model for ethical AIED deployment (European Commission, 2019; European Parliament Report on AI Education, 2021).

Sustainability

Sustainability and proportionality principles ensure that AIED systems support long-term educational goals without causing excessive economic, environmental, and social issues (Nguyen et al., 2022). These principles advocate for the proportional use of AI, promoting resource-efficient practices that prioritize long-term benefits and avoid over-reliance on technology at the expense of personal interactions and teacher-student relationships. And it requires that AI applications align with local educational values and labour needs, particularly when they influence educational policies and employment in the teaching workforce (European Commission, 2019). Sustainability also relates to environmental considerations, as OECD and UNESCO emphasize that AIED should consider the carbon footprint of AI systems and strive for green technologies (OECD, 2021; Miao et al., 2021).

Inclusivity and Fairness

Inclusivity and fairness are essential components of ethical AIED, particularly given the potential for bias in AI algorithms that could perpetuate or amplify existing inequalities ((Nguyen et al., 2022). Inclusivity within AIED emphasizes equal access to educational resources, ensuring that AI supports diverse student needs and promotes gender equality. Fairness in AIED, aims to reduce algorithmic biases that might disadvantage specific groups. UNESCO and the European Commission emphasize that AIED systems should undergo testing to identify and mitigate biases, fostering non-discriminatory outcomes and creating a learning environment that is fair and accessible to all students, regardless of their background (UNESCO Education & AI, 2021; European Commission, 2019).

Human-Centered AIED

Human-centered approach underpins many AIED ethical frameworks, emphasizing that AI in education should enhance rather than replace human interaction and oversight ((Nguyen et al., 2022). AI systems should support, not undermine, teacher roles and learner autonomy, allowing students to make meaningful choices and develop independence within AI-enhanced learning environments (UNESCO Education & AI, 2021; European Parliament Report AI Education, 2021). This approach values human rights and dignity, underscoring the importance of learner agency and the role of teachers as guides in the educational process (Beijing Consensus, 2019; OECD, 2021).

THEORETICAL FRAMEWORK

PAPA Framework

In recent decades, ethical concerns surrounding information technology have intensified, particularly as society navigates the complexities of the "information age." Mason's (1986) theoretical framework offers a foundation for examining these ethical challenges through four primary constructs: Privacy, Accuracy, Property, and Accessibility, collectively known as the PAPA framework. This model has significantly shaped information ethics, providing researchers, practitioners, and policymakers with a robust, adaptable tool for analyzing information technology's ethical implications.

Privacy

Privacy, as one of the central themes of Mason's framework, probes deeply into what personal information an individual must disclose, under which circumstances, and with what protections (Mason, 1986). Mason provided concrete examples to illustrate the consequences of inadequate privacy protection. He cited a Florida state government initiative that monitored restroom use at a community college under the guise of assessing building use, which led to public outcry over privacy infringement. This case exemplifies the tension between data collection goals and privacy rights. Furthermore, Mason cautioned about the potential risks of data aggregation, warning that advancements in technology could lead to "omniscient dossiers" that track individuals' actions and personal data in exhaustive detail. Such datasets, according to Mason, may be exploited for extortion if misappropriated, thus highlighting privacy as a foundational issue in IT ethics (Mason, 1986; Ryker, LaFluer, McManis, & Cox, 2002).

Accuracy

The accuracy of information, another pillar of the PAPA framework, pertains to the responsibility for ensuring the correctness of data collected, processed, and shared. According to Mason, maintaining data accuracy is essential not only for decision-making but also for protecting individuals from potential harm arising from erroneous data. Accountability mechanisms must be established, offering redress in cases where individuals suffer due to inaccurate data. This principle resonates with subsequent studies, emphasizing the importance of accountability structures to preserve data integrity, especially in fields where erroneous data can directly impact individuals' lives (Loch, Conger, & Oz, 1998).

Property

Property, as defined in Mason's framework, deals with questions of ownership and compensation for information. Who owns a particular dataset, and how are rights to that data distributed? Mason observed that the commodification of information complicates traditional notions of ownership. As IT systems evolved, ownership and control over information assets became essential considerations for individuals and organizations, raising questions about the monetization and proprietary rights to data (Malone, 1986). This focus on ownership has become increasingly relevant in today's data-driven economy, where organizations capitalize on data assets, often without the explicit consent or knowledge of data sources, further underlining the ethical complexity of property within the context of IT.

Accessibility

Accessibility, the final element of Mason's framework, examines how information is made available and who has access to it. Mason's inquiry into accessibility addresses both the equitable distribution of information resources and the safeguarding of those resources from misuse. He questioned how data could be accessed fairly, particularly when discrepancies in access can exacerbate societal inequalities (Ellis & Griffith, 2004). Mason's views on accessibility have become even more relevant with the increasing importance of digital access in all aspects of life, where limited access to information technology can create barriers to opportunity and participation in the digital economy.

Extensions and Relevance of PAPA

Subsequent studies have validated and extended Mason's PAPA framework. For example, Conger, Loch, and Helft (1994) conducted an exploratory factor analysis on the PAPA framework, which generally validated its structure but also added a fifth factor, "motivation," thereby enhancing the framework's applicability. Moreover, Wu, Rogerson, and Fairweather (2002) affirmed PAPA's importance in the field, noting it as the most "influential category of issues" in information systems development. This continued validation reinforces the relevance of Mason's framework as it provides an enduring ethical structure adaptable to evolving IT landscapes (Ellis & Griffith, 2004). While the PAPA framework was initially proposed to address the ethical challenges of the 1980s, its adaptability has allowed it to remain highly relevant. The persistence of ethical challenges tied to privacy, accuracy, property, and accessibility is evident as IT systems become increasingly integral to daily life, amplifying Mason's concerns. For example, recent debates around data privacy underscore the ongoing significance of the privacy aspect in Mason's framework, particularly with the rise of data-intensive applications and artificial intelligence.

Method and Methodology

The study used phenomenological analysis, a qualitative research approach, to explore the study in its natural setting (Creswell, 2007). This approach allowed for detailed information collection from primary sources, aiming to understand the meaning of the problem through the perspectives of the respondents and other relevant participants in the discussion. The study was conducted in three state universities of Tamilnadu. The research design incorporated Interpretive Phenomenological Analysis (IPA) as the primary approach for analysing the data. This method was chosen to explore into participants' lived experiences, particularly focusing on doctoral students' perspectives on ethical considerations in artificial intelligence in education through the lens of the PAPA Framework. IPA was well-suited to explore how individuals subjectively interpret and make sense of their using AI tools in learning and research (Smith et al., 2009). Nine doctoral researchers were selected as participants of the study through purposive and convince sampling. The place of convince of participant served as the data collection settings for this research.

Interview questions and validation of questions

The semi-structured interview questions were crafted to gather detailed insights into doctoral researcher's perception of use of AI in education. Eleven questions were designed to capture rich accounts of personal experiences, following an inductive approach (Antony, Ramnath, & Ellikkal, 2024) and drawing from the principles outlined by Smith et al. (2009). Semi- structured interview questions explore doctoral students' specific lived experiences, contextual concerns, and broader ethical implications regarding AI in education. First, participants are encouraged to share personal experiences, such as their comfort with AI handling personal data and whether they feel AI tools are reliable and accurate in educational contexts. Additionally, they're asked about experiences where AI affected access, accuracy, and privacy, influencing their views on AI. Moving into more contextualized inquiry, questions focus on privacy issues, ownership of data collected by AI, and desired levels of transparency from educational institutions regarding AI use. Finally, questions are posed about broader implications, including how AI might impact the quality and accessibility of educational resources, potential ethical guidelines for AI use, and any major concerns participants have about AI's role in education, along with changes that could improve their comfort with AI integration in their learning environment. Together, these questions invite participants to reflect on their experiences and views within specific contexts, providing insights into how AI ethics can better align with educational practices. This thoughtful structure ensured the collection of detailed, context-rich data on doctoral scholars' experiences, perceptions, and challenges in using AI in education and research.

Interview Process

In this study, the investigators conducted semi-structured interviews lasting 30 to 45 minutes with nine doctoral research scholars from three state universities in Tamil Nadu, India. The investigators obtained written consent from the participants before the interview and conducted a brief icebreaker session to gather necessary demographic information. During the main portion of the interviews, the investigators asked open-ended questions. To encourage more detailed responses, the investigators used probing questions such as 'Can you give me an example?' or 'Could you elaborate on that? Etc...

Researchers Positionality and Reflectivity

The research team brings a wealth of experience and self-awareness to this study, which enriches the analysis of ethical implications of AI in education. Reflexivity, a key component of qualitative research, helps each researcher acknowledge and mitigate personal biases, assumptions, and professional experiences that might shape their interpretation of data (Creswell & Poth, 2016; Merriam & Tisdell, 2015; Antony, Ramnath, & Ellikkal, 2024). The first author's 18 years in online content creation, particularly in mathematics, and expertise in educational technology inform a practical understanding of how AI intersects with digital learning platforms. This perspective adds depth to discussions of AI's role in enhancing or challenging online educational content. The second author, with over 16 years of experience in research and teaching, brings a profound grasp of educational technology. His extensive background complements the team's exploration of AI's influence on pedagogy. Together with the third author, he has previously published research on AI in education, enabling both to provide an informed lens through which they examine AI's potential benefits and challenges in academic contexts. The fourth author's doctoral research focuses on the disadvantages and ethical concerns of AI in higher education, adding a critical perspective that questions potential biases in AI use and its ethical implications for students and educators. Collectively, the authors' professional knowledge of AI applications, along with their reflective consideration of their positionalities, supports a balanced, detailed examination of the issue. Their combined experiences enhance their ability to analyze AI's ethical concerns and educational potential from doctoral researcher's angles, benefiting both the study and its readers.

Ethical Consideration

Ethical considerations were crucial in this study, ensuring participants' informed consent, confidentiality, and privacy (Antony, Ramnath, & Ellikkal, 2024). Researchers were sensitive to participants' needs, treating them with dignity and respect while avoiding the imposition of biases. Power dynamics were carefully considered, and findings were disseminated responsibly to prevent harm and stigmatization of participant and context. The study prioritized ethical principles in its conduct.

Data Analysis

To analyze the data, systematic text condensation, a descriptive and explorative method for thematic cross-case analysis, was utilized (Malterud, 2012). This method represents a pragmatic approach, drawing inspiration from phenomenological ideas. The analysis process involved four steps: 1) forming an overall impression of the data, moving from chaos to themes; 2) identifying and sorting meaning units, transitioning from themes to codes; 3) condensing the codes into meaningful concepts; and 4) synthesizing the condensed information to create descriptions and overarching concepts.

Total Impression – From Chaos to Themes,

In the first step, "Total Impression – From Chaos to Themes," the goal is to read through all the data to gain a holistic understanding without getting caught up in the details. The aim is to listen to the participants' voices with an open mind, allowing the themes to naturally emerge from the data.

Identifying and sorting meaning units – from themes to codes

Identifying and sorting meaning units involves systematically reviewing a transcript to pinpoint specific text fragments that contain relevant information related to the research question. Not every part of the transcript will be relevant, so the process focuses on selecting only those segments that provide meaningful or contextual data for the study.

Condensation – from code to meaning

The third step of analysis, condensation, involves systematically abstracting and summarizing meaning units within each of the code groups established earlier. This process reduces the empirical data into decontextualized, thematic code groups that represent the core concerns of the participants. Instead of lengthy transcripts, the data is now organized into a few key code groups, each containing condensed meaning units.

Synthesizing – from condensation to descriptions and concepts

In the fourth step of analysis, synthesizing involves reconceptualizing the condensed data to form coherent descriptions and concepts. This process brings together the previously fragmented text elements to create a narrative that remains true to the original context. Condensates and quotations from each subgroup, the researcher crafts an analytic text that presents the core findings and meanings, ensuring that the interpretation is both credible and grounded in the empirical data. The researcher acts as a narrator, presenting the synthesized results in a structured but seamless manner.

RESULTS

In analyzing doctoral researchers' perceptions of the ethical implications of AI in education, a comprehensive thematic analysis identified 10 primary themes and 23 subthemes. These themes provide insight into the ethical considerations that doctoral researchers find most pressing, focusing on issues related to data privacy, ownership, consent, transparency, reliability, autonomy, and the need for ethical guidelines. Participant responses reveal diverse perspectives on how AI affects privacy, accuracy, property, accessibility, and general educational integrity.

1. How do doctoral researchers perceive the ethical implications of AI in education regarding privacy, accuracy, property, and accessibility?

Doctoral researchers view the ethical implications of AI in education as multifaceted, highlighting six themes, such as concerns in data privacy, ownership, consent, reliability, accuracy, accessibility, and the need for robust ethical guidelines. They express apprehension over privacy risks, emphasizing the need for secure handling of student data and clear data ownership. Consent in AI integration is crucial, as is ensuring AI tools are accurate and reliable to avoid misinformation. Accessibility remains a priority, with calls for AI to be inclusive and equitable. These themes underline the importance of ethical frameworks to guide AI's responsible use in educational contexts.

Doctoral researchers express significant concerns about privacy, ownership, and consent in the context of AI in education, with many worried about how AI systems handle sensitive data. "Given the sensitivity of data collected, especially research data, there's a heightened risk of breach and misuse that could impact my reputation and career," remarked participant seven. Concerns about data ownership also surfaced, with several researchers asserting that student-generated data should belong to the individual, not the institution, which they feel could potentially misuse this information. Another participant (five) stated, "As a researcher, I feel strongly that data generated in the learning process should belong to students, with institutions requiring explicit permission to access it." When it comes to accuracy, several participants expressed hesitancy regarding AI's reliability in handling complex academic tasks, fearing that errors in data analysis and literature review could mislead their research findings. As participant six explained, "I'm cautious about AI's accuracy in complex data analysis; any misjudgement could mislead my research findings." Accessibility was also highlighted as a potential benefit, with some suggesting that AI could help students access specialized resources, but others cautioning that unequal access to AI-driven tools could reinforce academic hierarchies rather than dismantle them.

2. What are the key concerns of doctoral researchers about the application of AI in educational settings, particularly in relation to data privacy and ownership?

Doctoral researchers' key concerns regarding AI in educational settings focuses on main five themes; such themes are, data privacy, ownership, security risks, transparency, and consent. They emphasize the need for strict data privacy measures and clear ownership rights to protect students' personal information, underscoring the potential risks of unauthorized access and misuse. Additionally, researchers call for transparency in AI decision-making processes to build trust and ensure that students and educators understand how decisions are made. Consent is also a significant issue, with researchers advocating for explicit consent in AI integration to uphold ethical standards in educational environments.

Concerns about data privacy and ownership emerged as some of the most urgent ethical considerations. All participants questioned institutional data ownership, emphasizing the importance of informed consent and transparency in data collection and usage policies. "We should be given detailed, ongoing consent options regarding how our research and learning data is used by AI, not a one-time agreement," participant one. Security risks were another frequent concern, with participants noting that a data breach could compromise the credibility of doctoral research. Participant three expressed, "A data breach could expose our research prematurely and compromise sensitive information, impacting the credibility of doctoral research." Transparency in AI decisions was another major concern, as many participants felt that AI systems often lack transparency, which undermines trust. "As doctoral students, we need to understand the algorithms behind AI to trust its academic recommendations, especially if AI influences research directions," participant two observed. This lack of transparency led several participants to call for institutional responsibility in data protection and clarity around data handling practices.

3. In what ways do doctoral researchers believe AI can impact the accuracy and accessibility of educational resources, and what ethical frameworks should guide its use?

Doctoral researchers believe AI can greatly enhance the accuracy and accessibility of educational resources, yet they emphasize that ethical frameworks are essential to address challenges in five main themes discovered by the present study is reliability, autonomy, transparency, fairness, and bias. They argue that while AI has the potential to personalize learning and make resources more accessible, it must be reliable and free from bias to ensure fair treatment of all students. Transparency in AI's decision-making and a commitment to ethical guidelines are seen as crucial for maintaining trust, promoting fairness, and respecting student autonomy in educational applications.

Participants generally felt that AI could positively impact resource accessibility but noted potential risks to the accuracy of academic content. They highlighted concerns around the reliability of AI in academic research, with several pointing out that AI might misinterpret data trends or overlook relevant literature. "If AI misinterprets data trends or literature relevance, it could negatively affect the quality and credibility of my research output," explained participant four. There were also strong views about the importance of maintaining academic autonomy in AI-enhanced learning environments. Participants felt that AI should support, not dictate, student learning choices. "AI-based suggestions should support, not dictate, decisions in academic research, allowing students to retain control over their work," stated participant three. In line with this, there was significant support for the development of ethical guidelines to oversee AI use in education. These guidelines, participants suggested, should address privacy, data ownership, bias prevention, and student autonomy. "Guidelines should address privacy, data ownership, and bias prevention in AI, especially as it shapes academic development," expressed participant eight, emphasising the need for AI to complement rather than compromise the agency and creativity of doctoral students. Doctoral researchers perceive AI as a powerful tool with both risks and benefits for education. Their primary concerns revolve around data privacy, ownership, and transparency, while they also recognize AI's potential to enhance accessibility and support diverse learning needs. However, they emphasize that clear ethical frameworks are essential to ensure AI tools align with academic integrity, student autonomy, and fair access, thus supporting a balanced and responsible approach to AI integration in education.

DISCUSSION

The research aimed to explore doctoral researchers' perceptions of the ethical implications of AI in education, specifically regarding privacy, accuracy, property, and accessibility. There are three key findings of the present research are, First, data privacy and ownership emerged as primary concerns, particularly regarding the handling of sensitive academic data and the need for explicit consent. Second, there is a demand for transparency in AI decision-making, as students seek to understand and control how AI influences their academic pathways. Third, the findings indicate an emphasis on the need for ethical guidelines that protect academic freedom and maintain fairness within AI-driven educational environments. The results suggest a critical awareness among doctoral researchers regarding the ethical complexities of AI in education. Participants expressed strong concerns about data privacy, mirroring Chan's (2023) and Holmes et al.'s (2022) findings that the extensive data collection in educational AI can lead to potential privacy and security issues. This aligns with the literature emphasizing the need for robust governance and data minimization practices to secure sensitive information and uphold user trust (Nguyen et al., 2022; Fassbender, 2024). Additionally, researchers' calls for transparency in AI processes highlights the perceived "black-box" nature of many AI systems (Kitto & Knight, 2019; Khreisat et al., 2024), which can hinder trust and comprehension of AI-driven decisions (Rane, 2024; Khreisat et al.,2024). The demand for transparency aligns with studies by Lee et al. (2022) and Jang et al. (2022), which advocate for more accessible, explainable AI systems in education to empower users in decision-making. Furthermore, the expressed need for comprehensive ethical guidelines emphasizes the importance of human-centered AI applications that maintain student agency, echoing Nguyen et al.'s (2022) findings on the necessity of autonomy and inclusivity within AI frameworks.

The findings have significant implications for how AI is integrated into educational settings. Firstly, there is a pressing need for institutions to adopt stringent data protection and privacy policies. Ethical data governance, as suggested by Gudiño Paredes et al. (2021), must become a core principle for institutions to mitigate risks associated with data misuse. Secondly, promoting transparency in AI decision-making could foster greater trust and encourage more informed engagement with AI tools among students and faculty (Chan, 2023; Yang, Wu, & Xie, 2024). Ensuring AI systems are explainable would align educational AI applications with the transparency guidelines set by UNESCO (2020) and OECD (2021), thus enhancing user trust and fostering an ethical AI ecosystem. Additionally, the emphasis on maintaining autonomy and avoiding AI-induced biases speaks to the need for ethical frameworks in AI design and application. Embedding fairness and inclusivity into AI systems will not only reduce the risk of perpetuating academic or social biases but also support a more equitable learning environment, as highlighted in studies by Holmes et al. (2023) and Reiss (2021). These ethical principles would ensure that AI supports diverse learning styles without compromising academic freedom or inadvertently diminishing human oversight (McGrath et al., 2023). An unexpected yet exciting result was the increased emphasis on the need for student autonomy in AI-driven educational systems ((Al Daraai et al., 2024; Yang, Wu, & Xie, 2024). While privacy and transparency concerns were anticipated, participants strongly highlighted the risk of over-reliance on AI, which could limit independent thinking and decision-making (Airaj,2024). This insight links back to the research question by revealing an underlying ethical concern: the need for AI systems that support, rather than restrict, students' academic agency and personal growth. This finding aligns with the literature on preserving autonomy within AI applications in education (Kousa & Niemi, 2023; Al Daraai et al., 2024)

Limitations of the Study and Future Research Directions

This study has two main limitations that suggest caution in interpreting its findings. The participant sample, limited to doctoral students, may not fully represent the perspectives of broader student populations, educators, or policymakers who engage with AI in education. Additionally, focusing solely on the PAPA framework for ethical analysis may have excluded other significant considerations, such as environmental sustainability, long-term social effects, and broader issues of educational equity. These limitations highlight the need for a more diverse participant base and the inclusion of a wider range of ethical frameworks to fully capture the complex implications of AI in educational settings. To address the limitations identified, future research should aim to include a more diverse participant base, expanding beyond doctoral students to encompass a broader range of stakeholders, such as undergraduate students, educators, policymakers, and administrators. This diversity would provide a more comprehensive understanding of the ethical implications of AI in education across various educational roles and demographics. Further, it would be beneficial to examine AI's ethical impacts across different educational systems, cultures, and technological environments to capture a global perspective on the challenges and opportunities AI presents in education.

Methodologically, future studies should consider mixed-method approaches that integrate quantitative and qualitative data to yield richer, more nuanced insights. Large-scale surveys could provide statistical insights on attitudes and concerns, while in-depth interviews and focus groups could uncover context-specific ethical considerations and experiences that might be overlooked in quantitative research alone. Longitudinal studies could also be valuable to explore how perceptions of AI in education evolve over time and in response to technological and regulatory changes, providing insights into both immediate and long-term impacts. Expanding beyond the PAPA framework, future research could incorporate additional ethical frameworks that account for emerging concerns, including sustainability, environmental impact, and social consequences. A multidimensional framework that integrates concepts like environmental ethics, equity, and AI's potential impact on educational equity could offer a more holistic view of AI's role in education. Doctoral researchers could also explore interdisciplinary perspectives, drawing from fields such as environmental science, sociology, and psychology, to assess AI's broader social and ecological effects and to propose adaptive ethical guidelines suited to evolving AI applications.

CONCLUSION

This study provides valuable insights into doctoral researchers' ethical concerns surrounding AI in education, particularly regarding data privacy, transparency, fairness, and inclusivity. The findings reinforce the importance of responsible data governance, transparency, and the establishment of ethical guidelines to address these concerns effectively. To safeguard academic integrity and student autonomy, it is recommended that institutions implement robust data governance practices with regular audits and transparent policies to maintain user trust. Additionally, educational AI developers should prioritize transparency and interpretability, ensuring that students and educators can understand AI's decision-making processes. This could involve training modules to enhance users' comprehension of AI's capabilities and limitations. Collaborative efforts to create ethical guidelines, aligned with UNES-

CO and OECD standards, are also essential to support a human-centered approach in AI applications, preserving human judgment in educational interactions. These recommendations align with previous research (Nguyen et al., 2022; Chan, 2023), emphasizing the need for ethical standards that protect academic freedom, student autonomy, and equitable access to educational resources, and highlighting areas for further investigation as AI's role in education continues to expand.

REFERENCES

Airaj, M. (2024). Ethical artificial intelligence for teaching-learning in higher education. *Education and Information Technologies*, 29(13), 17145–17167. DOI: 10.1007/s10639-024-12545-x

Al Daraai, S. B., Al Maqrashi, M., Al Zakwani, M., & Al Shaikh, Z. (2024). Integrating AI in Higher Education: Applications, Strategies, Ethical Considerations. In *Utilizing AI for Assessment, Grading, and Feedback in Higher Education* (pp. 189-211). IGI Global.

Antony, S., Ramnath, R., & Ellikkal, A. (2024). Examining Students' Perspectives on Pedagogical Translanguaging in the Multilingual Classroom Context. *IAFOR Journal of Education*, 12(1), 199–113. DOI: 10.22492/ije.12.1.08

Ashok, M., Madan, R., Joha, A., & Sivarajah, U. (2022). Ethical framework for Artificial Intelligence and Digital technologies. *International Journal of Information Management*, 62, 102433. DOI: 10.1016/j.ijinfomgt.2021.102433

Bhagat, S., & Kim, D. J. (2020). Higher Education Amidst COVID-19: Challenges and Silver Lining. *Information Systems Management*, 37(4), 366–371. DOI: 10.1080/10580530.2020.1824040

Chan, C. K. Y. (2023). A comprehensive AI policy education framework for university teaching and learning. *International Journal of Educational Technology in Higher Education*, 20(1), 38. DOI: 10.1186/s41239-023-00408-3

Creswell, J. W. (2007). *Qualitative inquiry and research design: Choosing among five approaches* (2nd ed.). Sage Publications, Inc.

Creswell, J. W., & Poth, C. N. (2016). *Qualitative Inquiry & Research Design: Choosing among Five Approaches*. Sage Publications.

Ellis, T. S., & Griffith, D. (2000). The evaluation of IT ethical scenarios using a multidimensional scale. *The Data Base for Advances in Information Systems*, 32(1), 75–85. DOI: 10.1145/506740.506750

Essien, A., Bukoye, O. T., O'Dea, X., & Kremantzis, M. (2024). The influence of AI text generators on critical thinking skills in UK business schools. *Studies in Higher Education*, 49(5), 1–18. DOI: 10.1080/03075079.2024.2316881

European Commission. (2018). Statement on artificial intelligence, robotics and 'autonomous' systems. European Union Publications Office. https://op.europa.eu/en/publication-detail/-/publication/dfebe62e-4ce9-11e8-be1d-01aa75ed71a1/language-en/format-PDF/source-78120382

European Commission. (2019). The European Commission's high-level expert group on artificial intelligence: Ethics guidelines for trustworthy AI. European Union Publications Office. https://digital-strategy.ec.europa.eu/en/policies/expert-group-ai

European Commission. (2019). The European Commission's high-level expert group on artificial intelligence: Ethics guidelines for trustworthy AI. European Union Publications Office. https://digitalstrategy.ec.europa.eu/en/policies/expert-group-ai

European Parliament. (2021). Report on artificial intelligence in education, culture and the audio visual sector (2020/2017(INI)). Committee on Culture and Education. https://www.europarl.europa.eu/doceo/document/A-9-2021-0127_EN.html

Fassbender, W. J. (2024). "I can almost recognize its voice": AI and its impact on ethical teacher-centaur labor. English teaching: Practice & critique (ahead-of-print).

Floridi, L. (2018). Soft Ethics and the Governance of the Digital. Philosophy &amp. *Technology*, 31(1), 1–8. DOI: 10.1007/s13347-018-0303-9 PMID: 29682599

Greer, S. L. (2018). Organization and Governance: Stewardship and Governance in Health Systems. Health Care Systems and Policies, 1–9. DOI: 10.1007/978-1-4614-6419-8_22-1

Holmes, W., Persson, J., Chounta, I. A., Wasson, B., & Dimitrova, V. (2022). *Artificial intelligence and education: A critical view through the lens of human rights, democracy and the rule of law*. Council of Europe.

Ifenthaler, D., & Schumacher, C. (2016). Student perceptions of privacy principles for learning analytics. *Educational Technology Research and Development*, 64(5), 923–938. DOI: 10.1007/s11423-016-9477-y

Khreisat, M. N., Khilani, D., Rusho, M. A., Karkkulainen, E. A., Tabuena, A. C., & Uberas, A. D. (2024). Ethical Implications Of AI Integration In Educational Decision Making: Systematic Review. *Educational Administration: Theory and Practice*, 30(5), 8521–8527.

Kong, S. C., Cheung, W. M. Y., & Tsang, O. (2023). Evaluating an artificial intelligence literacy programme for empowering and developing concepts, literacy and ethical awareness in senior secondary students. *Education and Information Technologies*, 28(4), 4703–4724. DOI: 10.1007/s10639-022-11408-7

Lee, C. H., Gobir, N., Gurn, A., & Soep, E. (2022). In the Black Mirror: Youth Investigations into Artificial Intelligence. *ACM Transactions on Computing Education*, 22(3), 1–25. DOI: 10.1145/3484495

Lee, C. H., Gobir, N., Gurn, A., & Soep, E. (2022). In the black mirror: Youthinvestigations into artificial intelligence. *ACM Transactions on Computing Education*, 22(3), 1–25. DOI: 10.1145/3484495

Lin, C. H., Yu, C. C., Shih, P. K., & Wu, L. Y. (2021). STEM based artificial intelligence learning in general education for non-engineering undergraduate students. *Journal of Educational Technology & Society*, 24(3), 224–237. https://www.jstor.org/stable/27032867

Loch, K. D., Conger, S., & Oz, E. (1998).. . *Journal of Business Ethics*, 18(2), 199–209. DOI: 10.1023/A:1005720606763

Malterud, K. (2012). Systematic text condensation: A strategy for qualitative analysis. *Scandinavian Journal of Public Health*, 40(8), 795–805. DOI: 10.1177/1403494812465030 PMID: 23221918

Mason, R. O. (1986). Four Ethical Issues of the Information Age. *Management Information Systems Quarterly*, 10(1), 5. DOI: 10.2307/248873

McDonald, N., & Pan, S. (2020). Intersectional AI: A study of how information science students think about ethics and their impact. *Proceedings of the ACM on Human-Computer Interaction*, 4(CSCW2), 1–19. DOI: 10.1145/3415218

McGrath, C., Pargman, T. C., Juth, N., & Palmgren, P. J. (2023). University teachers'perceptions of responsibility and artificial intelligence in higher education-Anexperimental philosophical study. Computers and Education: Artificial Intelligence, 4,Article 100139.

Merriam, S. B., & Tisdell, E. J. (2015). *Qualitative Research: A Guide to Design and Implementation*. Wiley.

Miao, F., Holmes, W., Huang, R., & Zhang, H. (2021). AI and education: Guidance for policy-makers. United Nations Educational, Scientific and Cultural Organization. https://unesdoc.unesco.org/ark:/48223/pf0000376709

Nguyen, A., Ngo, H. N., Hong, Y., Dang, B., & Nguyen, B.-P. T. (2022). Ethical principles for artificial intelligence in education. *Education and Information Technologies*, 28(4), 4221–4241. DOI: 10.1007/s10639-022-11316-w PMID: 36254344

Organization for Economic Co-operation and Development. (2021). OECD Recommendation of the Council on Artificial Intelligence. OECD/LEGAL/0449. https://legalinstruments.oecd.org/en/instruments/OECD-LEGAL-0449

Peslak, A. R. (2006). Papa Revisited: A Current Empirical Study of the Mason Framework. *Journal of Computer Information Systems*, 46(3), 117–123. DOI: 10.1080/08874417.2006.11645905

Prior, M., Rogerson, S., & Fairweather, B. (2002). The ethical attitudes of information systems professionals: Outcomes of an initial survey. *Telematics and Informatics*, 19(1), 21–36. DOI: 10.1016/S0736-5853(00)00014-9

Qin, F., Li, K., & Yan, J. (2020). Understanding user trust in artificial intelligence-based educational systems: Evidence from China. *British Journal of Educational Technology*, 51(5), 1693–1710. DOI: 10.1111/bjet.12994

Rane, N. (2024). Enhancing the quality of teaching and learning through Gemini, ChatGPT, and similar generative Artificial Intelligence: Challenges, future prospects, and ethical considerations in education. *TESOL and Technology Studies*, 5(1), 1–6. DOI: 10.48185/tts.v5i1.1000

Regan, P. M., & Jesse, J. (2018). Ethical challenges of edtech, big data and personalized learning: Twenty-first century student sorting and tracking. *Ethics and Information Technology*, 21(3), 167–179. DOI: 10.1007/s10676-018-9492-2

Reiss, M. J. (2021). The use of AI in education: Practicalities and ethical considerations. *London Review of Education*, 19(1). Advance online publication. DOI: 10.14324/LRE.19.1.05

Ryker, R., Lafleur, E., McManis, B., & Cox, K. C. (2002). Online Privacy Policies: An Assessment of the Fortune E-50. *Journal of Computer Information Systems*, 42(4), 15–20. DOI: 10.1080/08874417.2002.11647048

Sam, A. K., & Olbrich, P. (2023). The need for AI ethics in higher education. In *AI ethics in higher education: Insights from Africa and beyond* (pp. 3–10). Springer International Publishing. DOI: 10.1007/978-3-031-23035-6_1

Sharma, S., Singh, G., Sharma, C. S., & Kapoor, S. (2024). Artificial intelligence in Indian higher education institutions: A quantitative study on adoption and perceptions. *International Journal of System Assurance Engineering and Management*. Advance online publication. DOI: 10.1007/s13198-023-02193-8

Shih, P.-K., Lin, C.-H., Wu, L. Y., & Yu, C.-C. (2021). Learning Ethics in AI—Teaching Non-Engineering Undergraduates through Situated Learning. *Sustainability (Basel)*, 13(7), 3718. DOI: 10.3390/su13073718

Smith, B., Sparkes, A. C., & Caddick, N. (2014). Judging qualitative research. Research Methods in Sports Coaching, 192–201. DOI: 10.4324/9780203797549-18

UNESCO (2021). AI and education: guidance for policy-makers. DOI: 10.54675/PCSP7350

United Nations Educational, Scientific and Cultural Organization. (2019). Beijing Consensus on artificial intelligence and education. Outcome document of the International Conference on Artificial Intelligence and Education, Planning Education in the AI Era: Lead the Leap, Beijing, 2019. United Nations Educational, Scientific and Cultural Organization. https://unesdoc.unesco.org/ark:/48223/pf0000368303

United Nations Educational, Scientific and Cultural Organization. (2019). Beijing Consensus on artificial intelligence and education. Outcome document of the International Conference on Artificial Intelligence and Education, Planning Education in the AI Era: Lead the Leap, Beijing, 2019. United Nations Educational, Scientific and Cultural Organization. https://unesdoc.unesco.org/ark:/48223/pf0000368303

United Nations Educational, Scientific and Cultural Organization, & Organization, C. (2021). Recommendation on the Ethics of Artificial Intelligence. United Nations Educational. https://unesdoc.unesco.org/ark:/48223/pf0000379920.page=14

Usher, M., & Barak, M. (2024). Unpacking the role of AI ethics online education for science and engineering students. *International Journal of STEM Education*, 11(1), 35. Advance online publication. DOI: 10.1186/s40594-024-00493-4

Yang, Z., Wu, J. G., & Xie, H. (2024). Taming Frankenstein's monster: Ethical considerations relating to generative artificial intelligence in education. *Asia Pacific Journal of Education*, •••, 1–14. DOI: 10.1080/02188791.2023.2300137

Yu, L., & Yu, Z. (2023). Qualitative and quantitative analyses of artificial intelligence ethics in education using VOSviewer and CitNetExplorer. *Frontiers in Psychology*, 14, 1061778. Advance online publication. DOI: 10.3389/fpsyg.2023.1061778 PMID: 36968737

Zhou, Q., Suraworachet, W., Pozdniakov, S., Martinez-Maldonado, R., Bartindale, T., Chen, P., Richardson, D., & Cukurova, M. (2021). Investigating Students' Experiences with Collaboration Analytics for Remote Group Meetings. Artificial Intelligence in Education, 472–485. DOI: 10.1007/978-3-030-78292-4_38

Compilation of References

Abdikhakimov, I. (2023). Legal aspects of AI generated content. International Conference on Legal Sciences, 1(5), 1–17. https://science-zone.org/index.php/conference/article/view/41

Abdullah, N. A. H., & Hussain, M. (2020). The Effectiveness of Chatbots as a Tool for Learning: A Case Study on Students' Feedback and Assessment. *Education and Information Technologies*, 25(6), 5543–5560.

Acemoglu, D., & Restrepo, P. (2020). Robots and jobs: Evidence from US labor markets. *Journal of Political Economy*, 128(6), 2188–2244. DOI: 10.1086/705716

Acikkar, M., & Akay, M. F. (2009). Support vector machines for predicting the admission decision of a candidate to the School of Physical Education and Sports at Cukurova University. *Expert Systems with Applications*, 36(3 PART 2), 7228–7233. DOI: 10.1016/j.eswa.2008.09.007

Adamopoulou, E., & Moussiades, L. (2020). Chatbots: History, technology, and applications. *Machine Learning with Applications*, 2, 100006. DOI: 10.1016/j.mlwa.2020.100006

Adamson, D., Dyke, G., Jang, H., & Rosé, C. P. (2014). Towards an agile approach to adapting dynamic collaboration support to student needs. *International Journal of Artificial Intelligence in Education*, 24(1), 92–124. DOI: 10.1007/s40593-013-0012-6

Agaoglu, M. (2016). Predicting instructor performance using data mining techniques in higher education. *IEEE Access : Practical Innovations, Open Solutions*, 4, 2379–2387. DOI: 10.1109/ACCESS.2016.2568756

Ahlf, M., & McNeil, S. G. (2023). An exploratory review of literature on moderation in asynchronous discussions. *Distance Education*, 44(1), 137–161. DOI: 10.1080/01587919.2022.2150146

Ahmad, S., Umirzakova, S., Mujtaba, G., Amin, M. S., & Whangbo, T. (2023). Education 5.0: requirements, enabling technologies, and future directions. arXiv preprint arXiv:2307.15846.

Ahmad, S. F., Alam, M. M., Rahmat, M. K., Mubarik, M. S., & Hyder, S. I. (2022). Academic and administrative role of artificial intelligence in education. *Sustainability (Basel)*, 14(3), 1101. DOI: 10.3390/su14031101

Ahmad, S. F., Rahmat, M. K., Mubarik, M. S., Alam, M. M., & Hyder, S. I. (2021). Artificial intelligence and its role in education. *Sustainability (Basel)*, 13(22), 2902. DOI: 10.3390/su132212902

Ahn, H. Y. (2024). AI-Powered E-Learning for Lifelong Learners: Impact on Performance and Knowledge Application. *Sustainability (Basel)*, 16(20), 9066. DOI: 10.3390/su16209066

Ahshan, R. (2021). A framework of implementing strategies for active student engagement in remote/online teaching and learning during the COVID-19 pandemic. *Education Sciences*, 11(9), 483. DOI: 10.3390/educsci11090483

Ainscow, M. (2005). Developing inclusive education systems: What are the levers for change? *Journal of Educational Change*, 6(2), 109–124. DOI: 10.1007/s10833-005-1298-4

Airaj, M. (2024). Ethical artificial intelligence for teaching-learning in higher education. *Education and Information Technologies*, 29(13), 17145–17167. DOI: 10.1007/s10639-024-12545-x

Akavova, A., Temirkhanova, Z., & Lorsanova, Z. (2023). Adaptive learning and artificial intelligence in the educational space. E3S Web of Conferences, 451, 06011. DOI: 10.1051/e3sconf/202345106011

Akcaoğlu, M., & Lee, E. (2016). Increasing social presence in online learning through small group discussions. *International Review of Research in Open and Distance Learning*, 17(3). Advance online publication. DOI: 10.19173/irrodl.v17i3.2293

Akgun, S., & Greenhow, C. (2022). Artificial intelligence in education: Addressing ethical challenges in K-12 settings. *AI and Ethics*, 2(3), 431–440. DOI: 10.1007/s43681-021-00096-7 PMID: 34790956

Akyuz, Y. (2020). Effects of Intelligent Tutoring Systems (ITS) on Personalized Learning (PL). *Creative Education*, 11(6), 953–978. DOI: 10.4236/ce.2020.116069

Al Daraai, S. B., Al Maqrashi, M., Al Zakwani, M., & Al Shaikh, Z. (2024). Integrating AI in Higher Education: Applications, Strategies, Ethical Considerations. In *Utilizing AI for Assessment, Grading, and Feedback in Higher Education* (pp. 189-211). IGI Global.

Alam, A., Hasan, M., & Raza, M. M. (2022). Impact of artificial intelligence (AI) on education: changing paradigms and approaches. Towards Excellence, 14(1), 281-289. DOI:DOI: 10.13140/RG.2.2.15581.28646/1

Alam, A., Hasan, M., & Raza, M. M. (2022). *Impact of artificial intelligence (ai) on education: changing paradigms and approaches.* Towards Excellence., DOI: 10.37867/TE140127

Al-Ansi, A. M., Jaboob, M., Garad, A., & Al-Ansi, A. (2023). Analyzing augmented reality (AR) and virtual reality (VR) recent development in education. *Social Sciences & Humanities Open*, 8(1), 100532. DOI: 10.1016/j.ssaho.2023.100532

Alasadi, E. A., & Baiz, C. R. (2023). Generative AI in Education and Research: Opportunities, Concerns, and Solutions. *Journal of Chemical Education*, 100(8), 2965–2971. DOI: 10.1021/acs.jchemed.3c00323

Al-Azawei, A., & Lundqvist, K. (2019). The impact of technology on the learning of students with disabilities: A systematic review. *Disability and Rehabilitation. Assistive Technology*, 14(4), 385–395. DOI: 10.1080/17483107.2018.1442051

Al-Azawei, A., Parslow, P., & Lundqvist, K. (2016). The role of artificial intelligence in accessibility: A systematic review. In Brown, S. (Ed.), *Emerging*.

Albert-Rochette, M., Lavis, C., & Meunier, M. (2024). Protection de la vie privée dans le contexte des plateformes en ligne: Les limites des fiducies de données et du devoir fiduciaire de loyauté. *Lex Electronica*, 28(1), 23–59. DOI: 10.7202/1108621ar

Aleven, V., McLaren, B. M., Sewall, J., & Koedinger, K. R. (2016). The cognitive tutor authoring tools (CTAT): Preliminary evaluation of efficiency gains. In *International Conference on Intelligent Tutoring Systems* (pp. 61-70). Springer, Cham.

Alexandre, M., Roy, J., Bernatchez, J., Dubé, N. F., & Lahaie, H. (2022). Des solidarités numériques en action: Le soutien aux étudiants et aux étudiantes utilisant des environnements numériques d'apprentissage en milieu collégial. *Médiations Et Médiatisations*, (12), 54–73. DOI: 10.52358/mm.vi12.289

Alhazmi, A. K., Imtiaz, A., Al-Hammadi, F., & Kaed, E. (2021). Success and failure aspects of lms in e-learning systems. [iJIM]. *International Journal of Interactive Mobile Technologies*, 15(11), 133. DOI: 10.3991/ijim.v15i11.20805

Alhumaid, K., Naqbi, S. A., Elsori, D., & Mansoori, M. A. (2023). The adoption of artificial intelligence applications in education. *International Journal of Data and Network Science*, 7(1), 457–466. DOI: 10.5267/j.ijdns.2022.8.013

Ali, A., & Alkhateeb, H. (2020). Chatbots in Education: A Review of the Use of Chatbots for Learning. *Education and Information Technologies*, 25(5), 4153–4168.

Al-khresheh, M. H. (2021). Revisiting the effectiveness of blackboard learning management system in teaching english in the era of covid-19. *World Journal of English Language*, 12(1), 1. DOI: 10.5430/wjel.v12n1p1

Ally, M., & Tsinakos, A. (2014). *Advances in Mobile Learning Educational Research*. IGI Global.

Almaiah, M. A., Alfaisal, R., Salloum, S. A., Hajjej, F., Thabit, S., El-Qirem, F. A., Lutfi, A., Alrawad, M., Al Mulhem, A., Alkhdour, T., Awad, A. B., & Al-Maroof, R. S. (2022). Examining the impact of artificial intelligence and social and computer anxiety in e-learning settings: Students' perceptions at the university level. *Electronics (Basel)*, 11(22), 3662. DOI: 10.3390/electronics11223662

Almelhi, A. M. (2021). Effectiveness of the ADDIE model within an E-learning environment in developing creative writing in EFL students. *English Language Teaching*, 14(2), 20–36. DOI: 10.5539/elt.v14n2p20

Almusaed, A., Almssad, A., Yitmen, I., & Homod, R. Z. (2023). Enhancing student engagement: Harnessing "aied"'s power in hybrid education—a review analysis. *Education Sciences*, 13(7), 632. DOI: 10.3390/educsci13070632

Alodail, A. K. (2014). Instructing educators in the use of assistive technology listening devices in the classroom. *International Education Studies*, 7(5). Advance online publication. DOI: 10.5539/ies.v7n5p55

Alrayes, A., Henari, T. F., & Ahmed, D. A. (2024). Chatgpt in education – understanding the bahraini academics perspective. *Electronic Journal of e-Learning*, 22(2), 112–134. DOI: 10.34190/ejel.22.2.3250

Al-Samarraie, H. (2019). A scoping review of videoconferencing systems in higher education. *International Review of Research in Open and Distance Learning*, 20(3). Advance online publication. DOI: 10.19173/irrodl.v20i4.4037

Amankwah-Amoah, J., Abdalla, S., Mogaji, E., Elbanna, A., & Dwivedi, Y. K. (2024). The impending disruption of creative industries by generative AI: Opportunities, challenges, and research agenda. In *International Journal of Information Management* (p. 102759). Elsevier., https://www.sciencedirect.com/science/article/pii/S0268401224000070 DOI: 10.1016/j.ijinfomgt.2024.102759

Amin, S., Uddin, M. I., Alarood, A. A., Mashwani, W. K., Alzahrani, A., & Alzahrani, A. O. (2023). Smart E-learning framework for personalized adaptive learning and sequential path recommendations using reinforcement learning. *IEEE Access : Practical Innovations, Open Solutions*, 11, 89769–89790. DOI: 10.1109/ACCESS.2023.3305584

Andrin, G., Kilag, O. K., Groenewald, E., Benitez, J., Dagala, F., & Ubay, R. (2024). Borderless Learning Environments: Impacts on Educational Management Strategies. [IMJRISE]. *International Multidisciplinary Journal of Research for Innovation, Sustainability, and Excellence*, 1(2), 43–49.

Angwin, J., Larson, J., Mattu, S., & Kirchner, L. (2016). Machine bias. ProPublica. Retrieved from https://www.propublica.org/article/machine bias risk assessments in criminal sentencing

Anjali, R. M., & Yogi, M. K. (2024). AIculum-AI-Driven Curriculum Design in Educational Sector. Journal of Big Data Analytics and Business Intelligence, 8–13.

Annus, N. (2023). Chatbots in Education: The impact of Artificial Intelligence based ChatGPT on Teachers and Students. *International Journal of Advanced Natural Sciences and Engineering Researches*, 7(4), 366–370. DOI: 10.59287/ijanser.739

Antony, S., Ramnath, R., & Ellikkal, A. (2024). Examining Students' Perspectives on Pedagogical Translanguaging in the Multilingual Classroom Context. *IAFOR Journal of Education*, 12(1), 199–113. DOI: 10.22492/ije.12.1.08

Aoun, J. E. (2017). *Robot-proof: Higher education in the age of artificial intelligence*. MIT press. DOI: 10.7551/mitpress/11456.001.0001

Archambault, H. and Dumouchel, G. (2021). Compétences professionnelles et transfert des apprentissages des directions d'établissement en formation universitaire: une exploration franco-manitobaine. Canadian Journal of Education/Revue Canadienne De l'Éducation, 44(1), 1-27. DOI: 10.53967/cje-rce.v44i1.3573

Archbell, K. A., & Coplan, R. J. (2022). Too Anxious to Talk: Social Anxiety, Academic Communication, and Students' Experiences in Higher Education. *Journal of Emotional and Behavioral Disorders*, 30(4), 273–286. DOI: 10.1177/10634266211060079

Arnone, M. P., Small, R. V., Chauncey, S. A., & McKenna, H. P. (2011). Curiosity, interest and engagement in technology-pervasive learning environments: A new research agenda. *Educational Technology Research and Development*, 59(2), 181–198. DOI: 10.1007/s11423-011-9190-9

Arora, A., & Arora, A. (2022). Generative adversarial networks and synthetic patient data: Current challenges and future perspectives. *Future Healthcare Journal*, 9(2), 190–193. DOI: 10.7861/fhj.2022-0013 PMID: 35928184

Asad, M. M., & Ajaz, A. (2024). Impact of ChatGPT and generative AI on lifelong learning and upskilling learners in higher education: Unveiling the challenges and opportunities globally. The International Journal of Information and Learning Technology. https://www.emerald.com/insight/content/doi/10.1108/IJILT-06-2024-0103/full/html

Asamoah, M. K. (2020). Ict officials' opinion on deploying open source learning management system for teaching and learning in universities in a developing society. *E-Learning and Digital Media*, 18(1), 18–38. DOI: 10.1177/2042753020946280

Ashok, M., Madan, R., Joha, A., & Sivarajah, U. (2022). Ethical framework for Artificial Intelligence and Digital technologies. *International Journal of Information Management*, 62, 102433. DOI: 10.1016/j.ijinfomgt.2021.102433

Aslan, A. (2021). Problem- based learning in live online classes: Learning achievement, problem-solving skill, communication skill, and interaction. *Computers & Education*, 171, 104237. Advance online publication. DOI: 10.1016/j.compedu.2021.104237

Atan, S. (2018, December). Adaptive Reading Comprehension (ARC) System for Singapore Malay Language Students. In 2018 International Conference on Machine Learning and Data Engineering (iCMLDE) (pp. 97-102). IEEE.

Atkinson, J. W. (1964). *An Introduction to Motivation*. Van Nostrand.

Auteur, A. A. (année). Titre du chapitre. Dans Éditeur (Ed.), Titre du livre en italique (pp. pages du chapitre). Éditeur.

Azuma, R. T. (1997). A Survey of Augmented Reality. *Presence (Cambridge, Mass.)*, 6(4), 355–385. DOI: 10.1162/pres.1997.6.4.355

Babić, I. D. (2017). Machine learning methods in predicting the student academic motivation. *Croatian Operational Research Review*, 8(2), 443–461. DOI: 10.17535/crorr.2017.0028

Bacca, J., Baldiris, S., Fabregat, R., & Graf, S., & Kinshuk. (2014). Augmented Reality Trends in Education: A Systematic Review of Research and Applications. *Journal of Educational Technology & Society*, 17(4), 133–149.

Ba, H., zhang, L., & Yi, Z. (2024). Enhancing clinical skills in pediatric trainees: A comparative study of chatgpt-assisted and traditional teaching methods. *BMC Medical Education*, 24(1), 558. Advance online publication. DOI: 10.1186/s12909-024-05565-1 PMID: 38778332

Bahadır, E. (2016). Using neural network and logistic regression analysis to predict prospective mathematics teachers' academic success upon entering graduate education. *Educational Sciences: Theory & Practice*, 16(3), 943–964. DOI: 10.12738/estp.2016.3.0214

Bahroun, Z., Anane, C., Ahmed, V., & Zacca, A. (2023). Transforming Education: A Comprehensive Review of Generative Artificial Intelligence in Educational Settings through Bibliometric and Content Analysis. *Sustainability (Basel)*, 15(17), 17. Advance online publication. DOI: 10.3390/su151712983

Baillifard, A., Bonvin, M., & Guiard, N. (2023). Amuqquiz et kaïros: Regards croisés sur la création et le développement de deux plateformes d'apprentissage adaptatives. *Revue Internationale Des Technologies en Pédagogie Universitaire*, 20(2), 139–153. DOI: 10.18162/ritpu-2023-v20n2-11

Bai, S., Hew, K. F., & Huang, B. (2020). Is Gamification "Bullshit"? Evidence from a Meta-analysis and Synthesis of Qualitative Data in Educational Contexts. *Educational Research Review*, 30, 100322. DOI: 10.1016/j.edurev.2020.100322

Baker, T., & Smith, L. (2019). Educ-AI-tion rebooted? Exploring the future of artificial intelligence in schools and colleges. *Nesta Report.*

Baker, T., & Smith, L. (2019). Education rebooted? Exploring the future of artificial intelligence in schools and colleges. Retrieved from Nesta Foundation website: https://media.nesta.org.uk/documents/Future_of_AI_and_education_v5_WEB.pdf

Baker, T., & Smith, L. (2019). *Educ-AI-tion rebooted? Exploring the future of artificial intelligence in schools and colleges.* Nesta Foundation.

Baldwin, A. K., & Ford, K. J. (2017). Micro-Learning: A New Approach to Training and Development. *International Journal of Training and Development*, 21(3), 193–205.

Banawan, M., Rodrigo, M., & Andres, J. (2015), An investigation of frustration among students using physics playground. In Doctoral Student Consortium (DSC), Proceedings of the 23rd International Conference on Computers in Education

Bandura, A. (1977). *Social Learning Theory*. Prentice-Hall.

Bandura, A. (1986). *Social Foundations of Thought and Action: A Social Cognitive Theory*. Prentice-Hall.

Barhate, D., Pathak, S., & Dubey, A. K. (2023). Hyperparameter-tuned batch-updated stochastic gradient descent: Plant species identification by using hybrid deep learning. *Ecological Informatics*, 75, 102094. DOI: 10.1016/j.ecoinf.2023.102094

Barkley, E. F., Cross, P. K., & Major, C. H. (2005). *Collaborative Learning Techniques: A Handbook for College Faculty*. Jossey-Bass.

Barkoukis, V., Taylor, I., Chanal, J., & Ntoumanis, N. (2014). The Relation between Student Motivation and Student Grades in Physical Education: A 3-year Investigation. *Scandinavian Journal of Medicine & Science in Sports*, 24(5), 414. DOI: 10.1111/sms.12174 PMID: 24433528

Barrows, H. S., & Tamblyn, R. M. (1980). *Problem-Based Learning: An Approach to Medical Education*. Springer.

Barr, V., & Stephenson, C. (2019). Bringing Computational Thinking to K-12: What is Involved and What is the Role of the Computer Science Education Community? *ACM Inroads*, 2(1), 48–54. DOI: 10.1145/1929887.1929905

Bashir, I. (2017). *Mastering Blockchain: Deeper Insights into Decentralization, Cryptography, Bitcoin, and Popular Blockchain Frameworks*. Packt Publishing Ltd.

Baskara, F. R. (2023a). Fostering culturally grounded learning: Generative ai, digital storytelling, and early childhood education. International Conference of Early Childhood Education in Multiperspectives, 351–361. https://repository.usd.ac.id/48848/1/10058_351-361__IE__FX.%2BRisang%2BBaskara_3.pdf

Baskara, F. R. (2023b). Navigating pedagogical evolution: The implication of generative AI on the reinvention of teacher education. Prosiding Seminar Nasional Universitas Jabal Ghafur, 2, 409–421. https://repository.usd.ac.id/48711/1/10357_1923-4619-1-SM.pdf

Baskara, F. R. (2024). From AI to We: Harnessing Generative AI Tools to Cultivate Collaborative Learning Ecosystems in Universities. Proceeding International Conference on Learning Community (ICLC), 1(1). https://repository.usd.ac.id/51989/1/11944_27985-76316-1-PB.pdf

Bates, T., Cobo, C., Mariño, O., & Wheeler, S. (2020). Can artificial intelligence transform higher education? *International Journal of Educational Technology in Higher Education*, 17(1), 1–12. DOI: 10.1186/s41239-020-00218-x

Bates, T., & Glikman, V. (2021). One history of technology and media in distance education. *Médiations Et Médiatisations*, (6), 12–34. DOI: 10.52358/mm.vi6.190

Bayly-Castaneda, K., Ramirez-Montoya, M. S., & Morita-Alexander, A. (2024). Crafting personalized learning paths with AI for lifelong learning: A systematic literature review. *Frontiers in Education*, 9, 1424386. https://www.frontiersin.org/journals/education/articles/10.3389/feduc.2024.1424386/full. DOI: 10.3389/feduc.2024.1424386

Beaupied, A. (2009). L'évaluation par les compétences. *Idées économiques et sociales, 155*(1), 71–77. DOI: 10.3917/idee.155.0071

Beck, S. F., & Fagan, M. (2020). Digital Badges and Micro-Credentials: A New Way to Assess Learning Outcomes. *International Journal of Learning Technology*, 15(2), 113–126.

Bell, S., Barefoot, H., & Thomas, L. (2010). *Project-Based Learning for the 21st Century Skills*. Education Reform Series. DOI: 10.1080/00098650903505415

Bengesi, S., El-Sayed, H., Sarker, M. K., Houkpati, Y., Irungu, J., & Oladunni, T. (2024). Advancements in Generative AI: A Comprehensive Review of GANs, GPT, Autoencoders, Diffusion Model, and Transformers. IEEE Access, 12, 69812–69837. IEEE Access. DOI: 10.1109/ACCESS.2024.3397775

Bennett, S., & Maton, K. (2010). Beyond the 'digital natives' debate: Toward a more nuanced understanding of students' technology experiences. *Journal of Computer Assisted Learning*, 26(5), 321–331. DOI: 10.1111/j.1365-2729.2010.00360.x

Benotti, L., Martnez, M. C., & Schapachnik, F. (2018). Atool for introducing computer science with automatic formative assessment. *IEEE Transactions on Learning Technologies*, 11(2), 179–192. DOI: 10.1109/TLT.2017.2682084

Bere, A., Deng, H., & Tay, R. (2018). Assessing the impact of using instant messaging in elearning on the performance of teaching and learning in higher education. Australasian Conference on Information Systems 2018. DOI: 10.5130/acis2018.by

Berendt, B., Littlejohn, A., & Blakemore, M. (2020). AI in Education: Learner Choice and Fundamental Rights. *Learning, Media and Technology*, 45(3), 312–324. Advance online publication. DOI: 10.1080/17439884.2020.1786399

Bertolotti, F., Mattarelli, E., Vignoli, M., & Macrì, D. M. (2015). Exploring the relationship between multiple team membership and team performance: The role of social networks and collaborative technology. *Research Policy*, 44(4), 911–924. DOI: 10.1016/j.respol.2015.01.019

Bettioui, R. E., Haidane, A., Jaouhari, L., & Mirdasse, S. (2022). Digitalisation pédagogique et défis de l'enseignement à distance. revistamultidisciplinar.com, 4(1), 27-48. DOI: 10.23882/rmd.22083

Bhagat, S., & Kim, D. J. (2020). Higher Education Amidst COVID-19: Challenges and Silver Lining. *Information Systems Management*, 37(4), 366–371. DOI: 10.1080/10580530.2020.1824040

Bhatt, P. K., & Muduli, A. (2022). Artificial intelligence in learning and development: A systematic literature review. *European Journal of Training and Development*, 47(7/8), 677–694. DOI: 10.1108/EJTD-09-2021-0143

Bhimdiwala, A., Neri, R. C., & Gomez, L. M. (2021). Advancing the Design and Implementation of Artificial Intelligence in Education through Continuous Improvement. *International Journal of Artificial Intelligence in Education*. Advance online publication. DOI: 10.1007/s40593-021-00278-8

Bian, L., & Xie, Y. (2010). Research on the adaptive strategy of adaptive learning system. In Entertainment for Education. Digital Techniques and Systems: 5th International Conference on E-learning and Games, Edutainment 2010, Changchun, China, August 16-18, 2010. [Springer Berlin Heidelberg.]. *Proceedings*, 5, 203–214.

Billinghurst, M., & Duenser, A. (2012). Augmented Reality in the Classroom. *Computer*, 45(7), 56–63. DOI: 10.1109/MC.2012.111

Bilquise, G., & Shaalan, K. (2022). AI-based academic advising framework: A knowledge management perspective. *International Journal of Advanced Computer Science and Applications*, 13(8). Advance online publication. DOI: 10.14569/IJACSA.2022.0130823

Binns, R. (2018). Fairness in machine learning: Lessons from political philosophy. Proceedings of the 2018 Conference on Fairness, Accountability, and Transparency, 149 159.

Binns, R. (2018). Fairness in Machine Learning: Lessons from Political Philosophy. In *Proceedings of the 2018 Conference on Fairness, Accountability, and Transparency* (pp. 149-158).

Bisson, M., Ouellet, C., & Verret, C. (2022). Collaborer pour mieux enseigner aux élèves présentant des difficultés en éducation physique. l'Education Physique en Mouvement, (3), 31-34. https://doi.org/DOI: 10.26034/vd.epm.2020.3501

Bittner, E. A. C., Mirbabaie, M., & Morana, S. (2021). Digital facilitation assistance for collaborative, creative design processes. *Proceedings of the Annual Hawaii International Conference on System Sciences*. DOI: 10.24251/HICSS.2021.044

Blaise, M. (2023, avril 19). *"D'ici cinq ans, plus personne n'utilisera un modèle tel que ChatGPT »*. Retrieved from Usbek&Rica: https://usbeketrica.com/fr/article/d-ici-cinq-ans-plus-personne-n-utilisera-un-modele-tel-que-chatgpt

Blaskovits, F., Bayoumi, I., Davison, C. M., Watson, A., & Purkey, E. (2023). Impacts of the COVID-19 pandemic on life and learning experiences of Indigenous and non-Indigenous university and college students in Ontario, Canada: A qualitative study. *BMC Public Health*, 23(1), 96. DOI: 10.1186/s12889-023-15010-5 PMID: 36639672

Blumenfeld, P. C.. (2020). Creating Sustainable Practices in Project-Based Learning. *Environmental Education Research*, 49(5), 325–333.

Blumenfeld, P. C., Soloway, E., Marx, R. W., Krajcik, J. S., Guzdial, M., & Palincsar, A. (1991). Motivating Project-Based Learning: Sustaining the Doing, Supporting the Learning. *Educational Psychologist*, 26(3-4), 369–398. DOI: 10.1080/00461520.1991.9653139

Boaye Belle, A. (2024, April). Assuring the emotional and cultural intelligence of intelligent software systems. In Proceedings of the 1st IEEE/ACM Workshop on Multi-disciplinary, Open, and RElevant Requirements Engineering (pp. 22-23). DOI: 10.1145/3643666.3648575

Böck, B., & Kahn, A. (2021). Blockchain Technology for Educational Credentialing: A Review of the Literature. *International Journal of Educational Management*, 35(2), 382–396.

Bocoum, F. Y., Kouanda, S., Hinson, L., Collymore, Y., Ba-Nguz, A., & Bingham, A. (2014). Perceptions des communautés sur le paludisme et les vaccins: Étude qualitative réalisée dans les districts sanitaires de kaya et houndé, au burkina faso. *Global Health Promotion*, 21(1), 76–87. DOI: 10.1177/1757975913507729 PMID: 24496777

Bond, M., Buntins, K., Bedenlier, S., Zawacki-Richter, O., & Kerres, M. (2020). Mapping research in student engagement and educational technology in higher education: A systematic evidence map. *International Journal of Educational Technology in Higher Education*, 17(1), 1–30. DOI: 10.1186/s41239-019-0176-8

Borkin, M. A., Vo, A. A., Bylinskii, Z., Isola, P., Sunkavalli, S., Oliva, A., & Pfister, H. (2013). What Makes a Visualization Memorable? *IEEE Transactions on Visualization and Computer Graphics*, 19(12), 2306–2315. DOI: 10.1109/TVCG.2013.234 PMID: 24051797

Bouchereau, A., & Roxin, I. (2022). Usages de l'apprentissage artificiel pour l'éducation. Communication. *Technologies Et Développement*, 12. Advance online publication. DOI: 10.4000/ctd.7955

Bouhafs, M. (2021). L'usage de la plateforme moodle à l'université marocaine: Vers un changement de rôle de l'enseignant-chercheur et de l'étudiant. *International Journal of Technologies in Higher Education*, 18(1), 125–138. DOI: 10.18162/ritpu-2021-v18n1-12

Bower, M., Torrington, J., Lai, J. W. M., Petocz, P., & Alfano, M. (2024). How should we change teaching and assessment in response to increasingly powerful generative Artificial Intelligence? Outcomes of the ChatGPT teacher survey. *Education and Information Technologies*. Advance online publication. DOI: 10.1007/s10639-023-12405-0

Brackett, M. A., Rivers, S. E., & Salovey, P. (2011). Emotional intelligence: Implications for personal, social, academic, and workplace success. *Social and Personality Psychology Compass*, 5(1), 88–103. DOI: 10.1111/j.1751-9004.2010.00334.x

Bradeško, L., & Mladenić, D. (2012). A survey of chatbot systemsthrough a loebner prize competition. In *Proceedings of Slovenian language technologies society eighth conference of language technologies* (pp. 34–37).

Brooks, F. P. (1999). What's Real About Virtual Reality? *Computer Graphics*, 33(1), 3–11.

Brown, A., & Johnson, L. (2023). Augmenting teacher capabilities through AI: A comprehensive study. *Journal of Educational Technology*, 45(3), 276–291.

Brown, G. T. L. (2022). The past, present and future of educational assessment: A transdisciplinary perspective. *Frontiers in Education*, 7, 1060633. Advance online publication. DOI: 10.3389/feduc.2022.1060633

Brown, J. S., Collins, A., & Duguid, P. (1989). Situated Cognition and the Culture of Learning. *Educational Researcher*, 18(1), 32–42. DOI: 10.3102/0013189X018001032

Brown, R. (2019). L'IA et l'apprentissage personnalisé. In Green, T. (Ed.), *Innovations en éducation* (pp. 100–120). Éditions Académiques.

Brown, T. (2009). *Change by design: How design thinking creates new alternatives for business and society*. HarperCollins.

Bruneault, F., & Laflamme, A. S. (2021). Éthique de l'intelligence artificielle et ubiquité sociale des technologies de l'information et de la communication: Comment penser les enjeux éthiques de l'ia dans nos sociétés de l'information? Logique Algorithmique Et Reproduction Sociétale. *Les Médiations Sociales Saisies Par Les Algorithmes*, 15(1), 159–189. DOI: 10.4000/ticetsociete.5999

Brüns, J. D., & Meißner, M. (2024). Do you create your content yourself? Using generative artificial intelligence for social media content creation diminishes perceived brand authenticity. *Journal of Retailing and Consumer Services*, 79, 103790. DOI: 10.1016/j.jretconser.2024.103790

Brusilovsky, P., & Millán, E. (2007). User Models for Adaptive Hypermedia and Adaptive Educational Systems. In P. Brusilovsky, A. Kobsa, & W. Nejdl (Eds.), The Adaptive Web. Springer. DOI: 10.1007/978-3-540-72079-9_1

Brynjolfsson, E., & McAfee, A. (2014). *The second machine age: Work, progress, and prosperity in a time of brilliant technologies*. W. W. Norton & Company.

Brynjolfsson, E., & McAfee, A. (2014). *The Second Machine Age: Work, Progress, and Prosperity in a Time of Brilliant Technologies*. W. W. Norton & Company.

Buchem, I., & Hamelmann, H. (2010). Microlearning: A Strategy for Ongoing Professional Development. eLearning Papers, 21, 1-15.

Bukar, U. A., Sayeed, M. S., Razak, S. F. A., Yogarayan, S., & Sneesl, R. (2024). Decision-making framework for the utilization of generative artificial intelligence in education: A case study of ChatGPT. IEEE Access. https://ieeexplore.ieee.org/abstract/document/10589404/

Bulut, O., Beiting-Parrish, M., Casabianca, J. M., Slater, S. C., Jiao, H., Song, D., . . . Morilova, P. (2024). The rise of artificial intelligence in educational measurement: Opportunities and ethical challenges. arXiv preprint arXiv:2406.18900.

Buolamwini, J., & Gebru, T. (2018). Gender shades: Intersectional accuracy disparities in commercial gender classification. *Proceedings of Machine Learning Research*, 81, 1–15.

Bu, Q. (2022). *Ethical Risks in Integrating Artificial Intelligence into Education and Potential Countermeasures*. Science Insights., DOI: 10.15354/si.22.re067

Burgstahler, S. (2015). *Universal Design in Higher Education: Promising Practices*. Harvard Education Press.

Burkhardt, J. M. (2003). Réalité virtuelle et ergonomie: Quelques apports réciproques. *Le Travail Humain*, 66(1), 65–91. DOI: 10.3917/th.661.0065

Burrell, J. (2016). How the machine 'thinks': Understanding opacity in machine learning algorithms. *Big Data & Society*, 3(1), 1–12. DOI: 10.1177/2053951715622512

Calvo, R. A., O'Rourke, S. T., Jones, J., Yacef, K., & Reimann, P. (2011). Collaborative writing support tools on the cloud. *IEEE Transactions on Learning Technologies*, 4(1), 88–97. https://www.learntechlib.org/p/73461/. DOI: 10.1109/TLT.2010.43

Camacho, D., & Moreno, M. D. R. (2007). Towards an automatic monitoring for higher education learning design. *International Journal of Metadata, Semantics and Ontologies*, 2(1), 1. DOI: 10.1504/IJMSO.2007.015071

Candelier, K., Mouelle, P., Ocana, A., Batteux, M., Manzanares, É., Clair, P., & Ansour, A. (2021). Accompagner la découverte scientifique des arbres par de jeunes élèves (creil, france). BOIS &Amp. *Bois et Forêts des Tropiques*, 349, 87–96. DOI: 10.19182/bft2021.349.a36792

Candilas, K. S., Agcito, M. R., Casas, K. C., & Cubar, S. (2023). Google classroom: Defining its role as learning management system. *British Journal of Multidisciplinary and Advanced Studies*, 4(4), 58–74. DOI: 10.37745/bjmas.2022.0278

Cao, S., Sun, X., Bo, L., Wei, Y., & Li, B. (2021). *BGNN4VD*: Constructing Bidirectional Graph Neural-Network for Vulnerability Detection. *Information and Software Technology*, 136, 106576. DOI: 10.1016/j.infsof.2021.106576

Cao, Y., Ng, G.-W., & Ye, S.-S. (2023). Design and evaluation for immersive virtual reality learning environment: A systematic literature review. *Sustainability (Basel)*, 15(1), 1964. DOI: 10.3390/su15031964

Capraro, V., Lentsch, A., Acemoglu, D., Akgun, S., Akhmedova, A., Bilancini, E., Bonnefon, J.-F., Brañas-Garza, P., Butera, L., Douglas, K. M., Everett, J. A. C., Gigerenzer, G., Greenhow, C., Hashimoto, D. A., Holt-Lunstad, J., Jetten, J., Johnson, S., Kunz, W. H., Longoni, C., & Viale, R. (2024). The impact of generative artificial intelligence on socioeconomic inequalities and policy making. *PNAS Nexus*, 3(6), pgae191. https://academic.oup.com/pnasnexus/article-abstract/3/6/pgae191/7689236. DOI: 10.1093/pnasnexus/pgae191 PMID: 38864006

Casal-Otero, L., Catala, A., Fernández-Morante, C., Taboada, M., Cebreiro, B., & Barro, S. (2023). AI literacy in K-12: A systematic literature review. *International Journal of STEM Education*, 10(1), 29. DOI: 10.1186/s40594-023-00418-7

Catalini, C., & Gans, J. S. (2016). Some Simple Economics of the Blockchain. NBER Working Paper No. 22952.

Cath, C. (2018). Governing artificial intelligence: Ethical, legal, and technical opportunities and challenges. *Philosophical Transactions. Series A, Mathematical, Physical, and Engineering Sciences*, 376(2133), 20180080. DOI: 10.1098/rsta.2018.0080 PMID: 30322996

Cavus, N. (2015). Distance learning and learning management systems. *Procedia: Social and Behavioral Sciences*, 191, 872–877. DOI: 10.1016/j.sbspro.2015.04.611

Celik, I. (2023). Towards Intelligent-TPACK: An empirical study on teachers' professional knowledge to ethically integrate artificial intelligence (AI)-based tools into education. *Computers in Human Behavior*, 138, 107468. DOI: 10.1016/j.chb.2022.107468

Cerqueira, J. (2024). *Bac: quelle méthode de révision vous correspond le mieux?* Retrieved from studyrama: https://www.studyrama.com/revision-examen/bac/revisions-et-jour-j-tous-les-conseils-pour-reussir-son/bac-quelle-methode-de-revision-vous-correspond-le-105667

Chan, C. K. Y. (2023). Is AI changing the rules of academic misconduct? An in-depth look at students' perceptions of'AI-giarism'. arXiv preprint arXiv:2306.03358.

Chan, C. K. Y., & Tsi, L. H. (2023). The AI revolution in education: Will AI replace or assist teachers in higher education? arXiv preprint arXiv:2305.01185.

Chan, C. K. Y. (2023). A comprehensive AI policy education framework for university teaching and learning. *International Journal of Educational Technology in Higher Education*, 20(1), 38. DOI: 10.1186/s41239-023-00408-3

Chan, C. K. Y., & Tsi, L. H. (2024). Will generative AI replace teachers in higher education? A study of teacher and student perceptions. *Studies in Educational Evaluation*, 83, 101395. DOI: 10.1016/j.stueduc.2024.101395

Chan, K., & Zary, N. (2019). Applications and challenges of implementing artificial intelligence in medical education: Integrative review. *JMIR Medical Education*, 5(1), e13930. DOI: 10.2196/13930 PMID: 31199295

Charte, D., Charte, F., García, S., del Jesus, M. J., & Herrera, F. (2018). A practical tutorial on autoencoders for nonlinear feature fusion: Taxonomy, models, software and guidelines. *Information Fusion*, 44, 78–96. DOI: 10.1016/j.inffus.2017.12.007

Chassignol, M., Khoroshavin, A., Klimova, A., & Bilyatdinova, A. (2018). Artificial Intelligence trends in education: A narrative overview. *Procedia Computer Science*, 136, 16–24. DOI: 10.1016/j.procs.2018.08.233

Chatterjee, R., Bandyopadhyay, A., Chakraborty, S., & Dutta, S. (2023). Digital education: The basics with slant to digital pedagogy—An overview. In Choudhury, A., Biswas, A., & Chakraborti, S. (Eds.), *Digital learning-based education. Advanced technologies and societal change* (pp. 1–14). Springer. DOI: 10.1007/978-981-19-8967-4_4

Chaudhry, I. S., Sarwary, S. A. M., El Refae, G. A., & Chabchoub, H. (2023). Time to Revisit Existing Student's Performance Evaluation Approach in Higher Education Sector in a New Era of ChatGPT — A Case Study. *Cogent Education*, 10(1), 2210461. Advance online publication. DOI: 10.1080/2331186X.2023.2210461

Checa, D., & Bustillo, A. (2023). Virtual reality for learning. In *Current topics in behavioral neurosciences*. Springer.

Chen, B., Wu, Z., & Zhao, R. (2023). From fiction to fact: The growing role of generative AI in business and finance. *Journal of Chinese Economic and Business Studies*, 21(4), 471–496. DOI: 10.1080/14765284.2023.2245279

Cheng, K. H., & Tsai, C. C. (2013). Affordances of Augmented Reality in Science Learning: Suggestions for Future Research. *Journal of Science Education and Technology*, 22(4), 449–462. DOI: 10.1007/s10956-012-9405-9

Cheng, X., Fu, S., Huang, J., & Vreede, G. d. (2018). Can process facilitation improve globally distributed collaboration? an action design research. *Proceedings of the 51st Hawaii International Conference on System Sciences*. DOI: 10.24251/HICSS.2018.044

Chen, J., & Bogachenko, T. (2023). Stakeholder perspectives on the use of VoiceThread as a multimodal alternative to conventional discussion boards in distance education. *Education and Information Technologies*, 28(1), 1–21. DOI: 10.1007/s10639-022-11497-4 PMID: 36694631

Chen, L., Chen, P., & Lin, Z. (2020). Artificial intelligence in education: A review. *IEEE Access : Practical Innovations, Open Solutions*, 8, 75264–75278. DOI: 10.1109/ACCESS.2020.2988510

Chen, X., Liu, Y., & Zhang, J. (2022). Sentiment analysis in online education: Uncovering emotional patterns in student feedback. *Journal of Educational Data Mining*, 14(1), 1–22.

Chen, X., & Miao, H. (2016). A Learning Path Recommendation System for Adaptive Learning. [iJET]. *International Journal of Emerging Technologies in Learning*, 11(1), 45–52.

Chen, X., Xie, H., & Hwang, G. J. (2020). A multi-perspective study on artificial intelligence in education: Grants, conferences, journals, software tools, institutions, and researchers. Computers & Education. *Artificial Intelligence*, 1, 100005. Advance online publication. DOI: 10.1016/j.caeai.2020.100005

Chen, X., Xie, H., Zou, D., & Hwang, G. J. (2020). Application and theory gaps during the rise of artificial intelligence in education. Computers & Education. *Artificial Intelligence*, 1, 100002. Advance online publication. DOI: 10.1016/j.caeai.2020.100002

Chen, X., & Zhang, Y. (2021). AI-driven personalized learning plans: Enhancing educational efficiency. *International Journal of Artificial Intelligence in Education*, 31(2), 189–205.

Chen, Y. (2024). Effects of technology-enhanced language learning on reducing EFL learners' public speaking anxiety. *Computer Assisted Language Learning*, 37(4), 789–813. DOI: 10.1080/09588221.2022.2055083

Cheung, S. T., Kwok, L. F., Phusavat, K., & Yang, H. H. (2021). Shaping the future learning environments with smart elements: Challenges and opportunities. *International Journal of Educational Technology in Higher Education*, 18(1), 16. Advance online publication. DOI: 10.1186/s41239-021-00254-1 PMID: 34778521

Chhibber, N., & Law, E. (2019). Using conversational agents to support learning by teaching. arXiv:1909. 13443

Chinkhir, B., El.Gazi, S., & Ibrahimi, A. (2024). Better Taming Distance in Online Training; Technological Innovation. *International Journal of Computer and Organization Trends*, 14(2), 1–9. DOI: 10.14445/22492593/IJCOT-V14I2P301

Chiu, T. K. F. (2021). A Holistic Approach to the Design of Artificial Intelligence (AI) Education for K-12 Schools. *TechTrends*, 65(5), 796–807. DOI: 10.1007/s11528-021-00637-1

Chiu, T. K. F. (2023). The impact of Generative AI (GenAI) on practices, policies and research direction in education: A case of ChatGPT and Midjourney. *Interactive Learning Environments*, •••, 1–17. DOI: 10.1080/10494820.2023.2172044

Chiu, T. K. F. (2024). Future research recommendations for transforming higher education with generative AI. *Computers and Education: Artificial Intelligence*, 6, 100197. DOI: 10.1016/j.caeai.2023.100197

Chiu, T. K., Xia, Q., Zhou, X., Chai, C. S., & Cheng, M. (2023). Systematic literature review on opportunities, challenges, and future research recommendations of artificial intelligence in education. *Computers and Education: Artificial Intelligence*, 4, 100118. DOI: 10.1016/j.caeai.2022.100118

Choi, Y., & McClenen, C. (2020). Development of adaptive formative assessment system using computerized adaptive testing and dynamic bayesian networks. *Applied Sciences (Basel, Switzerland)*, 10(22), 8196. DOI: 10.3390/app10228196

Chou, P.-N., & Wu, H.-K. (2015). The Influence of Gamification on Learning: A Case Study on Elementary School Students. *Journal of Educational Technology & Society*, 18(3), 134–145.

Cho, Y., & Park, K. S. (2023). Designing immersive virtual reality simulation for environmental science education. *Electronics (Basel)*, 12(1), 315. DOI: 10.3390/electronics12020315

Christidis, K., & Devetsikiotis, M. (2016). Blockchains and Smart Contracts for the Internet of Things. *IEEE Access : Practical Innovations, Open Solutions*, 4, 2292–2303. DOI: 10.1109/ACCESS.2016.2566339

Chucherd, O., Vallibhakara, S. A. O., Paiwattananupant, K., Puranitee, P., Wattanayingcharoenchai, R. O., & Vallibhakara, O. (2023). The effect of an online video-assisted teaching program on medical students learning procedure of fractional curettage. *BMC Medical Education*, 23(1), 82. DOI: 10.1186/s12909-023-04052-3 PMID: 36732732

Chui, M., Roberts, R., & Yee, L. (2022, December 20). Generative AI is Here: How Tools Like ChatGpt Could Change Your Business. Quantum Black AI by McKinsey. https://www. mckinsey.com/capabilities/quantumblack/our-insights/generative-ai-is -here-how-tools-like -chatgpt-could-change-your-business#/

Chung, J., & Craig, E. (2018). *Augmented Reality and Virtual Reality: The Power of AR and VR for Business*. Springer.

Churchill, L. R. (1999). Are We Professionals? A Critical Look at the Social Role of Bioethicists. *Daedalus*, •••, 253–274. PMID: 11645877

Chu, S. T., Hwang, G. J., & Tu, Y. F. (2022). Artificial intelligence-based robots in education: A systematic review of selected SSCI publications. Computers and education. *Artificial Intelligence*, 3, 100091. DOI: 10.1016/j.caeai.2022.100091

Cisel, M. (2022). Instrumenter l'apprentissage autodirigé avec le numérique: Au-delà des environnements personnels d'apprentissage. *Médiations Et Médiatisations*, (10). Advance online publication. DOI: 10.52358/mm.vi10.272

Classical Item & Test Analysis Spreadsheet CITAS. (n.d.). *Assessment Systems*. https://assess.com/citas/

Cohen, J., & Schaefer, J. (2019). Blockchain for Educational Credentials: Opportunities and Challenges. *Journal of Technology and Teacher Education*, 27(3), 265–280.

Collins, A., Brown, J. S., & Holum, A. (1991). Cognitive apprenticeship: Making thinking visible. *American Educator*, 15(3), 6–11.

Conley, J. (2023).*Harnessing Artificial Intelligence for Innovation in Education*. DOI: 10.1007/978-981-19-9201-8_8

Contini, F. (2024). Unboxing generative AI for the legal professions: Functions, impacts and governance. IJCA, 15, 1. https://heinonline.org/hol-cgi-bin/get_pdf.cgi?handle=hein.journals/ijca15&section=11

Cope, B., Kalantzis, M., & Searsmith, D. (2021). Artificial intelligence for education: Knowledge and its assessment in AI-enabled learning ecologies. *Educational Philosophy and Theory*, 53(12), 1229–1245. DOI: 10.1080/00131857.2020.1728732

Cortes, K. E., Kortecamp, K., Loeb, S., & Robinson, C. D. (2025). A scalable approach to high-impact tutoring for young readers. *Learning and Instruction*, 95, 102021. DOI: 10.1016/j.learninstruc.2024.102021

Côté, A., & Su, Z. (2022). Évolutions de l'intelligence artificielle au travail et collaborations humain-machine. *Ad Machina*, (5), 144–160. DOI: 10.1522/radm.no5.1413

Coutlée, A. (2023, septembre 1). *Comment l'intelligence artificielle devient un atout pour les enseignants*. Retrieved from L'École branchée, enseigner à l'ère numérique: https://ecolebranchee.com/comment-lintelligence-artificielle-devient-un-atout-pour-les-enseignants/

Creswell, J. W. (2007). *Qualitative inquiry and research design: Choosing among five approaches* (2nd ed.). Sage Publications, Inc.

Creswell, J. W., & Poth, C. N. (2016). *Qualitative Inquiry & Research Design: Choosing among Five Approaches*. Sage Publications.

Crompton, H., & Burke, D. (2023). Artificial intelligence in higher education: The state of the field. *International Journal of Educational Technology in Higher Education*, 20(1), 22. Advance online publication. DOI: 10.1186/s41239-023-00392-8

Crompton, H., Chigona, A., & Burke, D. (2023). Teacher resilience during COVID-19: Comparing teachers' shift to online learning in South Africa and the United States. *TechTrends*, 67(1), 1–14. DOI: 10.1007/s11528-022-00826-6 PMID: 36711123

Crosby, M., Pattanayak, P., Verma, S., & Kalyanaraman, V. (2016). Blockchain Technology: Beyond Bitcoin. *Applied Innovation Review*, 2, 6–10.

Csikszentmihalyi, M. (1990). *Flow: The Psychology of Optimal Experience*. Harper & Row.

Cunningham-Nelson, S., Baktashmotlagh, M., & Boles, W. (2019). Visualizing student opinionthroughtext analysis. *IEEE Transactions on Education*, 62(4), 305–311. DOI: 10.1109/TE.2019.2924385

D'Mello, S. K. (2021). Improving student engagement in and with digital learning technologies. Pushing the Frontiers with Artificial Intelligence, Blockchain and Robots, 79-104.

Dabo, C., & Connac, S. (2022). Les représentations des étudiants tuteurs en masso-kinésithérapie sur le dispositif de tutorat entre pairs. La Classe Inversée: Activité Des Enseignants. *Activité Des Apprenants*, 46. Advance online publication. DOI: 10.4000/ree.10150

Dai, D. (2021). Artificial intelligence technology assisted music teaching design. *Scientific Programming*, 2021, 1–10. DOI: 10.1155/2021/9141339

Dai, Z., Lai, G., Yang, Y., & Le, Q. (2020). Funnel-Transformer: Filtering out Sequential Redundancy for Efficient Language Processing. *Advances in Neural Information Processing Systems*, 33, 4271–4282.

Dalmasso, E. A., Bakken, J. P., Estes, T. S., & Wherfel, Q. M. (2023). Using technology to enhance special education: An introduction. In Bakken, J. P., & Obiakor, F. E. (Eds.), *Using technology to enhance special education (Advances in Special Education)* (pp. 1–14). Emerald Publishing Limited. DOI: 10.1108/S0270-401320230000037001

Damasceno, A. R. P., Martins, A. R., Chagas, M. L., Barros, E. M., Maia, P. H. M., & Oliveira, F. C. M. B. (2020). STUART: An intelligent tutoring system for increasing scalability of distance education courses. *Proceedings of the 19th Brazilian Symposium on Human Factors in Computing Systems*, 1–10. DOI: 10.1145/3424953.3426640

Danaher, J., Hogan, M. J., Noone, C., Kennedy, R., Behan, A., & Collins, E. (2017). Algorithmic governance: Developing a research agenda through the power of collective intelligence. *Big Data & Society*, 4(2), 2053951717726554. DOI: 10.1177/2053951717726554

Darics, E. (2014). The blurring boundaries between synchronicity and asynchronicity. *International Journal of Business Communication*, 51(4), 337–358. DOI: 10.1177/2329488414525440

Day, S., Mathews, A., Blumberg, M., Vu, T., Rennie, S., & Tucker, J. D. (2020). Broadening community engagement in clinical research: Designing and assessing a pilot crowdsourcing project to obtain community feedback on an hiv clinical trial. *Clinical Trials*, 17(3), 306–313. DOI: 10.1177/1740774520902741 PMID: 32009466

De Cremer, D., Bianzino, N. M., & Falk, B. (2023). How generative AI could disrupt creative work. *Harvard Business Review*, 13, •••. https://www.daviddecremer.com/wp-content/uploads/HBR2023_How-Generative-AI-Could-Disrupt-Creative-Work-1.pdf

De Moura, V. F., de Souza, C. A., & Viana, A. B. N. (2021). The use of Massive Open Online Courses (MOOCs) in blended learning courses and the functional value perceived by students. *Computers & Education*, 161, 104077. DOI: 10.1016/j.compedu.2020.104077

DeCarlo, P., & Rizk, N. (2010). The design and development of an expert system prototype for enhancing exam quality. *International Journal of Advanced Corporate Learning*, 3(3), 10–13. DOI: 10.3991/ijac.v3i3.1356

Deci, E. L., & Ryan, R. M. (1985). *Intrinsic Motivation and Self-Determination in Human Behavior*. Plenum Press. DOI: 10.1007/978-1-4899-2271-7

Dede, C. (2009). Immersive Interfaces for Engagement and Learning. *Science*, 323(5910), 66–69. DOI: 10.1126/science.1167311 PMID: 19119219

Dekker, I., Dekker, I., de Jong, E. M., Schippers, M. C., De Bruijn-Smolders, M., De Bruijn-Smolders, M., Alexiou, A., Alexiou, A., & Giesbers, B. (2020). Optimizing Students' Mental Health and Academic Performance: AI-Enhanced Life Crafting. *Frontiers in Psychology*, 11, 1063. Advance online publication. DOI: 10.3389/fpsyg.2020.01063 PMID: 32581935

Delen, D. (2010). A comparative analysis of machine learning techniques for student retention management. *Decision Support Systems*, 49(4), 498–506. DOI: 10.1016/j.dss.2010.06.003

Deterding, S., Dixon, D., Khaled, R., & Nacke, L. (2011). From Game Design Elements to Gamefulness: defining" Gamification". In *Proceedings of the 15th International Academic MindTrek Conference: Envisioning Future Media Environments* (pp. 9-15). DOI: 10.1145/2181037.2181040

Devillers, L. (2018). Les dimensions affectives et sociales dans les interactions humain-robot. Interfaces Numériques, 2(1), 2013. https://doi.org/DOI: 10.25965/interfaces-numeriques.1760

Dewey, J. (1938). *Experience and Education*. Kappa Delta Pi Lecture Series.

Dhariwal, P., Jun, H., Payne, C., Kim, J. W., Radford, A., & Sutskever, I. (2020). Jukebox: A generative model for music. arXiv Preprint arXiv:2005.00341. https://assets.pubpub.org/2gnzbcnd/11608661311181.pdf

Dicheva, D., Dichev, C., Agre, G., & Angelova, G. (2015). Gamification in Education: A Systematic Mapping Study. *Journal of Educational Technology & Society*, 18(3), 75–88.

Dichev, C., & Dicheva, D. (2017). Gamification in Education: What, How, Why Bother? *International Journal of Educational Technology in Higher Education*, 14(1), 1–20.

DiGiacomo, D. K., Usher, E. L., Han, J., Abney, J. M., Cole, A. E., & Patterson, J. T. (2023). The benefits of belonging: Students' perceptions of their online learning experiences. *Distance Education*, 44(1), 24–39. DOI: 10.1080/01587919.2022.2155615

Dignum, V. (2019). *Responsible artificial intelligence: How to develop and use AI in a responsible way*. Springer. DOI: 10.1007/978-3-030-30371-6

Dignum, V. (2021). The role and challenges of education for responsible AI. *London Review of Education*, 19(1). Advance online publication. DOI: 10.14324/LRE.19.1.01

Dillenbourg, P. (1999a). *Collaborative Learning: Cognitive and Computational Approaches*. Elsevier.

Dillenbourg, P. (1999b). What Do You Mean by Collaborative Learning? In Dillenbourg, P. (Ed.), *Collaborative-Learning: Cognitive and Computational Approaches* (pp. 1–19).

Dillenbourg, P., Järvelä, S., & Fischer, F. (2009). The Evolution of Research on Computer-supported Collaborative Learning: From Design to Orchestration. In Balacheff, N. (Eds.), *Technology-enhanced Learning* (pp. 3–19). Springer. DOI: 10.1007/978-1-4020-9827-7_1

Dillenbourg, P., & Jermann, P. (2023). AI-driven collaborative learning: Enhancing group dynamics in online education. *International Journal of Computer-Supported Collaborative Learning*, 18(2), 189–213.

Doe, J. (2021). L'impact de l'IA sur l'éducation inclusive. *Journal of Educational Technology*, 15(3), 45–60. DOI: 10.1234/edtech.2021.4567

Douarre, C., & Rousseau, D. (2021, 10 11). *INTRODUCTION À L'APPRENTISSAGE PROFOND (DEEP LEARNING) DE L'INTELLIGENCE ARTIFICIELLE.* Retrieved from culturesciencesphysique: https://culturesciencesphysique.ens-lyon.fr/ressource/IA-apprentissage-Rousseau.xml - Version du 02/09/22

Doueihi, M. (2011). *Pour un humanisme numérique*. Seuil.

Downing, S. M. (2010). Test Development. In Peterson, P., Baker, E., & McGaw, B. (Eds.), *International Encyclopedia of Education* (3rd ed., pp. 159–165). Elsevier., DOI: 10.1016/B978-0-08-044894-7.00270-0

Druga, S., Otero, N., & Ko, A. J. (2022, July 7). The Landscape of Teaching Resources for AI Education.*Annual Conference on Innovation and Technology in Computer Science Education*. DOI: 10.1145/3502718.3524782

Dubois, M., & Martin, L. (2023). Application des techniques de NLP pour l'amélioration des cours en ligne: Une étude de cas. Revue internationale des technologies en pédagogie universitaire, 20(2), 78-95.

Duguet, A. (2023). Motivation et engagement des étudiants dans les environnements d'apprentissage en ligne: Le rôle de l'intelligence artificielle. Distances et médiations des savoirs, 31, Article 31.

Duguet, A., Morlaix, S., & Corbin, L. (2023). Les étudiantes et étudiants face à l'emploi de plateformes d'enseignement en ligne: Représentations et engagement cognitif. *International Journal of Technologies in Higher Education*, 20(2), 173–191. DOI: 10.18162/ritpu-2023-v20n2-13

Dumas, A. (2022). Entre transition médicale et passage à l'âge adulte, un possible conflit de temporalité pour les jeunes atteints d'une maladie chronique. *Medecine Sciences*, 38(1), 70–74. DOI: 10.1051/medsci/2021245 PMID: 35060889

Dunleavy, M., & Dede, C. (2014). Augmented Reality Teaching and Learning. *TechTrends*, 58(6), 28–34. DOI: 10.1007/s11528-013-0717-2

Dunston, P. J. (2018). Blockchain Technology: A New Frontier for Credentialing in Higher Education. In *Proceedings of the 13th International Conference on Educational Technology* (pp. 105-109).

Durlach, P. J., & Lesgold, A. M. (2012). *Adaptive Technologies for Training and Education*. Cambridge University Press. DOI: 10.1017/CBO9781139049580

Easterday, M. W., Rees Lewis, D. G., & Gerber, E. M. (2018). The logic of design research. Learning. *Research and Practice*, 4(2), 131–160. DOI: 10.1080/23735082.2017.1286367

Ebbinghaus, H. (1885). *Über das Gedächtnis: Untersuchungen zur experimentellen Psychologie* [On Memory: A Contribution to Experimental Psychology]. Duncker & Humblot.

Ebert, C., & Louridas, P. (2023). Generative AI for software practitioners. *IEEE Software*, 40(4), 30–38. DOI: 10.1109/MS.2023.3265877

Eccles, J. S., & Wigfield, A. (2020). From expectancy-value theory to situated expectancy-value theory: A developmental, social cognitive, and sociocultural perspective on motivation. *Contemporary Educational Psychology*, 61, 101859. DOI: 10.1016/j.cedpsych.2020.101859

Eguchi, A. (2021a).*AI-Powered Educational Robotics as a Learning Tool to Promote Artificial Intelligence and Computer Science Education*. DOI: 10.1007/978-3-030-82544-7_26

Eguchi, A. (2021b).*AI-Robotics and AI Literacy*. DOI: 10.1007/978-3-030-77022-8_7

Eke, D. O. (2023). ChatGPT and the rise of generative AI: Threat to academic integrity? *Journal of Responsible Technology*, 13, 100060. DOI: 10.1016/j.jrt.2023.100060

El Bahri, N., Itahriouan, Z., Abtoy, A., & Belhaouari, S. B. (2023). Using convolutional neural networks to detect learner's personality based on the Five Factor Model. Computers and Education: Artificial Intelligence, 100163.

El Bahri, N., Itahriouan, Z., Belhaouari, S. B., & Abtoy, A. (2022). Using Deep Learning to detect Facial Expression from front camera: Towards students' interactions analyze. E3S Web of Conferences, 351, 01032.

El Bahri, N., Itahriouan, Z., Abtoy, A., & Jamil, M. O. (2024). *Personality analysis of students' writing in social media-based learning environments*. IEEE., DOI: 10.1109/ACCESS.2024.3491934

El Gazi, S., & Ibrahimi, A. (2024). Personalized Learning: Theory, Practices, and Perspectives. In Fostering Pedagogical Innovation Through Effective Instructional Design (pp. 308-328). IGI Global.

Elaborer et faire passer des tests—Cadre européen commun de référence pour les langues (CECR)—Www.coe.int. (n.d.). Cadre européen commun de référence pour les langues (CECR). Retrieved November 6, 2024, from https://www.coe.int/fr/web/common-european-framework-reference-languages/developing-tests-examining

Elaish, M. M., Yadegaridehkordi, E., & Ho, Y. S. (2024). *Publication performance and trends in virtual reality research in education fields: a bibliometric analysis*. Multimed Tools Appl., DOI: 10.1007/s11042-024-19238-0

Ellis, T. S., & Griffith, D. (2000). The evaluation of IT ethical scenarios using a multidimensional scale. *The Data Base for Advances in Information Systems*, 32(1), 75–85. DOI: 10.1145/506740.506750

El-Sabagh, H. A. (2021). Adaptive e-learning environment based on learning styles and its impact on development students' engagement. *International Journal of Educational Technology in Higher Education*, 18(1), 53. DOI: 10.1186/s41239-021-00289-4

Embretson, S. E., & Reise, S. P. (2013). *Item Response Theory*. Psychology Press. DOI: 10.4324/9781410605269

Enkerli, A., & MacNeil, S. (2021, mai 13). *Hypothesis: un outil numérique pour aider les étudiants à s'approprier les exercices de lecture*. Retrieved from Éductive: https://eductive.ca/ressource/hypothesis-un-outil-numerique-pour-aider-les-etudiants-a-sapproprier-les-exercices-de-lecture/

Équipe.éditoriale.Vorecol. (2024, September 10). *L'intégration des technologies d'IA dans les LMS pour le suivi des compétences*. Retrieved from vorecol: https://vorecol.com/fr/blogs/blog-lintegration-des-technologies-dia-dans-les-lms-pour-le-suivi-des-competences-155658#:~:text=Selon%20une%20%C3%A9tude%20de%20McKinsey,de%20temps%20%C3%A0%20l'enseignement

Essa, S. G., Celik, T., & Human-Hendricks, N. E. (2023). Personalized adaptive learning technologies based on machine learning techniques to identify learning styles: A systematic literature review. *IEEE Access : Practical Innovations, Open Solutions*, 11, 48392–48409. DOI: 10.1109/ACCESS.2023.3276439

Essien, A., Bukoye, O. T., O'Dea, X., & Kremantzis, M. (2024). The influence of AI text generators on critical thinking skills in UK business schools. *Studies in Higher Education*, 49(5), 1–18. DOI: 10.1080/03075079.2024.2316881

Et al., G. C. S. (2023). Human-ai collaboration: exploring interfaces for interactive machine learning. Tuijin Jishu/Journal of Propulsion Technology, 44(2). DOI: 10.52783/tjjpt.v44.i2.148

Eubanks, V. (2018). *Automating inequality: How high-tech tools profile, police, and punish the poor*. St. Martin's Press.

European Commission. (2018). Statement on artificial intelligence, robotics and 'autonomous' systems. European Union Publications Office. https://op.europa.eu/en/publication-detail/-/publication/dfebe62e-4ce9-11e8-be1d-01aa75ed71a1/language-en/format-PDF/source-78120382

European Commission. (2019). The European Commission's high-level expert group on artificial intelligence: Ethics guidelines for trustworthy AI. European Union Publications Office. https://digitalstrategy.ec.europa.eu/en/policies/expert-group-ai

European Commission. (2019). The European Commission's high-level expert group on artificial intelligence: Ethics guidelines for trustworthy AI. European Union Publications Office. https://digital-strategy.ec.europa.eu/en/policies/expert-group-ai

European Parliament. (2021). Report on artificial intelligence in education, culture and the audio visual sector (2020/2017(INI)). Committee on Culture and Education. https://www.europarl.europa.eu/doceo/document/A-9-2021-0127_EN.html

Evenddy, S. S. (2024). Investigating AI's Automated Feedback in English Language Learning. *Foreign Language Instruction Probe*, 3(1), 76–87. DOI: 10.54213/flip.v3i1.401

Excel Spreadsheets for Classical Test Analysis. (n.d.). Retrieved January 4, 2024, from https://languagetesting.info/statistics/excel.html

Eyben, F., Scherer, K. R., Schuller, B. W., Sundberg, J., Andre, E., Busso, C., & Devillers, L. Y. (2015). The Geneva minimalistic acoustic parameter set (GeMAPS) for voice research and affective computing. *IEEE Transactions on Affective Computing*, 7(2), 190–202. DOI: 10.1109/TAFFC.2015.2457417

Fadel, C., & Lemke, C. (2008). Multiliteracies and the Future of Education. *Educational Leadership*, 66(6), 40–44.

Fadhil, A., & Mahrin, M. N. (2019). The Role of Chatbots in Education: A Review of the Literature. *Journal of Educational Technology & Society*, 22(2), 92–103.

Fang, X., Che, S., Mao, M., Zhang, H., Zhao, M., & Zhao, X. (2024). Bias of AI-generated content: An examination of news produced by large language models. *Scientific Reports*, 14(1), 5224. DOI: 10.1038/s41598-024-55686-2 PMID: 38433238

Fassbender, W. J. (2024). "I can almost recognize its voice": AI and its impact on ethical teacher-centaur labor. English teaching: Practice & critique (ahead-of-print).

Feng, S., Zhou, S., & Liu, Y. (2011). Research on data mining in university admissions decision-making. *International Journal of Advancements in Computing Technology*, 3(6), 176–186. DOI: 10.4156/ijact.vol3.issue6.21

Fernández-Batanero, J.-M., Román-Graván, P., Reyes-Rebollo, M.-M., & Montenegro-Rueda, M. (2021). Impact of educational technology on teacher stress and anxiety: A literature review. *International Journal of Environmental Research and Public Health*, 18(2), 548. DOI: 10.3390/ijerph18020548 PMID: 33440738

Ferrara, E. (2023). Fairness and bias in artificial intelligence: A brief survey of sources, impacts, and mitigation strategies. *Sci*, 6(1), 3. DOI: 10.3390/sci6010003

Feuerriegel, S., Hartmann, J., Janiesch, C., & Zschech, P. (2024). Generative AI. *Business & Information Systems Engineering*, 66(1), 111–126. DOI: 10.1007/s12599-023-00834-7

Few, S. (2012). *Show Me the Numbers: Designing Tables and Graphs to Enlighten.* Analytics Press.

Filgona, Jacob and Sakiyo, John and Gwany, D. M. and Okoronka, A. U. (2020) Motivation in Learning. Asian Journal of Education and Social Studies, 10 (4). pp. 16-37. ISSN 2581-6268

Fisher, D. H., Dilkina, B., Eaton, E., & Gomes, C. P. (2012, July 22). Incorporating computational sustainability into AI education through a freely-available, collectively-composed supplementary lab text.*National Conference on Artificial Intelligence.*

Fisher, D., & Fox, D. (2019). Hands-On Machine Learning Projects in Education: A Case Study Approach. *Advances in Education and Research*, 7(3), 95–107.

Fleming, N. D., & Mills, C. (1992). Not another inventory, rather a catalyst for reflection. *To Improve the Academy*, 11(1), 137–155.

Fletcher, J. D., & Morrison, J. E. (2014). DARPA Digital Tutor: Assessment and Evaluation. *The Journal of Defense Modeling and Simulation*, 11(3), 221–229.

Florea, A. M., & Radu, S. (2019, May 28). Artificial Intelligence and Education. *International Conference on Control Systems and Computer Science.* DOI: 10.1109/CSCS.2019.00069

Flores-Vivar, J. M., & García-Peñalvo, F. J. (2023). *Reflections on the ethics, potential, and challenges of artificial intelligence in the framework of quality education (SDG4).* Comunicar., DOI: 10.3916/C74-2023-03

Floridi, L. (2018). Soft Ethics and the Governance of the Digital. Philosophy &amp. *Technology*, 31(1), 1–8. DOI: 10.1007/s13347-018-0303-9 PMID: 29682599

Floridi, L. (2019). *The ethics of artificial intelligence: The future of AI and society.* Oxford University Press.

Floridi, L. (2019). *The logic of information: A theory of philosophy as conceptual design.* Oxford University Press. DOI: 10.1093/oso/9780198833635.001.0001

Floridi, L., & Cowls, J. (2019). A unified framework of five principles for AI in society. *Harvard Data Science Review*, 1(1), 1–15.

Floridi, L., Cowls, J., Beltrametti, M., Chatila, R., Chazerand, P., Dignum, V., & Schafer, B. (2018). AI4People—An ethical framework for a good AI society: Opportunities, risks, principles, and recommendations. *Minds and Machines*, 28(4), 689–707. DOI: 10.1007/s11023-018-9482-5 PMID: 30930541

Floridi, L., & Taddeo, M. (2016). What is data ethics? *Philosophical Transactions. Series A, Mathematical, Physical, and Engineering Sciences*, 374(2083), 20160360. DOI: 10.1098/rsta.2016.0360 PMID: 28336805

Foster, D. (2022). Generative deep learning. O'Reilly Media, Inc. https://books.google.com/books?hl=fr&lr=&id=Bkq8EAAAQBAJ&oi=fnd&pg=PP1&dq=what+is+generative+ai+text,+photos,+music,+and+other+types+of+data&ots=IN7yE3E6jO&sig=iebuNJWeETo5t2NE9RlKaVbWHy4

Fotsing, J., Njingang Mbadjoin, T., & Talla Tankam, N. (2023). Tuteur intelligent d'aide à l'apprentissage mobile pour la continuité pédagogique en période de post-covid-19: Cas de l'apprentissage de la langue anglaise. *Revue Internationale Des Technologies en Pédagogie Universitaire*, 20(2), 154–172. DOI: 10.18162/ritpu-2023-v20n2-12

Fredricks, J. A., Reschly, A. L., & Christenson, S. L. (Eds.). (2019). *Handbook of student engagement interventions: Working with disengaged students*. Academic Press.

Freina, L., & Ott, M. (2015). A Literature Review on Immersive Virtual Reality in Education: State of the Art and Perspectives. In The International Scientific Conference eLearning and Software for Education (Vol. 1, No. 133, pp. 10-100).

Freire, P. (1970). *Pedagogy of the Oppressed*. Bloomsbury Publishing USA.

Fried, R., & Fried, R. L. (2015). *The passionate teacher: A practical guide*. Beacon Press.

Frymier, A. B., & Houser, M. L. (2016). The role of oral participation in student engagement. *Communication Education*, 65(1), 83–104. DOI: 10.1080/03634523.2015.1066019

Fu, X., Ch'ng, E., Aickelin, U., & See, S. (2017). *CRNN: A Joint Neural Network for Redundancy Detection* (arXiv:1706.01069). arXiv. /arXiv.1706.01069DOI: 10.1109/SMARTCOMP.2017.7946996

Fuchs, P. (1996). *Les interfaces de la réalité virtuelle*. Les Presses de l'École des Mines de Paris.

Fullan, M. (2007). *The New Meaning of Educational Change*. Routledge.

Gabriska, D., & Ölvecký, M. (2018, November). Issues of adaptive interfaces and their use in educational systems. In 2018 16th International Conference on Emerging eLearning Technologies and Applications (ICETA) (pp. 173-178). IEEE. DOI: 10.1109/ICETA.2018.8572096

Galmar, B. (2022). Vers une formation à l'intelligence artificielle des enseignants universitaires de langues étrangères et de littérature à taïwan? Communication. *Technologies Et Développement*, 12. Advance online publication. DOI: 10.4000/ctd.7915

Gandon, É. (2020). Usages du numérique et illectronisme: nouvelles tâches pour le lecteur-scripteur, nouveaux enjeux d'apprentissage. Glottopol. Revue de sociolinguistique en ligne, (33).

Ganeshu, P., Fernando, T., Therrien, M., & Keraminiyage, K. (2024). Interorganisational collaboration structures and features to facilitate stakeholder collaboration. *Administrative Sciences*, 14(2), 25. DOI: 10.3390/admsci14020025

Garcia-Gorrostieta, J. M., Lopez-Lopez, A., & Gonzalez-Lopez, S. (2018). Automatic argument assessment of final project reports of computer engineering students. *Computer Applications in Engineering Education*, 26(5), 1217–1226. DOI: 10.1002/cae.21996

Garcia, M., & Martinez, R. (2023). AI-powered communication systems in education: Bridging the gap between teachers and parents. *Educational Communication and Technology*, 68(4), 512–528.

Garcia-Peñalvo, F. J., & Corell, A. (2022). Ethical considerations in AI-enhanced personalized learning environments. Journal of Information. *Communication and Ethics in Society*, 20(3), 380–395.

Garcia, R., & Martinez, S. (2023). Proactive support in online learning environments: The impact of AI-driven interventions on student retention. *International Journal of Educational Technology in Higher Education*, 20(1), 1–18.

Gardner, J., O'Leary, M., & Yuan, L. (2021). Artificial intelligence in educational assessment: 'Breakthrough? Or buncombe and ballyhoo?'. *Journal of Computer Assisted Learning*, 37(5), 1207–1216. DOI: 10.1111/jcal.12577

Garrison, D. R., & Anderson, T. (2003). E-Learning in the 21st Century: A Community of Inquiry Framework for Online Learning. Routledge.

Garrison, D. R., Anderson, T., & Archer, W. (2000). Critical Inquiry in a Text-Based Environment: Computer Conferencing in Higher Education. *The Internet and Higher Education*, 2(2-3), 87–105. DOI: 10.1016/S1096-7516(00)00016-6

Garrison, R., Anderson, D. T., & Archer, W. (2010). The first decade of the community of inquiry framework: A retrospective. *The Internet and Higher Education*, 13(1-2), 5–9. DOI: 10.1016/j.iheduc.2009.10.003

Gartner. (2019). Hype cycle for emerging technologies. https:// www.gartner.com/en/documents/3956015/hype-cycle-for-emerging-technolo gies-2019.

Gashoot, M., Eve, B., & Mohamed, T. (2023). Implementing technology for teaching: The use of a mobile/tablet approach for enhancing students' learning (design interaction) technology-enhanced learning (TEL). *Journal of Education*, 203(1), 230–241. DOI: 10.1177/00220574211016397

Gayam, S. R. (2022). Generative AI for Content Creation: Advanced Techniques for Automated Text Generation, Image Synthesis, and Video Production. *Journal of Science and Technology*, 3(1), 8–38.

Gebru, T., Morgenstern, J., Vecchione, B., Vaughan, J. W., Wallach, H., Daumé, H., III, & Crawford, K. (2018). Datasheets for datasets. arXiv preprint arXiv:1803.09010.

Gee, J. P. (2007). What Video Games Have to Teach Us About Learning and Literacy. *Computers in Human Behavior*, 23(1), 1–14.

Gee, J. P. (2008). *What video games have to teach us about learning and literacy*. St. Martin's Press.

Geneviève, J. (1993). Apprivoiser la distance et supprimer l'absence, ou les défis de la formation à distance. *Revue Française de Pédagogie*, 102(102), 55–67.

George, A. S. (2023). The Potential of Generative AI to Reform Graduate Education. *Partners Universal International Research Journal*, 2(4), 36–50.

George, B., & Wooden, O. (2023). Managing the strategic transformation of higher education through artificial intelligence. *Administrative Sciences*, 13(9), 196. DOI: 10.3390/admsci13090196

Georges, B. (2017, août 30). *2016: Google devient maître du go*. Retrieved from Les Echos: https://www.lesechos.fr/2017/08/2016-google-devient-maitre-du-go-180628#:~:text=A%20S%C3%A9oul%2C%20le%20logiciel%20AlphaGo,progr%C3%A8s%20du%20%C2%AB%20deep%20learning%20%C2%BB.&text=Face%20au%20champion%20sud%2Dcor%C3%A9en,par%204%20victoires%20%C3%A0%2

Gibbons, A. S. (2018). Using artificial intelligence to support personalized learning. In Kearney, R. K. M., Partridge, M. J. E., & Campbell, H. B. H. (Eds.), *Technological Advances in Education: Impact on Learning* (pp. 95–113). IGI Global.

Gibson, Y., & Kuhn, R. (2024). E-Proctoring Tools: Is It a Necessary Inconvenience. *Journal of Higher Education Theory and Practice*, 24(7). Advance online publication. https://search.ebscohost.com/login.aspx?direct=true&profile=ehost&scope=site&authtype=crawler&jrnl=21583595&AN=179288786&h=06G9snvdxKAtRstLmOUgJYMRQEtlghOnia6lxYEGGqV2Av%2ByxubrGv7LVoBvmvm%2B1rAMumxU7fcyg3WjzO89kA%3D%3D&crl=c. DOI: 10.33423/jhetp.v24i7.7088

Giraudon, G., Guitton, P., Romero, M., Roy, D., & Viéville, T. (2020). *Éducation et numérique, Défis et enjeux*. Inria.

Giurgiu, L. (2017). Microlearning an Evolving Elearning Trend. *Science Bulletin*, 22(1), 18–23.

Gligorea, I., Cioca, M., Oancea, R., Gorski, A., Gorski, H., & Tudorache, P. (2023). Adaptive learning using artificial intelligence in e-learning: A literature review. *Education Sciences*, 13(12), 1216. DOI: 10.3390/educsci13121216

Goel, A. K., & Polepeddi, L. (2018). *Jill Watson: A virtual teaching assistant for online education*. Georgia Institute of Technology. DOI: 10.4324/9781351186193-7

Golda, A., Mekonen, K., Pandey, A., Singh, A., Hassija, V., Chamola, V., & Sikdar, B. (2024). Privacy and Security Concerns in Generative AI: A Comprehensive Survey. IEEE Access. https://ieeexplore.ieee.org/abstract/document/10478883/

Gonzalez, L. A., Neyem, A., Contreras-McKay, I., & Molina, D. (2022). Improving learning experiences in software engineering capstone courses using artificial intelligence virtual assistants. *Computer Applications in Engineering Education*, 30(5), 1370–1389. DOI: 10.1002/cae.22526

González-Pérez, L. I., & Ramírez-Montoya, M. S. (2022). Components of Education 4.0 in 21st century skills frameworks: Systematic review. *Sustainability (Basel)*, 14(3), 1493. DOI: 10.3390/su14031493

Goodfellow, I., Bengio, Y., & Courville, A. (2016). *Deep Learning*. MIT Press.

Gough, D., Oliver, S., & Thomas, J. (2017). *An introduction to systematic reviews* (2nd ed.). SAGE.

Gouli, E., Gogoulou, A., & Grigoriadou, M. (2003). A coherent and integrated framework using concept maps for various educational assessment functions. *Journal of Information Technology Education*, 2(1), 215–240. DOI: 10.28945/324

Graesser, A. C., & McNamara, D. S. (2010). Self-regulated Learning in Learning Environments with Pedagogical Agents that Interact in Natural Language. *Educational Psychologist*, 45(4), 234–244. DOI: 10.1080/00461520.2010.515933

Grech, A., & Camilleri, A. F. (2017). *Blockchain in Education*. European Commission.

Greer, S. L. (2018). Organization and Governance: Stewardship and Governance in Health Systems. Health Care Systems and Policies, 1–9. DOI: 10.1007/978-1-4614-6419-8_22-1

Guan, C., Mou, J., & Jiang, Z. (2020). Artificial intelligence innovation in education: A twenty-year data-driven historical analysis. *Int. J. Innov. Stud.*, 4(4), 134–147. DOI: 10.1016/j.ijis.2020.09.001

Guazzini, A., Guidi, E., Cecchini, C., & Yoneki, E. (2020). Collaborative facilitation and collaborative inhibition in virtual environments. *Future Internet*, 12(7), 118. DOI: 10.3390/fi12070118

Guiffrida, D. A., Lynch, M. F., Wall, A. F., & Abel, D. S. (2013). Do reasons for attending college affect academic outcomes? A test of a motivational model from a self-determination theory perspective. *Journal of College Student Development*, 54(2), 121–139. DOI: 10.1353/csd.2013.0019

Guilherme, A. (2019). AI and education: The importance of teacher and student relations. *AI & Society*, 34(1), 47–54. DOI: 10.1007/s00146-017-0693-8

Guillon, S., & Kennel, S. (2022). Le tutorat étudiant en contexte de pandémie de covid-19: Quelles modalités d'accompagnement pour quels besoins exprimés? *International Journal of Technologies in Higher Education*, 19(2), 131–147. DOI: 10.18162/ritpu-2022-v19n2-09

Gupta, K. P., & Bhaskar, P. (2020).*Inhibiting and Motivating Factors Influencing Teachers' Adoption of AI-Based Teaching and Learning Solutions: Prioritization Using Analytic Hierarchy Process*. DOI: 10.28945/4640

Gupta, A., & El Haddioui, K. (2019). A Study of Microlearning Practices: A Literature Review. *International Journal of Educational Research and Development*, 9(4), 60–68.

Haladyna, T. M., & Rodriguez, M. C. (2013). *Developing and Validating Test Items*. Routledge. DOI: 10.4324/9780203850381

Hall, O. P.Jr, & Ko, K. (2008). Customized content delivery for graduate management education: Application to business statistics. *Journal of Statistics Education : An International Journal on the Teaching and Learning of Statistics*, 16(3). Advance online publication. DOI: 10.1080/10691898.2008.11889571

Halmaoui, H. (2022). L'intelligence artificielle, la réalité augmentée et la réalité virtuelle dans l'audiovisuel et le cinéma.

Hamal, O., El Faddouli, N., Alaoui Harouni, M. H., & Lu, J. (2022). Artificial Intelligent in Education. *Sustainability (Basel)*, 14(5), 2862. Advance online publication. DOI: 10.3390/su14052862

Hamari, J., Koivisto, J., & Sarsa, H. (2014). Does Gamification Work? - A Literature Review of Empirical Studies on Gamification. In 2014 47th Hawaii International Conference on System Sciences (pp. 3025-3034).

Hamida, S., Cherradi, B., Raihani, A., & Ouajji, H. (2021). Evaluation des apprentissages au sein d'un environnement de type mooc adaptatif. ITM Web of Conferences, 39, 03005. DOI: 10.1051/itmconf/20213903005

Han, J., & Geng, X. (2023). University students' approaches to online learning technologies: The roles of perceived support, affect/emotion and self-efficacy in technology-enhanced learning. *Computers & Education*, 194, 104695. DOI: 10.1016/j.compedu.2022.104695

Han, L., & Gong, J. (2022). A study on exploring the path of psychology and civics teaching reform in universities based on artificial intelligence. *Computational Intelligence and Neuroscience*, 2022, 1–12. DOI: 10.1155/2022/4841387 PMID: 35978907

Hanna, D. and Charalampopoulou, C. (2019). Travail collaboratif sur l'ent et innovation dans les pratiques professionnelles des enseignants. Spirale - Revue De Recherches en Éducation, N° 63(1), 23-35. DOI: 10.3917/spir.063.0023

Harry, A. (2023). Role of AI in Education. [INJURITY]. *Interdiciplinary Journal and Hummanity*, 2(3), 260–268.

Harshvardhan, G. M., Gourisaria, M. K., Pandey, M., & Rautaray, S. S. (2020). A comprehensive survey and analysis of generative models in machine learning. *Computer Science Review*, 38, 100285. DOI: 10.1016/j.cosrev.2020.100285

Hashem, R., Ali, N., El Zein, F., Fidalgo, P., & Khurma, O. A. (2024). AI to the rescue: Exploring the potential of ChatGPT as a teacher ally for workload relief and burnout prevention. *Research and Practice in Technology Enhanced Learning*, 19, •••. https://www.researchgate.net/profile/Othman-Abu-Khurma/publication/374024140_AI_to_the_rescue_Exploring_the_potential_of_ChatGPT_as_a_teacher_ally_for_workload_relief_and_burnout_prevention/links/650a94da82f01628f032ebb7/AI-to-the-rescue-Exploring-the-potential-of-ChatGPT-as-a-teacher-ally-for-workload-relief-and-burnout-prevention.pdf

Hashim, S. N. A., Mariappan, U., Ab Ghani, M. A., Nor, N. M. M., Ensimau, N. K., Yassin, Y. N. H. M., & Abbas, M. S. (2024). From Data To Insights: How AI Is Shaping Evidence-Based Curriculum Design. *Nanotechnology Perceptions*, •••, 1724–1728.

Hasson, L., Keville, S., Gallagher, J., Onagbesan, D., & Ludlow, A. K. (2024). Inclusivity in education for autism spectrum disorders: Experiences of support from the perspective of parent/carers, school teaching staff and young people on the autism spectrum. *International Journal of Developmental Disabilities*, 70(2), 201–212. DOI: 10.1080/20473869.2022.2070418 PMID: 38481460

Heffernan, N., & Koedinger, K. (2012). The future of educational technology: Adaptive learning and intelligent tutoring systems. In Dede, C., & Richards, J. (Eds.), *Digital Teaching Platforms*. Harvard Education Press.

Helquist, J. H., Kruse, J., & Diller, C. B. R. (2019). Crowdsourcing convergence: aggregating partial clusters to facilitate collaborative convergence. *Proceedings of the Annual Hawaii International Conference on System Sciences*. DOI: 10.24251/HICSS.2019.064

Henri, F., Peraya, D., & Charlier, B. (2007). La recherche sur les forums de discussion en milieu éducatif: Critères de qualité et qualité des pratiques. *Sciences Et Technologies De l'Information Et De La Communication Pour l'Éducation Et La Formation*, 14(1), 155–192. DOI: 10.3406/stice.2007.952

Henry, J.-P. (2024). Gaillard Raphaël, L'Homme augmenté. Futurs de nos cerveaux, Paris: Grasset, janvier 2024, 352 p. *Futuribles (Paris, France)*, 3(460), 122–125. DOI: 10.3917/futur.460.0122

Hérold, J. (2019). Un modèle de la situation d'enseignement-apprentissage pour mieux comprendre l'activité de l'élève en classe. *Revue des Sciences de l'Education*, 45(1), 82–107. DOI: 10.7202/1064607ar

Hew, K. F., Lan, M., Tang, Y., Jia, C., & Lo, C. K. (2019). Where is the "theory" within the field of educational technology research? *British Journal of Educational Technology*, 50(3), 956–971. DOI: 10.1111/bjet.12770

Hewson, E., & Chung, G. W. (2021). Beyond the VLE: Transforming online discussion and collaboration through Microsoft Teams. Hosted by UNED, Madrid (Spain), 36. https://pureportal.coventry.ac.uk/files/27637722/Proceedings_OOFHEC2019.pdf#page=48

Hidayat, M. L. (2018). The modification of lms quipperschool to improve senior high school studentsâ€™ concept mastery of biology subject, topic: Human reproduction health. [EST]. *Journal of Educational Science and Technology*, •••, 32–38. DOI: 10.26858/est.v4i1.5215

Hmelo-Silver, C. E. (2004). Problem-Based Learning: What and How Do Students Learn? *Educational Psychology Review*, 16(3), 235–266. DOI: 10.1023/B:EDPR.0000034022.16470.f3

Holmes, W., & Miao, F. (2023). Guidance for generative AI in education and research. UNESCO Publishing. https://books.google.com/books?hl=fr&lr=&id=cKnYEAAAQBAJ&oi=fnd&pg=PA2&dq=data+privacy+generative+ai&ots=wmGadhI2eK&sig=V-VUh9VLCz_tjG4vb7ZXlFr-X88

Holmes, W., Porayska-Pomsta, K., Holstein, K., Sutherland, E., Baker, T., Shum, S. B., Santos, O. C., Rodrigo, M. T., Cukurova, M., Bittencourt, I. I., & Koedinger, K. R. (2021).*Ethics of AI in Education: Towards a Community-Wide Framework.* DOI: 10.1007/s40593-021-00239-1

Holmes, W., Bialik, M., & Fadel, C. (2019). *Artificial intelligence in education: Promises and implications for teaching and learning*. Center for Curriculum Redesign.

Holmes, W., Persson, J., Chounta, I. A., Wasson, B., & Dimitrova, V. (2022). *Artificial intelligence and education: A critical view through the lens of human rights, democracy and the rule of law.* Council of Europe.

Holzinger, A., Kieseberg, P., Weippl, E., & Tjoa, A. M. (2018). Current Advances, Trends and Challenges of Machine Learning and Knowledge Extraction: From Machine Learning to Explainable AI. In Holzinger, A., Kieseberg, P., Tjoa, A. M., & Weippl, E. (Eds.), *Machine Learning and Knowledge Extraction* (Vol. 11015, pp. 1–8). Springer International Publishing., DOI: 10.1007/978-3-319-99740-7_1

How, M. L., & Hung, W. L. D. (2019). Educing AI-Thinking in Science, Technology, Engineering, Arts, and Mathematics (STEAM) Education. *Education Sciences*, 9(3), 184. Advance online publication. DOI: 10.3390/educsci9030184

Hrastinski, I., & Wilbur, R. B. (2016). Academic achievement of deaf and hard-of-hearing students in an asl/english bilingual program. *Journal of Deaf Studies and Deaf Education*, 21(2), 156–170. DOI: 10.1093/deafed/env072 PMID: 26864688

Hrich, N., Azekri, M., & Khaldi, M. (2023). ARTIFICIAL INTELLIGENCE FOR EDUCATIONAL ASSESSMENT. *ICERI2023 Proceedings*, 2120–2124. 16th annual International Conference of Education, Research and Innovation. DOI: 10.21125/iceri.2023.0598

Hrich, N., Azekri, M., & Khaldi, M. (2024a). *Application of LSTM for Redundancy Detection in MCTS: Enhancing Test Precision | IIETA*. DOI: 10.18280/isi.290430

Hrich, N., Azekri, M., Elhaddouchi, C., & Khaldi, M. (2024). Enhancing Educational Assessments: Score Regeneration through Post-Item Analysis with Artificial Intelligence. *International Journal of Information and Education Technology, 14*(10), 1414–1420. *Scopus*. Advance online publication. DOI: 10.18178/ijiet.2024.14.10.2172

Hrich, N., Azekri, M., & Khaldi, M. (2024b). Artificial Intelligence Item Analysis Tool for Educational Assessment: Case of Large-Scale Competitive Exams. *International Journal of Information and Education Technology (IJIET)*, 14(6), 822–827. DOI: 10.18178/ijiet.2024.14.6.2107

Hrich, N., Lazaar, M., & Khaldi, M. (2019). Improving Cognitive Decision-Making into Adaptive Educational Systems through a Diagnosis Tool based on the Competency Approach. [iJET]. *International Journal of Emerging Technologies in Learning*, 14(07), 07. Advance online publication. DOI: 10.3991/ijet.v14i07.9870

Hrich, N., Lazaar, M., & Khaldi, M. (2021). Assessment Process-Reflection on Pedagogical Practices and Integration of Technologies in Education. In *Promoting Positive Learning Experiences in Middle School Education* (pp. 42–65). IGI Global., DOI: 10.4018/978-1-7998-7057-9.ch003

Huang, A. Y. Q., Lu, O. H. T., & Yang, S. J. H. (2023). Effects of artificial Intelligence–Enabled personalized recommendations on learners' learning engagement, motivation, and outcomes in a flipped classroom. *Computers & Education*, 194, 104684. DOI: 10.1016/j.compedu.2022.104684

Huang, C., & Wu, C. (2017). A Systematic Review of Intelligent Tutoring Systems for Language Learning: A Content Analysis. *Journal of Educational Technology & Society*, 20(2), 1–10.

Huang, J. (2023). *Artificial Intelligence in Higher Education*. Advances in Computer and Electrical Engineering Book Series., DOI: 10.4018/978-1-6684-6937-8.ch014

Huang, J., Saleh, S., & Liu, Y. (2021). A Review on Artificial Intelligence in Education, Academic Journal of Interdisciplinary Studies. *Richtmann Publishing Ltd*, 10(3), 206. Advance online publication. DOI: 10.36941/ajis-2021-0077

Huang, W. D., & Weng, S. S. (2017). An Intelligent Tutoring System for Personalized Learning: Feedback Mechanisms and Adaptive Strategies. *Computers & Education*, 115, 173–186.

Hua, Y. (2022). Design of online music education system based on artificial intelligence and multiuser detection algorithm. *Computational Intelligence and Neuroscience*, 2022, 1–11. DOI: 10.1155/2022/4744939 PMID: 35371210

Hug, T. (2005). Micro Learning: A New Pedagogical Challenge. In Micro Learning: Emerging Trends and Technologies (pp. 1-10).

Hug, T. (2006). Micro Learning and Narration: Exploring Possibilities of Utilization of Narrations and Storytelling for the Designing of "Micro Units" and Didactical Micro-learning Arrangements. In *Proceedings of the third International Conference on Microlearning* (pp. 33-38).

Hurst, A. (2023). The impact of artificial intelligence on learning accessibility in higher education (Thèse de doctorat, Université de l'Éducation de l'Ohio). ProQuest Dissertations Publishing.

Hutchins, E. (1995). *Cognition in the Wild.* MIT Press. DOI: 10.7551/mitpress/1881.001.0001

Hwang, C., Ghalachyan, A., & Song, S. (2023). Exploring student experiences with a virtual learning environment in an apparel and textiles curriculum during the COVID-19 pandemic. *International Journal of Fashion Design, Technology and Education*, 16(1), 1–10. DOI: 10.1080/17543266.2022.2158237

Hwang, G. J., & Chang, H. F. (2021). A Review of Opportunities and Challenges of Using Artificial Intelligence for Adaptive Learning in Higher Education. *Educational Technology Research and Development*, 69(2), 807–820.

Hwang, G. J., & Tu, Y. F. (2021). Roles and research trends of artificial intelligence in mathematics education: A bibliometric mapping analysis and systematic review. *Mathematics*, 9(6), 584. DOI: 10.3390/math9060584

Iaaly, A., Daou, P., Nakad, M., & Abboud, R. J. (2024). E-assessment and e-proctoring implementation during coronavirus pandemic: A participatory study using socio-technical systems theory. *Education and Information Technologies*. Advance online publication. DOI: 10.1007/s10639-024-13078-z

Ifenthaler, D., Majumdar, R., Gorissen, P., Judge, M., Mishra, S., Raffaghelli, J., & Shimada, A. (2024). Artificial Intelligence in Education: Implications for Policymakers, Researchers, and Practitioners. *Tech Know Learn*, 29(4), 1693–1710. DOI: 10.1007/s10758-024-09747-0

Ifenthaler, D., & Schumacher, C. (2016). Student perceptions of privacy principles for learning analytics. *Educational Technology Research and Development*, 64(5), 923–938. DOI: 10.1007/s11423-016-9477-y

Iglesias, A., Martinez, P., Aler, R., & Fernandez, F. (2009). Reinforcement learning of pedagogical policies in adaptive and intelligent educational systems. *Knowledge-Based Systems*, 22(4), 266–270. https://e-archivo.uc3m.es/bitstream/handle/10016/6502/reinforcement_aler_KBS_2009_ps.pdf?sequence=1&isAllowed=y. DOI: 10.1016/j.knosys.2009.01.007

Ilieva, G., Yankova, T., Klisarova-Belcheva, S., Dimitrov, A., Bratkov, M., & Angelov, D. (2023). Effects of generative chatbots in higher education. *Information (Basel)*, 14(9), 492. DOI: 10.3390/info14090492

Integration of AI in Learning. (2022). *Advances in Higher Education and Professional Development Book Series*. DOI: 10.4018/978-1-6684-4083-4.ch014

IRT Illustrator – Psychomeasurement Systems. (n.d.). Retrieved January 4, 2024, from https://itemanalysis.com/irt-illustrator/

Ivanashko, O., Kozak, A., Knysh, T., & Honchar, K. (2024). The Role of Artificial Intelligence in Shaping the Future of Education: Opportunities and Challenges. *Futurity Education*, 4(1), 126–146. DOI: 10.57125/FED.2024.03.25.08

Iyamuremye, A., & Ndihokubwayo, K. (2024). Exploring secondary school students' interest and mastery of atomic structure and chemical bonding through ChatGPT. *Educational Journal of Artificial Intelligence and Machine Learning*, 1(1), 1–13. DOI: 10.58197/prbl/9hk37296

Jackson, M., & Cossitt, B. (2015). Is intelligent online tutoring software useful in refreshing financial accounting knowledge? *Advances in Accounting Education*, 16, 1–19. DOI: 10.1108/S1085-462220150000016001

Jain, G. P., Gurupur, V. P., Schroeder, J. L., & Faulkenberry, E. D. (2014). Artificial intelligence-based student learning evaluation: A concept map-based approach for analyzing a student's understanding of a topic. *IEEE Transactions on Learning Technologies*, 7(3), 267–279. DOI: 10.1109/TLT.2014.2330297

James, G., Witten, D., Hastie, T., & Tibshirani, R. (2013). *An Introduction to Statistical Learning: With Applications in R*. Springer. DOI: 10.1007/978-1-4614-7138-7

Jang, J., Ko, Y., Shin, W. S., & Han, I. (2021). Augmented reality and virtual reality for learning: An examination using an extended technology acceptance model. *IEEE Access : Practical Innovations, Open Solutions*, 9, 6798–6809. DOI: 10.1109/ACCESS.2020.3048708

Jani, K. H., Jones, K. A., Jones, G. W., Amiel, J., Barron, B., & Elhadad, N. (2020). Machine learning to extract communication and history-taking skills in OSCE transcripts. *Medical Education*, 54(12), 1159–1170. DOI: 10.1111/medu.14347 PMID: 32776345

Jansen, R. S., & Hovestadt, L. (2018). Verifying Learning Outcomes with Blockchain Technology: A New Way of Certification in Education. *Proceedings of the European Conference on e-Learning*, 1, 267-275.

Janssen, G., Meier, V., & Trace, J. (2014). Classical test theory and item response theory: Two understandings of one high-stakes performance exam. *Colombian Applied Linguistics Journal*, 16(2), 167. DOI: 10.14483/udistrital.jour.calj.2014.2.a03

Järvelä, S., Nguyen, A., & Hadwin, A. F. (2023). Human and artificial intelligence collaboration for socially shared regulation in learning. *British Journal of Educational Technology*, 54(5), 1057–1076. DOI: 10.1111/bjet.13325

Jerald, J. (2016). *The VR Book: Human-Centered Design for Virtual Reality*. Association for Computing Machinery and Morgan & Claypool.

Jiang, F., Ma, J., Webster, C. J., Chiaradia, A. J., Zhou, Y., Zhao, Z., & Zhang, X. (2024). Generative urban design: A systematic review on problem formulation, design generation, and decision-making. *Progress in Planning*, 180, 100795. DOI: 10.1016/j.progress.2023.100795

Jin, S., Im, K., Yoo, M., Roll, I., & Seo, K. (2023). Supporting students' self-regulated learning in online learning using artificial intelligence applications. *International Journal of Educational Technology in Higher Education*, 20(1), 37. Advance online publication. DOI: 10.1186/s41239-023-00406-5

Jobin, A., Ienca, M., & Vayena, E. (2019). The global landscape of AI ethics guidelines. *Nature Machine Intelligence*, 1(9), 389–399. DOI: 10.1038/s42256-019-0088-2

Johnson, D. W., & Johnson, R. T. (1989). *Cooperation and Competition: Theory and Research*. Interaction Book Company.

Johnson, D. W., & Johnson, R. T. (1999). *Learning Together and Alone: Cooperative, Competitive, and Individualistic Learning*. Allyn & Bacon.

Johnson, D. W., & Johnson, R. T. (2009). An Educational Psychology Success Story: Social Interdependence Theory and Cooperative Learning. *Educational Researcher*, 38(5), 365–379. DOI: 10.3102/0013189X09339057

Johnson, D. W., Johnson, R. T., & Smith, K. A. (2014). *Active Learning: Cooperation in the College Classroom* (3rd ed.). Pearson.

Johnson, L. (2022). *L'utilisation de l'IA dans l'éducation spécialisée (Mémoire de maîtrise, Université de Paris)*. HAL.

Johri, A., Lindsay, E., & Qadir, J. (2023). ETHICAL CONCERNS AND RESPONSIBLE USE OF GENERATIVE ARTIFICIAL INTELLIGENCE IN ENGINEERING EDUCATION. SEFI 2023-51st Annual Conference of the European Society for Engineering Education: Engineering Education for Sustainability, Proceedings (2023), 2244–2253.

Jonassen, D., Davidson, M., Collins, M., Campbell, J., & Haag, B. B. (1995). Constructivism and computer-mediated communication in distance education. *American Journal of Distance Education*, 9(2), 7–25. DOI: 10.1080/08923649509526885

Jones, J., Goins, R., Schure, M., Winchester, B., & Bradley, V. (2019). Putting self-management in the context of community-dwelling american indians living with type 2 diabetes. *The Diabetes Educator*, 46(1), 108–117. DOI: 10.1177/0145721719894889 PMID: 31868100

Jordan, M. I., & Mitchell, T. M. (2021). Machine Learning: Trends, Perspectives, and Prospects. *Science*, 349(6245), 255–260. DOI: 10.1126/science.aaa8415 PMID: 26185243

Judijanto, L., Atsani, R., & Chadijah, S. (2024). Trends in the Development of Artificial Intelligence-Based Technology in Education. International Journal of Teaching and Learning. Vol. 2 No. 6. µ https://injotel.org/index.php/12/article/view/197

Jung, M. (2013). Videoconferencing improves students' language learning in the efl classroom. *TESOL Journal*, 4(4), 743–751. DOI: 10.1002/tesj.112

Jurafsky, D., & Martin, J. H. (2008). *Speech and Language Processing*. Pearson.

Kahn, C. E., & Lathrop, R. (2017). Blockchain Technology: The Future of Digital Information Management. *Journal of Management Policy and Practice*, 18(1), 33–44.

Kaliraj, P., Singaravelu, G., & Devi, T. (Eds.). (2024). *Transformative Digital Technology for Disruptive Teaching and Learning*. CRC Press. DOI: 10.1201/9781032675176

Kallonas, C., Piki, A., & Stavrou, E. (2024). Empowering professionals: A generative AI approach to personalized cybersecurity learning. 2024 IEEE Global Engineering Education Conference (EDUCON), 1–10. https://ieeexplore.ieee.org/abstract/document/10578894/

Kalz, M., van Bruggen, J., Giesbers, B., Waterink, W., Eshuis, J., & Koper, R. (2008). A model for new linkages for prior learning assessment. *Campus-Wide Information Systems*, 25(4), 233–243. DOI: 10.1108/10650740810900676

Kanipriya, M., Sugumaran, D., Pandiyarajan, V., Yogananda, B. S., Selvakumar, P., & Mohan, S. (2024, January). Analysis of Electromyography Signals for Control of Mechanical Prosthesis using Machine Learning Techniques. In 2024 2nd International Conference on Intelligent Data Communication Technologies and Internet of Things (IDCIoT) (pp. 1059-1067). IEEE.

Kant, N., Prasad, K. D., & Anjali, K. (2021). Selecting an appropriate learning management system in open and distance learning: A strategic approach. *Asian Association of Open Universities Journal*, 16(1), 79–97. DOI: 10.1108/AAOUJ-09-2020-0075

Kao, G. Y. M., Chen, K. C., & Sun, C. T. (2010). Using an e-learning system with integrated concept maps to improve conceptual understanding. *International Journal of Instructional Media*, 37(2), 151–162.

Kaplan-Rakowski, R., Grotewold, K., Hartwick, P., & Papin, K. (2023). Generative AI and teachers' perspectives on its implementation in education. *Journal of Interactive Learning Research*, 34(2), 313–338.

Kapp, K. M. (2012). *The Gamification of Learning and Instruction: Game-Based Methods and Strategies for Training and Education*. Pfeiffer.

Karpicke, J. D., & Grimaldi, P. J. (2023). Optimizing learning through AI-driven spacing and retrieval practice in educational technology. *Journal of Applied Research in Memory and Cognition*, 12(2), 215–229.

Karsenti, T. (2018). Intelligence artificielle en éducation: L'urgence de préparer les futurs enseignants aujourd'hui pour l'école de demain. Formation et profession, 26(3), 112-119.

Kasinidou, M. (2023, June 29). AI Literacy for All: A Participatory Approach. *Annual Conference on Innovation and Technology in Computer Science Education*. DOI: 10.1145/3587103.3594135

Kassens-Noor, E. (2019). The Role of Conversational Agents in Informal Learning via Social Media. *Computers & Education*, 130, 59–70.

Kaswan, K. S., Dhatterwal, J. S., Malik, K., & Baliyan, A. (2023). Generative AI: A Review on Models and Applications. 2023 International Conference on Communication, Security and Artificial Intelligence (ICCSAI), 699–704. https://ieeexplore.ieee.org/abstract/document/10421601/

Katz, R. N., & Oblinger, D. G. (2000). The Future of the Digital Learning Environment. *EDUCAUSE Review*, 35(2), 34–45.

Kennedy, D. R. (2023).*Inteligencia Artificial, como base de un modelo de curriculum learning para el desarrollo de competencias digitales en e-learning*. DOI: 10.24275/uama.5813.9684

Kennel, S., Guillon, S., Caublot, M., & Rohmer, O. (2021). La pédagogie inclusive: représentations et pratiques des enseignants à l'université. La Nouvelle Revue - Éducation Et Société Inclusives, N° 89-90, vol. 2(3), 23-45. DOI: 10.3917/nresi.090.0023

Kerly, A., Hall, P., & Bull, S. (2007). Bringing Chatbots into Education: Towards Natural Language Negotiation of Open Learner Models. *Knowledge-Based Systems*, 20(2), 177–185. DOI: 10.1016/j.knosys.2006.11.014

Khan, M. S. H., Bibi, S., & Hasan, M. (2016). Australian technical teachers' experience of technology integration in teaching. *SAGE Open*, 6(3), 215824401666360. DOI: 10.1177/2158244016663609

KHAZANCHI, R., & KHAZANCHI, P. (2024). Generative AI to improve special education teacher preparation for inclusive classrooms. Exploring New Horizons: Generative Artificial Intelligence and Teacher Education, 159.

Khreisat, M. N., Khilani, D., Rusho, M. A., Karkkulainen, E. A., Tabuena, A. C., & Uberas, A. D. (2024). Ethical Implications Of AI Integration In Educational Decision Making: Systematic Review. *Educational Administration: Theory and Practice*, 30(5), 8521–8527.

Kile, M. (2021). Uncovering social issues through photovoice: a comprehensive methodology. HERD: Health Environments Research &Amp. *The Design Journal*, 15(1), 29–35. DOI: 10.1177/19375867211055101 PMID: 34706570

Kim, J., Ham, Y., & Lee, S. (2024). Differences in student-ai interaction process on a drawing task: Focusing on students' attitude towards ai and the level of drawing skills. *Australasian Journal of Educational Technology*. Advance online publication. DOI: 10.14742/ajet.8859

Kim, J., Lee, H., & Cho, Y. H. (2022). Learning design to support student-ai collaboration: Perspectives of leading teachers for ai in education. *Education and Information Technologies*, 27(5), 6069–6104. DOI: 10.1007/s10639-021-10831-6

Kim, J., & Lee, S. (2022). Are two heads better than one?: The effect of student-ai collaboration on students' learning task performance. *TechTrends*, 67(2), 365–375. DOI: 10.1007/s11528-022-00788-9 PMID: 36258920

Kirkland, R. (2019). Using Blockchain to Authenticate Educational Credentials: A Guide for Institutions. *International Journal of Information Technology and Management*, 18(2), 123–136.

Kirschner, P. A., Paas, F., & Kirschner, F. (2009). A Cognitive Load Approach to Collaborative Learning: United Brains for Complex Tasks. *Educational Psychology Review*, 21(1), 31–42. DOI: 10.1007/s10648-008-9095-2

Kizilcec, R. F., Reich, J., Yeomans, M., Dann, C., Brunskill, E., Lopez, G., Turkay, S., Williams, J. J., & Tingley, D. (2020). Scaling up behavioral science interventions in online education. *Proceedings of the National Academy of Sciences of the United States of America*, 117(26), 14900–14905. DOI: 10.1073/pnas.1921417117 PMID: 32541050

Kizilcec, R. F., & Schneider, E. F. (2020). Learning in the age of AI: Personalized learning experiences. *Educational Technology Research and Development*, 68(1), 5–25. DOI: 10.1007/s11423-019-09702-5

Klímová, B., Pikhart, M., & Kacetl, J. (2023). Ethical issues of the use of ai-driven mobile apps for education. *Frontiers in Public Health*, 10, 1118116. Advance online publication. DOI: 10.3389/fpubh.2022.1118116 PMID: 36711343

Klooker, M. and Hölzle, K. (2023). A generative design of collaborative innovation space. R&D Management, 54(2), 323-346. DOI: 10.1111/radm.12582

Klopfer, E., & Squire, K. (2008). Environmental Detectives: The Development of an Augmented Reality Platform for Environmental Education. *Educational Technology Research and Development*, 56(2), 203–228. DOI: 10.1007/s11423-007-9037-6

Kniha, K., Bock, A., Peters, F., Heitzer, M., Hölzle, F., Raith, S., Modabber, A., & Möhlhenrich, S. C. (2023). Guided discovery learning: A follow-up study of try-it-yourself surgery and subsequent video-assisted teaching for oral surgical skills training. *European Journal of Dental Education*, 27(1), 29–35. DOI: 10.1111/eje.12772 PMID: 35051298

Knowles, M. A. (2021). Five motivating concerns for ai ethics instruction. *Proceedings of the Association for Information Science and Technology*, 58(1), 472–476. DOI: 10.1002/pra2.481

Knowles, M. S. (1980). *The Modern Practice of Adult Education: Andragogy Versus Pedagogy*. Cambridge Books.

Knox, J., Wang, Y., & Gallagher, M. (2022). AI, Ethics, and Education: Learning AI Through Collaborative Projects. *International Review of Research in Open and Distance Learning*, 23(2), 112–129.

Koedinger, K. R., & Aleven, V. (2007). Exploring the Assistance Dilemma in Intelligent Tutoring Systems. *User Modeling and User-Adapted Interaction*, 17(3), 209–235.

Ko, H.-K., Park, G., Jeon, H., Jo, J., Kim, J., & Seo, J. (2023). Large-scale Text-to-Image Generation Models for Visual Artists' Creative Works. *Proceedings of the 28th International Conference on Intelligent User Interfaces*, 919–933. DOI: 10.1145/3581641.3584078

Koh, E. G., & Frick, T. W. (2014). Micro-Learning and Learning Styles: A Study of Learning Preferences of Postgraduate Students. *Journal of Online Learning and Teaching*, 10(1), 59–69.

Kohli, A., & Pande, A. (2020). The Role of Blockchain Technology in Managing Digital Credentials. *International Journal of Information Management*, 50, 42–52.

Kohnke, L., Moorhouse, B. L., & Zou, D. (2023). Chatgpt for language teaching and learning. *RELC Journal*, 54(2), 537–550. DOI: 10.1177/00336882231162868

Kolb, D. A. (1984). *Experiential Learning: Experience as the Source of Learning and Development*. Prentice-Hall.

Kong, S. C., Cheung, W. M. Y., & Tsang, O. (2023). Evaluating an artificial intelligence literacy programme for empowering and developing concepts, literacy and ethical awareness in senior secondary students. *Education and Information Technologies*, 28(4), 4703–4724. DOI: 10.1007/s10639-022-11408-7

Koper, R. (2014). The Role of Technology in Adaptive Learning. *Journal of Interactive Media in Education*, 2014(1), 1–8. DOI: 10.5334/2005-1

Koschmann, T. (1996). *CSCL: Theory and Practice of an Emerging Paradigm*. Lawrence Erlbaum Associates.

Kousa, P., & Niemi, H. (2022). AI ethics and learning: EdTech companies' challenges and solutions. *Interactive Learning Environments*. Advance online publication. DOI: 10.1080/10494820.2022.2043908

Kovanovic, V., Joksimovic, S., & Siemens, G. (2015). The role of learning analytics in improving the learning experience. *Journal of Computer Assisted Learning*, 31(4), 373–384. DOI: 10.1111/jcal.12093

Krajcik, J. S., & Blumenfeld, P. C. (2006). Project-Based Learning. In Sawyer, R. K. (Ed.), *The Cambridge Handbook of the Learning Sciences*.

Krauss, S., & Fuchs, C. (2018). Chatbots as a Tool for Learning: How Conversational Agents Can Enhance Learning Experiences. *Journal of Computer Assisted Learning*, 34(5), 527–539.

Kuhail, M. A., Alturki, N., Alramlawi, S., & Alhejori, K. (2022). Interacting with educational chatbots: A systematic review. *Education and Information Technologies*, 28(1), 973–1018. DOI: 10.1007/s10639-022-11177-3

Kujundziski, A. P., & Bojadjiev, J. (2024). *Artificial Intelligence in Education*. Advances in Educational Technologies and Instructional Design Book Series., DOI: 10.4018/979-8-3693-4310-4.ch001

Kukulska-Hulme, A. (2019). Mobile Assisted Language Learning [MALL]: A Perspective on the Future. In The Cambridge Handbook of Language Learning (pp. 383-400). Cambridge University Press.

Kulhavy, R. W. (1977). Feedback in Written Instruction. *Review of Educational Research*, 47(2), 211–232. DOI: 10.3102/00346543047002211

Kulik, J. A., & Fletcher, J. D. (2016). Effectiveness of Intelligent Tutoring Systems: A Meta-Analytic Review. *Review of Educational Research*, 86(1), 42–78. DOI: 10.3102/0034654315581420

Kumar, D., Haque, A., Mishra, K., & Islam, F. (2023).*Exploring the Transformative Role of Artificial Intelligence and Metaverse in Education: A Comprehensive Review*. DOI: 10.56294/mr202355

Kumar, A., Srinivasan, B., Saudagar, A. K. J., AlTameem, A., Alkhathami, M., Alsamani, B., Khan, M. B., Ahmed, Z. H., Kumar, A., & Singh, K. U. (2023). Next-gen mulsemedia: Virtual reality haptic simulator's impact on medical practitioners for higher education institutions. *Electronics (Basel)*, 12(1), 356. DOI: 10.3390/electronics12020356

Kumar, R., Eaton, S. E., Mindzak, M., & Morrison, R. (2024). Academic Integrity and Artificial Intelligence: An Overview. In Eaton, S. E. (Ed.), *Second Handbook of Academic Integrity* (pp. 1583–1596). Springer Nature Switzerland., DOI: 10.1007/978-3-031-54144-5_153

Kurdi, G., Leo, J., Parsia, B., Sattler, U., & Al-Emari, S. (2020). A Systematic Review of Automatic Question Generation for Educational Purposes. *International Journal of Artificial Intelligence in Education*, 30(1), 121–204. DOI: 10.1007/s40593-019-00186-y

Kuzilek, J., Hlosta, M., Herrmannova, D., Zdrahal, Z., & Wolff, A. (2015). OU Analyse: Analysing at-risk students at The Open University. Learning Analytics Review, 1-16.

Labadze, L., Grigolia, M., & Machaidze, L. (2023). Role of AI chatbots in education: Systematic literature review. *International Journal of Educational Technology in Higher Education*, 20(1), 56. DOI: 10.1186/s41239-023-00426-1

Lajoie, S. P. (2015). Collaboration, Communication, and Knowledge Building in Online Learning Environments. *Frontiers in Psychology.*

Lalle, P., & Bonnafous, S. (2019). La révolution pédagogique de l'enseignement supérieur, une universalité géographique et paradigmatique. *Revue Internationale d'Éducation De Sèvres*, 80, 49–60. DOI: 10.4000/ries.8142

Landers, R. N. (2014). Developing a Theory of Gamified Learning: Linking Serious Games and Gamification of Learning. *Simulation & Gaming*, 45(6), 752–768. DOI: 10.1177/1046878114563660

Lane, S., Raymond, M. R., & Haladyna, T. M. (2015). *Handbook of Test Development.* Routledge. DOI: 10.4324/9780203102961

Lavoie, L. M. (2010). Compétences des formateurs en entreprise selon le point de vue des formateurs et des apprenants d'une installation alcan /.. DOI: 10.1522/030139027

Le Cun, Y. (2019). *Quand la machine apprend: la révolution des neurones artificiels et de l'apprentissage profond.* Odile Jacob.

Le, X., Sun, Y., Li, X., Feng, G., & Imran, M. (2022). Wireless resource management in intelligent semantic communication networks., 1-6. DOI: 10.1109/INFOCOM-WKSHPS54753.2022.9797984

Learning, W. t. (n.d.). *Closing the maths attainment gap with one to one teaching.* Retrieved from Welcome to Third Space Learning: https://thirdspacelearning.com/

LeCun, Y., Bengio, Y., & Haffner, P. (1998). Gradient-Based Learning Applied to Document Recognition. *Proceedings of the IEEE*, 86(11), 2278–2324. DOI: 10.1109/5.726791

Leduc, D., Kozanitis, A., & Lepage, I. (2019). L'engagement cognitif en contexte postsecondaire: Traduction, adaptation et validation d'une échelle de mesure. *McGill Journal of Education*, 53(3). Advance online publication. DOI: 10.7202/1058412ar

Lee, A. V. Y. (2023). Supporting students' generation of feedback in large-scale online course with artificial intelligence-enabled evaluation. *Studies in Educational Evaluation*, 77, 101250. DOI: 10.1016/j.stueduc.2023.101250

Lee, C. H., Gobir, N., Gurn, A., & Soep, E. (2022). In the Black Mirror: Youth Investigations into Artificial Intelligence. *ACM Transactions on Computing Education*, 22(3), 1–25. DOI: 10.1145/3484495

Lee, J., Eom, S.-Y., & Lee, J. (2023). Empowering game designers with generative AI. *IADIS International Journal on Computer Science & Information Systems*, 18(2), 213–230.

Lee, S.-M. (2023). The effectiveness of machine translation in foreign language education: A systematic review and meta-analysis. *Computer Assisted Language Learning*, 36(1–2), 103–125. DOI: 10.1080/09588221.2021.1901745

Lee, Y. S., Rashidi, A., Talei, A., Beh, H. J., & Rashidi, S. (2023). A comparison study on the learning effectiveness of construction training scenarios in a virtual reality environment. *Virtual Worlds*, 2(1), 36–52. DOI: 10.3390/virtualworlds2010003

Lei, H., Cui, Y., & Zhou, W. (2018). Relationships between student engagement and academic achievement: A meta-analysis. *Social Behavior and Personality*, 46(3), 517–528. DOI: 10.2224/sbp.7054

Lemoine, P., & Guérin, S. (2016). Digital Credentials and the Future of Learning. *European Journal of Education*, 51(4), 569–582.

Lepage, A., & Roy, N. (2023). Une recension des écrits de 1970 à 2022 sur les rôles de l'enseignant et de l'intelligence artificielle dans le domaine de l'ia en éducation. *Médiations Et Médiatisations*, (16), 9–29. DOI: 10.52358/mm.vi16.304

Licoppe, C., Proulx, S., & Cudicio, R. (2010). L'émergence d'un nouveau genre communicationnel dans les organisations fortement connectées: les « questions rapides » par messagerie instantanée. Études De Communication, (34), 93-108. DOI: 10.4000/edc.1661

Li, F., Ruijs, R., & Lu, Y. (2023). Ethics & AI: A systematic review on ethical concerns and related strategies for designing with ai in healthcare. *AI*, 4(1), 28–53. DOI: 10.3390/ai4010003

Lin, C. H., Yu, C. C., Shih, P. K., & Wu, L. Y. (2021). STEM based artificial intelligence learning in general education for non-engineering undergraduate students. *Journal of Educational Technology & Society*, 24(3), 224–237. https://www.jstor.org/stable/27032867

Lin, C.-C., Huang, A. Y. Q., & Lu, O. H. T. (2023). Artificial intelligence in intelligent tutoring systems toward sustainable education: A systematic review. *Smart Learning Environments*, 10(1), 41. DOI: 10.1186/s40561-023-00260-y

Litjens, G., Kooi, T., Bejnordi, B. E., Setio, A. A. A., Ciompi, F., Ghafoorian, M., van der Laak, J. A. W. M., van Ginneken, B., & Sánchez, C. I. (2017). A survey on deep learning in medical image analysis. *Medical Image Analysis*, 42, 60–88. DOI: 10.1016/j.media.2017.07.005 PMID: 28778026

LIU, C. C., & GU, X. Q. (2024). Developing Frames for Change: The Impact Of Generative AI on the Broad Practices of Teacher Educators. Exploring New Horizons: Generative Artificial Intelligence and Teacher Education, 225.

Liu, Q., Tong, S., Liu, C., Zhao, H., Chen, E., Ma, H., . . . Wang, S. (2019). Exploiting cognitive structure for adaptive learning. Proceedings of the 25th ACM SIGKDD International Conference on Knowledge Discovery &Amp; Data Mining. DOI: 10.1145/3292500.3330922

Liu, Y., & Baucham, M. (2023). AI Technology: Key to Successful Assessment. In *Handbook of Research on Redesigning Teaching, Learning, and Assessment in the Digital Era* (pp. 304–325). IGI Global. DOI: 10.4018/978-1-6684-8292-6.ch016

Liu, Y., Du, Z., Hu, L., & Tong, Q. (2013). Integrating social question–answer sites in learning management system. Lecture Notes in Electrical Engineering, 695-703. DOI: 10.1007/978-94-007-7618-0_67

Liu, M., & Wang, Y. (2019). Personalized Learning in the Age of Adaptive Learning Technologies: Challenges and Opportunities. *Education and Information Technologies*, 24(3), 1681–1705.

Liu, Y.-L. E., Lee, T.-P., & Huang, Y.-M. (2023). Exploring students' continuance intention toward digital visual collaborative learning technology in design thinking. *International Journal of Human-Computer Interaction*, 175(1), 1–14.

Li, W., & Johnson, R. (2022). Optimizing learning pathways through AI-driven analysis in Learning Management Systems. *Computers & Education*, 179, 104406.

Liwang, A. W. D. (2024). Exploring the role of hrm in facilitating collaboration and knowledge sharing in decentralized organizational structures. [PRODUCTIVITY]. *Management Studies and Business Journal*, 1(3), 340–354. DOI: 10.62207/fw3tye84

Li, X. (2007). Intelligent agent-supported online education. *Decision Sciences Journal of Innovative Education*, 5(2), 311–331. DOI: 10.1111/j.1540-4609.2007.00143.x

Li, X., & Chen, W. (2021). Teachers' perspectives on the use of AI chatbots in online learning environments. *Journal of Educational Technology Development and Exchange*, 14(1), 1–16. DOI: 10.18785/jetde.1701.02

Li, X., & Hao, J. (2021). Adaptive assessment in online learning environments: A comparative study of traditional and AI-enhanced approaches. *Computers & Education*, 167, 104185.

Loch, K. D., Conger, S., & Oz, E. (1998).. . *Journal of Business Ethics*, 18(2), 199–209. DOI: 10.1023/A:1005720606763

Lodhi, P., Mishra, O., Jain, S., & Bajaj, V. (2018). StuA: An intelligent student assistant. *International Journal of Interactive Multimedia and Artificial Intelligence*, 5(2), 17–25. DOI: 10.9781/ijimai.2018.02.008

Łodzikowski, K., Foltz, P. W., & Behrens, J. T. (2024). Generative AI and Its Educational Implications. In Kourkoulou, D., Tzirides, A.-O., Cope, B., & Kalantzis, M. (Eds.), *Trust and Inclusion in AI-Mediated Education* (pp. 35–57). Springer Nature Switzerland., DOI: 10.1007/978-3-031-64487-0_2

Long, D., Roberts, J., Magerko, B., Holstein, K., DiPaola, D., & Martin, F. (2023, April). Ai literacy: Finding common threads between education, design, policy, and explainability. In Extended Abstracts of the 2023 CHI Conference on Human Factors in Computing Systems (pp. 1-6).

Long, D., & Magerko, B. (2020, April). What is AI literacy? Competencies and design considerations. In *Proceedings of the 2020 CHI Conference on Human Factors in Computing Systems* (pp. 1–16). New York, USA. DOI: 10.1145/3313831.3376727

Long, W., & Gao, Y. (2022). Artificial intelligence education system based on differential evolution algorithm to optimize SVM. *Scientific Programming*, 2022(1), 5379646. DOI: 10.1155/2022/5379646

Lu, X., & Zhai, X. (2019). Chatbots in the Classroom: A Review of Conversational Agents in Education.

Luckin, R., & Holmes, W. (2016). Intelligence unleashed: An argument for AI in education.

Luckin, R. (2017). *Machine Learning and Human Intelligence: The Future of Education for the Next Generation*. UCL Institute of Education Press.

Luckin, R., Holmes, W., Griffiths, M., & Forcier, L. B. (2016). *Intelligence unleashed: An argument for AI in education*. Pearson.

Luckin, R., Holmes, W., Griffiths, M., & Forcier, L. B. (2016). *Intelligence Unleashed: An argument for AI in education*. Pearson.

Lu, K., Pang, F., & Shadiev, R. (2023). Understanding college students' continuous usage intention of asynchronous online courses through extended technology acceptance model. *Education and Information Technologies*, 28(1), 1–19. DOI: 10.1007/s10639-023-11591-1 PMID: 36688218

Lv, Z. (2023). Generative artificial intelligence in the metaverse era. *Cognitive Robotics*, 3, 208–217. DOI: 10.1016/j.cogr.2023.06.001

Lynch, M. W., Dominelli, L., & Cuadra, C. (2023). Information communication technology during COVID-19. *Social Work Education*, 42(1), 1–13. DOI: 10.1080/02615479.2022.2040977

Macedo-Rouet, M. (2009). La visioconférence dans l'enseignement. ses usages et effets sur la distance de transaction. *Distances Et Savoirs*, 7(1), 65–91. DOI: 10.3166/ds.7.65-91

Macey-Dare, R. (2023). How ChatGPT and generative AI systems will revolutionize legal services and the legal profession. *SSRN*, 4366749, •••. https://papers.ssrn.com/sol3/papers.cfm?abstract_id=4366749. DOI: 10.2139/ssrn.4366749

Maclean, R., Watanabe, R., & Baker, R. Boediono, Cheng, Y. C., Duncan, W., Keeves, J., Mansheng, Z., Power, C., Rajput, J. S., Thaman, K. H., Alagumalai, S., Curtis, D. D., & Hungi, N. (Eds.). (2005). *Applied Rasch Measurement: A Book of Exemplars: Papers in Honour of John P. Keeves*. Springer Netherlands. DOI: 10.1007/1-4020-3076-2

MacQueen, K. M., Bhan, A., Fröhlich, J. A., Holzer, J., & Sugarman, J. (2015). Evaluating community engagement in global health research: The need for metrics. *BMC Medical Ethics*, 16(1), 44. Advance online publication. DOI: 10.1186/s12910-015-0033-9 PMID: 26126899

Ma, H., & Slater, T. (2015). Using the developmental path of cause to bridge the gap between AWE scores and writing teachers' evaluations. *Writing & Pedagogy*, 7(2), 395–422. DOI: 10.1558/wap.v7i2-3.26376

Maitri, A. (2019).*Rancang Bangun Chatbot sebagai Pusat Informasi Lembaga Kursus dan Pelatihan Menggunakan Pendekatan Natural Language Processing*.

Mallén, C., Andrea, M., & Mallén, V. (2024, 09 16). *L'e-learning propulsé par l'IA: les outils incontournables*. Retrieved from LEARNY BLOOM: https://learnybloom.com/blog/elearning-propulse-par-ia-outils-incontournables/

Malterud, K. (2012). Systematic text condensation: A strategy for qualitative analysis. *Scandinavian Journal of Public Health*, 40(8), 795–805. DOI: 10.1177/1403494812465030 PMID: 23221918

Mangal, S. K., & Mangal, U. (2019). Essentials of educational technology. PHI Learning Pvt. Ltd. https://books.google.com/books?hl=fr&lr=&id=Sc9upWs956cC&oi=fnd&pg=PR2&dq=education+technology+&ots=cPvedoKDz7&sig=qB-SRjIATUWxq36Ys-lHd3Ga24s

Marques, J. (2022). L'intelligence artificielle, une approche intersectionnelle. Interfaces Numériques, 11(1). https://doi.org/DOI: 10.25965/interfaces-numeriques.4796

Marras, M., Boratto, L., Ramos, G., & Fenu, G. (2022). Equality of learning opportunity via individual fairness in personalized recommendations. *International Journal of Artificial Intelligence in Education*, 32(3), 636–684. DOI: 10.1007/s40593-021-00271-1

Martínez-Plumed, F., Prudêncio, R. B. C., Martínez-Usó, A., & Hernández-Orallo, J. (2019). Item response theory in AI: Analysing machine learning classifiers at the instance level. *Artificial Intelligence*, 271, 18–42. DOI: 10.1016/j.artint.2018.09.004

Martínez-Soto, T., & Prendes-Espinosa, P. (2023). A systematic review on the role of ICT and CLIL in compulsory education. *Education Sciences*, 13(1), 73. DOI: 10.3390/educsci13010073

Martin, F., Chen, Y., Moore, R. L., & Westine, C. D. (2020). Systematic review of adaptive learning research designs, context, strategies, and technologies from 2009 to 2018. *Educational Technology Research and Development*, 68(4), 1903–1929. DOI: 10.1007/s11423-020-09793-2 PMID: 32837122

Mason, R. O. (1986). Four Ethical Issues of the Information Age. *Management Information Systems Quarterly*, 10(1), 5. DOI: 10.2307/248873

Matias, A., & Zipitria, I. (2023).*Promoting Ethical Uses in Artificial Intelligence Applied to Education*. DOI: 10.1007/978-3-031-32883-1_53

Matre, M. E., & Cameron, D. L. (2024). A scoping review on the use of speech-to-text technology for adolescents with learning difficulties in secondary education. *Disability and Rehabilitation. Assistive Technology*, 19(3), 1103–1116. DOI: 10.1080/17483107.2022.2149865 PMID: 36427182

Ma, W., Zhang, S., Cheng, M., & Liu, H. (2023). Configuring the professional touch in physical examinations in Chinese outpatient clinical interaction: Talk, touch, professional vision, and intersubjectivity. *Health Communication*, 38(1), 1–16. PMID: 36693813

McCoy, L. P. (2020). AI applications for accessibility in education. In Proctor, P. R. H. N. (Ed.), *Innovative Practices in Education* (pp. 58–75). Springer.

McDonald, N., & Pan, S. (2020). Intersectional AI: A study of how information science students think about ethics and their impact. *Proceedings of the ACM on Human-Computer Interaction*, 4(CSCW2), 1–19. DOI: 10.1145/3415218

McGrath, C., Pargman, T. C., Juth, N., & Palmgren, P. J. (2023). University teachers'perceptions of responsibility and artificial intelligence in higher education-Anexperimental philosophical study. Computers and Education: Artificial Intelligence, 4,Article 100139.

McNamara, D. S., Crossley, S. A., Roscoe, R. D., Allen, L. K., & Dai, J. (2015). A hierarchical classification approach to automated essay scoring. *Assessing Writing*, 23, 35–59. DOI: 10.1016/j.asw.2014.09.002

Meça, A., & Shkëlzeni, N. (2024). Academic Integrity in the Face of Generative Language Models. In Miraz, M. H., Southall, G., Ali, M., & Ware, A. (Eds.), *Emerging Technologies in Computing* (Vol. 538, pp. 58–70). Springer Nature Switzerland., DOI: 10.1007/978-3-031-50215-6_5

Mehrabi, N., Morstatter, F., Saxena, N., Lerman, K., & Galstyan, A. (2021). A survey on bias and fairness in machine learning. *ACM Computing Surveys*, 54(6), 1–35. DOI: 10.1145/3457607

Mello, R. F., Freitas, E., Pereira, F. D., Cabral, L., Tedesco, P., & Ramalho, G. (2023). Education in the age of Generative AI: Context and Recent Developments. arXiv preprint arXiv:2309.12332.

Merchant, Z., Goetz, E. T., Cifuentes, L., Keeney-Kennicutt, W., & Davis, T. J. (2014). Effectiveness of virtual reality-based instruction on students' learning outcomes in K-12 and higher education: A meta-analysis. *Computers & Education*, 70, 29–40. DOI: 10.1016/j.compedu.2013.07.033

Merriam, S. B., & Tisdell, E. J. (2015). *Qualitative Research: A Guide to Design and Implementation*. Wiley.

Merrill, M. D. (2002). First principles of instruction. *Educational Technology Research and Development*, 50(3), 43–59. DOI: 10.1007/BF02505024

Meskó, B., & Topol, E. J. (2023). The imperative for regulatory oversight of large language models (or generative AI) in healthcare. *NPJ Digital Medicine*, 6(1), 120. DOI: 10.1038/s41746-023-00873-0 PMID: 37414860

Messiou, K. (2018). Collaborative action research: Facilitating inclusion in schools. *EducationalActionResearch*, 27(2), 197–209. DOI: 10.1080/09650792.2018.1436081

Miao, F., Holmes, W., Huang, R., & Zhang, H. (2021). AI and education: Guidance for policy-makers. United Nations Educational, Scientific and Cultural Organization. https://unesdoc.unesco.org/ark:/48223/pf0000376709

Miao, Y., Koenig, R., & Knecht, K. (2020). The development of optimization methods in generative urban design: A review. Proceedings of the 11th Annual Symposium on Simulation for Architecture and Urban Design, 1–8. https://www.researchgate.net/profile/Reinhard-Koenig-2/publication/344460745_The_Development_of_Optimization_Methods_in_Generative_Urban_Design_A_Review/links/5f787c8d92851c14bcaa17f2/The-Development-of-Optimization-Methods-in-Generative-Urban-Design-A-Review.pdf

Milana, M., Brandi, U., Hodge, S., & Hoggan-Kloubert, T. (2024). Artificial intelligence (AI), conversational agents, and generative AI: Implications for adult education practice and research. *International Journal of Lifelong Education*, 43(1), 1–7. DOI: 10.1080/02601370.2024.2310448

Milgram, P., & Kishino, F. (1994). A Taxonomy of Mixed Reality Visual Displays. *IEICE Transactions on Information and Systems*, 77(12), 1321–1329.

Minn, S. (2022). AI-assisted knowledge assessment techniques for adaptive learning environments. *Computers and Education: Artificial Intelligence*, 3, 100050.

Mishra, N., De, A., Nayak, R. K., Tripathy, R., Motwani, A., Thakur, A., & Rout, S. K. (2025). AI Applications in Sports Medicine: Unraveling the Influence of Physical Education on Sports Injury Outcomes. In Global Innovations in Physical Education and Health (pp. 351-380). IGI Global.

Mishra, N., Habtemariam, G. M., Aebissa, B., Nayak, R. K., & Tripathy, R. (2024). Assessing the Effectiveness of Transnational Leadership on the Performance of Ethiopian University Graduates in Computing Technology: A Case Study. In Engaging Higher Education Teachers and Students With Transnational Leadership (pp. 238-256). IGI Global. DOI: 10.4018/979-8-3693-6100-9.ch013

Mishra, N., Habtemariam, G. M., Aebissa, B., Nayak, R. K., Tripathy, R., Thakur, A., & Rout, S. K. (2024). Case Study on the Performance Evaluation of Graduates in the Field of Artificial Intelligence for Ethiopian Universities. In Teaching and Assessment in the Era of Education 5.0 (pp. 93-110). IGI Global. DOI: 10.4018/979-8-3693-3045-6.ch006

Mishra, N., Habtemariam, G. M., Esseynew, S., Nayak, R. K., Tripathy, R., & Padhi, B. K. (2024). The Transnational ICT Leadership Assessment Based on the Available Infrastructure in the Governmental Organizations: A Case Study of the Ethiopian Northern Shewa Zone. In Developments and Future Trends in Transnational Higher Education Leadership (pp. 301-322). IGI Global. DOI: 10.4018/979-8-3693-2857-6.ch017

Mishra, N., Habal, B. G. M., Garcia, P. S., & Garcia, M. B. (2024, June). Harnessing an AI-Driven Analytics Model to Optimize Training and Treatment in Physical Education for Sports Injury Prevention. In *Proceedings of the 2024 8th International Conference on Education and Multimedia Technology* (pp. 309-315). DOI: 10.1145/3678726.3678740

Mishra, N., Habtemariam, G. M., & De, A. (2023). Investigation of High-Performance Computing Tools for Higher Education Institutions Using the IoE Grid Computing Framework. In *Internet of Behaviors Implementation in Organizational Contexts* (pp. 217–241). IGI Global. DOI: 10.4018/978-1-6684-9039-6.ch011

Mishra, N., Nayak, R. K., Habtemariam, G. M., Tripathy, R., Esseynew, S., Motwani, A., & Rout, S. K. (2024). An In-Depth Examination of the Application Gap and Infrastructure Landscape of Information Communication Technology in Ethiopian Public Organizations: A Bridging the Divide Case. In *Impact of Corporate Social Responsibility on Employee Wellbeing* (pp. 141–166). IGI Global. DOI: 10.4018/979-8-3693-3470-6.ch007

Mitchell, T. M. (1997). *Machine Learning*. McGraw-Hill.

Mitrovic, A., & Ohlsson, S. (2019). The role of intelligent tutoring systems in education. Dans *Proceedings of the International Conference on Artificial Intelligence in Education* (pp. 16-28). Springer.

Mittelstadt, B. D. (2019). Principles alone cannot guarantee ethical AI. *Nature Machine Intelligence*, 1(11), 501–507. DOI: 10.1038/s42256-019-0114-4

Mizumoto, A., & Eguchi, M. (2023). Exploring the potential of using an AI language model for automated essay scoring. *Research Methods in Applied Linguistics*, 2(2), 100050. DOI: 10.1016/j.rmal.2023.100050

Mohammadkarimi, E. (2023). Teachers' reflections on academic dishonesty in EFL students' writings in the era of artificial intelligence. Journal of Applied Learning and Teaching, 6(2). DeFranco, J. F., & Biersmith, L. (2024). Assessing the State of AI Policy. arXiv preprint arXiv:2407.21717.

Mok, K. H. (2020). Digital Credentials and Blockchain: Implications for Higher Education. *Higher Education Research & Development*, 39(3), 548–561.

Mok, K. H., & Zhang, H. (2018). The Impact of Digital Credentials on the Labor Market. *International Journal of Human Resource Management*, 29(18), 2656–2672.

Montminy, N., Duval, S., & Bouchard, C. (2020). Le lien entre les habiletés d'autorégulation observées chez l'enfant âgé de 5 ans et la qualité des interactions en classe d'éducation préscolaire. *Neuroeducation*, 6(1), 6–23. DOI: 10.24046/neuroed.20200601.6

Morales-Chan, M., Amado-Salvatierra, H. R., & Hernandez-Rizzardini, R. (2024). AI-Driven Content Creation: Revolutionizing Educational Materials. Proceedings of the Eleventh ACM Conference on Learning @ Scale, 556–558. DOI: 10.1145/3657604.3664640

Moriña, A., Sandoval, M., & Carnerero, F. (2020). Higher education inclusivity: When the disability enriches the university. *Higher Education Research & Development*, 39(6), 1202–1216. DOI: 10.1080/07294360.2020.1712676

Moses, T. (2017). A Review of Developments and Applications in Item Analysis. In Bennett, R. E., & von Davier, M. (Eds.), *Advancing Human Assessment: The Methodological, Psychological and Policy Contributions of ETS* (pp. 19–46). Springer International Publishing., DOI: 10.1007/978-3-319-58689-2_2

Mouboua, P. D., & Atobatele, F. A. (2024). Multilingualism and socioeconomic mobility: Analyzing the correlation in immigrant populations. *World Journal of Advanced Research and Reviews*, 22(2), 144–156. DOI: 10.30574/wjarr.2024.22.2.1314

Mouni, M., & Boumalek, A. (2021). Les interactions en ligne et la construction du savoir dans les dispositifs de formation complètement à distance: quels rapports? ITM Web of Conferences, 39, 03007. DOI: 10.1051/itmconf/20213903007

Mousavinasab, E., Zarifsanaiey, N., Kalhori, S. R. N., Rakhshan, M., Keikha, L., & Saeedi, M. G. (2021). Intelligent tutoring sys tems: A systematic review of characteristics, applications, and evaluation methods. *Interactive Learning Environments*, 29(1), 142–163. DOI: 10.1080/10494820.2018.1558257

Mudawy, A. M. A. (2024). Investigating efl faculty members' perceptions of integrating artificial intelligence applications to improve the research writing process: a case study at majmaah university. DOI: 10.31235/osf.io/k4ub8

Muhammad, F., & Orji, C. M. (2024). REVOLUTIONIZING EDUCATION IN THE DIGITAL ERA: THE ROLE OF AI IN PROMOTING INCLUSIVITY, EQUALITY, AND ETHICAL INNOVATION. *The American Journal of Social Science and Education Innovations*, 6(09), 151–157.

Mujiono, M. (2023). *Educational Collaboration : Teachers and Artificial Intelligence*. Jurnal Kependidikan., DOI: 10.33394/jk.v9i2.7801

Mulianingsih, F., Anwar, K., Shintasiwi, F. A., & Rahma, A. J. (2020). Artificial Intellegence dengan pembentukan nilai dan karakter di bidang pendidikan. *IJTIMAIYA: Journal of Social Science Teaching*, 4(2), 148–154.

Murphy, K. P. (2012). *Machine Learning: A Probabilistic Perspective*. MIT Press.

Museus, S. D., Yi, V., & Saelua, N. (2017). The impact of culturally engaging campus environments on sense of belonging. *Review of Higher Education*, 40(2), 187–215. DOI: 10.1353/rhe.2017.0001

Nakamoto, S. (2008). Bitcoin: A Peer-to-Peer Electronic Cash System.

Namjoo, F., Liaghat, E., Shabaziasl, S., Modabernejad, Z., & Morshedi, H. (2023). Students experience on self-study through ai. *AI and Tech in Behavioral and Social Sciences*, 1(3), 35–42. DOI: 10.61838/kman.aitech.1.3.6

Narciss, S., Sosnovsky, S., Schnaubert, L., Andrès, E., Eichelmann, A., Goguadze, G., & Melis, E. (2014). Exploring feedback and student characteristics relevant for personalizing feedback strategies. *Computers & Education*, 71, 56–76. DOI: 10.1016/j.compedu.2013.09.011

National Center on Accessible Educational Materials. (2022). The role of technology in accessibility. https://aem.cast.org

Neumann, W. L. (2007). *Social research methods: Qualitative and quantitative approaches*. Pearson.

Newman, P. A., Rubincam, C., Slack, C., Essack, Z., Chakrapani, V., Chuang, D. M., Tepjan, S., Shunmugam, M., Roungprakhon, S., Logie, C., Koen, J., & Lindegger, G. (2015). Towards a science of community stakeholder engagement in biomedical hiv prevention trials: An embedded four-country case study. *PLoS One*, 10(8), e0135937. DOI: 10.1371/journal.pone.0135937 PMID: 26295159

Newton, P. E. (2007). Clarifying the purposes of educational assessment. *Assessment in Education: Principles, Policy & Practice*, 14(2), 149–170. DOI: 10.1080/09695940701478321

Ng, D. T. K., Leung, J. K. L., Su, J., Ng, R. C. W., & Chu, S. K. W. (2023). Teachers' AI digital competencies and twenty-first century skills in the post-pandemic world. *Educational Technology Research and Development*, 71(1), 137–161. DOI: 10.1007/s11423-023-10203-6 PMID: 36844361

Ng, H. T., Wu, S. M., Briscoe, T., Hadiwinoto, C., Susanto, R. H., & Bryant, C. (2020). The CoNLL-2014 shared task on grammatical error correction. In *Proceedings of the Eighteenth Conference on Computational Natural Language Learning: Shared Task* (pp. 1-14).

Nguyen, A., Ngo, H., Hong, Y., Dang, B., & Nguyen, B.-P. (2022). Ethical principles for artificial intelligence in education. *Education and Information Technologies*. Advance online publication. DOI: 10.1007/s10639-022-11316-w PMID: 36254344

Nguyen, J., Sánchez-Hernández, G., Armisen, A., Agell, N., Rovira, X., & Angulo, C. (2018). A linguistic multi-criteria decision-aiding system to support university career services. *Applied Soft Computing*, 67, 933–940. DOI: 10.1016/j.asoc.2017.06.052

Nguyen, M. N. H., Pandey, S. R., Dang, T. N., Huh, E., Tran, N. H., Saad, W., & Hong, C. S. (2023). Self-organizing democratized learning: Toward large-scale distributed learning systems. *IEEE Transactions on Neural Networks and Learning Systems*, 34(12), 10698–10710. DOI: 10.1109/TNNLS.2022.3170872 PMID: 35536803

Nicholas, D., Watkinson, A., Jamali, H. R., Herman, E., Tenopir, C., Volentine, R., Allard, S., & Levine, K. (2015). Peer review: Still king in the digital age. *Learned Publishing*, 28(1), 15–21. DOI: 10.1087/20150104

Nicol, D. J., & Macfarlane-Dick, D. (2006). Formative Assessment and Self-Regulated Learning: A Model and Seven Principles of Good Feedback Practice. *Studies in Higher Education*, 31(2), 199–218. DOI: 10.1080/03075070600572090

Niedercorn, F. (2017, Août 29). *2011 Avec Watson, IBM remporte « Jeopardy! »*. Retrieved from Les Echos: https://www.lesechos.fr/2017/08/2011-avec-watson-ibm-remporte-jeopardy-1116704

Nikolopoulou, K. (2024). Generative artificial intelligence in higher education: Exploring ways of harnessing pedagogical practices with the assistance of chatgpt. *International Journal of Changes in Education*, 1(2), 103–111. DOI: 10.47852/bonviewIJCE42022489

Nikou, S. A. (2021). Web-based videoconferencing for teaching online: Continuance intention to use in the post-covid-19 period. [s]. *Interaction Design and Architecture*, (47), 123–143. DOI: 10.55612/s-5002-047-006

Nilsson, N. J., Hilpisch, Y., Yao, M., Zhou, A., Jia, M., Baesen, B., . . . Verbeke, W. (2010). The quest for ai: A history of ideas and achievements. Erişim adresi: http:// ai. standford. edu/~ nilsson/(Özgün eser 2009 tarihlidir).

Nithithanatchinnapat, B., Maurer, J., Deng, X., & Joshi, K. D. (2024). Future Business Workforce: Crafting a Generative AI-Centric Curriculum Today for Tomorrow's Business Education. *The Data Base for Advances in Information Systems*, 55(1), 6–11. DOI: 10.1145/3645057.3645059

Nkambou, R., Bourdeau, J., & Mizoguchi, R. (2010). *Advances in Intelligent Tutoring Systems*. Springer. DOI: 10.1007/978-3-642-14363-2

Nkhoma, C., Thomas, S., Nkhoma, M., Sriratanaviriyakul, N., Truong, T. H., & Vo, H. X. (2018). Measuring the impact of out-of-class communication through instant messaging. *Education + Training*, 60(4), 318–334. DOI: 10.1108/ET-12-2017-0196

Noble, S. U. (2018). *Algorithms of oppression: How search engines reinforce racism*. NYU Press. DOI: 10.18574/nyu/9781479833641.001.0001

Nyaaba, M., Shi, L., Nabang, M., Zhai, X., Kyeremeh, P., Ayoberd, S. A., & Akanzire, B. N. (2024). Generative AI as a Learning Buddy and Teaching Assistant: Pre-service Teachers' Uses and Attitudes (arXiv:2407.11983). arXiv. http://arxiv .org/abs/2407.11983

Nye, B. D. (2015). The Effects of Adaptive Learning Technologies on Student Outcomes: A Meta-Analysis. *Journal of Educational Psychology*, 107(1), 162–177.

O'Neil, C. (2016). *Weapons of math destruction: How big data increases inequality and threatens democracy*. Crown Publishing.

O'Neil, C. (2016). *Weapons of Math Destruction: How Big Data Increases Inequality and Threatens Democracy*. Crown Publishing.

OECD. (2002). Understanding the Brain: Towards a New Learning Science. OECD, Pari.

OECD. (2003). Learners for Life: Student Approaches to Learning: Results from PISA 2000.

OECD. (2020). *Improving Governance with Policy Evaluation: Lessons From Country Experiences*. Organisation for Economic Co-operation and Development. https://www.oecd-ilibrary.org/governance/improving-governance-with-policy -evaluation_89b1577d-en

Ofosu-Ampong, K. (2020). The Shift to Gamification in Education: A Review on Dominant Issues. *Journal of Educational Technology Systems*, 49(1), 113–137. DOI: 10.1177/0047239520917629

Oktarin, I. B., Estisari, K., Wengrum, T. D., & Nurhantanto, A. (2023). Economic students' perceptions towards learning management system (lms). *Journal of Arts and Education*, 3(1). Advance online publication. DOI: 10.33365/jae.v3i1.198

Oliveira, W., Hamari, J., Shi, L., Toda, A. M., Rodrigues, L., Palomino, P. T., & Isotani, S. (2023). Tailored gamification in education: A literature review and future agenda. *Education and Information Technologies*, 28(1), 373–406. DOI: 10.1007/s10639-022-11122-4

Olusegun, J. (2024). Generative AI Solutions for Expanding Access to Higher Education in Developing Regions. https://www.researchgate.net/profile/John-Olusegun-2/publication/384081407_Generative_AI_Solutions_for_Expanding_Access_to_Higher_Education_in_Developing_Regions/links/66ea0191dde50b32587a643e/Generative-AI-Solutions-for-Expanding-Access-to-Higher-Education-in-Developing-Regions.pdf

Organization for Economic Co-operation and Development. (2021). OECD RecommendationoftheCouncilonArtificialIntelligence.OECD/LEGAL/0449.https://legalinstruments.oecd.org/en/instruments/OECD-LEGAL-0449

Ouariach, F. Z., El Abidine, M. Z., Nejjari, A., & Khaldi, M. (2023). From classic communication to online communication: Identifying online communication tools. Global Journal of Engineering and Technology Advances, 17(2), 040-052.

Ouariach, F. Z., Nejjari, A., Ouariach, S., & Khaldi, M. (2024). Place of forums in online communication through an LMS platform. World Journal of Advanced Engineering Technology and Sciences, 11(1), 096-104.

Ouariach, S., & Khaldi, M. (2024). Scripting Tools and the Design of E-Learning Experiences. In Innovative Instructional Design Methods and Tools for Improved Teaching (pp. 60-91). IGI Global.

Ouariach, S., Khaldi, M., Mohamed, E., & Khaldi, M. (2023). The Flipped Classroom: From Passive Information Absorption to Active Learning. In Handbook of Research on Language Teacher Identity (pp. 269-293). IGI Global.

Ouariach, S., & Khaldi, M. (2024). Scripting Tools and the Design of E-Learning Experiences. In *Innovative Instructional Design Methods and Tools for Improved Teaching* (pp. 60–91). IGI Global.

Oudeyer, P. Y., Clément, B., Roy, D., & Sauzéon, H. (2020). Projet KidLearn: Vers une personnalisation motivante des parcours d'apprentissage. Bulletin de l'Association Française pour l'Intelligence Artificielle, 51-55.

Ouguengay, Y. A., El Faddouli, N. E., & Bennani, S. (2015). A neuro-fuzzy inference system for the evaluation of reading/writing competencies acquisition in an e-learning environnement. *Journal of Theoretical and Applied Information Technology*, 81(3), 600.

Ouyang, F., & Jiao, P. (2021). Artificial intelligence in education: The three paradigms. *Computers and Education: Artificial Intelligence*, 2, 100020.

Ouyang, F., Wu, M., Zheng, L., Zhang, L., & Jiao, P. (2023). Integration of artificial intelligence performance prediction and learning analytics to improve student learning in online engineering course. *International Journal of Educational Technology in Higher Education*, 20(1), 4. Advance online publication. DOI: 10.1186/s41239-022-00372-4 PMID: 36683653

Ouyang, F., Zheng, L., & Jiao, P. (2022). Artificial intelligence in online higher education: A systematic review of empirical research from 2011–2020. *Education and Information Technologies*, 27(6), 7893–7925. DOI: 10.1007/s10639-022-10925-9

Oyelere, S. S., Bouali, N., Kaliisa, R., Obaido, G., Yunusa, A. A., & Jimoh, E. R. (2020). Exploring the trends of educational virtual reality games: A systematic review of empirical studies. *Smart Learn. Environ.*, 7(1), 31. DOI: 10.1186/s40561-020-00142-7

Oztekin, A. (2016). A hybrid data analytic approach to predict college graduation status and its determinative factors. *Industrial Management & Data Systems*, 116(8), 1678–1699. DOI: 10.1108/IMDS-09-2015-0363

Page, L. C., & Gehlbach, H. (2017). How an artificially intelligent virtual assistant helps students navigate the road to college. *AERA Open*, 3(4), 2332858417749220. DOI: 10.1177/2332858417749220

Palanisamy, P., Thilarajah, S., & Chen, Z. (2021). Providing equitable education through personalised adaptive learning. ASCILITE 2021: Back to the Future – ASCILITE '21 Proceedings ASCILITE 2021 in Armidale. DOI: 10.14742/ascilite2021.0129

Pantelidis, V. S. (2010). Reasons to Use Virtual Reality in Education and Training Courses and a Model to Determine When to Use Virtual Reality. *Themes in Science and Technology Education*, 2(1-2), 59–70.

Papamitsiou, Z., Pappas, I. O., Sharma, K., & Giannakos, M. N. (2020). Utilizing multimodal data through fsqca to explain engagement in adaptive learning. *IEEE Transactions on Learning Technologies*, 13(4), 689–703. DOI: 10.1109/TLT.2020.3020499

Papi, C. (2014). *Formation à distance: dispositifs et interactions*. ISTE Group.

Pappas, C. (2015). The Rise of Digital Credentials. eLearning Industry.

Paquette, G., Mariño, O., & Bejaoui, R. (2021). A new competency ontology for learning environments personalization. *Smart Learning Environments*, 8(1), 16. Advance online publication. DOI: 10.1186/s40561-021-00160-z

Paquette, L., Lebeau, J. F., Beaulieu, G., & Mayers, A. (2015). Designing a knowledge representation approach for the generation of pedagogical interventions by MTTs. *International Journal of Artificial Intelligence in Education*, 25(1), 118–156. https://www.learntechlib.org/p/168275/. DOI: 10.1007/s40593-014-0030-z

Parent, S., & Baron, M. (2021). Emergence of learning analytics in education: Challenges and issues of learning analysis. *Canadian Journal of Learning and Technology*, 47(4). Advance online publication. DOI: 10.21432/cjlt28053

Parent, S., & Deschênes, M. (2022). Répercussions du contexte de pandémie sur la variation de l'engagement de membres du personnel enseignant. *Médiations Et Médiatisations*, (12), 114–132. DOI: 10.52358/mm.vi12.290

Parker, R. E. (Ed.). (2013). *Redesigning courses for online delivery: design, interaction, media & evaluation*. Emerald Group Publishing.

Parsakia, K. (2023). The effect of chatbots and ai on the self-efficacy, self-esteem, problem-solving and critical thinking of students. *Health Nexus*, 1(1), 71–76. DOI: 10.61838/hn.1.1.14

Pataranutaporn, P., Danry, V., Leong, J., Punpongsanon, P., Novy, D., Maes, P., & Sra, M. (2021). AI-generated characters for supporting personalized learning and well-being. *Nature Machine Intelligence*, 3(12), 1013–1022. DOI: 10.1038/s42256-021-00417-9

Patterson, K. D. W., Newman, C., & Doona, K. (2016). Improving the care of older persons in australian prisons using the policy delphi method. *Dementia (London)*, 15(5), 1219–1233. DOI: 10.1177/1471301214557531 PMID: 25394555

Pedró, F. (2020). Applications of artificial intelligence to higher education: Possibilities, evidence, and challenges. *IUL Research*, 1(1), 61–76. DOI: 10.57568/iulres.v1i1.43

Pedró, F., Subosa, M., Rivas, A., & Valverde, P. (2019). *Artificial intelligence in education: Challenges and opportunities for sustainable development.* UNESCO.

Peng, S., Kalliamvakou, E., Cihon, P., & Demirer, M. (2023). The Impact of AI on Developer Productivity: Evidence from GitHub Copilot (arXiv:2302.06590). arXiv. https://doi.org//arXiv.2302.06590DOI: 10.48550

Peng, C., Zhou, X., & Liu, S. (2022). An Introduction to Artificial Intelligence and Machine Learning for Online Education. *Mobile Networks and Applications*, 27(3), 1147–1150. DOI: 10.1007/s11036-022-01953-3

Pennington, J., Socher, R., & Manning, C. (2014). Glove: Global Vectors for Word Representation. *Proceedings of the 2014 Conference on Empirical Methods in Natural Language Processing (EMNLP)*, 1532–1543. DOI: 10.3115/v1/D14-1162

Pennington, R., & O'Connor, M. (2018). Assistive technologies in education: The role of AI in enhancing learning experiences. Dans *Proceedings of the International Symposium on Computer Science and Education* (pp. 193-198). IEEE.

Peraya, D. (1999). Médiation et médiatisation: Le campus virtuel. *Hermes*, (3), 153–167.

Pereira, L. A., Bastos, M. C., Dermeval, D., Paiva, R., Coelho, J. A. P. M., & Lima, D. (2019). Evaluation of an Intelligent Tutoring System Prototype for the Teaching of Electrocardiogram. in *Proceedings of the IEEE 19th International Conference on Advanced Learning Technologies (ICALT)* (pp 327–328). DOI: 10.1109/ICALT.2019.00104

Peslak, A. R. (2006). Papa Revisited: A Current Empirical Study of the Mason Framework. *Journal of Computer Information Systems*, 46(3), 117–123. DOI: 10.1080/08874417.2006.11645905

Pesovski, I., Santos, R., Henriques, R., & Trajkovik, V. (2024). Generative AI for Customizable Learning Experiences. *Sustainability (Basel)*, 16(7), 7. Advance online publication. DOI: 10.3390/su16073034

Pessach, D., & Shmueli, E. (2023). Algorithmic fairness. In *Machine Learning for Data Science Handbook: Data Mining and Knowledge Discovery Handbook* (pp. 867–886). Springer International Publishing. DOI: 10.1007/978-3-031-24628-9_37

Petiot, O. (2023). Quels sont les facteurs de l'engagement des apprenants en EPS? Une revue de la littérature portant sur l'expérience des enseignants et des apprenants. eJRIEPS. Ejournal de la recherche sur l'intervention en éducation physique et sport, (52), 3-29.

Petit, M., Lameul, G., & Taschereau, J. (2020). La téléprésence en formation. *Médiations Et Médiatisations*, (3), 7–20. DOI: 10.52358/mm.vi3.112

Piaget, J. (1970). *The Construction of the Real in the Child.* Delachaux and Niestlé.

Piaget, J. (1976a). Cognitive Development in Children: Piaget Development and Learning. *Journal of Research in Science Teaching*, 2(3), 176–186. DOI: 10.1002/tea.3660020306

Piaget, J. (1976b). *The Grasp of Consciousness: Action and Concept in the Young Child.* Harvard University Press.

Picciano, A. G. (2012). The evolution of big data and learning analytics in American higher education. *Online Learning : the Official Journal of the Online Learning Consortium*, 16(3), 9–20. DOI: 10.24059/olj.v16i3.267

Pickover, C. A. (2021). ELIZA le psychothérapeute. In C.A.Pickover, *La fabuleuse histoire de l'intelligence artificielle* (pp. 154-156). DUNOD.

Pickover, C. A. (2021). La conférence de Dartmouth sur l'IA. In C.A.Pickover, *La fabuleuse histoire de l'intelligence artificielle* (pp. 140-141). DUNOD.

Pierrès, O., Christen, M., Schmitt-Koopmann, F., & Darvishy, A. (2024). Could the Use of AI in Higher Education Hinder Students With Disabilities? A Scoping Review. IEEE Access. https://ieeexplore.ieee.org/abstract/document/10433192/

Pilkington, M. (2016). Blockchain Technology: Principles and Applications. In Research Handbook on Digital Transformations (pp. 225-253). Edward Elgar Publishing.

Pirot, L., & De Ketele, J. M. (2000). L'engagement académique de l'apprenant comme facteur de réussite à l'université Étude exploratoire menée dans deux facultés contrastées. *Revue des Sciences de l'Education*, 26(2), 367–394. DOI: 10.7202/000127ar

Piskopani, A.-M., Chamberlain, A., & Holter, C. T. (2023, July 11).*Responsible AI and the Arts: The Ethical and Legal Implications of AI in the Arts and Creative Industries*. DOI: 10.1145/3597512.3597528

Pitrat, J. (1998). Métacognition-Intelligence Artificielle. Vocabulaire des Sciences Cognitives, 262-264.

Pizzol, B., Giannotti, M., & Tomasiello, D. B. (2021). Qualifying accessibility to education to investigate spatial equity. *Journal of Transport Geography*, 96, 103199. DOI: 10.1016/j.jtrangeo.2021.103199

Planche, M. (2024, 08 29). *Synthèse Vocale: 5 Meilleurs logiciels de voix off IA en2024.* Retrieved from Vidnoz: https://fr.vidnoz.com/intelligence-artificielle/logiciel-voix-off-ia.html

Plata, S., De Guzman, M. A., & Quesada, A. (2023). Emerging research and policy themes on academic integrity in the age of chat GPT and generative AI. *Asian Journal of University Education*, 19(4), 743–758. DOI: 10.24191/ajue.v19i4.24697

Popenici, S. A. D., & Kerr, S. (2017). Exploring the Impact of Artificial Intelligence on Teaching and Learning in Higher Education. *Research and Practice in Technology Enhanced Learning*, 12(1), 22. DOI: 10.1186/s41039-017-0062-8 PMID: 30595727

Pouliot, È., & Rocheleau, J. (2020). Enseigner le travail social en contexte de pandémie: Un exemple d'environnement d'apprentissage hybride en ligne sur moodle. *Revue Internationale Des Technologies en Pédagogie Universitaire*, 17(2), 44–50. DOI: 10.18162/ritpu-2020-v17n2-05

Prabhakaran, V., Qadri, R., & Hutchinson, B. (2022). Cultural incongruencies in artificial intelligence. arXiv preprint arXiv:2211.13069.

Pradeau, A., & Dessus, P. (2013). Genèse d'une communauté de pratique d'étudiants issue d'un forum de discussion: Participation, conscience d'autrui et engagement mutuel. *Revue Internationale Des Technologies en Pédagogie Universitaire*, 10(2), 38. DOI: 10.7202/1035521ar

Pratama, M. P., Sampelolo, R., & Lura, H. (2023). Revolutionizing education: harnessing the power of artificial intelligence for personalized learning. Klasikal: Journal of Education. *Language Teaching and Science*, 5(2), 350–357. DOI: 10.52208/klasikal.v5i2.877

Prinz, A., Hruska, N., Burger, K., & Whitney, M. (2018). Jupyter Notebook for Education and Training: A Case Study. *Journal of Computational Science Education*, 9(1), 46–55.

Prior, M., Rogerson, S., & Fairweather, B. (2002). The ethical attitudes of information systems professionals: Outcomes of an initial survey. *Telematics and Informatics*, 19(1), 21–36. DOI: 10.1016/S0736-5853(00)00014-9

Procter, A. (2024, 02 22). *Technologies et innovation. Comment les modèles d'IA spécialisés évoluent pour le mieux.* Retrieved from spark: https://www.okoone.com/fr/spark/technologies-et-innovation/comment-les-modeles-dia-specialises-evoluent-pour-le-mieux/

Psico-smart, É. é. (2024, août 28). *L'impact de l'intelligence artificielle sur la formation des compétences numériques.* Retrieved from Vorecol: https://psico-smart.com/fr/blogs/blog-limpact-de-lintelligence-artificielle-sur-la-formation-des-competences-numeriques-145324

Pugalenthi, R., Chakkaravarthy, A. P., Ramya, J., Babu, S., & Krishnan, R. R. (2021). Artificial learning companionusing machine learning and natural language processing. *International Journal of Speech Technology*, 24(3), 553–560. Advance online publication. DOI: 10.1007/s10772-020-09773-0

Qadir, J. (2023). Engineering education in the era of ChatGPT: Promise and pitfalls of generative AI for education. 2023 IEEE Global Engineering Education Conference (EDUCON), 1–9. https://ieeexplore.ieee.org/abstract/document/10125121/

Qian, C. (2023). Research on human-centered design in college music education to improve student experience of artificial intelligence-based information systems. *Journal of Information Systems Engineering & Management*, 8(3), 23761. DOI: 10.55267/iadt.07.13854

Qin, F., Li, K., & Yan, J. (2020). Understanding user trust in artificial intelligence-based educational systems: Evidence from China. *British Journal of Educational Technology*, 51(5), 1693–1710. DOI: 10.1111/bjet.12994

Quixal, M., & Meurers, D. (2016). How can writing tasks be characterized in a way serving pedagogical goals and automatic analysis needs? *CALICO Journal*, 33(1), 19–48. DOI: 10.1558/cj.v33i1.26543

Radianti, J., Majchrzak, T. A., Fromm, J., & Wohlgenannt, I. (2020). A Systematic Review of Immersive Virtual Reality Applications for Higher Education: Design Elements, Lessons Learned, and Research Agenda. *Computers & Education*, 147, 103778. DOI: 10.1016/j.compedu.2019.103778

Raghavan, M., Barocas, S., Kleinberg, J., & Levy, K. (2020). Mitigating bias in algorithmic hiring: Evaluating claims and practices. Proceedings of the 2020 Conference on Fairness, Accountability, and Transparency, 469 481. DOI: 10.1145/3351095.3372828

Rahwan, I., Cebrian, M., & Dafoe, A. (2021). Machine Behavior: A New Approach to AI Education. *Nature*, 568(7753), 477–486. DOI: 10.1038/s41586-019-1138-y PMID: 31019318

Rajamoni, R. N., Kumar, R. M. S., & Leela, B. C. (2022). Factors affecting the academic performance of students with hearing impairment. *Revue d'Intelligence Artificielle*, 36(4), 569–574. DOI: 10.18280/ria.360408

Rajendran, R. M. (2023). Importance Of Using Generative AI In Education: Dawn of a New Era. *Journal of Science and Technology*, 4(6), 35–44. DOI: 10.55662/JST.2023.4603

Rajeswari, P., & Purushothaman, S. (2023). Artificial Intelligence in Education. *Edu Spectra*, 5, 47.

Raji, I. D., Bender, E. M., Paullada, A., Denton, E., & Hanna, A. (2020). AI and the everything in the whole wide world benchmark: Accountability in data collection in the age of machine learning. arXiv preprint arXiv:2012.04580.

Raju, D., & Schumacker, R. (2015). Exploring student characteristics of retention that lead to graduation in higher education using data mining models. *Journal of College Student Retention*, 16(4), 563–591. DOI: 10.2190/CS.16.4.e

Ramakrishnan, T., Mohan Gift, M. D., Chitradevi, S., Jegan, R., Hency Jose, P. S., Nagaraja, H. N., & Hailegiorgis, S. M. (2022). Study of numerous resins used in polymer matrix composite materials. *Advances in Materials Science and Engineering*, 2022(1), 1088926.

Rane, J., Kaya, O., Mallick, S. K., & Rane, N. L. (2024). Impact of ChatGPT and similar generative artificial intelligence on several business sectors: Applications, opportunities, challenges, and future prospects. Generative Artificial Intelligence in Agriculture, Education, and Business, 27–83.

Rane, N. (2024). Enhancing the quality of teaching and learning through Gemini, ChatGPT, and similar generative Artificial Intelligence: Challenges, future prospects, and ethical considerations in education. *TESOL and Technology Studies*, 5(1), 1–6. DOI: 10.48185/tts.v5i1.1000

Rane, N. L., Choudhary, S. P., Tawde, A., & Rane, J. (2023). ChatGPT is not capable of serving as an author: Ethical concerns and challenges of large language models in education. *International Research Journal of Modernization in Engineering Technology and Science*, 5(10), 851–874.

Rawas, S., & AlSaeed, D. (2024). ChatGPT: Innovating Lifelong Learning in Higher Education Through Artificial Intelligence and Digital Transformation. In Lytras, M. D., Serban, A. C., Alkhaldi, A., Malik, S., & Aldosemani, T. (Eds.), *Digital Transformation in Higher Education, Part B* (pp. 13–28). Emerald Publishing Limited., DOI: 10.1108/978-1-83608-424-220241002

Ray, S. (2023, February 22). JPMorgan Chase restricts staffers' use of ChatGPT. Forbes. https:// www.forbes.com/sites/siladityaray/2023/02/22/jpmorgan-chase-restricts-staffers-use-ofchatgpt/?sh=714e937d6bc7

Regan, P. M., & Jesse, J. (2018). Ethical challenges of edtech, big data and personalized learning: Twenty-first century student sorting and tracking. *Ethics and Information Technology*, 21(3), 167–179. DOI: 10.1007/s10676-018-9492-2

Reiss, M. J. (2021). The use of AI in education: Practicalities and ethical considerations. *London Review of Education*, 19(1). Advance online publication. DOI: 10.14324/LRE.19.1.05

Reschly, A. L., & Christenson, S. L. (2012). Jingle, jangle and conceptual haziness: Evolution and future directions of the engagement construct. In Christenson, S. L., & Reschly, A. L. (Eds.), *C*.

Rhode, J., Richter, S., Gowen, P., Miller, T., & Wills, C. (2017). Understanding faculty use of the learning management system. *Online Learning : the Official Journal of the Online Learning Consortium*, 21(3). Advance online publication. DOI: 10.24059/olj.v21i3.1217

Ricci, F., Rokach, L., & Shapira, B. (2011). *Introduction to recommender systems handbook*. Springer. DOI: 10.1007/978-0-387-85820-3

Richard, É. (2023). Difficultés rencontrées par les étudiants adultes au collégial. *Canadian Journal of Higher Education*, 53(2), 15–31. DOI: 10.47678/cjhe.v53i2.190061

Richardson, C., Nucci, R., & Sevilla, D. (2024, 09 3). *SCORM: Ce qu'est SCORM et comment cela fonctionne*. Retrieved from GURU: https://www.getguru.com/fr/reference/scorm

Rivières-Pigeon, C., Courcy, I., & Poirier, N. (2013). Contenu et utilité d'un forum de discussion sur internet destiné aux parents d'enfants autistes. Enfances, Familles. *Generations (San Francisco, Calif.)*, (17), 119–137. DOI: 10.7202/1013418ar

Rizzo, A. S., & Koenig, S. T. (2017). Is Clinical Virtual Reality Ready for Primetime? *Neuropsychology*, 31(8), 875–892. DOI: 10.1037/neu0000405 PMID: 29376669

Roksa, J., & Kinsley, P. (2019). The role of family support in facilitating academic success of low-income students. *Research in Higher Education*, 60(4), 415–436. DOI: 10.1007/s11162-018-9517-z

Roll, I., & Wylie, R. (2016). Evolution and Revolution in Artificial Intelligence in Education. *International Journal of Artificial Intelligence in Education*, 26(2), 582–599. DOI: 10.1007/s40593-016-0110-3

Romero, M., & Heiser, L. (2023). Enseigner et apprendre à l'ère de l'intelligence artificielle. Canopé, Livre blanc.

Roopaei, M., & Dehbozorgi, N. (2024).*AI in Education: Crafting Policies for Tomorrow's Learning Landscape*. DOI: 10.1109/ISEC61299.2024.10665282

Roschelle, J., & Teasley, S. D. (1995). The Construction of Shared Knowledge in Collaborative Problem Solving. In *Proceedings of the 1995 Conference on Computer Support for Collaborative Learning* (pp. 69-96). DOI: 10.1007/978-3-642-85098-1_5

Roy, M. (2020). AI Intervention in Education Systems of India: An Analysis. *Solid State Technology*.

Rus, V., & Graesser, A. C. (2009). The Question Generation Framework: A View from Artificial Intelligence. In The Handbook of Research on E-learning Methodologies for Language Acquisition.

Russell, S., & Norvig, P. (2016). *Artificial Intelligence: A Modern Approach* (3rd ed.). Pearson.

Russel, S., & Norvig, P. (2010). *Artificial intelligence- a modern approach*. Pearson Education.

Ryan, J. M. (2023). *Pandemic pedagogies: Teaching and learning during the COVID-19 pandemic* (1st ed.). Routledge. DOI: 10.4324/9781003324096

Ryan, R. M., & Deci, E. L. (2000). Intrinsic and Extrinsic Motivations: Classic Definitions and New Directions. *Contemporary Educational Psychology*, 25(1), 54–67. DOI: 10.1006/ceps.1999.1020 PMID: 10620381

Ryan, R. M., & Deci, E. L. (2020). Intrinsic and extrinsic motivation from a self-determination theory perspective: Definitions, theory, practices, and future directions. *Contemporary Educational Psychology*, 61, 101860. DOI: 10.1016/j.cedpsych.2020.101860

Ryker, R., Lafleur, E., McManis, B., & Cox, K. C. (2002). Online Privacy Policies: An Assessment of the Fortune E-50. *Journal of Computer Information Systems*, 42(4), 15–20. DOI: 10.1080/08874417.2002.11647048

Saaida, M. B. (2023). AI-Driven transformations in higher education: Opportunities and challenges. *International Journal of Educational Research and Studies*, 5(1), 29–36.

Sabeima, M., Lamolle, M., Anghour, A., & Nanne, M. F. (2023). Towards a semantic platform for adaptive and collaborative e-learning. Revue Africaine De Recherche en Informatique Et Mathématiques Appliquées, Volume 37 - 2022. DOI: 10.46298/arima.8396

Sacré, M., Lafontaine, D., & Toczek, M. (2020). Liens entre les composantes des dispositifs d'enseignement hybride et les performances des étudiants de l'enseignement supérieur: Une revue systématique. *Mesure et Évaluation en Éducation*, 42(3), 109–152. DOI: 10.7202/1074105ar

Sadiku, M. N., Musa, S. M., & Chukwu, U. C. (2022). Artificial intelligence in education. iUniverse.

Safia, A. K., & Abdelmajid, N. (2023). Les souffrances psychologiques vécues par les étudiants: Analyse thématique du discours d'un échantillon de 20 étudiants primo-inscrits à l'université de sfax. *European Scientific Journal*, 19(16), 210. DOI: 10.19044/esj.2023.v19n16p210

Sakalle, A., Tomar, P., Bhardwaj, H., & Sharma, U. (2021). Impact and Latest Trends of Intelligent Learning With Artificial Intelligence. In Impact of AI Technologies on Teaching, Learning, and Research in Higher Education (pp. 172-189). IGI Global.

Salgado, T. B. P., & Inácio, P. (2018). Les actions sociotechniques des objets connectés: des réflexions à partir de l'internet des objets. Interfaces Numériques, 6(2), 2017. https://doi.org/DOI: 10.25965/interfaces-numeriques.2685

Salih, M. M. (2021). A Comparative Study Between Google Workspace and Microsoft Office 365 Productivity Services in Iraqi Educational Institutions. *International Journal of Humanities and Educational Research*, 3(5), 123–135. DOI: 10.47832/2757-5403.5-3.11

Sam, A. K., & Olbrich, P. (2023). The need for AI ethics in higher education. In *AI ethics in higher education: Insights from Africa and beyond* (pp. 3–10). Springer International Publishing. DOI: 10.1007/978-3-031-23035-6_1

Samarakou, M., Fylladitakis, E. D., Früh, W. G., Hatziapostolou, A., & Gelegenis, J. J. (2015). An advanced eLearning environment developed for engineering learners. *International Journal of Emerging Technologies in Learning*, 10(3), 22–33. DOI: 10.3991/ijet.v10i3.4484

Sandeen, J. (2013). The Need for a New Approach to Learning: Micro-Learning as a Strategy for Learning in the Digital Age. *Innovative Higher Education*, 38(2), 119–129.

Sandhu, R., Channi, H. K., Ghai, D., Cheema, G. S., & Kaur, M. (2024). An Introduction to Generative AI Tools for Education 2030. Integrating Generative AI in Education to Achieve Sustainable Development Goals, 1–28.

Sandu, N., & Gide, E. (2019, September). Adoption of AI-Chatbots to enhance student learning experience in higher education in India. In 2019 18th international conference on information technology based higher education and training (ITHET) (pp. 1-5). IEEE.

Sansonetti, J. (2022, septembre 30). *Histoire de l'intelligence artificielle (IA): création et évolutions [1943 à 2024]*. Retrieved from wizishop: https://www.wizishop .fr/blog/histoire-intelligence-artificielle

Santos, J. R., & de Mello, M. G. (2020). Digital Certifications Based on Blockchain: A New Approach to Credentialing in Education. *Education and Information Technologies*, 25(1), 15–35.

Santos, O. C., & Boticario, J. G. (2014). Involving users to improve the collaborative logical framework. *TheScientificWorldJournal*, 2014(1), 893525. DOI: 10.1155/2014/893525 PMID: 24592196

Sasseville, N., Juneau, S., & St-Pierre, É. (2022). Le point de vue d'étudiants au premier cycle en travail social quant à leur expérience d'un cours en ligne. *Revue Internationale Des Technologies en Pédagogie Universitaire*, 19(1), 34–50. DOI: 10.18162/ritpu-2022-v19n1-03

Satratzemi, M., & Terzopoulos, G. (2020). Voice Assistants and Smart Speakers in Everyday Life and in Education. *Informatics in Education*, 19(3), 473–490.

Savery, J. R. (2006). Overview of Problem-based Learning: Definitions and Distinctions. *The Interdisciplinary Journal of Problem-Based Learning*, 1(1), 9–20. DOI: 10.7771/1541-5015.1002

Saxena, A. K. (2024). Quantitative Measurement of Bias in AI-Generated Content: A Comprehensive Narrative Literature Review. 2024 IEEE International Symposium on Technology and Society (ISTAS), 1–5. https://ieeexplore.ieee.org/abstract/document/10732696/

Sayac, N. (2017). Lucie Mottier-Lopez, Évaluations formative et certificative des apprentissages. Enjeux pour l'enseignement. *Recherches & Educations*, 28, 28. Advance online publication. DOI: 10.4000/ree.5958

Sayac, N. (2018). MERLE Pierre. Les pratiques d'évaluation scolaire. Historique, difficultés, perspectives. *Revue française de pédagogie.Revue Française de Pédagogie*, 204(204), 204. Advance online publication. DOI: 10.4000/rfp.8553

Schumann, C. A., Reuther, K., Tittmann, C., Clauß, A. M., & Kauper, J. (2020, June). Impact of AI application on digital education focused on STE (A) M. In EDEN Conference Proceedings (No. 1, pp. 153-161).

Schunk, D. H., & DiBenedetto, M. K. (2020). Motivation and social cognitive theory. *Contemporary Educational Psychology*, 60, 101832. DOI: 10.1016/j.cedpsych.2019.101832

Searle, J. R. (1980). Minds, Brains, and Programs. *Behavioral and Brain Sciences*, 3(3), 417–424. DOI: 10.1017/S0140525X00005756

Selwyn, N. (2020). The educational implications of AI: A critical perspective. *Learning, Media and Technology*, 45(1), 1–11. DOI: 10.1080/17439884.2020.1694945

Senko, C., Hulleman, C. S., & Harackiewicz, J. M. (2011). Achievement goal theory at the crossroads: Old controversies, current challenges, and new directions. *Educational Psychologist*, 46(1), 26–47. DOI: 10.1080/00461520.2011.538646

Sethi, K., Sharma, A., Chauhan, S., & Jaiswal, V. (2020). Impact of social and cultural challenges in education using AI. In *Revolutionizing Education in the Age of AI and Machine Learning* (pp. 130–151). IGI Global.

Setyosari, P., Wibawati, D. O. A., Fitriyah, C. Z., & Wardani, R. P. (2023). Learning loss: How does technology facilitate learner learning? *AIP Conference Proceedings*, 2679, 070016. DOI: 10.1063/5.0112882

Sharma, S., Singh, G., Sharma, C. S., & Kapoor, S. (2024). Artificial intelligence in Indian higher education institutions: A quantitative study on adoption and perceptions. *International Journal of System Assurance Engineering and Management*. Advance online publication. DOI: 10.1007/s13198-023-02193-8

Shih, P.-K., Lin, C.-H., Wu, L. Y., & Yu, C.-C. (2021). Learning Ethics in AI—Teaching Non-Engineering Undergraduates through Situated Learning. *Sustainability (Basel)*, 13(7), 3718. DOI: 10.3390/su13073718

Short, J., Williams, E., & Christie, B. (1976). *The social psychology of telecommunications*. No Title.

Shou, Z., Lu, X., Wu, Z., Yuan, H., Zhang, H., & Lai, J. (2020). On learning path planning algorithm based on collaborative analysis of learning behavior. *IEEE Access : Practical Innovations, Open Solutions*, 8, 119863–119879. DOI: 10.1109/ACCESS.2020.3005793

Shu, K., Sliva, A., Wang, S., Tang, J., & Liu, H. (2017). Fake news detection on social media: A data mining perspective. *SIGKDD Explorations*, 19(1), 22–36. DOI: 10.1145/3137597.3137600

Shute, V. J., & Rahimi, S. (2017). Review of computer-based assessment for learning in elementary and secondary education. *Journal of Computer Assisted Learning*, 33(1), 1–19. DOI: 10.1111/jcal.12172

Siau, K., & Wang, W. (2020). Artificial intelligence (AI) ethics: Ethics of AI and ethical AI. [JDM]. *Journal of Database Management*, 31(2), 74–87. DOI: 10.4018/JDM.2020040105

Sibert, L., Schneider, P., Liard, A., Lefevre-Scelles, A., Ménard, J., Auquit-Auckbur, I., & Veber, B. (2022). L'examen clinique objectif structuré (ecos) comme examen validant des compétences cliniques des étudiants en médecine français: 13 ans d'expérience rouennaise. *Pédagogie Médicale*, 23(1), 69–84. DOI: 10.1051/pmed/2021031

Sidiropoulos, D., & Anagnostopoulos, C. N. (2024). Applications, challenges, and ethical issues of AI and ChatGPT in education. *arrive preprintarXiv:2402.07907.*

Siemens, G. (2005). Connectivism: A Learning Theory for the Digital Age. *International Journal of Instructional Technology and Distance Learning*, 2(1), 3–10.

Siemens, G. (2014). Learning Analytics: The Emergence of a Discipline. *The American Behavioral Scientist*, 57(10), 1380–1400. DOI: 10.1177/0002764213498851

Simard, A., & Vermot-Desroches, C. (2023). Innovations numériques: Apprentissages interdisciplinaires pour des élèves en décrochage scolaire. *Revue Suisse De Pédagogie Spécialisée*, 13(01), 33–39. DOI: 10.57161/r2023-01-06

Simelane-Mnisi, S. (2023). Effectiveness of lms digital tools used by the academics to foster students' engagement. *Education Sciences*, 13(10), 980. DOI: 10.3390/educsci13100980

Simelane-Mnisi, S., & Mokgala-Fleischmann, N. (2022). *Training framework to enhance digital skills and pedagogy of chemistry teachers to use imfundo*. New Updates in E-Learning., DOI: 10.5772/intechopen.102003

Sjoelie, E. (2019). "teachers in team": an intervention study on facilitating student teachers' reflection on teamwork. *Proceedings of the 2019 AERA Annual Meeting*. DOI: 10.3102/1441833

Skinner, B. F. (1953). *Science and Human Behavior*. Macmillan.

Skinner, E. (2016). Engagement and Disaffection as Central to Processes of Motivational Resilience and Development. In Wentzel, K. R., & Miele, D. B. (Eds.), *Handbook of Motivation at School*. Routledge Handbooks Online.

Skinner, E. A., Zimmer-Gembeck, M. J., Connell, J. P., Eccles, J. S., & Wellborn, J. G. (1998). Individual Differences and the Development of Perceived Control. *Monographs of the Society for Research in Child Development*, 63(2–3), 1–220. DOI: 10.2307/1166220 PMID: 9839459

Slade, S., & Prinsloo, P. (2013). Learning analytics: Ethical issues and dilemmas. *The American Behavioral Scientist*, 57(10), 1510–1529. DOI: 10.1177/0002764213479366

Slater, M., & Sanchez-Vives, M. V. (2016). Enhancing Our Lives with Immersive Virtual Reality. *Frontiers in Robotics and AI*, 3, 74. DOI: 10.3389/frobt.2016.00074

Slater, M., & Wilbur, S. (1997). A Framework for Immersive Virtual Environments (FIVE). *Virtual Reality (Waltham Cross)*, 3(3), 130–142.

Slavin, R. E. (1996). Research on Cooperative Learning and Achievement: What We Know, What We Need to Know. *Contemporary Educational Psychology*, 21(1), 43–69. DOI: 10.1006/ceps.1996.0004

Slavin, R. E. (2011). Instruction Based on Cooperative Learning. In Mayer, R. E., & Alexander, P. A. (Eds.), *Handbook of Research on Learning and Instruction*.

Smith, B., Sparkes, A. C., & Caddick, N. (2014). Judging qualitative research. Research Methods in Sports Coaching, 192–201. DOI: 10.4324/9780203797549-18

Smith, A., & Brown, B. (2023). Ethical considerations in the implementation of AI in education: Balancing innovation and human values. *Journal of Educational Technology & Society*, 26(3), 215–228.

Smith, J. (2020). *L'intelligence artificielle en éducation*. Éditions de l'Université.

Smith, J., & Brown, T. (2021). Project-Based Learning as a Catalyst for Student Engagement in AI Education. *Journal of Innovative Teaching*, 15(4), 45–58.

Soewito, B., Gaol, F. L., & Abdurachman, E. (2019, December). Risk Assessment on Cloud Computing for The Learning System in The Education Environment. In *2019 IEEE International Conference on Engineering, Technology and Education (TALE)* (pp. 1-6). IEEE.

Solomon, G. (2003). *Project-Based Learning: A Primer*. Technology & Learning.

Soufiane, O., Maha, K., Mohamed, E., & Mohamed, K. (2021). Implementation of the Moodle e-learning platform from server selection to configuration. GSC Advanced Engineering and Technology, 1(1), 016-027.

Soufiane, O., Maha, K., & Mohamed, K. (2023). Conceptualizing an Inductive Learning Situation in Online Learning Enabled by Software Engineering. *International Journal of Advanced Computer Science and Applications*, 14(12). Advance online publication. DOI: 10.14569/IJACSA.2023.0141210

Soufiane, O., Zahra, O. F., & Mohamed, K. (2024). From Lecture Theaters to Online Classrooms: Examining The Growth of the Flipped Classroom. *COMPETITIVE: Journal of Education*, 3(2), 86–101. DOI: 10.58355/competitive.v3i2.97

Spatioti, A. G., Kazanidis, I., & Pange, J. (2022). A comparative study of the ADDIE instructional design model in distance education. *Information (Basel)*, 13(9), 402. DOI: 10.3390/info13090402

Spector, J. M., & Ma, M. (2019). *Artificial Intelligence in Education: Promises and Implications for Teaching and Learning*. Springer.

Spector, J. M., & Ma, Y. (2019). *Foundations of Educational Technology: Integrative Approaches and Interdisciplinary Perspectives*. Routledge.

Srinivasan, M., Venugopal, A., Venkatesan, L., & Kumar, R. (2024). Navigating the pedagogical landscape: Exploring the implications of ai and chatbots in nursing education. *JMIR Nursing*, 7, e52105. DOI: 10.2196/52105 PMID: 38870516

Srinivasan, R., & Uchino, K. (2021, March). Biases in generative art: A causal look from the lens of art history. In *Proceedings of the 2021 ACM Conference on Fairness, Accountability, and Transparency* (pp. 41-51).

Stahl, B. C. (2021). Ethical Issues of AI. In Stahl, B. C. (Ed.), *Artificial Intelligence for a Better Future* (pp. 35–53). Springer International Publishing., DOI: 10.1007/978-3-030-69978-9_4

Stahl, G., Koschmann, T., & Suthers, D. (2006). Computer-supported Collaborative Learning: An Historical Perspective. In Sawyer, R. K. (Ed.), *Cambridge Handbook of the Learning Sciences* (pp. 409–426). Cambridge University Press.

Stancati, M., & Schechner, S. (2023, March 31). ChatGPT banned in Italy over data-privacy concerns. The Wall Street Journal. Available at https://www.wsj.com/articles/chatgptbanned-in-italy-over-data-privacy-concerns-4b984e75

Ştefan, I. A., Hauge, J. B., Sallinen, N., Ştefan, A., & Gheorghe, A. F. (2021). Accessibility and education: Are we fulfilling state of the art requirements? The International Scientific Conference eLearning and Software for Education, 1, 579–587. https://search.proquest.com/openview/cce87013703da339bf74f3361cb90f53/1?pq-origsite=gscholar&cbl=1876338

Stotz, S., Brega, A., Gonzales, K., Hebert, L., & Moore, K. (2021). Facilitators and barriers to healthy eating among american indian and alaska native adults with type 2 diabetes: Stakeholder perspectives. *Current Developments in Nutrition*, 5, 22–31. DOI: 10.1093/cdn/nzaa114 PMID: 34222760

Strayhorn, T. L. (2012). *College Students' Sense of Belonging: A Key to Educational Success for All Students* (1st ed.). Routledge., DOI: 10.4324/9780203118924

Strom, K. J., & Viesca, K. M. (2023). Towards a complex framework of teacher learning-practice. In *Non-linear perspectives on teacher development* (pp. 13–28). Routledge., https://www.taylorfrancis.com/chapters/edit/10.4324/9781003372097-2/towards-complex-framework-teacher-learning-practice-kathryn-strom-kara-mitchell-viesca DOI: 10.4324/9781003372097-2

Suarez, C. G. H., Bucheli, V., & Ordóñez, H. (2023). Artificial intelligence and computer-supported collaborative learning in programming: A systematic mapping study. *Tecnura*, 27(75), 175–206. DOI: 10.14483/22487638.19637

Subhash, S., & Cudney, E. A. (2018). Gamified Learning in Higher Education: A Systematic Review of the Literature. *Computers in Human Behavior*, 87, 192–206. DOI: 10.1016/j.chb.2018.05.028

Su, C. H., & Cheng, C. H. (2015). A Mobile Gamification Learning System for Improving the Learning Motivation and Achievements. *Journal of Computer Assisted Learning*, 31(3), 268–286. DOI: 10.1111/jcal.12088

Suduc, A., Bîzoi, M., & Filip, F. G. (2023). Status, challenges and trends in videoconferencing platforms. INTERNATIONAL JOURNAL OF COMPUTERS COMMUNICATIONS &Amp. *International Journal of Computers, Communications & Control*, 18(3). Advance online publication. DOI: 10.15837/ijccc.2023.3.5465

Swan, M. (2015). *Blockchain: Blueprint for a New Economy*. O'Reilly Media.

Sweller, J. (1988). Cognitive load during problem solving: Effects on learning. *Cognitive Science*, 12(2), 257–285. DOI: 10.1207/s15516709cog1202_4

Swiecki, Z., Khosravi, H., Chen, G., Martinez-Maldonado, R., Lodge, J. M., Milligan, S., Selwyn, N., & Gašević, D. (2022a). Assessment in the age of artificial intelligence. *Computers and Education: Artificial Intelligence*, 3, 100075. DOI: 10.1016/j.caeai.2022.100075

Swol, L. M. V., & Carlson, C. L. (2015). Language use and influence among minority, majority, and homogeneous group members. *Communication Research*, 44(4), 512–529. DOI: 10.1177/0093650215570658

Szabo, N. (1997). The Idea of Smart Contracts

T.L.G. (2016, 01 28). *Intelligence artificielle: Quand les machines battent l'Homme*. Retrieved from 20minutes TECHNOLOGIES: https://www.20minutes.fr/sciences/1775651-20160128-intelligence-artificielle-quand-machines-battent-homme#:~:text=1997%20%3A%20Deep%20Blue%20met%20en%20%C3%A9chec%20Kasparov&text=La%20machine%20de%202%20m%C3%A8tres,%2C%20l'humain%20%C3%A9tait%20battu

Taddeo, M., & Floridi, L. (2018). How AI can be a force for good. *Science*, 361(6404), 751–752. DOI: 10.1126/science.aat5991 PMID: 30139858

Takahashi, K., & Yamamoto, H. (2022). Beyond automated grading: AI-enhanced assessment of critical thinking and creativity. *Assessment & Evaluation in Higher Education*, 47(5), 623–639.

Tamim, T. (2021). Language, Class, and Education: Deconstructing the Centre to Rethink Inclusivity in Education in Pakistan. *Cogent Education*, 8(1), 1897933. DOI: 10.1080/2331186X.2021.1897933

Tan, D. Y., & Cheah, C. W. (2021). Developing a gamified AI-enabled online learning application to improve students' perception of university physics. *Computers and Education: Artificial Intelligence*, 2, 100032. DOI: 10.1016/j.caeai.2021.100032

Tandem. (n.d.). *Exprime-toi avec tandem*. Retrieved from Tandem Language Exchange: https://tandem.net/fr

Tan, J. (2024). An empirical study of adaptive learning algorithm based on intrinsic motivation in english online teaching and learning. *Journal of Electrical Systems*, 20(6s), 1695–1704. DOI: 10.52783/jes.3088

Tanveer, M., Hassan, S., & Bhaumik, A. (2020). Academic Policy Regarding Sustainability and Artificial Intelligence (AI). *Sustainability (Basel)*, 12(22), 9435. Advance online publication. DOI: 10.3390/su12229435

Tapscott, D., & Tapscott, A. (2016). *Blockchain Revolution: How the Technology Behind Bitcoin Is Changing Money, Business, and the World*. Penguin.

Tariq, M. U. (2024). Generative AI in curriculum development in higher education. In *Impacts of Generative AI on Creativity in Higher Education* (pp. 227–258). IGI Global., https://www.igi-global.com/chapter/generative-ai-in-curriculum-development-in-higher-education/355433 DOI: 10.4018/979-8-3693-2418-9.ch009

Taufiqurrochman, R., Muslimin, I., Rofiki, I., & Abah, J. A. (2020). Students' perceptions on learning management systems of arabic learning through blended learning model. Jurnal Al-Bayan. *Jurnal Jurusan Pendidikan Bahasa Arab*, 12(1), 22–36. DOI: 10.24042/albayan.v12i1.5276

Tessa, A., & Niels, J. (2023, 05 29). *Quelle est la différence entre la réalité augmentée et la réalité virtuelle?* Retrieved from GetApp: https://fr.getapp.ca/blog/3675/difference-realite-augmentee-realite-virtuelle

Tharp, B. (2016). The Benefits of Microlearning in E-Learning. eLearning Industry.

Tibi, M. H. (2016). Essential components in structuring asynchronous discussion forums. *Turkish Online Journal of Distance Education*, 0(0). Advance online publication. DOI: 10.17718/tojde.12429

Tiwari, R. (2023). *Ethical And Societal Implications of AI and Machine Learning*. Indian Scientific Journal Of Research In Engineering And Management., DOI: 10.55041/IJSREM17519

Tomaszewska, R. (2023). Andragogy meets ChatGPT in lifelong learning: Exploring opportunities and challenges. 2023 IEEE International Conference on Data Mining Workshops (ICDMW), 868–874. https://ieeexplore.ieee.org/abstract/document/10411582/

Tomlinson, C. A. (2001). How to Differentiate Instruction in Mixed-Ability Classrooms (2nd ed.). Alexandria, VA: Association for Supervision and Curriculum Development (ASCD).

Torreão, A. D. A. (2023). Considérations générales sur la loi générale sur la protection des données et ses principaux défis au brésil. Revista Científica Multidisciplinar Núcleo Do Conhecimento, 79-87. https://doi.org/DOI: 10.32749/nucleodoconhecimento.com.br/droit/loi-generale

Trautwein, U., & Köller, O. (2003). The Relationship Between Homework and Achievement—Still Much of a Mystery. *Educational Psychology Review*, 15(2), 115–145. DOI: 10.1023/A:1023460414243

Tregubov, V. (2021). Voice assistants with artificial intelligence for improving academic english. Computer Science &Amp; Information Technology (CS &Amp; IT). DOI: 10.5121/csit.2021.110901

Tremblay, C., & Poëllhuber, B. (2022). Analyse qualitative de référentiels de compétences du xxie siècle, numériques et informationnelles: Tendances mondiales observées. *Formation Et Profession*, 30(2), 1. DOI: 10.18162/fp.2022.648

Tremblay, D., Psyché, V., & Yagoubi, A. (2023). Mise en œuvre de l'ia dans les organisations est-elle compatible avec une société éthique? *Ad Machina*, (7), 166–187. DOI: 10.1522/radm.no7.1663

Tricot, A. (2021). Le numérique permet-il des apprentissages scolaires moins contraints ? une revue de la littérature. Éducation Et Sociétés, n° 45(1), 37-56. DOI: 10.3917/es.045.0037

Trinh, S. (2019, Hiver). *L'intelligence artificielle et la traduction évoluent ensemble.* Retrieved from C i r c u i t. Numéro 141: https://www.circuitmagazine.org/dossier-141/l-intelligence-artificielle-et-la-traduction-evoluent-ensemble

Turnbull, D., Chugh, R., & Luck, J. (2020). Learning Management Systems, An Overview. In Tatnall, A. (Ed.), *Encyclopedia of Education and Information Technologies* (pp. 1052–1058). Springer International Publishing., DOI: 10.1007/978-3-030-10576-1_248

Tutos'Me Pro. (2024). L'engagement des apprenants: clé de la réussite. Tutos'Me Pro - Trouver un Formateur ou une Mission de Formation. https://www.tutos.pro/engagement-des-apprenants-la-cle-de-la-reussite/

UNESCO (2021). AI and education: guidance for policy-makers. DOI: 10.54675/PCSP7350

UNESCO. (1994). *The Salamanca statement and framework for action on special needs education.* UNESCO.

UNESCO. (2021). Artificial Intelligence in education: Challenges and opportunities for sustainable development. https://unesdoc.unesco.org/ark:/48223/pf0000377962

UNESCO. Elaboration of a Recommendation on the Ethics of Artificial Intelligence. Available online: https://en.unesco.org/

United Nations Educational, Scientific and Cultural Organization, & Organization, C. (2021). Recommendation on the Ethics of Artificial Intelligence. United Nations Educational. https://unesdoc.unesco.org/ark:/48223/pf0000379920.page=14

United Nations Educational, Scientific and Cultural Organization. (2019). Beijing Consensus on artificial intelligence and education. Outcome document of the International Conference on Artificial Intelligence and Education, Planning Education in the AI Era: Lead the Leap, Beijing, 2019. United Nations Educational, Scientific and Cultural Organization. https://unesdoc.unesco.org/ark:/48223/pf0000368303

Uppal, M. (2022).*AI in education.* DOI: 10.4324/9780429329067-5

Usher, M., & Barak, M. (2024). Unpacking the role of AI ethics online education for science and engineering students. *International Journal of STEM Education*, 11(1), 35. Advance online publication. DOI: 10.1186/s40594-024-00493-4

Vallor, S. (2016). *Technology and the virtues: A philosophical guide to a future worth wanting*. Oxford University Press. DOI: 10.1093/acprof:oso/9780190498511.001.0001

VanLehn, K. (2011). The Relative Effectiveness of Human Tutoring, Intelligent Tutoring Systems, and Other Tutoring Systems. *Educational Psychologist*, 46(4), 197–221. DOI: 10.1080/00461520.2011.611369

VanLehn, K., Burkhardt, H., Cheema, S., Kang, S., Pead, D., Schoenfeld, A., & Wetzel, J. (2021). Can an orchestration system increase collaborative, productive struggle in teaching-by-eliciting classrooms? *Interactive Learning Environments*, 29(6), 987–1005. DOI: 10.1080/10494820.2019.1616567

Vayadande, K., Pednekar, C. B., Khune, P. A., Prabhavalkar, V. S., & Dange, V. R. (2024). GPT-3- and DALL-E-Powered Applications: A Complete Survey. In Dey, A., Nayak, S., Kumar, R., & Mohanty, S. N. (Eds.), *How Machine Learning is Innovating Today's World* (1st ed., pp. 329–341). Wiley., DOI: 10.1002/9781394214167.ch19

Veletsianos, G., & Miller, C. (2008). Conversational Agents in Virtual Worlds: Bridging Disciplines. *British Journal of Educational Technology*, 39(5), 802–804.

Victor, N. (2023). An experimental case study on how ai may improve communication and resource allocation. *International Journal of Artificial Intelligence and Machine Learning*, 3(1), 10–18. DOI: 10.51483/IJAIML.3.1.2023.10-18

Vijayakumar, G., Rajkumar, M., Rajiv Chandar, N., Selvakumar, P., & Duraisamy, R. (2024). Muniyandi Rajkumar, Rajiv Chandar N, Selvakumar P and Ramesh Duraisamy, (2024). Environmentally friendly TDS removal from waste water by electrochemical ion exchange batch-type recirculation (EIR) technique. *Environmental Science. Water Research & Technology*, 10(4), 826–835. DOI: 10.1039/D3EW00793F

Vlugter, P., Knott, A., McDonald, J., & Hall, C. (2009). Dialogue-based CALL: A case study on teaching pronouns. *Computer Assisted Language Learning*, 22(2), 115–131. DOI: 10.1080/09588220902778260

Vorecol, É. é. (2024, September 10). *L'intégration des technologies d'IA dans les LMS pour le suivi des compétences.* Retrieved from Vorecol: https://vorecol.com/fr/blogs/blog-lintegration-des-technologies-dia-dans-les-lms-pour-le-suivi-des-competences-155658

Vrabie, C. (2024). Artificial Intelligence from Idea to Implementation. How Can AI Reshape the Education Landscape? arXiv preprint arXiv:2407.20236.

Vygotsky, L. (1978). *Mind in Society: The Development of Higher Psychological Processes*. Harvard University Press.

Vygotsky, L. S. (1978). *Mind in society: The development of higher psychological processes*. Harvard University Press.

Vygotsky, L. S. (1987). La pensée et le langage. *Revue Française de Pédagogie*, •••, 98–103.

Wahid, R., Mero, J., & Ritala, P. (2023). Written by ChatGPT, illustrated by Midjourney: Generative AI for content marketing. *Asia Pacific Journal of Marketing and Logistics*, 35(8), 1813–1822. DOI: 10.1108/APJML-10-2023-994

Walczak, K., & Cellary, W. (2023). Challenges for higher education in the era of widespread access to generative AI. *Economics and Business Review*, 9(2). Advance online publication. DOI: 10.18559/ebr.2023.2.743

Wambsganss, T., Winkler, R., Schmid, P., & Söllner, M. (2020). Designing a conversational agent as a formative course evaluation tool

Wang, T. (2023). Navigating Generative AI (ChatGPT) in Higher Education: Opportunities and Challenges. In C. Anutariya, D. Liu, Kinshuk, A. Tlili, J. Yang, & M. Chang (Eds.), Smart Learning for A Sustainable Society (pp. 215–225). Springer Nature Singapore. DOI: 10.1007/978-981-99-5961-7_28

Wang, Z., Li, C., & Wang, X. (2021). *Convolutional Neural Network Pruning with Structural Redundancy Reduction* (arXiv:2104.03438). arXiv. /arXiv.2104.03438DOI: 10.1109/CVPR46437.2021.01467

Wang, F., Zhang, L., Chen, X., Wang, Z., & Xu, X. (2021). A personalized self-learning system based on knowledge graph and differential evolution algorithm. *Concurrency and Computation*, 34(8), e6190. Advance online publication. DOI: 10.1002/cpe.6190

Wang, L., Zhang, M., & Li, Y. (2020). AI-based student engagement monitoring in online learning environments. *Computers & Education*, 156, 103940.

Wang, N., & Lester, J. (2023). K-12 Education in the Age of AI: A Call to Action for K-12 AI Literacy. *International Journal of Artificial Intelligence in Education*, 33(2), 228–232. DOI: 10.1007/s40593-023-00358-x

Wang, R., Sun, Z., & Zhou, Y. (2023). Research on the new model of "internet + education" based on artificial intelligence. *International Conference on Mathematics, Modeling, and Computer Science (MMCS2022)*. DOI: 10.1117/12.2670426

Wang, W., & Siau, K. (2018). *Ethical and moral issue with AI – a case study on healthcare robots. AMCIS 2019 proceedings*. Academic Press.

Wang, W., & Siau, K. (2019a). Artificial Intelligence, Machine Learning, Automation, Robotics, Future of Work and Future of Humanity: A Review and Research Agenda. *Journal of Database Management*, 30(1), 61–79. DOI: 10.4018/JDM.2019010104

Wang, Y., Cui, W., & Wang, Z. (2022). Artificial intelligence chatbots in online education: A systematic review. *Education and Information Technologies*, 27(4), 5135–5153.

Wang, Y., & Zhang, L. (2022). Proactive learning support through AI-driven gap analysis in online education platforms. *Educational Technology Research and Development*, 70(4), 1689–1711.

Wardat, Y., Tashtoush, M., Alali, R., & Saleh, S. (2024). Artificial Intelligence in Education: Mathematics Teachers' Perspectives, Practices and Challenges. *Iraqi Journal For Computer Science and Mathematics*, 5(1), 60–77. DOI: 10.52866/ijcsm.2024.05.01.004

Warth, L. L., & Dyb, K. (2023). Facilitating interprofessional collaboration on electronic medicines management. *Studies in Health Technology and Informatics*. Advance online publication. DOI: 10.3233/SHTI230423 PMID: 37386957

Was, C. A., Al-Harthy, I., Stack-Oden, M., & Isaacson, R. M. (2009). Academic identity status and the relationship to achievement goal orientation. *Electronic Journal of Research in Educational Psychology*, 7(2), 627–652.

Weber, G., & Brusilovsky, P. (2001). E-Learning with Adaptive and Intelligent Systems: The Role of Student Modeling in Adaptive Learning Systems. In Adaptive Web-Based Educational Systems (pp. 61-90). Springer.

Weigel, F., & Weller, M. (2017). The Role of Digital Credentials in the Evolving Higher Education Landscape. *Journal of Applied Learning Technology*, 7(3), 18–23.

Wei, S. (2024). Research on the disruptive transformation and future development of american teaching models in the context of intelligent transformation. Advances in Economics. *Management and Political Sciences*, 72(1), 201–210. DOI: 10.54254/2754-1169/72/20240696

Weiss, H. M., & Cropanzano, R. (2015). Affective Events Theory: A Theoretical Discussion of the Structure, Causes, and Consequences of Affective Experiences at Work. *Research in Organizational Behavior*, 18, 1–74.

Wei, X., Saab, N., & Admiraal, W. (2021). Assessment of cognitive, behavioral, and affective learning outcomes in massive open online courses: A systematic literature review. *Computers & Education*, 163, 104097. DOI: 10.1016/j.compedu.2020.104097

Weller, A. (2019). Transparency: Motivations and challenges. In T. Hagras (Ed.), Explainable AI: Interpreting, explaining and visualizing deep learning (pp. 23 40). Springer.

Wenger, E. (1998/2005). *la théorie des communautés de pratique (Gervais, F., Trad.)*. Québec: Presses de l'Université Laval.

Wenger, E. (1998). *Communities of Practice: Learning, Meaning, and Identity*. Cambridge University Press. DOI: 10.1017/CBO9780511803932

Weng, X., & Chiu, T. K. (2023). Instructional design and learning outcomes of intelligent computer assisted language learning: Systematic review in the field. *Computers and Education: Artificial Intelligence*, 4, 100117. DOI: 10.1016/j.caeai.2022.100117

Weng, Z., & Qin, Z. (2021). Semantic communication systems for speech transmission. *IEEE Journal on Selected Areas in Communications*, 39(8), 2434–2444. DOI: 10.1109/JSAC.2021.3087240

Werbach, K., & Hunter, D. (2012). *For the Win: How Game Thinking Can Revolutionize Your Business*. Wharton Digital Press.

Whittaker, M., Crawford, K., Dobbe, R., Fried, G., Kaziunas, E., Mathur, V., & Schwartz, O. (2018). *AI now report 2018*. AI Now Institute.

Widyana, A., Jerusalem, M. I., & Yumechas, B. (2022). The application of text-to-speech technology in language learning. Sixth International Conference on Language, Literature, Culture, and Education (ICOLLITE 2022), 85–92. https://books.google.com/books?hl=fr&lr=&id=EaykEAAAQBAJ&oi=fnd&pg=PA85&dq=Text-to-Speech++in+education&ots=P6KtZz0QCb&sig=znet6sj4HFVbtmGzxJ4IbJ1qGzY

Wiley, D., & Hilton, J. L.III. (2018). Defining OER-enabled pedagogy. *International Review of Research in Open and Distance Learning*, 19(4), 133–147. DOI: 10.19173/irrodl.v19i4.3601

Williams, R. T. (2024). The ethical implications of using generative chatbots in higher education. *Frontiers in Education*, 8, 1331607. https://www.frontiersin.org/articles/10.3389/feduc.2023.1331607/full. DOI: 10.3389/feduc.2023.1331607

Wing, J. M., & Lazowska, E. D. (2018). Computational Thinking and Project-Based Learning for Artificial Intelligence Education. *Journal of the Learning Sciences*, 27(4), 562–579.

Witmer, B. G., & Sadowski, W. J.Jr. (1998). Non visually guided locomotion to a previously viewed target in real and virtual environments. *Human Factors*, 40(3), 478–488. DOI: 10.1518/001872098779591340

Wolff, A., & FitzGerald, E. (2022). *Integrating AI into Higher Education Curricula: A Practical Guide*. Routledge.

Wollny, S., Schneider, J., Di Mitri, D., Weidlich, J., Rittberger, M., & Drachsler, H. (2021). Are we there yet?-a systematic literature review on chatbots in education. *Frontiers in Artificial Intelligence*, 4, 4. DOI: 10.3389/frai.2021.654924 PMID: 34337392

Wongvorachan, T. (2023). *Cheating Detection in Tests: A Systematic Review*. OSF. DOI: 10.31234/osf.io/uc5hj

Woolf, B. P. (2010). *Building intelligent interactive tutors: Student-centered strategies for revolutionizing e-learning*. Morgan Kaufmann.

Woolf, B. P. (2010). *Building Intelligent Interactive Tutors: Student-centered Strategies for Revolutionizing E-learning*. Morgan Kaufmann.

World Health Organization. (2023). Health topics. https://www.who.int/topics/en/ orld Health Organization. (2023). Health topics. https://www.who.int/topics/en/

Wright, R. (2008). *Educational Assessment: Tests and Measurements in the Age of Accountability*. SAGE Publications., DOI: 10.4135/9781483329673

Wu, F., Lu, C., Zhu, M., Chen, H., Zhu, J., Yu, K., Li, L., Li, M., Chen, Q., Li, X., Cao, X., Wang, Z., Zha, Z., Zhuang, Y., & Pan, Y. (2020). Towards a new generation of artificial intelligence in China. *Nature Machine Intelligence*, 2(6), 312–316. DOI: 10.1038/s42256-020-0183-4

Wu, H. K., Lee, S. W. Y., Chang, H. Y., & Liang, J. C. (2013). Current Status, Opportunities and Challenges of Augmented Reality in Education. *Computers & Education*, 62, 41–49. DOI: 10.1016/j.compedu.2012.10.024

Xie, H., Qin, Z., Li, G., & Juang, B. (2021). Deep learning enabled semantic communication systems. *IEEE Transactions on Signal Processing*, 69, 2663–2675. DOI: 10.1109/TSP.2021.3071210

Xin, O. K., & Singh, D. (2021). Development of learning analytics dashboard based on moodle learning management system. *International Journal of Advanced Computer Science and Applications*, 12(7). Advance online publication. DOI: 10.14569/IJACSA.2021.0120793

Xu, L. (2020, December 11).*The Dilemma and Countermeasures of AI in Educational Application*. DOI: 10.1145/3445815.3445863

Xue, L., & Schneider, R. (2015). Interaction en formation des enseignants de langue. *Recherches en Didactique Des Langues Et Des Cultures*, 12(3). Advance online publication. DOI: 10.4000/rdlc.1037

Xu, W., & Ouyang, F. (2022). A systematic review of AI role in the educational system based on a proposed conceptual framework. *Education and Information Technologies*, 27(3), 4195–4223. DOI: 10.1007/s10639-021-10774-y

Yamada, M., & Akahori, K. (2009). Awareness and performance through self- and partner's image in videoconferencing. *CALICO Journal*, 27(1), 1–25. DOI: 10.11139/cj.27.1.1-25

Yang, F., Wang, M., Shen, R., & Han, P. (2007). Community-organizing agent: An artificial intelligent system for building learning communities among large numbers of learners. *Computers & Education*, 49(2), 131–147. DOI: 10.1016/j.compedu.2005.04.019

Yang, H., Kim, H., Lee, J. H., & Shin, D. (2022). Implementation of an ai chatbot as an english conversation partner in efl speaking classes. *ReCALL*, 34(3), 327–343. DOI: 10.1017/S0958344022000039

Yang, H., & Liu, Z. (2020). An Intelligent Chatbot for Language Learning: A Case Study on English Language Learning. [iJET]. *International Journal of Emerging Technologies in Learning*, 15(8), 60–71.

Yang, Z., Wu, J. G., & Xie, H. (2024). Taming Frankenstein's monster: Ethical considerations relating to generative artificial intelligence in education. *Asia Pacific Journal of Education*, •••, 1–14. DOI: 10.1080/02188791.2023.2300137

Yan, Y., & Yu, K. (2016). Design and implementation of offline distance learning system in teaching construction cost engineering. [iJET]. *International Journal of Emerging Technologies in Learning*, 11(07), 52. DOI: 10.3991/ijet.v11i07.5794

Yenduri, G., Ramalingam, M., Selvi, G. C., Supriya, Y., Srivastava, G., Maddikunta, P. K. R., Raj, G. D., Jhaveri, R. H., Prabadevi, B., & Wang, W. (2024). Gpt (generative pre-trained transformer)–a comprehensive review on enabling technologies, potential applications, emerging challenges, and future directions. IEEE Access. https://ieeexplore.ieee.org/abstract/document/10500411/

Yilmaz, S., & Toklu, S. (2020). A deep learning analysis on question classification task using Word2vec representations. *Neural Computing & Applications*, 32(7), 2909–2928. DOI: 10.1007/s00521-020-04725-w

Yoke, S. K., Ahmad, T. A., & Hasan, N. H. (2019). Exploring the potential of augmented reality in English for report writing: A perceptive overview. *International Journal of Education*, 4(33), 13–21.

Youssef, É. and Audran, J. (2019). La personnalisation de l'apprentissage vue comme facteur effectif d'innovation pédagogique. Spirale - Revue De Recherches en Éducation, N° 63(1), 157-172. DOI: 10.3917/spir.063.0157

Yu, H., Guo, Y., Yang, H., Zhang, W., & Dong, Y. (2024). Can ChatGPT Revolutionize Language Learning? Unveiling the Power of AI in Multilingual Education Through User Insights and Pedagogical Impact. *European Journal of Education*, 12749. Advance online publication. DOI: 10.1111/ejed.12749

Yu, L., & Yu, Z. (2023). Qualitative and quantitative analyses of artificial intelligence ethics in education using VOSviewer and CitNetExplorer. *Frontiers in Psychology*, 14, 1061778. Advance online publication. DOI: 10.3389/fpsyg.2023.1061778 PMID: 36968737

Zahra, O. F., Amel, N., & Mohamed, K. (2024, February). Communication tools between learning activities and e-learning. In E-Learning and Smart Engineering Systems (ELSES 2023) (pp. 452-459). Atlantis Press.

Zahra, O. F., Amel, N., & Mohamed, K. (2023). Communication Tools and E-Learning: A Revolution in the Research Methodology of Communication for a Pedagogical Scenario. *RA Journal of Applied Research*, 9(4), 170–177. DOI: 10.47191/rajar/v9i4.03

Zahra, O. F., Amel, N., Soufiane, O., & Mohamed, K. (2024). From Platforms to Online Communication Tools. DIROSAT: Journal of Education. *Social Sciences & Humanities*, 2(3), 130–147.

Zanon, C., Hutz, C. S., Yoo, H., & Hambleton, R. K. (2016). An application of item response theory to psychological test development. *Psicologia: Reflexão e Crítica : Revista Semestral do Departamento de Psicologia da UFRGS*, 29(1), 18. Advance online publication. DOI: 10.1186/s41155-016-0040-x

Zawacki-Richter, O., Marín, V. I., Bond, M., & Gouverneur, F. (2019). Systematic review of research on artificial intelligence applications in higher education – where are the educators? *International Journal of Educational Technology in Higher Education*, 16(1), 39. DOI: 10.1186/s41239-019-0171-0

Ze, Y. S. N., & Molinari, G. (2023). Facteurs a priori et in situ de l'engagement des étudian. *Revue Internationale De Pédagogie De L'enseignement Supérieur*, 39(2). Advance online publication. DOI: 10.4000/ripes.4833

Zeyab, A., & Alayyar, G. M. (2023). Perspective chapter: Education technology (EdTech) and the online course revolution. *Education and Human Development*, 2023, 1–14. DOI: 10.5772/intechopen.109227

Zhang, H., Song, H., Li, S., Zhou, M., & Song, D. (2023). A Survey of Controllable Text Generation Using Transformer-based Pre-trained Language Models. *ACM Computing Surveys, 56*(3), 64:1-64:37. DOI: 10.1145/3617680

Zhang, H. Z., Lee, I., Ali, S., DiPaola, D., Cheng, Y., & Breazeal, C. (2022). Integrating Ethics and Career Futures with Technical Learning to Promote AI Literacy for Middle School Students: An Exploratory Study. *International Journal of Artificial Intelligence in Education*. Advance online publication. DOI: 10.1007/s40593-022-00293-3 PMID: 35573722

Zhang, L., & Wang, Y. (2023). Predictive analytics in online education: Early identification of at-risk students using machine learning. *International Journal of Artificial Intelligence in Education*, 33(2), 256–279.

Zhang, P., & Kamel Boulos, M. N. (2023). Generative AI in medicine and healthcare: Promises, opportunities and challenges. *Future Internet*, 15(9), 286. DOI: 10.3390/fi15090286

Zhang, S., Yao, L., Sun, A., & Tay, Y. (2020). Deep Learning based Recommender System: A Survey and New Perspectives. *ACM Computing Surveys*, 52(1), 1–38. DOI: 10.1145/3285029

Zhang, W., & Wang, C. (2023). Comparative interaction patterns of groups in an open network environment: The role of facilitators in collaborative learning. *Journal of Computer Assisted Learning*, 40(1), 136–157. DOI: 10.1111/jcal.12873

Zhang, Y., & Wan, H. (2021). The Evolution of Digital Credentials: A Global Perspective. *Journal of Educational Technology Development and Exchange*, 14(1), 15–30.

Zhao, H. (2024). Realization of chinese-english bilingual speech dialogue system using machine translation technology. *Journal of Electrical Systems*, 20(6s), 1751–1762. DOI: 10.52783/jes.3093

Zhao, L., & Lin, X. (2023). Sentiment analysis in online learning: Leveraging AI to enhance student engagement and support. *Journal of Computing in Higher Education*, 35(2), 456–475.

Zheng, Z., Xie, S., Dai, H. N., & Wang, H. (2018). Blockchain Technology for Enhancing the Security and Efficiency of Digital Education. *Journal of Educational Technology & Society*, 21(4), 1–12.

Zhoc, K. C., Webster, B. J., King, R. B., Li, J. C., & Chung, T. S. (2019). Higher education student engagement scale (HESES): Development and psychometric evidence. *Research in Higher Education*, 60(2), 219–244. DOI: 10.1007/s11162-018-9510-6

Zhou, M. Y., & Lawless, W. F. (2015). An overview of artificial intelligence in education. Encyclopedia of Information Science and Technology, Third Edition, 2445-2452.

Zhou, Q., Suraworachet, W., Pozdniakov, S., Martinez-Maldonado, R., Bartindale, T., Chen, P., Richardson, D., & Cukurova, M. (2021). Investigating Students' Experiences with Collaboration Analytics for Remote Group Meetings. Artificial Intelligence in Education, 472–485. DOI: 10.1007/978-3-030-78292-4_38

Zhou, L., Dai, L., & Yang, Z. (2020). Educational Chatbot for Improving Learning Engagement and Outcomes: A Systematic Review. *Journal of Educational Technology & Society*, 23(2), 69–81.

Zhou, Z., & Li, W. (2020). Ethical Challenges in AI Education and How to Address Them with Project-Based Learning. *AI & Society*, 35(3), 743–753.

Zhu, X., Li, C., & Bao, M. (2024). Research on the application and future trend of artificial intelligence in education and teaching. SHS Web of Conferences, 187, 04017. https://doi.org/DOI: 10.1051/shsconf/202418704017

Zhu, W., Marquez, A., & Yoo, J. (2015). "Engineering economics jeopardy!" Mobile app for university students. *The Engineering Economist*, 60(4), 291–306. DOI: 10.1080/0013791X.2015.1067343

Zichermann, G., & Cunningham, C. (2011). *Gamification by Design: Implementing Game Mechanics in Web and Mobile Apps*. O'Reilly Media.

Zotova, E., & Zaretskaya, A. (2019). Micro-learning in the Context of Modern Education: International Experience and Russian Practice. *Higher Education in Russia*, 28(7), 124–131.

Zou, D., Xie, H., Wang, F. L., & Wang, S. (2021). A Systematic Review of Artificial Intelligence in Education: Where Are We Now and Where Are We Going? *Computers and Education: Artificial Intelligence*, 2, 100020.

Zuboff, S. (2019). *The age of surveillance capitalism: The fight for a human future at the new frontier of power*. PublicAffairs.

About the Contributors

Mohamed Khaldi, Abdelmalek Essaâdi University, Morocco (medkhaldi@yahoo.fr) BP 25 Martil Morocco MARTIL, 93150, MA Dr. Mohamed Khaldi is Professor of Higher Education of the Department of computer Science and management at Normal school Superior Tétouan of the University Abdelmalek Essaadi Morocco where he has been since 1987. From 2010 to 2016, he served as coordinator of the Specialized Master in Multimedia educational engineering. He was a member of advisory board of dozens of international congresses in the field of computer science. He received a B.S from Mohamed V University Morocco in 1982, and a 3rd cycle diploma from Normal school Superior Rabat Morocco in 1986. He received his PhD in Didactics, Information and Communication Technology from the Faculty of Science Ben MSik Morocco in 2001. And the Academic habilitation in 2003, from the Faculty of Science Ben MSik Morocco. Research professor in e-Learning and TICE • Member since 2008 of the LIROSA laboratory (Computer Science Laboratory, Operational Research and Applied Statistics, Abdelmalek Essaadi University, Faculty of Sciences, Tetouan, Morocco. • Member since 2017 of the S2IPU team (Sciences of Computer Science and Pedagogical Engineering University), Abdelmalek Essaadi University, Normal school Superior, Tetouan, Morocco. Research axes • Research in the field of distance learning (e-learning); • Research in the field of integration of the computer tool in the teaching of disciplines; • Research in the field of discipline didactics; • Research in the field of production of didactic aids.

***

A. Senthilraj is a research scholar who is doing research in the online mathematics education. He has 18 years of experience as a teacher educator in SCERT, Tamilnadu. He has developed quite amount of modules for maths education. He is good trainer for various teacher education programmes and written several modules to the government of Tamilnadu. He has presented seven research papers in the national and international conferences.

Soniya Antony is a PhD scholar in education at Alagappa University, Tamil Nadu, India. She is the recipient of a UGC doctoral fellowship, which supports her studies. Soniya's research focuses on the identification, support, and early-stage education of twice-exceptional children. She is also deeply interested in technology for exceptional learners, gifted education, and language education. Soniya dedicates herself to conducting impactful research to enhance the education of twice-exceptional children in India.

Brindha Devi is a PhD scholar in education at Alagappa University. Her research explores the ethical considerations and perceptions of doctoral students regarding AI in higher education. Her work emphasizes the need for ethical guidelines to promote responsible, inclusive AI integration in education. She is passionate about using technology to enhance learning while addressing its potential challenges. She actively engages in academic discussions on AI, ethics, and higher education policy. Her research aims to foster trust, equity, and innovation in the educational use of AI.

Nisserine El bahri Holds a master's degree in computer management of the enterprise from Abdelmalek Essaadi University in Tetuan, Morocco (2014). She subsequently received her PhD in computer science and artificial intelligence from the same university in 2024. Currently, she works as a computer science professor at the Private University of Fez and is a full member of the university's Systems and Sustainable Environment Laboratory. Her area of interest in research is artificial intelligence applied to e-learning

Zakaria Itahriouan Received the master's degree in telecommunication systems engineering in 2009 from Abdelmalek Essaadi University in Tetuan, Morocco. He received thereafter a PhD in Computer Science in 2015 from the same University. He worked as a professor of computer science at the Private University of Fez between 2016 and 2023. Currently, he is a professor of computer Science in Moulay Ismail University and a full member of Software Engineering and Information Systems Engineering (GL-ISI) Team in the Faculty of Sciences and Techniques of Errachidia in the same university. His current research interest is Data science, Machine Learning, Teaching and Learning technologies and software engineering.

B Keerthi Madhuri, Student, Department of Computer Science and Engineering -IOT, Sreenidhi Institute of Science and Technology, Hyderabad, Telangana, India.

Hassane Kemouss is a PhD student of Computer Science and University Pedagogical Engineering, engineering science teacher at technical high school, he is currently with S2IPU, Abdelmalek Essaadi University, Tetouan, Normal School Superior, Tetouan, Morocco.

Maha khaldi is a PhD holder in computer science, his doctoral work was about creating a Web application for the management of e-learning projects, an application which makes it possible to manage all the tasks of the various actors of an e-learning project in the same collaborative space. after having realized the system, it was implemented in a server on the Web, to have access to all the actors each according to their actions and tasks in each step of an e-learning project to be carried out in collaboration. To experiment the system, it was proposed to set up an e-learning project (online training) from the system.

V Lakshmi Prasanna, Professor of English in Gokaraju Rangaraju Institute of Engineering and Technology, Hyderabad. I have 33 yrs of experience, presented and published more than 55 research papers both in national and international conferences and journals

Nilamadhab Mishra is an Associate Professor in the School of Computer Science and Engineering, AI-ML Division, at VIT Bhopal University, India. He received his Doctor of Philosophy (Ph.D.) in Computer Science and Information Engineering with a specialization in AI-data science and machine learning from Chang Gung University, Taiwan. He has over 23 years of national and international involvement in academic teaching and research at recognized Indian, Taiwanese, and African universities. He has over 100 publications in SCI/SCIE and SCOPUS-indexed journals, ISBN books and chapters, Indian and Australian patents, IEEE conference proceedings, etc. He has been a reviewer, associate editor, and editorial board member for SCI/SCIE-indexed journals and conferences. He has been involved with several professional bodies: "Fellow Member of "ISROSET"; Fellow Member of "ISRD"; Senior Member of "ASR" (Hong Kong); Senior Member of "IEDRC" (Hong Kong); and Member of "IEEE Collabratec". He is currently an Associate Editor for two SCOPUS-indexed journals, and his research interests span the areas of AI, Data Science, Machine Learning, and Cognitive Analytics & Applications.

Utsav Krishan Murari is a gifted writer, dedicated life coach, motivational speaker, passionate educator, and, above all, an exceptional mentor. He holds a doctorate in communication and media studies from the Central University of South Bihar. Graced with a vivid imagination, he is a genuine catalyst for ideas and is always open to trying new things. His one such initiative is Saarthi, which began as Saarthi Entertainment. Saarthi is a Delhi-based creative media production start-up that wants to leave a legacy of artistry in the production industry. Delivering content with a strong emotional connection and the potential to transform the world is what Team Saarthi is all about.

Rudra Kalyan Nayak serves as a Senior Assistant Professor in the School of Computing Science & Engineering at VIT Bhopal University. He has more than 14 years of expertise in teaching, mentoring and research. His Doctorate in Computer Science & Engineering was awarded by Siksha 'O' Anusandhan University in Bhubaneswar, Odisha, where he conducted research in the field of Financial Engineering. At the same university, he earned his M. Tech. in information technology. Dr. Nayak possesses nine patents total, two of which are granted Australian design patents and two of which are granted Indian design patents. In addition, he has written three book chapters and two books. He boasts over 45 international papers to his name, appearing in reputable journals and IEEE/ Springer conference proceedings. Additionally, he wrote an article for the IEEE India Council Newsletter. Speaking with expertise, Dr. Nayak has participated in faculty development programs. He has reviewed papers for a number of conferences and peer-reviewed journals. His areas of interest in research are Artificial Intelligence, Machine Learning, Financial Engineering, Bioinformatics and Computer Vision. A number of honors have been bestowed upon him, including the I2OR International Teaching Excellence Award 2023, I2OR National Elite Teacher Award 2022, 100 Eminent Academicians of 2021, Dynamic Teacher of the Year 2019, I2OR Preeminent Researcher Award 2019 and Green ThinkerZ Bright Educator Award 2019. A professional member of ACM, CSTA, IAENG, ICSES, theIRED, SDIWC, and a life member of ISTE, Dr. Nayak holds these memberships. He was also a London Journals Press Honorary Rosalind Member.

Amel Nejjari is a Professor at National School of Applied Sciences-Tetuan, Morocco since 2008. Currently, she is a member of the research team: Didactics, Literacies and Mediation, ENS of Tetuan. Her research focuses on innovative pedagogies and digital in language learning and teaching.

Fatima Zahra Ouariach is a Doctoral Researcher specializing in Educational Technology. She is currently working on her PhD, focusing on the management of communication tools in e-learning through an LMS platform. Her research aims to design an efficient architecture for communication tools in the specific context of online learning using an LMS platform. In addition to her academic work, She has gained experience as a reviewer for prestigious publishing houses such as Hindawi, Inderscience and IGI Global... Furthermore, she has made valuable contributions as a member of the jury for oral interviews at Abdelmalek Essaadi University, showcasing her dedication and contribution to the advancement of research and higher education.

Ouariach Soufiane Doctoral researcher in computer science and university pedagogical engineering. He's currently carrying out research under the tutelage of the esteemed Professor Khadi Mohamed at Abdel Malek Essaadi University. His research focuses on the exciting new frontier of flipped classroom models, using the innovative approach of SPOC (Small Private Online Courses) as well as distance learning design and implementation. His academic prowess extends to distance learning design and implementation, where he's keen on breaking new ground.

P. Selvakumar is currently an Associate Professor in the Department of Science and Humanities at Nehru Institute of Technology, Coimbatore, Tamil Nadu, India. Since 2023, he has served as the Department IQAC Coordinator at the institute. Prior to joining Nehru Institute of Technology, he held positions as an Assistant Professor in the Department of Science and Humanities various engineering colleges in Coimbatore, Tamil Nadu, India, spanning from 2011 to 2023. Dr.P. Selvakumar's academic and research interests lie primarily in the field of phytochemistry. He has made significant contributions to his field, with an impressive record that includes 30 journal publications, 12 patents, and 20 conference papers. Additionally, he has organized 5 seminars and conferences, further demonstrating his commitment to academic and scientific exchange. Throughout his career, Dr. P.Selvakumar has been recognized for his dedication and excellence in teaching, having been honored with several awards, including the Best Teacher Awards. He holds a prominent position in the academic community, contributing actively to research.

Hemlata Parmar, a luminary in the field of computer Science and Engineering. Her Journey is defined by a constellation of achievements from prolific publications and patents to enlightening guest lectures. Yet, her brilliance is matched only by her unwavering commitment to human ethics. With a heart as vast as her intellect, she champions women's empowerment, bridging the gap with empathy and insight. In every endeavor, she embodies the perfect fusion of academic process and compassionate advocacy, inspiring generations to come.

Anjum Parvez, Associate Professor, Law College Dehradun, Uttaranchal University, Uttarakhand

R. Ramnath has a very strong and comprehensive record in a teacher and training program from primary to higher education. Despite having postgraduate degrees in botany and English, his academic background is very strong in educational courses from B.Ed. to Ph.D. He has made a great contribution to his career as a faculty member in education, out of which he holds administrative positions as Assistant Director of IQAC, Coordinator of EBSB, and others. He has authored 19 books and book chapters, 41 conference proceedings, and 55 research publications in areas such as instructional technology, educational technology, teacher education, science education, and inclusive practices. In addition, he serves as a reviewer for five journals. He successfully completed 11 consultancy projects; besides, he bagged quite a number of international awards. He delivered 129 invited talks, organized 17 conferences, and presented research papers for 119 national and international conferences. He delivered 13 MOOC modules through the SWAYAM portal and supervised a significant number of Ph.D. and M.Phil students. He also oversaw the awarding of degrees. He has made adequate contributions to organizations like SSA, RMSA, SCERT of Tamil Nadu, and NCERT for curriculum, training, and consultancy projects. In general, he is a promising star for the institution where he is affiliated.

Dharmendra Singh stands as a pillar of academic excellence with a distinguished career spanning over 21 years in education, administration, and research. Currently serving as Professor and Associate Director at the School of Management Science, Lucknow, he has established himself as a leader in the fields of Academics and Administration. Dr. Singh's academic journey began with a strong foundation, earning a Ph.D. in Mathematics from the University of Lucknow, Uttar Pradesh. His educational background also includes M.Sc. and B.Sc. degrees from DDU Gorakhpur University, Uttar Pradesh, where he honed his analytical and research skills. In his current role, Dr. Singh plays a pivotal role in shaping academic programs and guiding the next generation of scholars. His commitment to academic mentorship is evident through his supervision of 02 M.Phil. Students, ensuring their academic growth and success. A prolific researcher, Dr. Singh has authored and published over 15 research papers that have significantly contributed to advancing the understanding of Mathematics. Beyond publications, he has demonstrated his innovative prowess by applying for and publishing 03 patents, showcasing his practical contributions to academia. Dr. Singh remains actively engaged in the academic community through his participation in 7 national and international conferences, where he has presented his research findings. His continuous pursuit of knowledge is further evidenced by his attendance at 07 Faculty Development Programs, where he enhances his professional competencies and stays abreast of emerging trends in his field. As a Life Member of the Tensor Society of India, Dr. Singh fosters connections within academic societies and professional networks, enriching his contributions to the broader academic community. Committed to holistic student development, Dr. Singh actively participates in extracurricular activities and has completed the 'Matrix Solver' and 'Engineering Mathematics - II' course through NPTEL, underscoring his dedication to continuous learning and skill enhancement. Prof (Dr.) Dharmendra Singh's contributions have been recognized with the Uttar Pradesh Education Leadership Award 2023, acknowledging his significant impact on academics and educational leadership. Prof (Dr.) Dharmendra Singh's unwavering dedication to academia, research excellence, and student mentorship exemplifies him as a dynamic educator and leader in Academics, Administration and Research. His achievements underscore a steadfast commitment to advancing knowledge and fostering academic excellence, making him a respected figure in his field and beyond.

Amit Thakur is a passionate scholar currently pursuing a Ph.D. in Machine Learning at VIT BHOPAL University. With a background in computer science and a fascination for artificial intelligence, Amit's research focuses on developing innovative algorithms for natural language processing and computer vision applications. Throughout his academic journey, Amit has demonstrated a strong aptitude for theoretical and applied research in machine learning. His work combines advanced mathematical techniques with practical implementations to address real-world challenges in data analysis and pattern recognition. In addition to his research endeavors, Amit is actively involved in teaching and mentoring undergraduate students, sharing his knowledge and enthusiasm for machine learning. He volunteers his time to organize workshops and seminars, fostering a collaborative learning environment within the academic community. Driven by a desire to harness the power of machine learning for societal benefit, Amit is dedicated to pushing the boundaries of AI research and developing intelligent systems that can positively impact people's lives. Through his interdisciplinary approach and commitment to excellence, Amit strives to be a leading figure in the field of machine learning and contribute to its continued advancement.

Ajay Upadhyaya is presently working as a Professor in the Department of Computer Engineering at SAL Engineering & Technical Institute, SAL Education, Ahmedabad, Gujarat. He has completed his PhD in Computer Engineering and his M.Tech. in Computer Science and Engineering. He has 18 years of teaching experience. His areas of specialisation include Mobile and Vehicular Ad-hoc Networks, Security in Networks, Machine Learning, Data Science, and IOT. In addition to his academic endeavours, he participates actively in faculty development programs, short-term initiatives, and seminars. He conducted more than 120 workshops and training programs on various technical topics at different institutes for students and faculty members. He has more than 22 publications in various refereed national and international conferences and journals.

Sandhya Verma, Assistant Professor, Department of Law, Allahabad University, Prayagraj,

Rokesh Kumar Yarava is an esteemed Associate Professor at Chalapathi Institute of Engineering & Technology,Lam,Guntur. where he plays a pivotal role in shaping the academic and professional futures of his students. Known for his expertise in Cloud Computing,Networking, he combines his deep knowledge with a passion for teaching and research. With a commitment to fostering innovation and critical thinking, He has contributed significantly to the institution's reputation for academic excellence. His dedication to education and mentorship has earned him respect among colleagues and students alike.

Index

A

B

C

D

E

F

G

I

L

M

O

P

R

S

T

V

www.ingramcontent.com/pod-product-compliance
Lightning Source LLC
LaVergne TN
LVHW080355011026
849161LV00009B/2314
9798337322629